Mathematical Engineering

Today, the development of high-tech systems is unthinkable without mathematical modeling and analysis of system behavior. As such, many fields in the modern engineering sciences (e.g. control engineering, communications engineering, mechanical engineering, and robotics) call for sophisticated mathematical methods in order to solve the tasks at hand.

The series Mathematical Engineering presents new or heretofore little-known methods to support engineers in finding suitable answers to their questions, presenting those methods in such manner as to make them ideally comprehensible and applicable in practice.

Therefore, the primary focus is—without neglecting mathematical accuracy—on comprehensibility and real-world applicability.

To submit a proposal or request further information, please use the PDF Proposal Form or contact directly: Dr. Thomas Ditzinger (thomas.ditzinger@springer.com)

Indexed by SCOPUS, zbMATH, SCImago.

Michael Robinson

Computational Homological Algebra

Springer

Michael Robinson
Department of Mathematics and Statistics
American University
Washington, D.C., USA

ISSN 2192-4732 ISSN 2192-4740 (electronic)
Mathematical Engineering
ISBN 978-3-032-08633-4 ISBN 978-3-032-08634-1 (eBook)
https://doi.org/10.1007/978-3-032-08634-1

This Springer imprint is published by the registered company Springer Nature Switzerland AG
The registered company address is: Gewerbestrasse 11, 6330 Cham, Switzerland

To Emerson Escolar and Yasu Hiraoka

Preface

Linear algebra is the study of linear maps between vector spaces. Its main insight is the *dimension theorem*, which provides the complete characterization of a single linear map up to vector space isomorphism. If one understands this characterization (revisited in detail as Theorem 1.2 in Chap. 1), one knows everything there is to know about a linear map. The broad success of linear algebra in applications is due to the dimension theorem and the algorithms that exploit it, like Gaussian elimination and QR factorizations.

Homological algebra is the study of what happens when linear maps are chained together, one after the next. These *sequences* of linear maps and the transformations between them—*sequence maps*—admit additional structure beyond what was present in the vector spaces from which they are built. Characterizing sequence maps is quite complicated, but for a special kind of sequence called a *chain complex*, a complete characterization is available. This characterization is algorithmic and intuitive. Chain complexes have broad application within mathematics.

Unlike linear algebra, homological algebra is little known outside of mathematics. This book is an attempt to reduce the barrier to entry for the key tools of homological algebra. Following Richard Hamming, "The purpose of computing is insight, not numbers" [1], this book develops the basic notions of homological algebra by emphasizing concrete, elementary, and computational examples in finite dimensional vector spaces.

This book is written for mid-division undergraduate students who have a solid background in linear algebra. Every important result in this book is mathematically proven, though readers who are less interested in proofs will still build insight by computing: working the examples and exercises. No abstract algebra is required, though in a few places remarks about it are made. No prerequisite knowledge of topology is needed, even though the most direct applications of homological algebra are to topology. Although Chaps. 4 and 5 are manifestly topological in nature, we proceed using simplicial complexes in an elementary way, covering all the background material as needed. Finally, although category theory is tightly bound into the story of homological algebra, the required material is introduced as needed. The category theory that is ultimately covered is elementary.

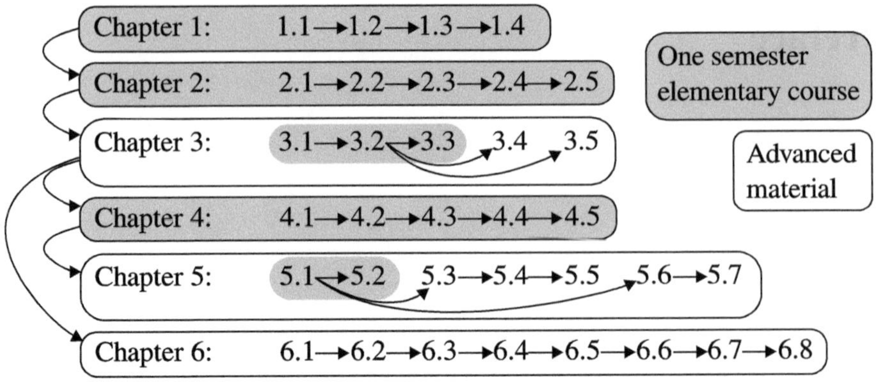

Fig. 1 Logical dependencies between the sections of the book. Shaded sections indicate the content of a one semester elementary course in computational homological algebra. The non-shaded portions consider more advanced material. It is possible to go directly to Chap. 6 from Chap. 3 if one is only interested in the algebra. The entire book could comfortably occupy a two-semester sequence

Figure 1 shows the logical dependencies of the sections of this book, with an arrow showing logical progression from one topic to the next. The book begins in Chap. 1 by remedying a gap in the typical linear algebra curriculum: the key properties of *vector space quotients*. The key turning point occurs in Chap. 2, where the Escolar-Hiraoka decomposition (first presented in 2016 in the paper [3]) is a unifying thread, both theoretically and computationally. It answers the question, "why chain complexes?" in a compelling and satisfying way. Specifically, a chain complex is the simplest kind of sequence that (1) has nontrivial sequence structure and (2) whose sequence maps have a finite presentation. If you allow for weaker constraints on the maps within the sequences, and therefore no longer have chain complexes, you lose that finite presentation. If you require a stronger constraint on the same maps, the resulting sequences are completely characterized by linear algebra.

The Escolar-Hiraoka decomposition provides a complete characterization of all chain maps. This means that to compute something or to prove something about chain maps, you merely need to decompose them and work with a finite list of simpler, indecomposable chain maps (Table 3.1 in Sect. 3.1 of Chapter 3). In no small part, this book is an appreciation of the computational insight provided by the Escolar-Hiraoka decomposition. Fortuitously, there are just 9 types of indecomposables (10 if you include the trivial one) in the Escolar-Hiraoka decomposition of a chain map. These types of indecomposables can be organized in a rather intuitive way. In fact, they are all pretty easy to guess, with one exception. That exception is a bit odd, but it turns out to be the key to understanding the connecting homomorphism in the Snake Lemma.

By the end of Chap. 3, the reader will be comfortable with algorithms for computing homology, induced maps on homology, and barcode decompositions of sequences. Upon this foundation, the reader is introduced to applications in combinatorial topology in Chaps. 4 and 5. While the combinatorics are in the forefront in Chap. 4, in Chap. 5 subdivision connects homology (algebra) with homotopy classes (topology) in a concrete way. The book ends in Chap. 6 with an elementary exploration of the derived category and derived functors.

This book aims to be the textbook I wanted to have when I taught a one-semester topics course on applied topology in Spring 2016. Although many of the students in my original course had taken linear algebra, they had not taken a course in abstract algebra nor a course in topology. At the time, I was unable to supply a fully satisfactory textbook. I was inspired by [5], which contains many examples and intellectual vistas but no proofs nor computational examples. I also drew from the classical introductions to homological algebra (for instance [4, 7, 9]), even though these are pitched to an audience with a solid background in abstract algebra. Newer texts with an applied topology focus [1, 2, 8] are aimed at an audience comfortable with topology. Additionally, because the derived category is *simply not taught at an elementary level* it seems rather mystical, even though it can be understood quite concretely.

My topics course covered the selection of Chaps. 1–5 shown in the shaded region of Fig. 1. Since my course did not include the discussion of chain homotopies and quasi-isomorphisms, it was possible to get to the Snake lemma (Theorem 3.2 in Sect. 3.3 of Chap. 3) about halfway through the semester. This schedule offered plenty of time to discuss simplicial complexes later in the semester (all of Chap. 4 and part of Chap. 5). A few students felt strong enough to give presentations at the end of the semester that touched on some of the introductory sections of Chap. 6.

Based upon the model of my original course, the book only treats sequences of finite dimensional vector spaces over $\mathbb{R}$, since this is what is usually encountered in applications. Some of the finer algebraic distinctions melt away under this limited setting. For finite dimensional vector spaces, the distinction between quasi-isomorphic chain complexes and chain homotopy-equivalent chain complexes vanishes, as does the difference between homology and cohomology. You can "get some of these back" if you look at chain complexes *of chain complexes*, which we do in Chap. 6. In this way, the algebra of modules both arises naturally and justifies some of the distinctions that are not visible with only vector spaces.

We can even discuss homotopy invariance using subdivisions in a rather "retro-futuristic" way, which later motivates the derived category in Chap. 6. By this, I mean that while combinatorial topology was popular in the past, it has been largely superseded by general topology in most mathematics curricula. On the other hand, combinatorial topology is increasing in popularity for practical applications because of its concrete and computational nature. Chapter 5 should convince the reader that abstract simplicial complexes are well suited for exploration with computers.

Because of the concrete and computational nature of the subject, the examples and exercises are the soul of this book. The reader is *strongly encouraged* to do (at least) the computational exercises, either by hand or with some automated aid, preferably using both. Nearly every exercise in the book is intended to be possible (if somewhat tedious) without the aid of a computer. Many of the examples were done "at the board" by the author during classes, and all should be "safe" for by-hand calculation. Consequently, the exercises in this book are too small for asymptotic scaling to matter, and chasing efficient algorithms complicates the intuition behind them. For this reason, computational *efficiency* is not a focus even though the book is titled "Computational"!

On the other hand, writing computational software to implement the algorithms (and some of the key constructions in the proofs) described in the book can be a means to greater insight. Using one's implementations tends to reveal subtle effects that depend on the scale of the computation beyond what the examples show. The author has implemented some of the key algorithms to ensure correctness of the examples and to facilitate typesetting. Although it somewhat defeats the purpose of doing one's own work, these implementations are available https://github.com/kb1dds/comphomalg to help a struggling reader. An errata page for this book is linked off the author's website at https://drmichaelrobinson.net/.

As a final personal note, I recall the meeting when Yasu Hiraoka announced the decomposition result that now forms the central theme of this book. Most of the talks in the meeting stayed close to the topological data analysis theme of the meeting, but Hiraoka's decomposition was both unexpected and spectacular. At the time I was impressed, but it took several years to come to a deep understanding. It was only after wrestling with the connection between the barcode decomposition of a sequence and the structure of a sequence map did I realize its central importance. Many years later, I am still greatly appreciative of his result!

Washington, D.C., USA Michael Robinson
August 2025

References

1. Hamming, R.: Numerical methods for scientists and engineers. Dover (1973)
2. Escolar, E.G., Hiraoka, Y.: Persistence modules on commutative ladders of finite type. Discrete & Computational Geometry **55**(1), 100–157 (2016)
3. Ghrist, R.: Elementary Applied Topology. CreateSpace Independent Publishing Platform (2014)
4. Gelfand, S.I., Manin, Y.I.: Methods of homological algebra. Springer (2003)
5. Hilton, P., Stammbach, U.: A course in homological algebra. Graduate Texts in Mathematics. Springer-Verlag, New York (1997). https://doi.org/10.1007/978-1-4419-8566-8
6. Weibel, C.A.: An Introduction to Homological Algebra. Cambridge Studies in Advanced Mathematics. Cambridge University Press (1994). 10.1017/CBO9781139644136
7. Carlsson, G., Vejdemo-Johansson, M.: Topological data analysis with applications. Cambridge University Press (2021)
8. Kaczynski, T., Mischaikow, K., Mrozek, M.: Computational homology. Springer (2004)
9. Edelsbrunner, H., Harer, J.: Computational Topology: An introduction. American Mathematical Society (2009)

Acknowledgments Many people read drafts of this book, worked its exercises and examples, provided useful suggestions for improvement, and supplied additional pointers. With apologies to anyone whom I have missed, the following people provided especially helpful inputs: Tate Altman, Uli Bauer, Dave D'Auria, Donna Dietz, Wendy Eldred, Ken Ewing, Olivia Freides, Zhi Gan, Scott Gerard, Barbara Giunti, Chad Guisti, Greg Henselman-Petrusek Roek, Steve Huntsman, Alan Hylton, Fangfei Lan, Jaehee Lee, Fabian Lenzen, Michael Lesnick, Zander Memon, Luke Miller, Natalie Tsuei, Matthew Wright, and Lori Ziegelmeier.

I gratefully acknowledge partial financial support for this book and the work that lead to it, which includes the following Federal contracts: N00014-15-1-2090, N00014-18-1-2541, and N00014-22-1-2659 from the Office of Naval Research, and HR0011-15-C-0050, N66001-15-C-4040, and HR0011-19-C-0072 from the Defense Advanced Research Projects Agency.

Finally, I want to thank my wife, Donna, for her continued love and support as this book has developed over the years. What was supposed to have been a project for only a year or so has taken nearly a decade to finish!

Competing Interests The author has no competing interests to declare that are relevant to the content of this manuscript.

Contents

Symbols

$(C^\bullet(X),\ d^\bullet),\ H^\bullet(X)$	The simplicial cochain complex and simplicial cohomology of an abstract simplicial complex X
$(C_\bullet(X),\ \partial_{X,\bullet}),\ H_\bullet(X)$	The simplicial chain complex and simplicial homology of an abstract simplicial complex X
$(f,\ g) : X \rightsquigarrow Y$	An equivalence class of simplicial rooves from an abstract simplicial complex X to another abstract simplicial complex Y
$(m_*)^\bullet$	Lower star indicates the induced map on the homology sequence
$/$	Quotient
$0_{V,W}$	The zero map $V \to W$
$:=$	The meaning of the expression on the left of this sign is defined by the expression on the right of this sign
$[f : \sigma]$	The sign of the permutation required to sort the vertices of $f(\sigma)$ according to some vertex partial order
$[v]_W$	A W-coset, or an element of the quotient space by W
$[v_0, \ldots, v_k]$	A dimension k simplex defined by the vertices listed, sorted according to a vertex partial order
$\setminus$	Set difference
β_k	The k-th Betti number (for a chain complex, which is implied)
$\bigoplus,\ \oplus$	Direct sum
$\bullet$	Refers to the set of all indices in a sequence
Kom(C)	The category of chain complexes whose terms are objects in **C**; we assume **C** = **Vec** if omitted
Seq(C)	The category of sequences whose terms are objects in **C**; we assume **C** = **Vec** if omitted
Set	The category of sets. Categories are set in **bold** type. A complete list of all categories used in this book is in the index
$\cdot$	Scalar multiplication (often omitted)

$\chi(V^\bullet)$	The Euler characteristic of a sequence $(V^\bullet,\ v^\bullet)$. The Euler characteristic can also be computed for an abstract simplicial complex
$\circ$	Composition of two functions or matrix multiplication
$\operatorname{cl}\sigma$	The closure of a simplex σ
$\operatorname{coimage}\ell$	The coimage of a linear map ℓ
$\operatorname{coker}\ell$	The cokernel of a linear map ℓ
$\cong$	Isomorphism
$\dim V$	Dimension of a vector space V
ℓ^T	The dual (or transpose) of a linear map ℓ
ℓ^{-1}	The inverse of ℓ
$\emptyset$	The empty set
$\operatorname{fr}A$	The frontier of a set A of simplices
$\operatorname{Hom}_{\mathbf{C}}(S,\ T)$	The class of morphisms in a category $\mathbf{C}$ with source S and target T
id_V	The identity map on V
$\operatorname{image}\ell$	The image of a linear map ℓ
$\operatorname{int}A$	The interior of a set A of simplices
$\ker\ell$	The kernel of a linear map ℓ
$\operatorname{lk}\sigma$	The link of a simplex σ
$\mathbb{R}$	The real numbers
$\mathbb{R}^{b,d},\ \mathbb{R}_{b,d}$	The bar with birth b and death d with increasing and decreasing indices, respectively. The indices b and d may be infinite
pr_A	The projection map onto a subspace A. If the domain is a product, we often use an index instead of the name of the space
$\operatorname{rank}\ell$	The rank of a linear map ℓ
σ^n	The shift functor on sequences of vector spaces
σ^T	The indicator function on a simplex σ
$\operatorname{star}\sigma$	The star of a simplex σ
$\operatorname{tr}m$	The trace of a matrix m
span	Span of a subset of a vector space
$\widetilde{C}_\bullet(X),\ \widetilde{H}_\bullet(X)$	The reduced simplicial chain complex and reduced simplicial homology of an abstract simplicial complex X
$a \cup b$	The cup product of two elements a and b of a simplicial cochain complex
$A \sqcup B$	Disjoint union of two sets A and B
$A \times B$	The cartesian product of A and B
$Bd(X)$	The barycentric subdivision of an abstract simplicial complex X
CX	The cone of an abstract simplicial complex X
$D(\mathbf{C})$	The derived category of a category $\mathbf{C}$
$f^\bullet \simeq_h g^\bullet$	A homotopy h between chain maps $f^\bullet$ and $g^\bullet$

$H^k(C^\bullet, c^\bullet)$ or $H_k(C_\bullet, c_\bullet)$	The index k homology of a chain complex with increasing or decreasing indices, respectively. Important: Homology is always a covariant functor, cohomology is always contravariant
$H_k(C^\bullet, c^\bullet)$ or $H^k(C_\bullet, c_\bullet)$	The index k cohomology of a chain complex with increasing or decreasing indices, respectively
$K(\mathbf{C})$	The homotopy category of a category $\mathbf{C}$
L^k	The Hodge k-Laplacian operator
$L_k F,\ R^k F$	Left and right k-th derived functors for a functor F
$m : A \to B$	A morphism m with source A and target B
$Mor(\mathbf{C})$	The class of morphisms in a category $\mathbf{C}$
$n \times m$	Dimensions of a linear map $\mathbb{R}^m \to \mathbb{R}^n$ as a matrix with n rows and m columns
$Ob(\mathbf{C})$	The class of objects in a category $\mathbf{C}$
PX	The prism of an abstract simplicial complex X
$Q_n((V^\bullet, v^\bullet))$	The n-th shifted quotient sequence functor applied to a sequence $(V^\bullet, v^\bullet)$
SX	The suspension of an abstract simplicial complex X
V^T	The dual space for a vector space V
Type n	A bar chain map in the Escolar-Hiraoka decomposition listed in Table 3.1

List of Figures

List of Tables

Chapter 1
Quotients of Vector Spaces

When linear maps are chained together in a sequence, the interaction between the images and kernels of adjacent linear maps matter. To disentangle these interactions, one needs to pay attention to which basis elements are present at which place in the sequence. It is easiest to think of "removing" portions of the vector spaces corresponding to certain basis elements once they are understood. This removal can be performed by the operation of a *vector space quotient*. The present chapter develops the fundamentals of quotients, their properties, and how to compute them. Special emphasis will be placed on effective algorithms for computing the quotient of two vector spaces, because this task will be done many times in the book. In this chapter we build the foundations of vector space quotients, with an eye towards their use throughout this book.

While the first few sections are likely to be review material, they are essential for setting the stage since they establish basic notation and computational conventions that will be used throughout the book. The reader is encouraged to consult the Appendix A as needed for a brief account of the basic facts of linear algebra. This book does not consider vector spaces in which the scalars come from a field other than $\mathbb{R}$. The centerpiece of this chapter is Theorem 1.2, which characterizes a single linear map and is proven with a constructive, algorithmic proof.

Specifically, in this chapter,

Section 1.1 Defines a *vector space quotient* and explore some of its properties through calculation, culminating in the classification result for linear maps, the dimension theorem (Theorem 1.2).

Section 1.2 Explores how quotients interact with linear maps, and how maps can be *induced*.

Section 1.3 Defines the *dual space* and recast Theorem 1.2 using them to obtain a more algorithmically useful version (Theorem 1.6).

The chapter ends with a section introducing *categories*. While category theory is not necessary for linear algebra, it is essential for what follows in the book. Histori-

M. Robinson, *Computational Homological Algebra*, Mathematical Engineering,
https://doi.org/10.1007/978-3-032-08634-1_1

cally, category theory and homological algebra were discovered concurrently. Category theory provides a language for building precise mathematical analogies. These analogies make it possible to "import" theory and techniques from one branch of mathematics into another. These importations will be performed repeatedly throughout the book, both to save effort and to highlight common themes. As an organizing principle, categories and *functors* allow one to transfer facts learned about one mathematical discipline to another. Additionally, categories allow for *diagrammatic reasoning* as a tool for calculation. The sooner the reader becomes acquainted with categories, the better! Therefore,

Section 1.4 Defines *categories* and *functors*, giving many examples of their usage.

1.1 Vector Space Quotients

A *linear map* is a function $\ell : V \to W$ from one vector space V to another W that preserves linear combinations. (See the Appendix A for a brief review of basic facts about linear algebra.) For every $v, w \in V$ and every $r, s \in \mathbb{R}$, then

$$\ell(r \cdot v + s \cdot w) = r \cdot \ell(v) + s \cdot \ell(w)$$

is the condition satisfied by linear maps ℓ. The vector space V is called the *domain* of ℓ and W is called the *codomain*[1] of ℓ.

Two linear maps will appear frequently enough in this book to warrant a definition.

Definition 1.1 The simplest linear map $0_{V,W} : V \to W$ is the one that assigns zero in W to every $v \in V$. This is called the *zero map*. For every vector space V, there is always the *identity map* $\mathrm{id}_V : V \to V$ defined by the equation $\mathrm{id}_V(x) := x$. These two linear maps are related in that the zero map is an identity for addition while the identity map is the identity for composition of linear maps.

Unless we state otherwise, the vector spaces throughout this book will be finite dimensional vector spaces over $\mathbb{R}$. Without losing any generality, we can write the elements of such a vector space V as *column vectors*. Each vector is then written as an ordered list of real numbers stacked vertically like

$$v = \begin{pmatrix} 1 \\ -2 \\ 3 \\ -4 \\ 5 \end{pmatrix}.$$

[1] Some authors prefer to call this the "range," though the term "range" seems to have several different definitions. Because authors who use "codomain" tend to use the same definition—the one used here—this book uses "codomain" to avoid confusion.

The length of the column is the *dimension* of the vector space. In this case, V has dimension 5. When written this way, linear maps can be written as matrices: a rectangular grid of real numbers. For instance, if V is a vector space of dimension 5 and W is a vector space of dimension 4, one example of a linear map ℓ is represented by the matrix

$$\ell = \begin{pmatrix} 2 & 3 & -1 & 0 & -7 \\ -7 & 9 & 8 & 8 & 1 \\ 0 & 4 & 5 & 1 & -2 \\ -8 & 2 & 6 & 5 & 8 \end{pmatrix},$$

which has 4 rows and 5 columns. The application of this linear map to the vector v above is the familiar row-into-column matrix-vector multiplication,

$$\ell(v) = \ell v = \begin{pmatrix} 2 & 3 & -1 & 0 & -7 \\ -7 & 9 & 8 & 8 & 1 \\ 0 & 4 & 5 & 1 & -2 \\ -8 & 2 & 6 & 5 & 8 \end{pmatrix} \begin{pmatrix} 1 \\ -2 \\ 3 \\ -4 \\ 5 \end{pmatrix}$$

$$= \begin{pmatrix} 2 \cdot 1 + 3 \cdot (-2) + (-1) \cdot 3 + 0 \cdot (-4) + (-7) \cdot 5 \\ (-7) \cdot 1 + 9 \cdot (-2) + 8 \cdot 3 + 8 \cdot (-4) + 1 \cdot 5 \\ 0 \cdot 1 + 4 \cdot (-2) + 5 \cdot 3 + 1 \cdot (-4) + (-2) \cdot 5 \\ (-8) \cdot 1 + 2 \cdot (-2) + 6 \cdot 3 + 5 \cdot (-4) + 8 \cdot 5 \end{pmatrix} = \begin{pmatrix} -42 \\ -28 \\ -7 \\ 26 \end{pmatrix}.$$

If the codomain of a linear map ℓ is equal to the domain of a linear map m, then the maps ℓ and m can be composed (as functions) to form a new linear map $(m \circ \ell)$ whose domain is the same as that of ℓ and whose codomain is the same as that of m. The usual row-into-column matrix multiplication is convenient because composition of two linear maps is exactly their matrix product. For instance, if X is a vector space of dimension 2 and $m : W \to X$ is represented by the matrix

$$m = \begin{pmatrix} 3 & 1 & -2 & 4 \\ 0 & 4 & -3 & -3 \end{pmatrix},$$

then the composition of m after the matrix ℓ from above, $m \circ \ell : V \to X$, is their matrix product

$$m\ell = \begin{pmatrix} 3 & 1 & -2 & 4 \\ 0 & 4 & -3 & -3 \end{pmatrix} \begin{pmatrix} 2 & 3 & -1 & 0 & -7 \\ -7 & 9 & 8 & 8 & 1 \\ 0 & 4 & 5 & 1 & -2 \\ -8 & 2 & 6 & 5 & 8 \end{pmatrix} = \begin{pmatrix} -33 & 18 & 19 & 26 & 16 \\ -4 & 18 & -1 & 14 & -14 \end{pmatrix}$$

Recall that the inner dimensions must match for two matrices to be multiplied; this is merely an indication that the codomain of ℓ is equal to the domain of m, both of which are W in this example. Let us verify that both $m\ell$ and $m \circ \ell$ give the same answer when applied to the vector v above. Specifically,

$$\begin{aligned}(m \circ \ell)v &= m(\ell(v)) = m\,(\ell v) = m \begin{pmatrix} -42 \\ -28 \\ -7 \\ 26 \end{pmatrix} \\ &= \begin{pmatrix} 3 & 1 & -2 & 4 \\ 0 & 4 & -3 & -3 \end{pmatrix} \begin{pmatrix} -42 \\ -28 \\ -7 \\ 26 \end{pmatrix} \\ &= \begin{pmatrix} -36 \\ -169 \end{pmatrix}\end{aligned}$$

and

$$\begin{aligned}(m\ell)v &= \begin{pmatrix} -33 & 18 & 19 & 26 & 16 \\ -4 & 18 & -1 & 14 & -14 \end{pmatrix} \begin{pmatrix} 1 \\ -2 \\ 3 \\ -4 \\ 5 \end{pmatrix} \\ &= \begin{pmatrix} -36 \\ -169 \end{pmatrix}.\end{aligned}$$

As a consequence, we do not need to distinguish between $m \circ \ell$ and $m\ell$. In this book, we prefer $m \circ \ell$ because it reminds us that matrices refer to the functions defining linear maps.

It should be clear from how matrix-matrix multiplication works that zero maps correspond to a matrix full of zeros.

For identity maps, it is useful to observe that an identity map has domain equal to its codomain. Linear maps which have the same domain and codomain correspond to *square matrices*: they have the same number of rows and columns. Since identity maps do not change any component of a vector they are applied to, identity maps are of the form

$$\begin{pmatrix} 1 & 0 & \cdots & 0 \\ 0 & 1 & \cdots & 0 \\ \vdots & \vdots & \ddots & \vdots \\ 0 & 0 & \cdots & 1 \end{pmatrix}$$

Such a matrix is also an example of a *diagonal matrix*, one in which the row and column locations of a nonzero entry must be equal.

Definition 1.2 Suppose that $\ell : V \to W$ is a linear map. The set of vectors that ℓ sends to zero in W is called the *kernel*, that is

$$\ker \ell := \{v \in V : \ell(v) = 0\}.$$

On the other hand, the *image* of ℓ is the subspace of W that ℓ produces as output, namely

$$\operatorname{image} \ell := \{w \in W : \text{there is a } v \in V \text{ such that } \ell(v) = w\}.$$

Some authors call the image the "range." Again, this book avoids the term "range."

A linear map $\ell : V \to W$ is injective if and only if $\ker \ell$ is trivial; ℓ is surjective if and only if $\operatorname{image} \ell = W$.

For instance, reusing the matrix m from before, the vector

$$\begin{pmatrix} 4 \\ 4 \end{pmatrix}$$

is in the image of m because

$$\begin{pmatrix} 4 \\ 4 \end{pmatrix} = m \begin{pmatrix} 1 \\ 1 \\ 0 \\ 0 \end{pmatrix} = \begin{pmatrix} 3 & 1 & -2 & 4 \\ 0 & 4 & -3 & -3 \end{pmatrix} \begin{pmatrix} 1 \\ 1 \\ 0 \\ 0 \end{pmatrix}.$$

On the other hand, the vector

$$\begin{pmatrix} 5/3 \\ 3 \\ 4 \\ 0 \end{pmatrix}$$

is in the kernel of m because

$$\begin{pmatrix} 3 & 1 & -2 & 4 \\ 0 & 4 & -3 & -3 \end{pmatrix} \begin{pmatrix} 5/3 \\ 3 \\ 4 \\ 0 \end{pmatrix} = \begin{pmatrix} 0 \\ 0 \end{pmatrix}.$$

Finding vectors that are *outside* the image of a linear map is considerably more difficult. Solving the problem of finding all vectors outside the image of a linear map is one of the main results (Theorem 1.2) of this section.

Lemma 1.1 *Both the image and the kernel of a linear map are subspaces.*

Proof Consider the linear map $\ell : V \to W$. A subset of W is a subspace if it is closed under linear combinations. Suppose that v and w are elements of the kernel of ℓ. Then

$$\ell(a \cdot v + b \cdot w) = a \cdot \ell(v) + b \cdot \ell(w) = a \cdot 0 + b \cdot 0 = 0,$$

for any $a, b \in \mathbb{R}$. Every linear combination of elements in the kernel is itself in the kernel. A similar computation shows that linear combinations of elements in the image are themselves in the image. □

There are various algorithms for finding the kernel of a linear map. For instance, the *method of free columns* is fairly intuitive and efficient. The reader is encouraged to recall the process from a standard reference, such as Strang (2006).

Example 1.1 Consider the linear map $m : \mathbb{R}^4 \to \mathbb{R}^4$ given by the matrix

$$m := \begin{pmatrix} -5 & 1 & 4 & -3 \\ 1 & 0 & -1 & 1 \\ 1 & 0 & -1 & 1 \\ 0 & 1 & -1 & 2 \end{pmatrix}.$$

This matrix can be row-reduced into the following form

$$\begin{pmatrix} 1 & 0 & -1 & 1 \\ 0 & 1 & -1 & 2 \\ 0 & 0 & 0 & 0 \\ 0 & 0 & 0 & 0 \end{pmatrix}.$$

Notice that there are two zero rows. We know that the kernel has dimension 2. The third and fourth columns are free. We start building out the basis for the kernel,

$$\begin{pmatrix} * \\ * \\ 1 \\ 0 \end{pmatrix} \text{ and } \begin{pmatrix} * \\ * \\ 0 \\ 1 \end{pmatrix}.$$

A little reflection yields the four unknown entries so that

$$\ker m = \operatorname{span} \left\{ \begin{pmatrix} 1 \\ 1 \\ 1 \\ 0 \end{pmatrix} \text{ and } \begin{pmatrix} -1 \\ -2 \\ 0 \\ 1 \end{pmatrix} \right\}.$$

Any vector subspace must contain the additive identity 0, which makes vector subspaces a generalization of the idea of a *plane through the origin*. What about planes that do not contain the origin? These are called *affine subspaces*. While affine subspaces do not usually play a role in linear algebra, they are crucial for understanding the structure of quotients.

The following example provides some insight: affine subspaces inherit algebraic structure from the ambient vector space.

Example 1.2 Consider the vector space $\mathbb{R}^2$ and the subspace spanned by $\begin{pmatrix} 1 \\ 2 \end{pmatrix}$. The set of all linear combinations consisting of this single vector is simple: it is the set of scalar multiples,

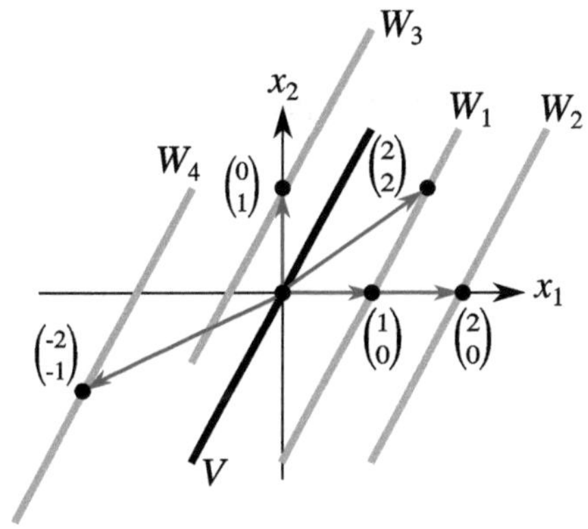

Fig. 1.1 The subspace V and some parallel affine subspaces (lines) in Example 1.2

$$V := \left\{ \begin{pmatrix} a \\ 2a \end{pmatrix} : a \in \mathbb{R} \right\},$$

represented graphically in Fig. 1.1. To obtain parallel affine subspaces (which are lines in this case), we merely need to add the same vector to each element of V. For instance, if we take

$$w := \begin{pmatrix} 1 \\ 0 \end{pmatrix},$$

and add it to every element of V, we obtain the subset

$$W_1 := \{v + w : v \in V\} = \left\{ \begin{pmatrix} 1 + a \\ 2a \end{pmatrix} : a \in \mathbb{R} \right\}.$$

We can interpret this as the sum of w with the *entire subspace V*

$$W_1 = V + w = V + \begin{pmatrix} 1 \\ 0 \end{pmatrix}.$$

This vector w we have added is not the only way to define W_1, though. We could also define it as

$$W_1 = V + \begin{pmatrix} 2 \\ 2 \end{pmatrix} = V + \begin{pmatrix} 1 \\ 0 \end{pmatrix} + \begin{pmatrix} 1 \\ 2 \end{pmatrix} = V + \begin{pmatrix} 1 \\ 0 \end{pmatrix},$$

since the vector $\begin{pmatrix} 1 \\ 2 \end{pmatrix}$ is already an element of V.

Like the elements of V, the set of affine subspaces parallel to V inherit vector addition and scalar multiplication from $\mathbb{R}^2$. For instance, scalar multiplication of w yields new parallel affine subspaces. In Fig. 1.1, $W_2 := V + 2 \cdot w$, which we can also interpret as scalar multiplication of the affine subspaces themselves,

$$W_2 = V + 2 \cdot w = 2 \cdot V + 2 \cdot w = 2 \cdot (V + w) = 2 \cdot W_1.$$

Vector addition of parallel affine subspaces also works so that we can write

$$\begin{aligned}
W_4 &:= -(W_3 + W_2) \\
&= -1 \cdot (W_3 + W_2) \\
&= (-1) \cdot W_3 + (-1) \cdot W_2 \\
&= (-1) \cdot V + \begin{pmatrix} 0 \\ -1 \end{pmatrix} + (-2) \cdot W_1 \\
&= V + \begin{pmatrix} 0 \\ -1 \end{pmatrix} - 2 \cdot V - 2 \cdot \begin{pmatrix} 1 \\ 0 \end{pmatrix} \\
&= V + \begin{pmatrix} -2 \\ -1 \end{pmatrix},
\end{aligned}$$

which is a affine subspace parallel to V.

The idea of the set of affine subspaces parallel to a subspace is formalized by the *quotient vector space.*

Definition 1.3 If V is a vector space and W is a subspace, the *quotient vector space* V/W is the vector space whose elements are the affine subspaces parallel to W. More precisely, the elements are subsets of the form

$$[v]_W := \{v + w : w \in W\},$$

where $v \in V$, and whose addition is given by

$$[u]_W + [v]_W := [u + v]_W = \{(u + v) + w : w \in W\},$$

and whose scalar multiplication is given by

$$a \cdot [v]_W := [a \cdot v]_W = \{(a \cdot v) + w : w \in W\}.$$

(We often leave out the $\cdot$ when writing scalar multiplication.) Each $[v]_W$ is called a W-*coset*. When no confusion can arise, we will often leave off the W subscript, writing $[v]$ instead of $[v]_W$.

W is itself a W-coset in V/W, since W and $[0]_W$ are equal as sets. We call this the *zero* W*-coset.*

Proposition 1.1 *For a vector space V and subspace W, V/W is also a vector space.*

Proof Even though V/W is not a subspace of V, it inherits associativity and the distributive properties from the operations on V. Suppose that $v, w, x \in V$. While

$$(v + w) + x = v + (w + x)$$

holds in V, we must show that

$$([v]+[w])+[x]=[v]+([w]+[x])$$

in V/W. To establish this fact, we use the definition of addition in V/W twice,

$$\begin{aligned}([v]+[w])+[x] &= [v+w]+[x] \\ &= [(v+w)+x] \\ &= [v+(w+x)] \\ &= [v]+[w+x] \\ &= [v]+([w]+[x]).\end{aligned}$$

The distributive properties work in the same way, namely if $a, b \in \mathbb{R}$,

$$a\cdot([v]+[w]) = a\cdot([v+w]) = [a\cdot(v+w)] = [a\cdot v + a\cdot w] = [a\cdot v]+[a\cdot w] = a\cdot[v]+a\cdot[w],$$

and

$$(a+b)\cdot[v] = [(a+b)\cdot v] = [a\cdot v + b\cdot v] = [a\cdot v]+[b\cdot v] = a\cdot[v]+b\cdot[v].$$

□

Example 1.3 Consider the following two subspaces of $\mathbb{R}^2$:

$$V := \operatorname{span}\left\{\begin{pmatrix}1\\0\end{pmatrix}\right\} = \left\{\begin{pmatrix}a\\0\end{pmatrix} : a \in \mathbb{R}\right\}$$

and

$$W := \operatorname{span}\left\{\begin{pmatrix}2\\-1\end{pmatrix}\right\} = \left\{\begin{pmatrix}2a\\-a\end{pmatrix} : a \in \mathbb{R}\right\}.$$

(See Fig. 1.2 for a graphical interpretation of this Example.)

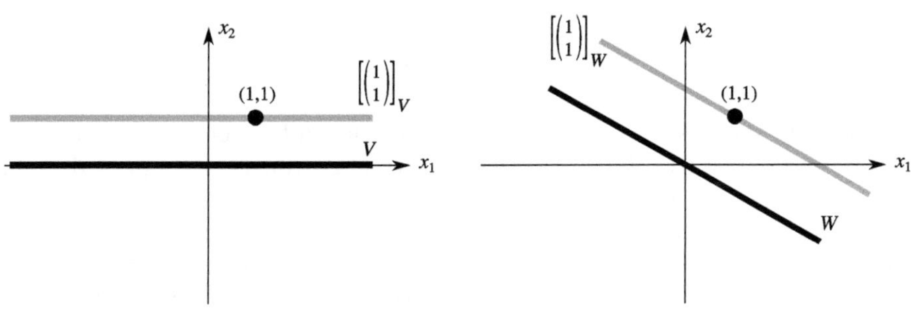

Fig. 1.2 V- and W-cosets for Example 1.3

The quotient $\mathbb{R}^2/V$ consists of cosets of the form

$$[x]_V = \begin{pmatrix} x_1 \\ x_2 \end{pmatrix} + \left\{ \begin{pmatrix} a \\ 0 \end{pmatrix} : a \in \mathbb{R} \right\} = \left\{ \begin{pmatrix} x_1 + a \\ x_2 \end{pmatrix} : a \in \mathbb{R} \right\},$$

where $x = \begin{pmatrix} x_1 \\ x_2 \end{pmatrix}$. When $a = -x_1$, we find that

$$\begin{pmatrix} 0 \\ x_2 \end{pmatrix} \in [x].$$

Therefore, for $y := \begin{pmatrix} y_1 \\ y_2 \end{pmatrix}$, $[x]_V = [y]_V$ if and only if $x_2 = y_2$. Thus, every coset of V in $\mathbb{R}^2$ is uniquely specified by a single number (namely x_2 in the above), so $\mathbb{R}^2/V$ is one-dimensional.

Addition of elements in $\mathbb{R}^2/V$ is given by

$$\begin{aligned}
[x]_V + [y]_V &= \left[\begin{pmatrix} x_1 \\ x_2 \end{pmatrix} \right]_V + \left[\begin{pmatrix} y_1 \\ y_2 \end{pmatrix} \right]_V \\
&= \left\{ \begin{pmatrix} x_1 + a \\ x_2 \end{pmatrix} : a \in \mathbb{R} \right\} + \left\{ \begin{pmatrix} y_1 + b \\ y_2 \end{pmatrix} : b \in \mathbb{R} \right\} \\
&= \left\{ \begin{pmatrix} x_1 + a + y_1 + b \\ x_2 + y_2 \end{pmatrix} : a, b \in \mathbb{R} \right\} \\
&= \left\{ \begin{pmatrix} (x_1 + y_1) + (a + b) \\ x_2 + y_2 \end{pmatrix} : a, b \in \mathbb{R} \right\} \\
&= \left\{ \begin{pmatrix} (x_1 + y_1) + c \\ x_2 + y_2 \end{pmatrix} : c \in \mathbb{R} \right\} \\
&= \left[\begin{pmatrix} x_1 + y_1 \\ x_2 + y_2 \end{pmatrix} \right]_V \\
&= [x + y]_V.
\end{aligned}$$

Similarly, the quotient $\mathbb{R}^2/W$ consists of cosets of the form

$$[x]_W = \left\{ \begin{pmatrix} x_1 + 2a \\ x_2 - a \end{pmatrix} : a \in \mathbb{R} \right\}.$$

The quotient space is still one-dimensional, which can be seen by observing that each coset has exactly one element whose second coordinate—as opposed to the first coordinate in the case of $\mathbb{R}^2/V$—is zero.

The quotient spaces $\mathbb{R}^2/V$ and $\mathbb{R}^2/W$ are *not* equal, because their cosets are different. This is especially visible in Fig. 1.2. However, they are isomorphic. The isomorphism established by many possible linear maps, one of which is the linear map $\ell : \mathbb{R}^2/V \to \mathbb{R}^2/W$ given by

$$\ell\left(\left[\begin{pmatrix} x_1 \\ x_2 \end{pmatrix}\right]_V\right) := \left[\begin{pmatrix} 2 & 0 \\ -1 & 1 \end{pmatrix}\begin{pmatrix} x_1 \\ x_2 \end{pmatrix}\right]_W .$$

To show that ℓ is an isomorphism, we need to verify that ℓ is surjective and that its kernel is trivial. The surjectivity of ℓ follows from the fact that the matrix

$$\begin{pmatrix} 2 & 0 \\ -1 & 1 \end{pmatrix}$$

is of full rank, since every representative for the quotient space $\mathbb{R}^2/W$ can be obtained as the image of a representative of the quotient $\mathbb{R}^2/V$. The kernel of ℓ consists of all V-cosets that are sent to the zero W-coset, that is

$$[0]_W = \left\{\begin{pmatrix} 2a \\ -a \end{pmatrix} : a \in \mathbb{R}\right\}.$$

By inspection of ℓ, all of these are obtained as images of elements of the zero V-coset, namely

$$[0]_V = \left\{\begin{pmatrix} a \\ 0 \end{pmatrix} : a \in \mathbb{R}\right\}.$$

Thus the entirety of the kernel of ℓ is the zero V-coset. Taken together, these two facts imply that ℓ is an isomorphism.

Quotients involving the trivial or the entire space are particularly simple.

Corollary 1.1 *For any vector space V, the space V/V is isomorphic to the trivial vector space.*

Proof Observe that if v and w are both elements of V, then $[v] = \{v + x : x \in V\} = V$ and the same is true for $[w]$. □

Corollary 1.2 *For any vector space V, the space $V/0$ (quotient with the trivial vector space) is isomorphic to V.*

Proof In this case, for any $v \in V$, $[v] = \{v\}$ the singleton set. The only difference between V and $V/0$ is that the latter's elements consist of singleton sets of the former. □

The kernel and image subspaces for a linear map $\ell : V \to W$ need not cover all of V nor W (respectively). The quotients are useful as well.

Definition 1.4 The *coimage* of ℓ is the quotient

$$\text{coimage}\, \ell := V/\ker \ell$$

and the *cokernel* of ℓ is the quotient

$$\operatorname{coker} \ell := W/\operatorname{image} \ell.$$

Proposition 1.2 *Consider a linear map $\ell : V \to W$. The following statements are equivalent:*

- *ℓ is injective,*
- *The kernel of ℓ is trivial, and*
- *The coimage of ℓ is isomorphic to V.*

Similarly, the following statements are equivalent:

- *ℓ is surjective,*
- *The cokernel of ℓ is trivial, and*
- *The image of ℓ is isomorphic to W.*

Proof Suppose that ℓ is an injective linear map. Then suppose that $v \in \ker \ell$ is an element of the kernel of ℓ. Then $0 = \ell(v) = \ell(v - 0) = \ell(v) - \ell(0)$ which means that $\ell(v) = \ell(0)$. But since ℓ is injective, this means that $v = 0$. On the other hand, if the kernel is trivial and $\ell(v) = \ell(w)$ then $0 = \ell(v) - \ell(w) = \ell(v - w)$. Since the kernel is trivial, this implies that $v - w = 0$, so $v = w$. Thus ℓ is injective.

On the other hand, surjectivity of ℓ is by definition that the image of ℓ is isomorphic to W.

Corollaries 1.1 and 1.2 make the requisite inferences to the coimage and cokernel. □

Exercise 1.1 Define the subspace

$$U := \operatorname{span}\left\{\begin{pmatrix}0\\1\end{pmatrix}\right\}$$

of $\mathbb{R}^2$.

1. Show that the quotient $\mathbb{R}^2/U$ is one dimensional.
2. Draw the zero U-coset and several typical nonzero cosets.
3. Prove that $\mathbb{R}^2/U$ is isomorphic to $\mathbb{R}^2/V$ from Example 1.3 by constructing an isomorphism.

Example 1.4 Consider the subspace V, where

$$V := \operatorname{span}\left\{\begin{pmatrix}1\\0\\2\end{pmatrix}\right\}.$$

The quotient $\mathbb{R}^3/V$ is 2-dimensional, because the linear map $\ell : \mathbb{R}^3/V \to \mathbb{R}^2$ given by

$$\ell\left(\left[\begin{pmatrix}x_1\\x_2\\x_3\end{pmatrix}\right]_V\right) := \begin{pmatrix}2 & 0 & -1\\0 & 1 & 0\end{pmatrix}\begin{pmatrix}x_1\\x_2\\x_3\end{pmatrix}$$

is of full rank and has kernel exactly equal to V, which is the zero V-coset.

Example 1.5 Consider the quotient $\mathbb{R}^2/W$ where

$$W := \operatorname{span}\left\{\begin{pmatrix}1\\0\\2\end{pmatrix}, \begin{pmatrix}0\\1\\1\end{pmatrix}\right\}.$$

The linear map $t : \mathbb{R}^2/W \to \mathbb{R}$

$$t\left(\left[\begin{pmatrix}x_1\\x_2\\x_3\end{pmatrix}\right]_W\right) := \begin{pmatrix}-2 & -1 & 1\end{pmatrix}\begin{pmatrix}x_1\\x_2\\x_3\end{pmatrix}$$

is of full rank and has kernel equal to W. Therefore, the quotient $\mathbb{R}^2/W$ is 1-dimensional.

Computing quotients is something that will happen many times in this book. To that end, we need an algorithm to compute quotients. The following two algorithms both compute a quotient V/W by constructing a basis for the cokernel of the injective map $\ell : W \to V$. In order to construct this basis, we will use the image of elements of W through ℓ. Algorithm 1 is a little easier to understand, but Algorithm 2 is faster. Both are well-suited for manual and computer-based calculations.

Algorithm 1 Computing a quotient V/W (slowly)

Require: A basis $\{v_1, \ldots, v_n\}$ for V, and a basis $\{w_1, \ldots, w_m\}$ for $W \subseteq V$ (not assuming that the basis of W is a subset of the basis for V)
Ensure: A set of vectors $S = \{z_1, \ldots, z_{n-m}\} \subset V$ so that $\{[z_1], \ldots, [z_{n-m}]\}$ is a basis for V/W

```
procedure SlowQuotient({v_i}, {w_i})
    if n = m then
        return S = {}.
    end if
    Initialize S := {}.
    for i = 1, ..., n do
        if {w_1, ..., w_m} ∪ S ∪ {v_i} is linearly independent then
            Append v_i to S.
        end if
    end for
    return S.
end procedure
```

Algorithm 1 is correct because if we use the S that is returned, then $\{w_1, \ldots, w_m\} \cup S$ is a basis for V. This set is linearly independent by construction. To show that it is a basis, we show that an arbitrary $x \in V$ can be written as a linear combination of $\{w_1, \ldots, w_m\} \cup S$. Start by writing

$$x = a_1 v_1 + \cdots + a_k v_k + a_{k+1} v_{k+1} + \cdots + a_n v_n$$

where without loss of generality we assume that $S = \{v_1, \ldots, v_k\}$. Each of the $v_{k+1}, \ldots, v_n$ can be written as a linear combination of $\{w_1, \ldots, w_m\} \cup S$ by construction. Thus x is a linear combination of $\{w_1, \ldots, w_m\} \cup S$.

Checking for linear independence at each step is inefficient. We can improve performance by combining the tests for independence into a single row reduction step, which results in Algorithm 2. This makes the interpretation of the algorithm as computing a cokernel a bit clearer as well.

Algorithm 2 Computing a quotient V/W (quickly)

Require: A basis $\{v_1, \ldots, v_n\}$ for V, and a basis $\{w_1, \ldots, w_m\}$ for $W \subseteq V$ (not assuming that the basis of W is a subset of the basis for V)
Ensure: A set of vectors $S = \{z_1, \ldots, z_{n-m}\} \subset V$ so that $\{[z_1], \ldots, [z_{n-m}]\}$ is a basis for V/W

procedure FASTQUOTIENT($\{v_i\}$, $\{w_i\}$)
  if $n = m$ **then**
    return $S = \{\}$.
  end if
  Form the matrix M with columns $\{w_1, \ldots, w_m, v_1, \ldots, v_n\}$.
  Compute M', the row echelon form of M.
  Initialize $S := \{\}$.
  for $i = m + 1, \ldots, n$ **do** ▷ iterating over rows of M'
    for $j = m + 1, \ldots, m + n$ **do** ▷ iterating of columns of M'
      if $M'_{i,j} \neq 0$ **then**
        Append v_{j-m} to S
        Exit the inner for loop (over columns).
      end if
    end for
  end for
  return S.
end procedure

The nested for loops in Algorithm 2 can be summarized as selecting the pivot columns that are not in W and then identifying which elements in the basis for V are implicated. Both for loops start at $m + 1$ since $\{w_1, \ldots, w_m\}$ was assumed be linearly independent.

Example 1.6 Consider the quotient of a 5-dimensional space by a 2-dimensional subspace spanned by

$$\begin{pmatrix} 3 \\ 8 \\ -6 \\ 3 \\ -4 \end{pmatrix}, \begin{pmatrix} -3 \\ -6 \\ -8 \\ 2 \\ 1 \end{pmatrix}.$$

Using Algorithm 2, we first form

$$M = \begin{pmatrix} 3 & -3 & 1 & 0 & 0 & 0 & 0 \\ 8 & -6 & 0 & 1 & 0 & 0 & 0 \\ -6 & -8 & 0 & 0 & 1 & 0 & 0 \\ 3 & 2 & 0 & 0 & 0 & 1 & 0 \\ -4 & 1 & 0 & 0 & 0 & 0 & 1 \end{pmatrix},$$

recognizing that $n = 5$, $m = 2$. Row reducing M yields the matrix M'

$$M' = \begin{pmatrix} 1 & 0 & 0 & 0 & 0 & 1/11 & -2/11 \\ 0 & 1 & 0 & 0 & 0 & 4/11 & 3/11 \\ 0 & 0 & \boxed{1} & 0 & 0 & 9/11 & 15/11 \\ 0 & 0 & 0 & \boxed{1} & 0 & 16/11 & 34/11 \\ 0 & 0 & 0 & 0 & \boxed{1} & 38/11 & 12/11 \end{pmatrix}$$

in which the pivot entries and columns selected by Algorithm 2 are marked with a box. Notice in particular that the search for pivots begins in row and column $m + 1 = 3$. Thus Algorithm 2 returns that

$$S = \left\{ \begin{pmatrix} 1 \\ 0 \\ 0 \\ 0 \\ 0 \end{pmatrix}, \begin{pmatrix} 0 \\ 1 \\ 0 \\ 0 \\ 0 \end{pmatrix}, \begin{pmatrix} 0 \\ 0 \\ 1 \\ 0 \\ 0 \end{pmatrix} \right\}.$$

Exercise 1.2 Compute a basis for the quotient of $\mathbb{R}^3$ by the 2-dimensional subspace spanned by

$$\begin{pmatrix} 6 \\ -6 \\ 6 \end{pmatrix}, \begin{pmatrix} 7 \\ 10 \\ 3 \end{pmatrix}$$

using both Algorithms 1 and 2. Do you get the same basis from both algorithms? What does this tell you?

Exercise 1.3 Compute a basis for the quotient of the subspace of $\mathbb{R}^3$ spanned by

$$\begin{pmatrix} 2 \\ -2 \\ 10 \end{pmatrix}, \begin{pmatrix} 8 \\ 0 \\ 3 \end{pmatrix}$$

using Algorithm 2.

Exercise 1.4 Compute the quotient of $\mathbb{R}^5$ by the 2-dimensional subspace spanned by

$$\begin{pmatrix} -9 \\ 3 \\ -4 \\ -5 \\ 5 \end{pmatrix}, \begin{pmatrix} -2 \\ -9 \\ 0 \\ 9 \\ 9 \end{pmatrix}$$

using Algorithms 1 or 2. (Hint: choose the ordering of your basis vectors carefully!)

Exercise 1.5 Compute the quotient of $\mathbb{R}^5$ by the 3-dimensional subspace spanned by

$$\begin{pmatrix} 9 \\ -3 \\ -7 \\ 0 \\ -10 \end{pmatrix}, \begin{pmatrix} 10 \\ -1 \\ 4 \\ 10 \\ 5 \end{pmatrix}, \begin{pmatrix} -7 \\ 1 \\ 6 \\ -6 \\ 7 \end{pmatrix}$$

using Algorithms 1 or 2. (Hint: choose the ordering of your basis vectors carefully!)

Definition 1.5 Suppose V and W are two vector spaces with bases $\{e_1, \ldots, e_n\}$ and $\{f_1, \ldots, f_m\}$ respectively. The *direct sum* of V and W is the $m + n$-dimensional vector space $V \oplus W$ whose basis is $\{e_1, \ldots, e_n, f_1, \ldots, f_m\}$, in which we assert that the basis elements of *both* V and W *together* are linearly independent.

The direct sum of a collection of spaces $\{V_i\}_{i=1}^n$ is usually written $\bigoplus_{i=1}^n V_i$.

The assertion part of the definition of direct sum is important and subtle. Consider the 1-dimensional vector space $\mathbb{R}$ spanned by the column vector (1). If we construct $\mathbb{R} \oplus \mathbb{R}$, notice that both vector spaces in the direct sum have the same basis, namely $\{(1)\}$. However, since the definition requires us to assert that the basis of $\mathbb{R} \oplus \mathbb{R}$ be of length 2, this means that we must distinguish between the two copies. We might write the basis of $\mathbb{R} \oplus \mathbb{R}$ as $\{(1), (1)'\}$ to avoid confusion. It is also fairly standard to add another component to each vector so that the basis is written

$$\mathbb{R} \oplus \mathbb{R} = \text{span}\left\{\begin{pmatrix}1\\0\end{pmatrix}, \begin{pmatrix}0\\1\end{pmatrix}\right\}$$

Lemma 1.2 *The direct sum $V \oplus W$ of two vector spaces W and V is well defined up to isomorphism; if different bases for V and W are used, an isomorphic direct sum would be obtained.*

Proof This follows because the dimension of $V \oplus W$ is equal to the sum of the dimensions of V and W and vector spaces of the same dimension are isomorphic. □

Exercise 1.6 Construct an isomorphism to show that $\mathbb{R}^3 \cong \mathbb{R}^2 \oplus \mathbb{R}$.

Exercise 1.7 Construct an isomorphism between $\mathbb{R} \oplus \mathbb{R}^2 \oplus \mathbb{R}^3$ and $\mathbb{R}^2 \oplus \mathbb{R}^4$.

Lemma 1.3 *If $\{v_1, \dots, v_n\}$ is a linearly independent subset of V, then*

$$\text{span}\{v_1, \dots, v_n\} \cong \text{span}\{v_1\} \oplus \cdots \oplus \text{span}\{v_n\} = \bigoplus_{k=1}^{n} \text{span}\{v_k\}$$

Proof Since $\{v_1, \dots, v_n\}$ is linearly independent, it is a basis for its span. On the other hand, $\{v_i\}$ is a basis for $\text{span}\{v_i\}$. Thus the direct sum of all such spans has $\{v_1, \dots, v_n\}$ as its basis. □

Example 1.7 Let U, V, and W be vector spaces and let $m_1 : U \to V, m_2 : U \to W$ be two linear maps. There is a unique linear map $m : U \to V \oplus W$ so that

$$m_1 = \text{pr}_1 \circ m$$

and

$$m_2 = \text{pr}_2 \circ m,$$

where pr_1 is the *projection map* $V \oplus W \to V$ given by

$$\mathrm{pr}_1(x, y) := x$$

and pr_2 is the projection map $V \oplus W \to W$ given by

$$\mathrm{pr}_2(x, y) := y.$$

The above construction can be summarized by saying that the following diagram *commutes*. which means that the composition of consecutive maps in the diagram is path independent,

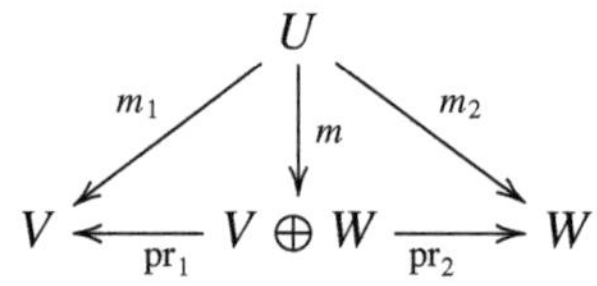

We will encounter many more commutative diagrams in this book (see Definition 1.6) as they are an efficient way to express complicated relationships between spaces and linear maps.

Definition 1.6 A *commutative diagram* is a collection of vector spaces and linear maps between them such that the composition of a sequence of consecutive linear maps depends only on the starting and ending vector spaces in the diagram.

Commutative diagrams are an efficient way to express a large, interrelated collection of matrix equations. Instead of writing the equations, we will often draw the diagram and assert that it must *commute*. However, there is no universally-agreed-upon way to indicate that a particular diagram is commutative or not. Always read the text surrounding any given diagram to determine if it is supposed to commute! Although arrows in a commutative diagram are often labeled, those that are not labeled can be inferred from the text.

Example 1.8 The following diagram is commutative,

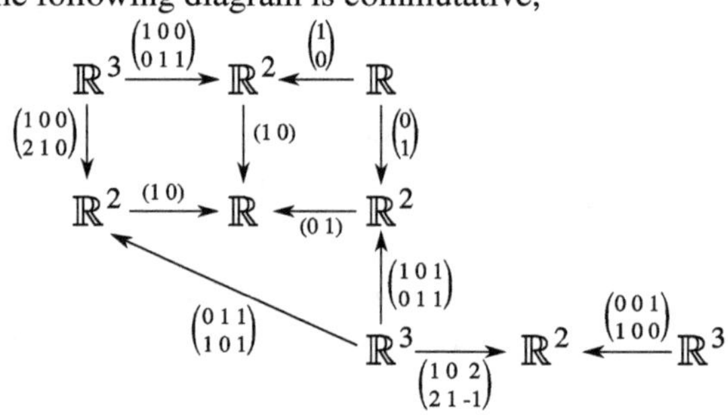

For instance, we can check that

$$(1\ 0)\begin{pmatrix}1&0&0\\0&1&1\end{pmatrix} = (1\ 0)\begin{pmatrix}1&0&0\\2&1&0\end{pmatrix}$$

Exercise 1.8 Here is another commutative diagram

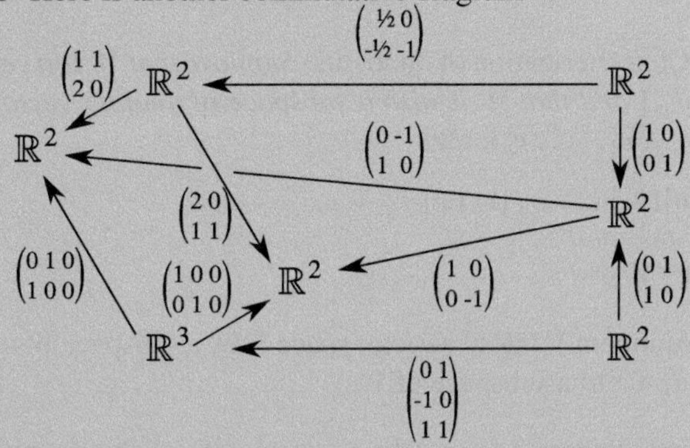

Verify that it is commutative by computation. (Hint: you need not check *every* path. How many matrix multiplications are sufficient?)

Commutative diagrams are called *commutative* because they generalize the idea of a commutative operation. For instance, consider the following diagram for a set V and two functions f and g.

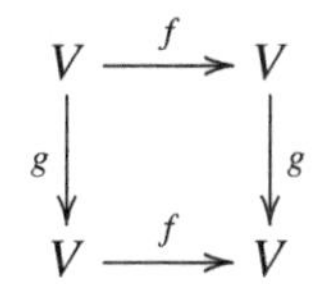

If this diagram commutes, then

$$g \circ f = f \circ g.$$

Exercise 1.9 Make the following diagram commute by choosing a matrix for the arrow labeled with a question mark.

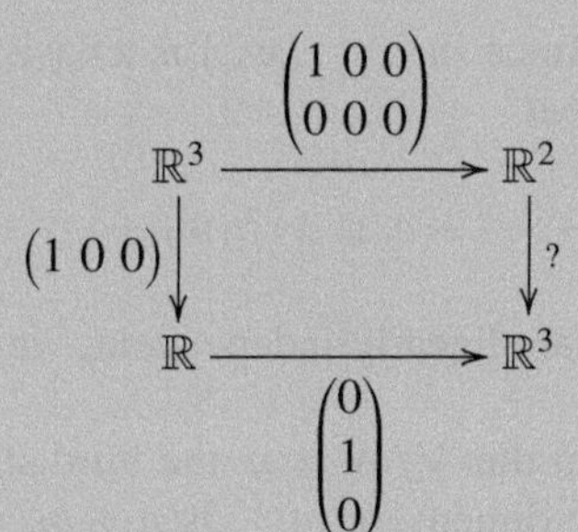

(Hint: There are many possible answers.)

With Lemma 1.3 in hand, we can characterize the quotients of a vector space in terms of bases.

Theorem 1.1 (Characterization of quotients) *Suppose that W is a vector space with basis $\{w_1, \ldots, w_m\}$ and that W is also a subspace of another vector space V with basis $\{w_1, \ldots, w_m, v_1, \ldots, v_n\}$, then*

1. $V/W = \mathrm{span}\{[v_1]_W, \ldots, [v_n]_W\}$,
2. $\dim V/W = n$, *and*
3. $V \cong W \oplus (V/W)$.

Although the quotient V/W of a vector space V by a subspace W *is not* a subspace of V, it *is isomorphic to* a subspace of V.

Proof 1. The first statement is merely a restatement of the definition of quotient space.
2. Since the $\{v_1, \ldots, v_n\}$ are a subset of a basis, they are automatically linearly independent. By Lemma 1.3, $\dim \mathrm{span}\{v_1, \ldots, v_n\} = n$. We claim that $\{[v_1]_W, \ldots, [v_n]_W\}$ is a linearly independent set of elements of V/W, since this implies the second statement

$$\dim V/W = \dim \mathrm{span}\{[v_1]_W, \ldots, [v_n]_W\} = \dim \mathrm{span}\{v_1, \ldots, v_n\} = n.$$

To show linear independence, suppose that there are $a_1, \ldots, a_n$ such that

$$[0]_W = a_1[v_1]_W + \cdots + a_n[v_n]_W.$$

This is equivalent to the statement that for all $z_1, \ldots, z_n \in W$,

$$a_1(v_1 + z_1) + \cdots + a_n(v_n + z_n) \in W.$$

Rearranging, this means that

$$a_1 v_1 + \cdots + a_n v_n \in W.$$

since W is a subspace. Since $\{w_1, \ldots, w_m\}$ is a basis for W, we can find coefficients $b_1, \ldots, b_m$ such that

$$a_1 v_1 + \cdots + a_n v_n = b_1 w_1 + \cdots + b_m w_m.$$

But since the v_i and w_i are linearly independent, we must conclude that $a_i = 0$ for all i.
3. Since we just proved that $\dim V/W = n$, the third statement merely follows by dimension counting: the dimension of V is $n + m$, while the dimension of W is m. □

Bases for quotient spaces are not canonical. First, there is the obvious flexibility of the basis choice that comes with every vector space. Second, there is flexibility in choosing the subspace representing the quotient itself.

To illustrate this, consider the quotient of $\mathbb{R}^3$ by the one dimensional subspace

$$V := \text{span}\left\{\begin{pmatrix}1\\0\\0\end{pmatrix}\right\}.$$

Two distinct bases for this space are

$$\left\{\left[\begin{pmatrix}0\\1\\0\end{pmatrix}\right]_V, \left[\begin{pmatrix}0\\0\\1\end{pmatrix}\right]_V\right\}$$

and

$$\left\{\left[\begin{pmatrix}0\\1\\1\end{pmatrix}\right]_V, \left[\begin{pmatrix}0\\1\\-1\end{pmatrix}\right]_V\right\}.$$

On the other hand, an isomorphic quotient space is $\mathbb{R}^3/W$ where

$$W := \text{span}\left\{\begin{pmatrix}0\\1\\0\end{pmatrix}\right\}.$$

Exercise 1.10 Consider the quotient of $\mathbb{R}^4$ by the subspace W spanned by the following two vectors,

$$\begin{pmatrix}-4\\4\\-2\\-1\end{pmatrix} \text{ and } \begin{pmatrix}-1\\2\\-3\\-2\end{pmatrix}.$$

1. Compute a basis for the quotient $\mathbb{R}^4/W$.
2. Find another basis for the quotient $\mathbb{R}^4/W$ using the same subspace W.
3. Find a third basis for a different quotient $\mathbb{R}^4/\mathbb{R}^2$ that is isomorphic to $\mathbb{R}^4/W$, but not identical to it.

Exercise 1.11 Find an isomorphism between $\mathbb{R}^5/A$ and $\mathbb{R}^4/B$ where

$$A := \operatorname{span}\left\{\begin{pmatrix}1\\0\\0\\0\\0\end{pmatrix}, \begin{pmatrix}0\\1\\0\\0\\0\end{pmatrix}, \begin{pmatrix}0\\0\\1\\0\\0\end{pmatrix}\right\}$$

and

$$B := \operatorname{span}\left\{\begin{pmatrix}1\\1\\1\\1\end{pmatrix}, \begin{pmatrix}1\\-1\\1\\1\end{pmatrix}\right\}.$$

Using Theorem 1.1 with the kernel and image of a linear map, we can observe that the linear map is compatible with a particular decomposition of its domain and codomain. The following theorem is justly famous, see for instance Strang (2006), where it is called the *dimension theorem.*

Theorem 1.2 (The dimension theorem) *For any linear map* $\ell : V \to W$,

1. $V \cong \ker \ell \oplus \operatorname{coimage} \ell$ *and*
2. $W \cong \operatorname{image} \ell \oplus \operatorname{coker} \ell$.

Proof This is an immediate consequence of Theorem 1.1 and the definition of coimage and cokernel (Definition 1.4) because

$$V \cong \ker \ell \oplus (V/\ker \ell) = \ker \ell \oplus \operatorname{coimage} \ell$$

and

$$W \cong \operatorname{image} \ell \oplus (W/\operatorname{image} \ell) = \operatorname{image} \ell \oplus \operatorname{coker} \ell.$$

□

1.2 Quotients and Linear Maps

Definition 1.7 If $f : V_1 \to V_2$ and $g : W_1 \to W_2$ are linear maps, then $(f \oplus g) : (V_1 \oplus W_1) \to (V_2 \oplus W_2)$ is given by

$$(f \oplus g)(v, w) := (f(v), g(w)).$$

As a matrix, $(f \oplus g)$ has the block diagonal form

$$\left(\begin{array}{c|c} f & 0_{W_1,V_2} \\ \hline 0_{V_1,W_2} & g \end{array}\right).$$

While simple in appearance, block form is very useful. Being able to rewrite linear maps in block form is fundamental to many constructions in this book. The key is to be careful with the bookkeeping!

Example 1.9 Consider the following 2×2 matrix, which is not in block diagonal form

$$\ell = \begin{pmatrix} 1 & 2 \\ 3 & 1 \end{pmatrix}.$$

We can factor the matrix into the product of a diagonal matrix and a full rank matrix.

$$\begin{pmatrix} 1 & 2 \\ 3 & 1 \end{pmatrix} = \begin{pmatrix} 1 & -2/5 \\ 3 & -1/5 \end{pmatrix} \begin{pmatrix} 1 & 0 \\ 0 & -5 \end{pmatrix}$$

We can reinterpret this calculation as factoring $\ell : \mathbb{R}^2 \to \mathbb{R}^2$ into two maps

$$\mathbb{R} \oplus \mathbb{R} \xrightarrow{(1)\oplus(-5)} \mathbb{R} \oplus \mathbb{R} \cong \mathbb{R}^2 \xrightarrow{m} \mathbb{R}^2,$$

where (1) and (-5) are 1×1 matrices and

$$m = \begin{pmatrix} 1 & -2/5 \\ 3 & -1/5 \end{pmatrix}.$$

Definition 1.8 If V is a vector space and $W \subseteq V$ is a subspace, then the linear map $p : V \to V/W$ given by

$$p(x) := [x]_W$$

is called the *canonical projection*.

Proposition 1.3 *The canonical projection is surjective and the kernel of the canonical projection $p : V \to V/W$ is equal to W.*

Proof Surjectivity follows from the definition of the quotient: if $[x]_W \in V/W$, then $p(x) = [x]_W$.

Since $[0]_W = [w]_W$ (as sets) for any $w \in W$, then $W \subseteq \ker p$. On the other hand, if $x \in \ker p$, this means that $[x]_W = [0]_W$, which implies $x \in W$ by definition. □

Suppose that $\{v_1, \ldots v_n\}$ is a basis for a vector space V and that $W = \operatorname{span}\{v_1, \ldots, v_k\}$ is spanned by the first k basis vectors. The canonical projection $p : V \to V/W$ has the block matrix form

$$0_{\mathbb{R}^k, \mathbb{R}^{n-k}} \oplus \mathrm{id}_{\mathbb{R}^{n-k}} = \left(\begin{array}{cccc|cccc} 0 & 0 & \cdots & 0 & 1 & 0 & \cdots & 0 \\ 0 & 0 & \cdots & 0 & 0 & 1 & \cdots & 0 \\ \vdots & \vdots & & \vdots & \vdots & \vdots & & \vdots \\ 0 & 0 & \cdots & 0 & 0 & 0 & \cdots & 1 \end{array}\right)$$

in which the first k columns appear to the left of the vertical bar and the last $(n - k)$ columns appear to the right of it.

Example 1.10 Let us find the canonical projection p_W for a given subspace W, using a method based on Algorithm 1. Consider the 1-dimensional subspace W given by

$$W = \operatorname{span}\left\{\begin{pmatrix}5\\9\\12\\0\end{pmatrix}\right\}.$$

If we run Algorithm 1 to obtain a basis for $\mathbb{R}^4/W$, we find that

$$\operatorname{span}\left\{\left[\begin{pmatrix}1\\0\\0\\0\end{pmatrix}\right]_W, \left[\begin{pmatrix}0\\1\\0\\0\end{pmatrix}\right]_W, \left[\begin{pmatrix}0\\0\\0\\1\end{pmatrix}\right]_W\right\}$$

is a basis for $\mathbb{R}^4/W$. Because of Theorem 1.1, we know that

$$\begin{pmatrix}5 & 1 & 0 & 0\\9 & 0 & 1 & 0\\12 & 0 & 0 & 0\\0 & 0 & 0 & 1\end{pmatrix}$$

is of full rank, since its columns correspond to the bases for W and for $\mathbb{R}^4/W$. We know that p_W should be surjective, and therefore will have rank 3. It also should have kernel precisely equal to W. This can be summarized by the matrix equation

$$p_W\begin{pmatrix}5 & 1 & 0 & 0\\9 & 0 & 1 & 0\\12 & 0 & 0 & 0\\0 & 0 & 0 & 1\end{pmatrix} = \begin{pmatrix}0 & 1 & 0 & 0\\0 & 0 & 1 & 0\\0 & 0 & 0 & 1\end{pmatrix}$$

Therefore

$$\begin{aligned}p_W &= \begin{pmatrix}0 & 1 & 0 & 0\\0 & 0 & 1 & 0\\0 & 0 & 0 & 1\end{pmatrix}\begin{pmatrix}5 & 1 & 0 & 0\\9 & 0 & 1 & 0\\12 & 0 & 0 & 0\\0 & 0 & 0 & 1\end{pmatrix}^{-1}\\ &= \begin{pmatrix}1 & 0 & -5/12 & 0\\0 & 1 & -3/4 & 0\\0 & 0 & 0 & 1\end{pmatrix}.\end{aligned}$$

It is easy to see that W lies in the kernel of this matrix. (The reader should check this!)

Exercise 1.12 In the previous example, show that W lies in the kernel of

$$\begin{pmatrix} 1 & 0 & -5/12 & 0 \\ 0 & 1 & -3/4 & 0 \\ 0 & 0 & 0 & 1 \end{pmatrix}.$$

The technique in Example 1.10 can be formalized into Algorithm 3 for constructing the canonical projection. The matrix q in Algorithm 3 is of full rank because of claim (3) of Theorem 1.1. Therefore, the inverse matrix q^{-1} is well defined.

Algorithm 3 Computing the canonical projection

Require: A basis $\{v_1, \ldots, v_n\}$ for V, and a basis $\{w_1, \ldots, w_m\}$ for $W \subseteq V$ (not assuming that the basis of W is a subset of the basis for V)

Ensure: An $(n-m) \times n$ matrix p defining the canonical projection $p : V \to V/W$

1: **procedure** CANONICALPROJECTION($\{v_i\}$,$\{w_j\}$)

2: Run Algorithms 1 or 2 to obtain a set of vectors $S = \{z_1, \ldots, z_{n-m}\}$ for which $\{[z_1]_W, \ldots, [z_{n-m}]_W\}$ spans V/W.

3: Form the matrix

$$q = \begin{pmatrix} w_1 & w_2 & \cdots & w_m & z_1 & z_2 & \cdots & z_{n-m} \end{pmatrix}$$

whose columns are the listed vectors.

4: Compute $p' = q^{-1}$, the inverse matrix for q.

5: Define p to be the last $n-m$ rows of p'.

6: **return** p.

7: **end procedure**

Exercise 1.13 Consider the following basis for $V \cong \mathbb{R}^2$,

$$V = \text{span}\left\{\begin{pmatrix} 1 \\ 0 \end{pmatrix}, \begin{pmatrix} 1 \\ 1 \end{pmatrix}\right\}.$$

Use Algorithm 3 to compute the canonical projection matrix for the subspace $W \subseteq$ given by

$$W = \text{span}\left\{\begin{pmatrix} 1 \\ -1 \end{pmatrix}\right\}.$$

Exercise 1.14 Use Algorithm 3 to compute the canonical projection matrix for the subspace $W \subseteq \mathbb{R}^3$ given by

$$W = \operatorname{span}\left\{ \begin{pmatrix} 2 \\ 4 \\ 0 \end{pmatrix}, \begin{pmatrix} 0 \\ 3 \\ 2 \end{pmatrix} \right\}.$$

Assume that the basis for V is the usual one, namely

$$V = \operatorname{span}\left\{ \begin{pmatrix} 1 \\ 0 \\ 0 \end{pmatrix}, \begin{pmatrix} 0 \\ 1 \\ 0 \end{pmatrix}, \begin{pmatrix} 0 \\ 0 \\ 1 \end{pmatrix} \right\}.$$

Exercise 1.15 Use Algorithm 3 to compute the canonical projection matrix for the subspace $W \subseteq \mathbb{R}^3$ given by

$$W = \operatorname{span}\left\{ \begin{pmatrix} 2 \\ -1 \\ 0 \end{pmatrix} \right\}.$$

Assume that the basis for V is the usual one, namely

$$V = \operatorname{span}\left\{ \begin{pmatrix} 1 \\ 0 \\ 0 \end{pmatrix}, \begin{pmatrix} 0 \\ 1 \\ 0 \end{pmatrix}, \begin{pmatrix} 0 \\ 0 \\ 1 \end{pmatrix} \right\}.$$

Exercise 1.16 Suppose that V is the vector space spanned by

$$V = \operatorname{span}\left\{ \begin{pmatrix} 0 \\ 1 \\ 0 \end{pmatrix}, \begin{pmatrix} 0 \\ 0 \\ -2 \end{pmatrix} \right\}.$$

Use Algorithm 3 to compute the canonical projection matrix for the subspace $W \subseteq V$ given by

$$W = \operatorname{span}\left\{ \begin{pmatrix} 0 \\ 3 \\ 3 \end{pmatrix} \right\}.$$

Proposition 1.4 *Suppose that $\ell : V \to W$ is a linear map and that $U \subseteq W$ is a subspace. Then there is a unique linear map $\ell' : V \to W/U$ given by $\ell' := p \circ \ell$.*

We say that ℓ' is the result of ℓ *descending* to the quotient W/U.

Proof The proof of this proposition follows because ℓ' is specified completely by the expression $p \circ \ell$. □

Proposition 1.4 is summarized by the diagram

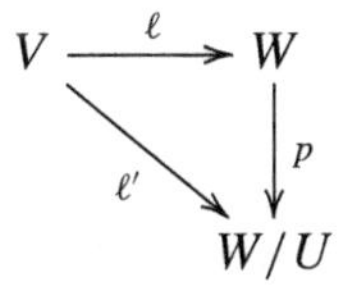

which is an example of a commutative diagram.

Proposition 1.5 *Suppose that $\ell : V \to W$ is a linear map and that $U \subseteq V$ is a subspace. There is a unique linear map $\ell' : V/U \to W$ that satisfies $\ell = \ell' \circ p$ if and only if $U \subseteq \ker \ell$.*

In a similar manner to Proposition 1.4, we say that ℓ' is the result of ℓ *descending* to the quotient V/U.

The proposition is summarized by the commutative diagram

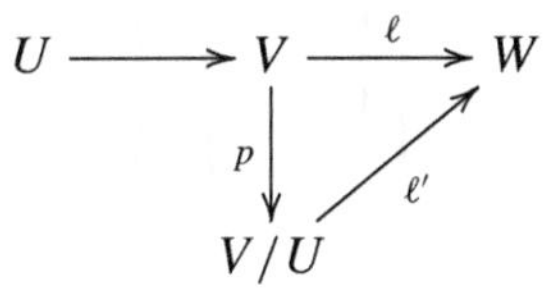

Notice that we have not labeled the arrow $U \to V$, since this can be inferred to be the injective linear map corresponding to the subset relationship $U \subseteq V$. That is, this map takes each element x of U to be the same x but now as an element of V.

Proof Suppose first that $U \subseteq \ker \ell$. Under this condition, we would need to define

$$\ell'([x]_U) := \ell(x), \tag{1.1}$$

which is a restatement of $\ell = \ell' \circ p$. We must show that this is well defined. Suppose that $x, y \in V$ such that $[x]_U = [y]_U$, which means $(x - y) \in U$. Since $U \subseteq \ker \ell$, this means

$$\begin{aligned} 0 &= \ell(x - y) \\ &= \ell(x) - \ell(y) \\ &= \ell'([x]_U) - \ell'([y]_U). \end{aligned}$$

Thus, different representatives of a U-coset have the same value under the proposed definition for ℓ'. It is well defined. Any other map satisfying the same properties as ℓ' must take the same values on V by Eq. 1.1. The resulting ℓ' map is unique.

Now suppose that there is a unique $\ell' : V/U \to W$ that satisfies $\ell = \ell \circ p$. Let $u \in U$ be given. Observe that

$$\ell(u) = (\ell' \circ p)(u) = \ell'(p(u)) = \ell'([u]_U) = \ell'([0]_U) = 0;$$

the final expression following from the linearity of ℓ'. Thus $u \in \ker \ell$. □

Example 1.11 Consider the matrix

$$m = \begin{pmatrix} 3 & 1 & -2 & 4 \\ 0 & 4 & -3 & -3 \end{pmatrix}$$

representing a linear map $m : \mathbb{R}^4 \to \mathbb{R}^2$. To compute the kernel of m, let us first apply a row operation,

$$\begin{pmatrix} 4 & -1 \\ 0 & 1 \end{pmatrix} \begin{pmatrix} 3 & 1 & -2 & 4 \\ 0 & 4 & -3 & -3 \end{pmatrix} = \begin{pmatrix} 12 & 0 & -5 & 19 \\ 0 & 4 & -3 & -3 \end{pmatrix}$$

Notice that this new matrix has the same kernel as m because the row operation matrix is of full rank, and the kernel can be read off from the last two columns,

$$\ker m = \operatorname{span} \left\{ \begin{pmatrix} 5/12 \\ 3/4 \\ 1 \\ 0 \end{pmatrix}, \begin{pmatrix} -19/12 \\ 3/4 \\ 0 \\ 1 \end{pmatrix} \right\}.$$

Let us now consider a 1-dimensional subspace V given by

$$V = \operatorname{span} \left\{ \begin{pmatrix} 1 \\ 0 \\ 0 \\ 0 \end{pmatrix} \right\}.$$

Since V is a cardinal plane, the quotient $\mathbb{R}^4/V$ can be constructed from the standard basis,

$$\operatorname{span} \left\{ \left[\begin{pmatrix} 0 \\ 1 \\ 0 \\ 0 \end{pmatrix} \right]_V, \left[\begin{pmatrix} 0 \\ 0 \\ 1 \\ 0 \end{pmatrix} \right]_V, \left[\begin{pmatrix} 0 \\ 0 \\ 0 \\ 1 \end{pmatrix} \right]_V \right\}$$

Notice that V is not a subspace of $\ker m$, which means that the hypothesis of Proposition 1.5 is not satisfied. If we try to compute a linear map $m' : \mathbb{R}^4/V \to \mathbb{R}^3$ using our basis for $\mathbb{R}^4/V$, we can nevertheless extract columns from m to obtain

$$m'_V = \begin{pmatrix} 1 & -2 & 4 \\ 4 & -3 & -3 \end{pmatrix}$$

Now, it should be the case that $(m'_V \circ p_V) = m$, where $p_V : \mathbb{R}^4 \to \mathbb{R}^4/V$ is the canonical projection. Given our basis for $\mathbb{R}^4/V$

$$p_V = \begin{pmatrix} 0 & 1 & 0 & 0 \\ 0 & 0 & 1 & 0 \\ 0 & 0 & 0 & 1 \end{pmatrix},$$

we can check if this is correct, by computing

$$\begin{aligned} m'_V \circ p_V &= \begin{pmatrix} 1 & -2 & 4 \\ 4 & -3 & -3 \end{pmatrix} \begin{pmatrix} 0 & 1 & 0 & 0 \\ 0 & 0 & 1 & 0 \\ 0 & 0 & 0 & 1 \end{pmatrix} \\ &= \begin{pmatrix} 0 & 1 & -2 & 4 \\ 0 & 4 & -3 & -3 \end{pmatrix}, \end{aligned}$$

which is definitely not m. This indicates that m did *not* descend to the quotient $\mathbb{R}^4/V$.

In contrast, consider the 1-dimensional subspace W given by

$$W = \operatorname{span}\left\{ \begin{pmatrix} 5 \\ 9 \\ 12 \\ 0 \end{pmatrix} \right\}.$$

Because $W \subset \ker m$, the hypotheses of Proposition 1.5 are satisfied. This means that m descends to a linear map $m' : \mathbb{R}^4/W \to \mathbb{R}^2$. To explore what this means, we need to compute a basis for $\mathbb{R}^4/W$. For instance, we can use Algorithm 1 to find that

$$\operatorname{span}\left\{ \left[\begin{pmatrix} 1 \\ 0 \\ 0 \\ 0 \end{pmatrix}\right]_W, \left[\begin{pmatrix} 0 \\ 1 \\ 0 \\ 0 \end{pmatrix}\right]_W, \left[\begin{pmatrix} 0 \\ 0 \\ 0 \\ 1 \end{pmatrix}\right]_W \right\}$$

is a basis for $\mathbb{R}^4/W$. The canonical projection for this quotient is a bit different from before, but was already computed in Example 1.10 as

$$p_W = \begin{pmatrix} 1 & 0 & -5/12 & 0 \\ 0 & 1 & -3/4 & 0 \\ 0 & 0 & 0 & 1 \end{pmatrix}$$

Using the basis we found for $\mathbb{R}^4/W$, m' is constructed by concatenating the first, second, and last columns of m

$$m'_W = \begin{pmatrix} 3 & 1 & 4 \\ 0 & 4 & -3 \end{pmatrix}$$

Let us see whether $m'_W \circ p_W = m$, where p_W is the canonical projection. We compute

$$\begin{aligned} m'_W \circ p_W &= \begin{pmatrix} 3 & 1 & 4 \\ 0 & 4 & -3 \end{pmatrix} \begin{pmatrix} 1 & 0 & -5/12 & 0 \\ 0 & 1 & -3/4 & 0 \\ 0 & 0 & 0 & 1 \end{pmatrix} \\ &= \begin{pmatrix} 3 & 1 & -2 & 4 \\ 0 & 4 & -3 & -3 \end{pmatrix} = m \end{aligned}$$

as desired.

Exercise 1.17 Consider the linear map $\ell : \mathbb{R}^3 \to \mathbb{R}^2$ given by the matrix

$$\ell := \begin{pmatrix} 1 & -1 & 2 \\ 1 & 1 & -3 \end{pmatrix}.$$

1. Using Proposition 1.4, find the map $\ell' : \mathbb{R}^3 \to \mathbb{R}^2/V$, where V is subspace spanned by

$$\begin{pmatrix} 1 \\ 1 \end{pmatrix}.$$

2. Does the subspace U spanned by

$$\begin{pmatrix} 1 \\ 5 \\ 2 \end{pmatrix}$$

satisfy the hypotheses of Proposition 1.5? If so, compute the associated map on the quotient $\mathbb{R}^3/U$.

3. Does the subspace U spanned by

$$\begin{pmatrix} 1 \\ 3 \\ 2 \end{pmatrix}$$

satisfy the hypotheses of Proposition 1.5? If so, compute the associated map on the quotient $\mathbb{R}^3/U$.

Exercise 1.18 Consider the linear map $m : \mathbb{R}^3 \to \mathbb{R}^3$ given by

$$m := \begin{pmatrix} 2 & 1 & -2 \\ 1 & -1 & -1 \\ 0 & -2 & 1 \end{pmatrix}$$

Suppose that U is the subspace given by

$$U := \text{span}\left\{ \begin{pmatrix} 1 \\ 1 \\ 1 \end{pmatrix} \right\},$$

and V is the subspace given by

$$V := \text{span}\left\{ \begin{pmatrix} 1 \\ 1 \\ 0 \end{pmatrix}, \begin{pmatrix} 2 \\ 0 \\ -1 \end{pmatrix} \right\}.$$

1. Apply Proposition 1.4 to obtain a map $m' : \mathbb{R}^3 \to \mathbb{R}^3/V$.
2. Observe that m' has a larger kernel than m. Compute the kernel, and
3. Verify that the subspace U is contained in that kernel.
4. Apply Proposition 1.5 to obtain a map $m'' : \mathbb{R}^3/U \to \mathbb{R}^3/V$ from m'.

Propositions 1.4 and 1.5 can be combined into a single, more useful result.

Theorem 1.3 (Conditions for descending to the quotient) *Suppose that $\ell : V \to W$ is a linear map with $U \subseteq V$ and $X \subseteq W$ such that $\ell(U) \subseteq X$. Then there is a unique* induced map *$\ell' : V/U \to W/X$ that makes the following diagram commute*

$$\begin{array}{ccc} U & & X \\ \downarrow & & \downarrow \\ V & \xrightarrow{\ \ell\ } & W \\ \downarrow{\scriptstyle p} & & \downarrow{\scriptstyle q} \\ V/U & \xrightarrow{\ \ell'\ } & W/X \end{array}$$

in which the unlabeled vertical maps are inclusions and $p : V \to V/U$, $q : W \to W/X$ are the canonical projections.

As before, we say that ℓ' is the result of ℓ *descending to the quotient*.

Proof First, use Proposition 1.5 to construct the map m in the diagram below

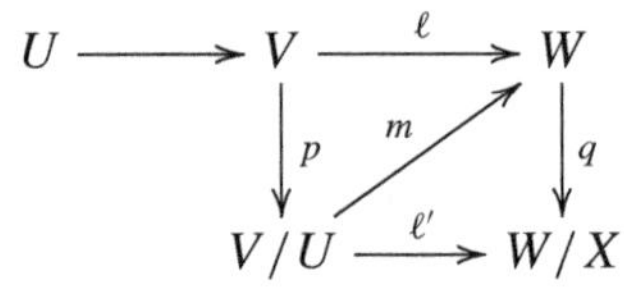

so that $\ell = m \circ p$. Then construct $\ell' := q \circ m$ using Proposition 1.4. The diagram commutes since $\ell' \circ p = (q \circ m) \circ p = q \circ (m \circ p) = q \circ \ell$. □

Exercise 1.19 Show that if the following diagram of vector spaces is commutative, then Theorem 1.3 induces a linear map $\ell' : V/f(U) \to W/g(X)$.

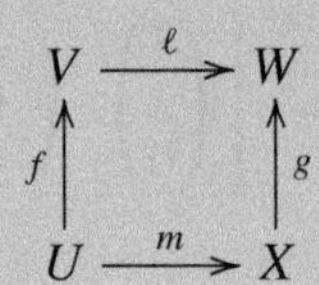

Theorem 1.3 is used frequently throughout the later chapters of this book, and it is often necessary to compute a matrix representation for the induced map on quotients in terms of given bases. Fortunately, the proof of Theorem 1.3 is constructive and straightforward. It can be recast as an algorithm.

Algorithm 4 Computing the map induced on quotients by subspaces

Require: A linear map $\ell : V \to W$ represented as a matrix, a basis $\{u_1, u_2, \ldots, u_m\} \subseteq V$ of a subspace $U \subseteq V$, and a basis $\{x_1, x_2, \ldots, x_n\} \subseteq W$ of a subspace $X \subseteq W$.
Ensure: The induced map $\ell' : V/U \to W/X$.

1: **procedure** INDUCEDLINEARMAP(ℓ, $\{u_i\}$, $\{x_j\}$)
2: Compute a basis $\{z_1, z_2, \ldots, z_k\}$ for V/U using Algorithms 1 or 2.
3: Form the matrix

$$\zeta := \begin{pmatrix} z_1 & z_2 & \cdots & z_k \end{pmatrix}.$$

4: Compute the canonical projection $q : W \to W/X$ using Algorithm 3.
5: Compute

$$m := q \circ \ell \circ \begin{pmatrix} u_1 & u_2 & \cdots & u_m \end{pmatrix}.$$

6: **if** $m \neq 0$ **then**
7: Produce an error. ▷ $\ell(U)$ is not a subset of X
8: **end if**
9: **return** the induced map $\ell' = q \circ \ell \circ \zeta$.
10: **end procedure**

Notice that the matrix in step (2) of Algorithm 4 defines an injective map $\zeta : V/U \to V$.

Example 1.12 Consider the diagram

$$\begin{array}{ccc} U=\mathbb{R} & & X=\mathbb{R}^2 \\ \begin{pmatrix}1\\1\end{pmatrix}\Big\downarrow & & \Big\downarrow\begin{pmatrix}1&0\\0&1\\0&0\end{pmatrix} \\ V=\mathbb{R}^2 & \xrightarrow{\begin{pmatrix}1&2\\3&4\\1&-1\end{pmatrix}} & W=\mathbb{R}^3 \\ (-1\ 1)\Big\downarrow & & \Big\downarrow(0\ 0\ 1) \\ V/U & \xrightarrow{\ \ell'\ } & W/X \end{array}$$

We can verify that this diagram satisfies the hypothesis of Theorem 1.3 because

$$\begin{pmatrix}1&2\\3&4\\1&-1\end{pmatrix}\begin{pmatrix}1\\1\end{pmatrix}=\begin{pmatrix}3\\7\\0\end{pmatrix},$$

which is in the image of the map $X \to W$. Notice also that the vertical maps compose to zero maps, which is to say that the U and X subspaces map to the zero cosets in the quotients U/V and W/X, respectively. This means that we can also verify the hypothesis of Theorem 1.3 by checking that

$$(0\ 0\ 1)\begin{pmatrix}1&2\\3&4\\1&-1\end{pmatrix}\begin{pmatrix}1\\1\end{pmatrix}=0.$$

Since both U/V and W/X are 1-dimensional spaces, the induced map ℓ' is a 1×1 matrix. To find the single entry in this matrix, let us start with a basis in U/V, for instance the column vector $\{(1)\}$. This vector has $\begin{pmatrix}1\\2\end{pmatrix}$ in V as a preimage, and we can take this vector as a basis for V/U as a subspace of V. Algorithm 4 will then have us define

$$\zeta := \begin{pmatrix}1\\2\end{pmatrix}.$$

Therefore, the induced map should take $\{(1)\}$ in U/V to

$$(0\ 0\ 1)\begin{pmatrix}1&2\\3&4\\1&-1\end{pmatrix}\begin{pmatrix}1\\2\end{pmatrix}=(-1).$$

Therefore, the induced map is simply the 1×1 matrix (-1).

If instead, the map $V \to W$ had the matrix

$$\begin{pmatrix} 1 & 2 \\ 3 & 4 \\ 0 & 0 \end{pmatrix},$$

then the induced map is still defined, but becomes the zero 1×1 matrix.

One can also start with a linear map between quotients.

Theorem 1.4 (Conditions for lifting a map from the quotient) *Suppose that $U \subseteq V$ and $X \subseteq W$ are two vector subspaces and that $\ell : V/U \to W/X$ is a linear map. There exists a linear map $\ell' : V \to W$ such that*

$$\begin{array}{ccc} V & \xrightarrow{\ell'} & W \\ \downarrow{\scriptstyle p} & & \downarrow{\scriptstyle q} \\ V/U & \xrightarrow{\ell} & W/X \end{array}$$

commutes when p and q are the canonical projections.

Proof Using Theorem 1.1, we can write $V \cong U \oplus (V/U)$ and $W \cong X \oplus (W/X)$. We then need only define

$$\ell' := 0_{U,X} \oplus \ell$$

to complete the proof. □

Notice that the linear map ℓ' is not unique in this case, because there may be other choices for the $U \to X$ map besides the zero map.

Exercise 1.20 Consider the map $m : \mathbb{R}^3 \to \mathbb{R}^3$ given by the matrix

$$m = \begin{pmatrix} -1 & 0 & 2 \\ 0 & 0 & -1 \\ -3 & 1 & 1 \end{pmatrix}.$$

Apply Proposition 1.3 and Algorithm 4 to compute the induced map when $U \subseteq \mathbb{R}^3$ is spanned by

$$\left\{ \begin{pmatrix} 1 \\ -1 \\ -1 \end{pmatrix}, \begin{pmatrix} 2 \\ -1 \\ -3 \end{pmatrix} \right\},$$

$V \subseteq \mathbb{R}^3$ is spanned by

$$\left\{ \begin{pmatrix} -2 \\ 1 \\ 0 \end{pmatrix} \right\},$$

and the induced map is constructed along $\mathbb{R}^3/U \to \mathbb{R}^3/V$.

Exercise 1.21 Consider the map $m : \mathbb{R}^3 \to \mathbb{R}^3$ given by the matrix

$$m = \begin{pmatrix} 2 & 0 & 0 \\ 0 & -3 & 1 \\ 1 & -1 & 0 \end{pmatrix}.$$

Apply Proposition 1.3 and Algorithm 4 to compute the induced map when $U \subseteq \mathbb{R}^3$ is spanned by

$$\left\{ \begin{pmatrix} 4 \\ -1 \\ 2 \end{pmatrix}, \begin{pmatrix} -1 \\ 0 \\ -1 \end{pmatrix} \right\},$$

$V \subseteq \mathbb{R}^3$ is spanned by

$$\left\{ \begin{pmatrix} 0 \\ 1 \\ 1 \end{pmatrix} \right\},$$

and the induced map is constructed along $\mathbb{R}^3/U \to \mathbb{R}^3/V$.

Exercise 1.22 Construct a map $m' : \mathbb{R}^3 \to \mathbb{R}^3$ such that it induces the map $m : \mathbb{R}^2 \to \mathbb{R}^2$ given by the matrix

$$m = \begin{pmatrix} -2 & 4 \\ -1 & 3 \end{pmatrix}$$

when the domain is constructed by a quotient with the subspace spanned by

$$\left\{ \begin{pmatrix} -7 \\ -12 \\ 4 \end{pmatrix} \right\}$$

and the codomain is constructed by a quotient with the subspace spanned by

$$\left\{ \begin{pmatrix} 4 \\ -9 \\ 2 \end{pmatrix} \right\}.$$

Exercise 1.23 Construct a map $m' : \mathbb{R}^3 \to \mathbb{R}^3$ such that it induces the map $m : \mathbb{R}^2 \to \mathbb{R}^2$ given by the matrix

$$m = \begin{pmatrix} 0 & 3 \\ 1 & 2 \end{pmatrix}$$

when the domain is constructed by a quotient with the subspace spanned by

$$\left\{ \begin{pmatrix} -13 \\ -3 \\ 1 \end{pmatrix} \right\}$$

and the codomain is constructed by a quotient with the subspace spanned by

$$\left\{ \begin{pmatrix} 2 \\ 0 \\ 3 \end{pmatrix} \right\}.$$

1.3 The Dual Vector Space

Definition 1.9 The *dual vector space* to a vector space V is the vector space V^T that consists of all linear maps $V \to \mathbb{R}$ with domain V and codomain $\mathbb{R}$. Addition in V^T is defined by the following recipe: if $f, g : V \to \mathbb{R}$ are elements of V^T, then $f + g$ is the linear map defined by

$$(f + g)(v) := f(v) + g(v).$$

Likewise, if $a \in \mathbb{R}$, then $a \cdot f$ is the linear map defined by

$$(a \cdot f)(v) := a \cdot f(v).$$

Example 1.13 Consider the vector space $V = \mathbb{R}^2$; its dual space V^T consists of all linear maps $\mathbb{R}^2 \to \mathbb{R}$. Since the $\mathbb{R}^2$ is spanned by two basis elements,

$$\left\{\begin{pmatrix}1\\0\end{pmatrix}, \begin{pmatrix}0\\1\end{pmatrix}\right\},$$

it is clear that any linear map $f : \mathbb{R}^2 \to \mathbb{R}$ must satisfy

$$f\left(a\begin{pmatrix}1\\0\end{pmatrix} + b\begin{pmatrix}0\\1\end{pmatrix}\right) = af\left(\begin{pmatrix}1\\0\end{pmatrix}\right) + bf\left(\begin{pmatrix}0\\1\end{pmatrix}\right).$$

From this, conclude that f is completely determined by exactly two real values: the value it takes on the first basis element, and the value it takes on the second. Therefore, since the dual vector space is a vector space, it must have dimension 2.

On the other hand, the dual vector space V is not equal to V^T in a basis-independent way. To see this, instead consider the basis for V that is spanned by

$$\left\{\begin{pmatrix}1\\-1\end{pmatrix}, \begin{pmatrix}1\\1\end{pmatrix}\right\},$$

for instance. The resulting linear maps have numerically different coefficients than from before.

Exercise 1.24 Consider the two linear maps given by

$$f\left(\begin{pmatrix}x\\y\\z\end{pmatrix}\right) := 2x - 3y + z,$$

and

$$g\left(\begin{pmatrix}x\\y\\z\end{pmatrix}\right) := -x + y - z.$$

What is their sum?

Exercise 1.25 Consider the vector space $V = \mathbb{R}^2$ and two bases defined as in Example 1.13, and the linear maps given by

$$\begin{aligned} f\left(a\begin{pmatrix}1\\0\end{pmatrix} + b\begin{pmatrix}0\\1\end{pmatrix}\right) &= af\left(\begin{pmatrix}1\\0\end{pmatrix}\right) + bf\left(\begin{pmatrix}0\\1\end{pmatrix}\right) \\ &= 2a + 7b. \end{aligned}$$

Rewrite this maps so that it is defined for the other basis in Example 1.13.

Duality is important in part because every linear map $\ell : V \to W$ between vector spaces induces a linear map ℓ^T, called the *dual map* or *transpose* of ℓ, between the dual spaces $\ell^T : W^T \to V^T$.

Definition 1.10 Given an element $f \in W^T$, which is a linear map $f : W \to \mathbb{R}$, define

$$\left(\ell^T f\right)(v) := f\left(\ell(v)\right)$$

where $v \in V$. This is a linear map $V \to \mathbb{R}$ and thus an element of V^T.

Example 1.14 Consider the linear map $\mathbb{R}^3 \to \mathbb{R}^2$ given by the 2×3 matrix

$$m := \begin{pmatrix} 1 & 3 & -4 \\ 1 & -2 & 0 \end{pmatrix}.$$

According to Definition 1.10, the dual map m^T is a linear map $(\mathbb{R}^2)^T \to (\mathbb{R}^3)^T$. Also, according to Example 1.13, every element f of $(\mathbb{R}^2)^T$ can be expressed as a linear combination

$$f\left(\begin{pmatrix} a \\ b \end{pmatrix}\right) = af\left(\begin{pmatrix} 1 \\ 0 \end{pmatrix}\right) + bf\left(\begin{pmatrix} 0 \\ 1 \end{pmatrix}\right) = af_1 + bf_2.$$

Definition 1.10 allows us to define a linear map in $(\mathbb{R}^3)^T$ by way of

$$\begin{aligned}
(m^T f)\left(\begin{pmatrix} x \\ y \\ z \end{pmatrix}\right) &= f\left(m\left(\begin{pmatrix} x \\ y \\ z \end{pmatrix}\right)\right) \\
&= f\left(\begin{pmatrix} 1 & 3 & -4 \\ 1 & -2 & 0 \end{pmatrix}\begin{pmatrix} x \\ y \\ z \end{pmatrix}\right) \\
&= f\left(\begin{pmatrix} x + 3y - 4z \\ x - 2y \end{pmatrix}\right) \\
&= (x + 3y - 4z) f\left(\begin{pmatrix} 1 \\ 0 \end{pmatrix}\right) + (x - 2y) f\left(\begin{pmatrix} 0 \\ 1 \end{pmatrix}\right) \\
&= (x + 3y - 4z) f_1 + (x - 2y) f_2 \\
&= \begin{pmatrix} f_1 & f_2 \end{pmatrix}\begin{pmatrix} 1 & 3 & -4 \\ 1 & -2 & 0 \end{pmatrix}\begin{pmatrix} x \\ y \\ z \end{pmatrix}.
\end{aligned}$$

By transposing, we can interpret the above equation as noting that $m^T(f)$ is given by

$$\begin{aligned} m^T(f) &= m^T \begin{pmatrix} f_1 \\ f_2 \end{pmatrix} \\ &= \begin{pmatrix} 1 & 1 \\ 3 & -2 \\ -4 & 0 \end{pmatrix} \begin{pmatrix} f_1 \\ f_2 \end{pmatrix}. \end{aligned}$$

Proposition 1.6 *For any pair of linear maps* $\ell : V \to W$ *and* $m : W \to X$, ℓ^T *and* m^T *are linear maps that satisfy the equation*

$$(m \circ \ell)^T = \ell^T \circ m^T.$$

Proof From Definition 1.10, both $(m \circ \ell)^T$ and $\ell^T \circ m^T$ are linear maps $X^T \to V^T$. Suppose that $f \in X^T$ and $v \in V$ are given. Since f is therefore a linear map $X \to \mathbb{R}$, what m^T does is produce another linear map $W \to \mathbb{R}$. Applying ℓ^T to $m^T(f)$ yields a linear map $V \to \mathbb{R}$. The commutative diagram below describes the situation,

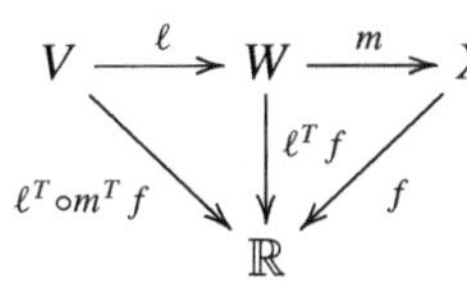

So that

$$\begin{aligned} \left(\ell^T \circ m^T f\right)(v) &= m^T f(\ell(v)) \\ &= f(m(\ell(v))) \\ &= f((m \circ \ell)(v)) \\ &= \left((m \circ \ell)^T f\right)(v), \end{aligned}$$

which establishes the result. □

Exercise 1.26 Prove the following statement or give a counterexample: "If $\ell : V \to W$ is a surjective linear map, then $\ell^T : W \to V$ is an injective linear map."

Suppose that a basis for V is $\{e_1, \ldots, e_n\}$ so that every element $v \in V$ can be written uniquely as a linear combination

$$v = \sum_{i=1}^{n} v^i e_i.$$

Caution: the superscript i is not being used as an exponent! The notation being used here is the *Einstein summation convention*: every variable indexing a sum must appear once as a superscript (indexing over coefficients) and once as a subscript (indexing over vectors). The roles of super- and subscripts are reversed for the elements of the dual space. The pairing of super- and subscripts is a useful visual check that helps avoid typographical errors. Additionally, the summation symbols can be omitted without ambiguity, since every paired index corresponds to a sum over that index. For complicated expressions with many summations, the visual clutter can be reduced considerably. (In this book, we will not omit summation symbols.)

Einstein summation convention is applicable when calculating within a specific basis and is most helpful when the usual matrix notation is awkward. In Chap. 2 and following, we will need other indices as well, which often leads to confusion when Einstein summation convention is also being used. Therefore, we will generally confine the use of Einstein summation to this chapter.

Using this decomposition for an arbitrary v define $e^k : V \to \mathbb{R}$ for $k = 1, \ldots, n$ by

$$e^k(v) = e^k\left(\sum_{i=1}^{n} v^i e_i\right) := v^k.$$

This is well defined by the uniqueness of the decomposition of v, and means that

$$e^k(e_i) = \begin{cases} 0 & \text{if } i \neq k \\ 1 & \text{if } i = k \end{cases} \tag{1.2}$$

by linear independence of the e_i.

Definition 1.11 The set $\{e^1, \ldots, e^n\}$ of linear maps is called the *dual basis* to $\{e_1, \ldots, e_n\}$. (We may take (1.2) as a definition, after extending by linearity.)

The following proposition establishes that the dual basis really is a basis for the dual vector space.

Theorem 1.5 (Self-duality of finite dimensional vector spaces) *If V is a finite dimensional real vector space, then $V \cong V^T$.*

Proof Since both spaces are finite dimensional, we show that the basis of V and the basis of V^T have the same number of elements, and thereby establish that they are isomorphic.

Each e^k is a linear map; let $w := \sum_{j=1}^{n} w^j e_j$ and observe that

$$\begin{aligned} e^k(a \cdot v + b \cdot w) &= e^k\left(a \cdot \sum_{i=1}^{n} v^i e_i + b \cdot \sum_{j=1}^{n} w^j e_j\right) \\ &= e^k\left(\sum_{i=1}^{n} (av^i + bw^i) e_i\right) \\ &= av^i + bw^i. \end{aligned}$$

Thus, each $e^k \in V^T$. It is sufficient to establish that $\{e^1, \dots, e^n\}$ is a basis since this has the same length n as the basis for V.

The $\{e^1, \dots, e^n\}$ are linearly independent. Suppose that there are b_k such that

$$\sum_{k=1}^{n} b_k e^k(v) = 0$$

for all $v \in V$. Then expanding v in the basis for V, we have

$$\begin{aligned} 0 &= \sum_{k=1}^{n} b_k e^k(v), \\ &= \sum_{k=1}^{n} b_k e^k \left(\sum_{i=1}^{n} v^i e_i \right), \\ &= \sum_{k=1}^{n} b_k v^k, \end{aligned}$$

which implies that $b_k = 0$ since the v^k are arbitrary.

The $\{e^1, \dots, e^n\}$ span V^T. Suppose that $\ell \in V^T$, then its value on an arbitrary $v \in V$ is given by

$$\begin{aligned} \ell(v) &= \ell \left(\sum_{k=1}^{n} v^k e_k \right), \\ &= \sum_{k=1}^{n} v^k \ell(e_k), \\ &= \sum_{k=1}^{n} e^k(v) \ell(e_k), \end{aligned}$$

which expresses ℓ as a linear combination of the $\{e^1, \dots, e^n\}$. □

Although Theorem 1.5 establishes that $V \cong V^T$, the proof does not establish a *canonical* choice of isomorphism $T : V \to V^T$. This is intentional, since there is no such canonical choice. One convenient choice of isomorphism is given by extending

$$T(e^i) := e_i,$$

by linearity. The element v in V given by

$$v = \sum_{i=1}^{n} v^i e_i$$

is transformed to the element of V^T given by

$$T(v) = \sum_{i=1}^{n} v^i T(e_i) = \sum_{i=1}^{n} v^i e^i.$$

The violation of the Einstein summation convention indicates something "interesting" is happening! Recalling that T takes each vector in V to one in V^T, we can interpret each element of the image of T as a linear map $V \to \mathbb{R}$. Specifically, if $v \in V$, then $T(v)$ is a linear map $V \to \mathbb{R}$. For instance, if

$$w = \sum_{i=1}^{n} w^i e_i$$

then

$$\begin{aligned}
(T(v))(w) &= \sum_{i=1}^{n} v^i e^i (w) \\
&= \sum_{i=1}^{n} v^i e^i \left(\sum_{j=1}^{n} w^j e_j \right) \\
&= \sum_{i=1}^{n} \sum_{j=1}^{n} v^i w^j e^i (e_j) \\
&= \sum_{i=1}^{n} v^i w^i.
\end{aligned}$$

If we write v and w as column vectors

$$v = \begin{pmatrix} v^1 \\ v^2 \\ \vdots \\ v^n \end{pmatrix}, \quad w = \begin{pmatrix} w^1 \\ w^2 \\ \vdots \\ w^n \end{pmatrix},$$

then $(T(v))(w)$ is their dot product, usually written

$$(T(v))(w) = v \bullet w = \begin{pmatrix} v^1 & v^2 & \cdots & v^n \end{pmatrix} \begin{pmatrix} w^1 \\ w^2 \\ \vdots \\ w^n \end{pmatrix}.$$

Therefore, T performs a vector *transpose* from a column vector to a row vector. In light of this, we will define

$$v^T := T(v).$$

If we additionally define

$$v_i^T := (v^T)_i = v^i$$

for all i, then we can resolve the violation of the Einstein summation convention in the following way

$$T(v) = \sum_{i=1}^{n} v^i T(e_i) = \sum_{i=1}^{n} v_i^T e^i.$$

This idea extends without any difficulty from vectors to linear maps, if we continue to use $T : V \to V^T$ as our choice of isomorphism. If $\ell : \mathbb{R}^m \to \mathbb{R}^n$ has the matrix representation in a basis $\{e_1, \ldots, e_n\}$

$$\ell = \begin{pmatrix} a_{11} & a_{12} & \cdots & a_{1m} \\ a_{21} & a_{22} & \cdots & a_{2m} \\ \vdots & \vdots & & \vdots \\ a_{n1} & a_{n2} & \cdots & a_{nm} \end{pmatrix}$$

then $\ell^T : \mathbb{R}^n \to \mathbb{R}^m$ has the matrix representation which is the transpose of that for ℓ in the basis $\{e^1, \ldots, e^n\}$

$$\ell^T = \begin{pmatrix} a_{11} & a_{21} & \cdots & a_{n1} \\ a_{12} & a_{22} & \cdots & a_{n2} \\ \vdots & \vdots & & \vdots \\ a_{1m} & a_{2m} & \cdots & a_{nm} \end{pmatrix}.$$

Exercise 1.27 Suppose that $\ell : \mathbb{R}^m \to \mathbb{R}^n$ is a linear map between finite dimensional vector spaces. Choose any bases for the domain and codomain of ℓ so that ℓ is written as a matrix. Show that $\ell \circ \ell^T$ and $\ell^T \circ \ell$ are represented by matrices that are

1. Square and
2. Symmetric. For example, if the matrix representation of $\ell^T \circ \ell$ is

$$\ell^T \circ \ell = \begin{pmatrix} a_{11} & a_{21} & \cdots & a_{m1} \\ a_{12} & a_{22} & \cdots & a_{m2} \\ \vdots & \vdots & & \vdots \\ a_{1m} & a_{2m} & \cdots & a_{mm} \end{pmatrix}$$

show that $a_{ji} = a_{ij}$ for all i and j.

Proposition 1.7 *If $f : V_1 \to V_2$ and $g : W_1 \to W_2$ are linear maps between finite dimensional spaces then $(f \oplus g)^T = f^T \oplus g^T$.*

Proof Since the block form of $(f \oplus g)$ is

$$\left(\begin{array}{c|c} f & 0_{W_1,V_2} \\ \hline 0_{V_1,W_2} & g \end{array}\right)$$

the block form of $(f \oplus g)^T$ is

$$\left(\begin{array}{c|c} f^T & (0_{V_1,W_2})^T \\ \hline (0_{W_1,V_2})^T & g^T \end{array}\right)$$

which is precisely $f^T \oplus g^T$ as soon as we notice that $(0_{V_1,W_2})^T = 0_{W_2^T,V_1^T} : W_2^T \to V_1^T$ and $(0_{W_1,V_2})^T = 0_{V_2^T,W_1^T} : V_2^T \to W_1^T$ are both zero maps. □

Theorem 1.6 (Duality between the fundamental subspaces of a linear map) *If $\ell : V \to W$ is a linear map between finite dimensional vector spaces, then*

$$\operatorname{coker} \ell \cong \ker \ell^T$$

and

$$\operatorname{coimage} \ell \cong \operatorname{image} \ell^T.$$

Proof Suppose that $f \in \ker \ell^T \subseteq W^T$. Then

$$(\ell^T f)(v) = f(\ell(v)) = 0$$

for all $v \in V$, which implies that $f(\operatorname{image} \ell) = 0$. In other words, $f : W \to \mathbb{R}$ is a linear function whose kernel contains $\operatorname{image} \ell$. By Proposition 1.4, this means that there is a unique $f' : W/\operatorname{image} \ell \to \mathbb{R}$ for which $f = f' \circ p$ for the canonical projection $p : W \to W/\operatorname{image} \ell$. Thus,

$$\ker \ell^T \cong (\operatorname{coker} \ell)^T \cong \operatorname{coker} \ell$$

in which the second isomorphism is a consequence of Theorem 1.5. Given this fact and Theorem 1.2, it follows that $\operatorname{coimage} \ell \cong \operatorname{image} \ell^T$. □

Example 1.15 Consider the matrix

$$\ell := \begin{pmatrix} 1 & 0 \\ 0 & 2 \\ 0 & 0 \end{pmatrix}.$$

The two columns of ℓ are linearly independent. The rank of ℓ is 2. As a result, the dimension of coker ℓ is 1. Moreover, since the last row of ℓ is all zeros, this means that any 3×1 vector with zeros in the first two rows lies in coker ℓ. Briefly,

$$\operatorname{coker} \ell = \operatorname{span} \left\{ \begin{pmatrix} 0 \\ 0 \\ 1 \end{pmatrix} \right\}.$$

On the other hand, notice that

$$\begin{pmatrix} 1 & 0 \\ 0 & 2 \\ 0 & 0 \end{pmatrix}^T \begin{pmatrix} 0 \\ 0 \\ 1 \end{pmatrix} = \begin{pmatrix} 1 & 0 & 0 \\ 0 & 2 & 0 \end{pmatrix} \begin{pmatrix} 0 \\ 0 \\ 1 \end{pmatrix} = \begin{pmatrix} 0 \\ 0 \end{pmatrix}.$$

Therefore, coker $\ell \subseteq \ker \ell^T$. On the other hand, given that the rank of ℓ^T is still 2, this means that the dimension of $\ker \ell^T$ is 1. We must conclude (in agreement with Theorem 1.6) that coker $\ell = \ker \ell^T$.

Proposition 1.8 *Every vector subspace can be realized as the kernel or the image of a linear map. That is, if A is a subspace of V, then there is a linear map $\ell : V \to W$ such that* $\ker \ell = A$. *Similarly, there is a linear map $m : U \to V$ such that* image $m = A$

Proof Suppose that $\{a_1, \ldots, a_m\}$ is a basis for A. Start by constructing a linear map $m : \mathbb{R}^m \to V$ by its matrix

$$m := \begin{pmatrix} a_1 \cdots a_m \end{pmatrix},$$

which simply means that the columns of m are given by the basis elements of A.

Since these columns are linearly independent by assumption, the image of m is precisely equal to A, thus according to Theorem 1.6,

$$A = \operatorname{image} m \cong \operatorname{coimage} m^T \cong V / \ker m^T.$$

Suppose that we find a basis for $\ker m^T$ of the following form,

$$V = \operatorname{span}\{v_1, \ldots, v_k\}.$$

where

$$k := \dim \ker m^T = \dim V - \dim A.$$

This construction ensures that for each $i = 1, \ldots, k$,

$$0 = m^T v_i = \begin{pmatrix} a_1 \cdots a_m \end{pmatrix}^T v_i.$$

Because of Proposition 1.6, this can be rewritten as

$$0 = v_i^T m = v_i^T \left(a_1 \cdots a_m\right),$$

so that we have the matrix equation

$$0 = \left(v_1 \cdots v_k\right)^T \left(a_1 \cdots a_m\right).$$

This implies that

$$\ell := \left(v_1 \cdots v_k\right)^T$$

has $A = \ker \ell$. □

Example 1.16 Suppose that we want a linear map $m : \mathbb{R}^3 \to \mathbb{R}$ to have a kernel that is spanned by

$$\begin{pmatrix} 1 \\ 0 \\ 0 \end{pmatrix} \text{ and } \begin{pmatrix} 0 \\ 1 \\ 1 \end{pmatrix}.$$

These two vectors are linearly independent, and

$$\begin{pmatrix} 0 \\ 1 \\ -1 \end{pmatrix}$$

is also linearly independent with both of them. Moreover, the transpose of this vector, namely the 1×3 matrix

$$m := (0\ 1\ -1)$$

has the property that

$$m \begin{pmatrix} 1 & 0 \\ 0 & 1 \\ 0 & 1 \end{pmatrix} = (0\ 1\ -1) \begin{pmatrix} 1 & 0 \\ 0 & 1 \\ 0 & 1 \end{pmatrix} = (0\ 0),$$

so this means that m has exactly the kernel we desired.

Example 1.17 Suppose that we want a linear map $\ell : \mathbb{R}^3 \to \mathbb{R}$ to have a kernel that is spanned by

$$\begin{pmatrix} 1 \\ -1 \\ 0 \end{pmatrix} \text{ and } \begin{pmatrix} 0 \\ 1 \\ 1 \end{pmatrix}.$$

The proof of Proposition 1.8 starts with the matrix

$$m = \begin{pmatrix} 1 & 0 \\ -1 & 1 \\ 0 & 1 \end{pmatrix}.$$

We then need to compute the kernel of m^T,

$$\ker m = \operatorname{span}\left\{\begin{pmatrix}1\\1\\-1\end{pmatrix}\right\},$$

so the matrix

$$\ell := \begin{pmatrix}1 & 1 & -1\end{pmatrix}$$

has the desired kernel.

Exercise 1.28 Using the construction in the proof of Proposition 1.8, construct a linear map $\mathbb{R}^3 \to \mathbb{R}^4$ whose kernel is spanned by the following two $\mathbb{R}^3$ vectors,

$$\begin{pmatrix}1\\2\\1\end{pmatrix} \text{ and } \begin{pmatrix}-1\\1\\2\end{pmatrix}.$$

Exercise 1.29 Using the construction in the proof of Proposition 1.8, construct a linear map $\mathbb{R}^4 \to \mathbb{R}^4$ whose kernel is spanned by the following two $\mathbb{R}^4$ vectors,

$$\begin{pmatrix}1\\1\\2\\4\end{pmatrix}, \text{ and } \begin{pmatrix}-1\\1\\3\\1\end{pmatrix}.$$

One of the most useful algorithmic consequences of Theorem 1.6 is an explicit formula for induced maps.

Theorem 1.7 (The projection formula) *Consider the following diagram*

$$\begin{array}{ccc} U & & X \\ \downarrow{\scriptstyle p} & & \downarrow{\scriptstyle q} \\ V & \xrightarrow{\ \ell\ } & W \end{array}$$

in which p and q are full rank maps satisfying $\ell(p(U)) \subseteq q(X)$. The map $m : U \to X$ given by

$$m := (q^T \circ q)^{-1} \circ q^T \circ \ell \circ p \tag{1.3}$$

makes the resulting diagram commute.

Dually, consider

$$\begin{array}{ccc} V & \xrightarrow{\ell} & W \\ \downarrow{\scriptstyle p} & & \downarrow{\scriptstyle q} \\ U & & X \end{array}$$

where p is of full rank and satisfies $\ker p \subseteq \ker \ell$. *The map* $m : U \to X$ *given by*

$$m := q \circ \ell \circ p^T \circ (p \circ p^T)^{-1} \tag{1.4}$$

makes the diagram commute.

Equations (1.3) and (1.4) are called the *projection formulas for the induced map.*

Proof Theorem 1.2 states that the first diagram can be equivalently rewritten as

$$\begin{array}{ccc} U & & X \\ \downarrow{\scriptstyle p} & & \downarrow{\scriptstyle q} \\ U \oplus (V/U) & \xrightarrow{\ell} & X \oplus (W/X), \end{array}$$

with the maps $U \to (V/U) \cong \operatorname{coker} p$ and $X \to (W/X) \cong \operatorname{coker} q$ being zero maps. According to Theorem 1.6, $(V/U) \cong \ker p^T$ and $(W/X) \cong \ker q^T$. The maps $p^T \circ p$ and $q^T \circ q$ are full rank with trivial kernel. They are isomorphisms. Therefore, if we try to solve for a map $m : U \to X$ making the diagram commute, we have

$$\begin{aligned} q \circ m &= \ell \circ p \\ q^T \circ q \circ m &= q^T \circ \ell \circ p \\ m &= (q^T \circ q)^{-1} \circ q^T \circ \ell \circ p. \end{aligned}$$

Dually, consider the second diagram

$$\begin{array}{ccc} V & \xrightarrow{\ell} & W \\ \downarrow{\scriptstyle p} & & \downarrow{\scriptstyle q} \\ U & & X \end{array}$$

The condition that $\ker p \subseteq \ker \ell$ implies that $\operatorname{coker} p^T \subseteq \ker \ell$ and consequently that $\operatorname{coimage} \ell \subseteq \operatorname{image} p^T$. With this in mind, we can solve for an induced map $m : U \to X$

$$\begin{aligned} m \circ p &= q \circ \ell \\ m \circ p \circ p^T &= q \circ \ell \circ p^T \\ m &= q \circ \ell \circ p^T \circ (p \circ p^T)^{-1}. \end{aligned}$$

□

Remark 1.1 If the map q is an isomorphism in Eq. (1.3), then it simplifies to

$$m = (q^T \circ q)^{-1} \circ q^T \circ \ell \circ p = q^{-1} \circ (q^T)^{-1} \circ q^T \circ \ell \circ p = q^{-1} \circ \ell \circ p.$$

A similar simplification occurs for Eq. (1.4) if p is an isomorphism.

Example 1.18 Consider the diagram

$$\begin{array}{ccc} \mathbb{R}^3 & \xrightarrow{\begin{pmatrix} 1 & -1 & 0 \\ 1 & 0 & 1 \\ 0 & 1 & 1 \end{pmatrix}} & \mathbb{R}^3 \\ {\scriptstyle \begin{pmatrix} 1 & -1 & 0 \\ 0 & 1 & 1 \end{pmatrix}} \Big\downarrow & & \Big\downarrow {\scriptstyle \begin{pmatrix} 1 & 1 & 1 \\ 1 & 1 & 0 \end{pmatrix}} \\ \mathbb{R}^2 & & \mathbb{R}^2 \end{array}$$

In Eq. (1.4), we need to compute

$$\begin{aligned} p \circ p^T &= \begin{pmatrix} 1 & -1 & 0 \\ 0 & 1 & 1 \end{pmatrix} \begin{pmatrix} 1 & -1 & 0 \\ 0 & 1 & 1 \end{pmatrix}^T \\ &= \begin{pmatrix} 1 & -1 & 0 \\ 0 & 1 & 1 \end{pmatrix} \begin{pmatrix} 1 & 0 \\ -1 & 1 \\ 0 & 1 \end{pmatrix} \\ &= \begin{pmatrix} 2 & -1 \\ -1 & 2 \end{pmatrix}, \end{aligned}$$

so that

$$(p \circ p^T)^{-1} = \begin{pmatrix} 2/3 & 1/3 \\ 1/3 & 2/3 \end{pmatrix}.$$

Then the map defined by Eq. (1.4) is simply

$$\begin{aligned} m &= q \circ \ell \circ p^T \circ (p \circ p^T)^{-1} \\ &= \begin{pmatrix} 1 & 1 & 1 \\ 1 & 1 & 0 \end{pmatrix} \begin{pmatrix} 1 & -1 & 0 \\ 1 & 0 & 1 \\ 0 & 1 & 1 \end{pmatrix} \begin{pmatrix} 1 & 0 \\ -1 & 1 \\ 0 & 1 \end{pmatrix} \begin{pmatrix} 2/3 & 1/3 \\ 1/3 & 2/3 \end{pmatrix} \\ &= \begin{pmatrix} 2 & 2 \\ 2 & 1 \end{pmatrix}. \end{aligned}$$

We can check that the resulting diagram commutes, namely that

$$
\begin{aligned}
m \circ p &= \begin{pmatrix} 2 & 2 \\ 2 & 1 \end{pmatrix} \begin{pmatrix} 1 & -1 & 0 \\ 0 & 1 & 1 \end{pmatrix} \\
&= \begin{pmatrix} 2 & 0 & 2 \\ 2 & -1 & 1 \end{pmatrix} \\
&= \begin{pmatrix} 1 & 1 & 1 \\ 1 & 1 & 0 \end{pmatrix} \begin{pmatrix} 1 & -1 & 0 \\ 1 & 0 & 1 \\ 0 & 1 & 1 \end{pmatrix} \\
&= q \circ \ell .
\end{aligned}
$$

If the requirement that $\ker p \subseteq \ker \ell$ is violated, then the map m defined by Eq. (1.4) may not result in a commutative diagram.

Example 1.19 In contrast to Example 1.18, consider the diagram (notice that exactly one minus sign has changed!)

$$
\begin{array}{ccc}
\mathbb{R}^3 & \xrightarrow{\begin{pmatrix} 1 & 1 & 0 \\ 1 & 0 & 1 \\ 0 & 1 & 1 \end{pmatrix}} & \mathbb{R}^3 \\
\begin{pmatrix} 1 & -1 & 0 \\ 0 & 1 & 1 \end{pmatrix} \Big\downarrow & & \Big\downarrow \begin{pmatrix} 1 & 1 & 1 \\ 1 & 1 & 0 \end{pmatrix} \\
\mathbb{R}^2 & & \mathbb{R}^2
\end{array}
$$

As in Example 1.18,

$$
(p \circ p^T)^{-1} = \begin{pmatrix} 2/3 & 1/3 \\ 1/3 & 2/3 \end{pmatrix},
$$

but notice that the ℓ map has trivial kernel while p clearly cannot! We may still attempt to construct the map defined by Eq. (1.4),

$$
\begin{aligned}
m &= q \circ \ell \circ p^T \circ (p \circ p^T)^{-1} \\
&= \begin{pmatrix} 1 & 1 & 1 \\ 1 & 1 & 0 \end{pmatrix} \begin{pmatrix} 1 & 1 & 0 \\ 1 & 0 & 1 \\ 0 & 1 & 1 \end{pmatrix} \begin{pmatrix} 1 & 0 \\ -1 & 1 \\ 0 & 1 \end{pmatrix} \begin{pmatrix} 2/3 & 1/3 \\ 1/3 & 2/3 \end{pmatrix} \\
&= \begin{pmatrix} 4/3 & 8/3 \\ 4/3 & 5/3 \end{pmatrix}.
\end{aligned}
$$

However, notice that the resulting diagram fails to commute, namely

$$
\begin{aligned}
m \circ p &= \begin{pmatrix} 4/3 & 8/3 \\ 4/3 & 5/3 \end{pmatrix} \begin{pmatrix} 1 & -1 & 0 \\ 0 & 1 & 1 \end{pmatrix} \\
&= \begin{pmatrix} 4/3 & 4/3 & 8/3 \\ 4/3 & 1/3 & 5/3 \end{pmatrix} \\
&\neq q \circ \ell \\
&= \begin{pmatrix} 1 & 1 & 1 \\ 1 & 1 & 0 \end{pmatrix} \begin{pmatrix} 1 & 1 & 0 \\ 1 & 0 & 1 \\ 0 & 1 & 1 \end{pmatrix} \\
&= \begin{pmatrix} 2 & 2 & 2 \\ 2 & 1 & 1 \end{pmatrix}.
\end{aligned}
$$

Exercise 1.30 Consider the diagram below

$$
\begin{array}{ccc}
\mathbb{R}^3 & & \mathbb{R}^2 \\
{\scriptstyle \begin{pmatrix} 3 & -2 & -1 \\ -3 & 2 & -0 \\ -2 & 2 & -0 \\ 4 & -1 & -1 \end{pmatrix}} \Big\downarrow & & \Big\downarrow {\scriptstyle \begin{pmatrix} 2 & 1 \\ 4 & 1 \\ 5 & 0 \\ -1 & -2 \end{pmatrix}} \\
\mathbb{R}^4 & \xrightarrow[\begin{pmatrix} -2 & -3 & 3 & -1 \\ -2 & -3 & 4 & -3 \\ -3 & -3 & 4 & -2 \\ -1 & 2 & -3 & 4 \end{pmatrix}]{} & \mathbb{R}^4
\end{array}
$$

1. Compute the induced map using Eq. (1.3).
2. Does the map you computed result in a commutative diagram? Why or why not?

Exercise 1.31 Consider the diagram below

$$
\begin{array}{ccc}
\mathbb{R}^3 & & \mathbb{R}^2 \\
\begin{pmatrix} -0 & 0 & 1 \\ -4 & 1 & -4 \\ 2 & 0 & -2 \\ -1 & 0 & 1 \end{pmatrix} \Big\downarrow & & \Big\downarrow \begin{pmatrix} -3 & 3 \\ 2 & -3 \\ -4 & 3 \\ -5 & 6 \end{pmatrix} \\
\mathbb{R}^4 & \xrightarrow[\begin{pmatrix} -2 & 3 & 2 & 1 \\ 2 & -3 & -2 & 0 \\ -2 & 3 & 2 & 2 \\ -4 & 6 & 4 & 1 \end{pmatrix}]{} & \mathbb{R}^4
\end{array}
$$

1. Compute the induced map using Eq. (1.3).
2. Does the map you computed result in a commutative diagram? Why or why not?

Remark 1.2 When implementing the projection formulas in software, it is efficient to compute kernels and cokernels using the singular value decomposition. Standard implementations of the singular value decomposition are available for many programming languages and are both fast and robust. The use of the singular value decomposition to identify a basis for the cokernel does require the programmer to select a small threshold for singular values, though this selection is usually not particularly critical.

1.4 Duality and Categories

Theorem 1.6 provides a rather direct interpretation of the "co" in cokernel (and coimage), namely that the cokernel (and coimage) of a linear map is isomorphic to the kernel (and image) of its dual. This indicates that the process of taking the dual "preserves the structure" of vector spaces and their associated linear maps. The notion of "preserves structure" seems intuitive enough, but it is helpful to be more formal. Such a formalization is essentially a study of mathematical analogies and has implications throughout mathematics. This formalization is called *category theory*. Historically, category theory and homological algebra grew together in the 1960s, each informing the development of the other.

Definition 1.12 A *category* $\mathbf{C}$ consists of a class of *objects* $Ob(\mathbf{C})$ and a class of morphisms $Mor(\mathbf{C})$ satisfying several conditions:

1. Every morphism $m \in Mor(\mathbf{C})$ has a unique *source object* and a unique *target object*. Both the source and target objects are in $Ob(\mathbf{C})$. We usually write

$$m : A \to B$$

to express the fact that the morphism m has source object A and target object B.
2. There is an operation $\circ$ called *composition*, that given a pair of morphisms $m_1 : A \to B$ and $m_2 : B \to C$, produces a unique morphism $(m_2 \circ m_1) : A \to C$ with the following properties:
   1. $\circ$ is associative: If $m_1 : A \to B$, $m_2 : B \to C$, and $m_3 : C \to D$, then

   $$m_3 \circ (m_2 \circ m_1) = (m_3 \circ m_2) \circ m_1,$$

   2. There is an *identity morphism* $\mathrm{id}_X : X \to X$ for every object $X \in Ob(\mathbf{C})$ such that if $m_1 : X \to A$, then

   $$\mathrm{id}_X \circ m_1 = m_1$$

   and if $m_2 : A \to X$, then

   $$m_2 \circ \mathrm{id}_X = m_2.$$

A pair of morphisms $m_1 : A \to B$ and $m_2 : B \to C$ is called *composable*. The class of morphisms with source object S and target object T is denoted $\mathrm{Hom}_{\mathbf{C}}(S, T)$, which suggests that the class of morphisms can be thought of as "homomorphisms from S to T in the category $\mathbf{C}$".

Nearly every mathematical concept you encounter will involve an instance of a category. Be on the lookout for them!

Example 1.20 The category **Set** has the class of sets as its objects. That is, every set is an object of **Set**. Each morphism of **Set** is a function from one set to another, and the composition of morphisms in **Set** is simply the composition of functions. Associativity of function composition is elementary. Because of this definition, every object A of **Set**—which is the set A—has an identity morphism id_A, which is the identity function

$$\mathrm{id}_A(x) := x.$$

Clearly, composition with the identity morphism of A leaves unchanged any other morphism whose source or target is A.

Many categories are easier to construct as a *subcategory*: one whose objects, morphisms, and composition come from a larger category. We say that a subcategory $\mathbf{C}$ of a larger category $\mathbf{D}$ is *full* if $\mathrm{Hom}_{\mathbf{C}}(A, B) = \mathrm{Hom}_{\mathbf{D}}(A, B)$ for every pair of objects A and B in **C**. In other words, a subcategory is full if it contains all relevant morphisms from the larger category.

Example 1.21 One the most important categories in this book is **Vec**, the *category of vector spaces*. The objects of **Vec** are finite dimensional vector spaces, and the morphisms are linear maps. Clearly, because vector spaces are themselves sets, and linear maps are functions between sets, much of the structure of **Set** is also present in **Vec**. We may interpret **Vec** as a subcategory of **Set** whenever it is convenient. However, observe that **Vec** is self-contained: composition of linear maps preserves linearity, and the identity function is a linear map as well.

Vec is a bit richer than **Set**, though. For one, **Vec** is not a full subcategory, since there are nonlinear functions between vector spaces that are contained in **Set** but not in **Vec**. For instance, the function $f : \mathbb{R} \to \mathbb{R}$ given by $f(x) = x^2$ is a morphism in **Set** but not **Vec**.

In addition to the identity map for each object in **Vec**, there is also the zero map—a morphism for each pair of objects. Composition with the zero map results in the zero map, which gives it a quality opposite the identity map. Notice that while the zero map is a morphism in **Set**, it does not exist for *every* pair of objects in **Set** since not every set is a vector space.

Categories generalize the idea of sets and functions, and therefore allow many of the same kinds of manipulations that work with sets to be carried out in other contexts. Most notions of mathematical objects as sets with some structure correspond naturally to a particular category. It is a good idea to try to rephrase element-based operations on sets as function-based ones, since they generalize more easily. Rather than thinking of a subset $A \subseteq B$, we can represent this situation as an injective function $i : A \to B$. This neatly generalizes to the notion of an injective linear map being a good way to think of a subspace, which is buttressed by Proposition 1.2.

As we have just seen, the morphisms can play the role of transformations between objects. This need not be the case, though. For instance, a *preordered set* can be thought of *as* a category.

Definition 1.13 Consider an arbitrary set P and a *preorder* $\leq$ on P. A *preorder* is a relation between elements in P that is

1. (Reflexive) $x \leq x$ for all $x \in P$, and
2. (Transitive) If $x \leq y$ and $y \leq z$, then $x \leq z$.

If we add an additional axiom

3. (Antisymmetric) $x \leq y$ and $y \leq x$ implies $x = y$,

we obtain a *partial order*. Alternatively, if we add the axiom

3'. (Symmetric) $x \leq y$ if and only if $y \leq x$.

The result is called an *equivalence relation*.

Instances of preorders, partial orders, and equivalence relations occur many times in this book.

Example 1.22 Recall that the *power set* of a set X is the set of all possible subsets of X. The power set is ordered by *inclusion*, which is to say that if A and B are subsets of X, then the relation $A \subseteq B$ is a partial order.

Definition 1.14 The *preorder category* $\mathbf{Pre}(P, \leq)$ of P) has the set P as its class of objects. The class of morphisms of $\mathbf{Pre}(P, \leq)$ consists of all ordered pairs $x \leq y$ of elements of P. We can interpret this to mean that a morphism $x \to y$ is only present if $x \leq y$.

Lemma 1.4 *The preorder category* $\mathbf{Pre}(P, \leq)$ *is a well defined category.*

Proof Everything follows from the fact that there is at most one morphism in $\mathrm{Hom}_{\mathbf{Pre}(P,\leq)}(x, y)$.

Transitivity of $\leq$ provides the composition of morphisms, since if $x \to y$ and $y \to z$, this means that $x \leq z$. Thus $x \to z$ must be a morphism. Moreover, since there is at most one morphism between x and z, this choice is unique.

Composition is associative simply because there is really no other option! Finally, for any given object x of $\mathbf{Pre}(P, \leq)$, the identity morphism $\mathrm{id}_x : x \to x$ is precisely the expression of the reflexivity axiom. □

Exercise 1.32 Draw a diagram of all objects and morphisms of $\mathbf{Pre}(P, \leq)$ where $P = \{a, b, c, d, e\}$ and $\leq$ is generated by the following

- $a \leq b$,
- $a \leq c$,
- $a \leq e$,
- $b \leq d$, and
- $c \leq d$.

On the other hand, we can also construct the category of all preorders.

Example 1.23 The *category of preorders* **Preorders** has every preordered set $(P, \leq)$ as an object. Each morphism $(P, \leq_P) \to (Q, \leq_Q)$ is an *order preserving function* $g : P \to Q$ such that if x and y are two elements of P satisfying $x \leq_P y$, then $g(x) \leq_Q g(y)$ in Q. These morphisms compose as functions on their respective sets.

To see that this defines a category, notice that the identity function is always order preserving and that the composition of two order preserving functions is another order preserving function. Associativity follows from the associativity of function composition.

Exercise 1.33 Suppose that $P = \{0, 1\}$ with $0 \leq_P 1$ generates a preordered set (there a few other easy facts about $\leq_P$ that follow). If $Q = \{a, b, c\}$ with $a \leq_Q c$ and $b \leq_Q c$ also generates a preordered set, construct all order preserving functions $P \to Q$ and all order preserving functions $Q \to P$.

Definition 1.15 We can restrict Definition 1.14 to define **Pos**, the full subcategory of **Preorders** in which each object is a partially ordered set $(P, \leq)$.

Definition 1.16 A pair of morphisms $p : A \to B$ and $q : B \to A$ are called *inverse morphisms* if $q \circ p = \mathrm{id}_A$ and $p \circ q = \mathrm{id}_B$. We usually write $q = p^{-1}$ or $p = q^{-1}$ in this case. We also declare that p and q (individually) are *isomorphisms*. Whenever an isomorphism exists between two objects A and B, we say that A and B are *isomorphic* objects.

Exercise 1.34 Show that vector space isomorphisms (which are bijective linear maps according to Proposition A.5 in Appendix A) are isomorphisms in **Vec**.

Exercise 1.35 Give an example of two non-isomorphic vector spaces that are isomorphic in **Set**. (Hint: this is not as difficult as it seems!)

Proposition 1.9 *Inverse morphisms are unique. Explicitly, suppose that* $p_1 : A \to B$ *and* $q_1 : B \to A$ *are inverse morphisms. For any other* $q_2 : B \to A$ *satisfying* $q_2 \circ p_1 = \mathrm{id}_A$ *and* $p_1 \circ q_2 = \mathrm{id}_B$*, it follows that* $q_2 = q_1$.

Proof This is simply calculation,

$$
\begin{aligned}
q_2 \circ p_1 &= \mathrm{id}_A = q_1 \circ p_1 \\
(q_2 \circ p_1) \circ q_1 &= (q_1 \circ p_1) \circ q_1 \\
q_2 \circ (p_1 \circ q_1) &= q_1 \circ (p_1 \circ q_1) \\
q_2 &= q_1,
\end{aligned}
$$

and similarly for $p_1 \circ q_2$. □

Isomorphic objects in a category are essentially indistinguishable, at least within the context of a given category. For instance, if A and B are isomorphic objects with $i : A \to B$ being an isomorphism, and if $m : B \to C$ is some arbitrary morphism, then $m \circ i : A \to C$ is also a morphism in that category. Similarly, if $m' : C \to B$ is a morphism, then $i^{-1} \circ m' : C \to A$ is also a morphism.

Proposition 1.10 *Isomorphism within an category is an equivalence relation. We therefore will often speak of* isomorphism classes.

Proof This is merely a verification of the axioms for an equivalence relation,

- (Reflexivity) First of all, each object in a category is isomorphic to itself, via its identity morphism.
- (Symmetry) If A and B are isomorphic, this means that there are isomorphisms $p : A \to B$ and $q : B \to A$ by definition.
- (Transitivity) If A and B are isomorphic by way of $i_1 : A \to B$ and if B and C are isomorphic by way of $i_2 : B \to C$, then clearly $i_2 \circ i_1 : A \to C$ is an isomorphism since its inverse is $i_1^{-1} \circ i_2^{-1}$. □

Exercise 1.36 Consider the preordered set $(P, \leq_P)$, in which $P = \{a, b, c, d, e\}$ and $\leq_P$ is generated by

- $a \leq_P b$,
- $b \leq_P c$,
- $d \leq_P c$, and
- $c \leq_P d$.

In the category $\mathbf{Pre}(P, \leq_P)$, determine all of the isomorphism classes of objects.

Example 1.24 Bijections between sets are isomorphisms in **Set**, since the definition of an inverse function for a bijection is precisely the definition of an inverse morphism.

Example 1.25 Isomorphisms in $\mathbf{Pre}(P, \leq)$ are a bit different from those in **Set**. Recall that if $A \to B$ is a morphism in $\mathbf{Pre}(P, \leq)$, this means that $A \leq B$. Since there can be at most one morphism $\mathbf{Pre}(P, \leq)$ between any ordered pair of objects, an inverse would simply be a morphism $B \to A$ if it exists. This implies that $B \leq A$. Thus, isomorphic objects in $\mathbf{Pre}(P, \leq)$ are elements A and B with $A \leq B$ and $B \leq A$. Isomorphic objects are therefore those elements that cannot by distinguished by the order. If $(P, \leq)$ is a *partial order* instead of a preorder, then we can additionally conclude that $A = B$, meaning that each isomorphism class contains only one object.

If two categories are related to one another, then results concerning one category can be transferred to the other. These relationships take the role of formalized analogies between categories. These formal analogies are called *functors*, and transform one category into another. They come in two kinds, depending on how they transform morphisms.

Definition 1.17 A *functor* F from a category $\mathbf{C}$ to another $\mathbf{D}$ assigns each object $x \in \mathbf{C}$ to an object $F(x) \in \mathbf{D}$ and a morphism $F(m) \in Mor(\mathbf{D})$ to each morphism $m \in Mor(\mathbf{C})$ according to the following:

1. $F(\mathrm{id}_{\mathbf{C}}) = \mathrm{id}_{\mathbf{D}}$, and
2. Either $F(f \circ g) = F(f) \circ F(g)$ for every pair of composable morphisms (we say that F is *covariant* in this case) or $F(f \circ g) = F(g) \circ F(f)$ (we say that F is *contravariant*).

If the adjective *covariant* or *contravariant* is not specified, the kind of functor should be clear from context.

We will encounter many examples of functors in this book; they are what allows us to study the structure of a complicated category by way of analogy with a simpler one.

Example 1.26 We constructed the category **Vec** of vector spaces as a subcategory of the category of sets **Set**. Subcategories can be thought of as a functor that *forgets* some structure.

In the case of **Vec**, a functor $F : \mathbf{Vec} \to \mathbf{Set}$ can be constructed by

$$F(V) := V$$

which reinterprets a vector space V as the set V, and

$$F(\ell : V \to W) := (\ell : V \to W)$$

which reinterprets the linear map ℓ as simply a function. Since composition of linear maps and functions works the "same way" in both categories, it is clear that F is a covariant functor.

Example 1.27 We can also construct a covariant functor $G : \mathbf{Set} \to \mathbf{Vec}$ that builds a vector space by specifying its basis. Concretely, if A is a set, then $G(A)$ is the vector space whose basis is A. An arbitrary element v of $G(A)$ can be written

$$v = \sum_{a \in A} v_a a.$$

If $f : A \to B$ is a function between two sets, $G(f) : G(A) \to G(B)$ is a linear map constructed by the action of f on the two bases. For each $a \in A$, define

$$(G(f))\,(a) := f(a)$$

where a is interpreted as basis element of the vector space $G(A)$ on the left side and an element of the set A on the right. Then we can extend by linearity

$$(G(f)) \left(\sum_{a \in A} v_a a \right) := \sum_{a \in A} v_a \,(G(f))\,(a) = \sum_{a \in A} v_a f(a),$$

observing that $f(a) \in B$ can be interpreted as an element of the basis for $G(B)$.

G preserves the identity morphisms, since for any set A the linear map $G(\mathrm{id}_A)$ is the identity map, as seen by its action on the basis

$$(G(\mathrm{id}_A))\,(a) := \mathrm{id}_A(a) = a.$$

Finally, G preserves composition. If $f : A \to B$ and $g : B \to C$ are two arbitrary functions, then for $a \in A$,

$$\begin{aligned}(G(g \circ f))\,(a) &= (g \circ f)(a)\\ &= g(f(a))\\ &= g((G(f))\,(a))\\ &= (G(g))\,((G(f))\,(a))\\ &= (G(g) \circ G(f))\,(a).\end{aligned}$$

Example 1.28 (A generalization of the construction that appears in Adámek et al. (1990, Sect. 3.3) and Rydeheard and Burstall (1988, p. 54)) Consider the category of *relations* **Rel** whose objects consist of two sets X, Y, and a subset $R \subseteq X \times Y$. We call R a *relation between* X *and* Y. A morphism $(X, Y, R) \to (X', Y', R')$ in **Rel** is defined by a pair of functions $f : X \to X'$, $g : Y \to Y'$ such that $(f(x), g(y)) \in R'$ whenever $(x, y) \in R$.

Composition of morphisms is simply the composition of the corresponding pairs of functions, which means that **Rel** satisfies the axioms for a category.

Example 1.29 There is a covariant functor $PosRep$: **Rel** $\to$ **Pos**, called the *poset representation* of a relation, that takes each (X, Y, R) to a collection $PosRep$ (X, Y, R) of subsets of the power set of X, for which $A \in PosRep(X, Y, R)$ if there is a $y \in Y$ such that $(x, y) \in R$ for every $x \in A$. The elements of $PosRep(X, Y, R)$ are ordered by subset inclusion.

$PosRep$ translates morphisms in **Rel** into order preserving maps among partially ordered sets. Specifically, the morphism $(X, Y, R) \to (X', Y', R')$ implemented by $f : X \to X'$ and $g : Y \to Y'$ is transformed into the function that takes $A \in PosRep(X, Y, R)$ to $f(A)$. By definition there is a $y \in Y$ such that $(x, y) \in R$ for every $x \in A$. Therefore, $(f(x), g(y)) \in R'$ by construction. Since each $x' \in f(A)$ is given by $x' = f(x)$ for some x, this means that $(x', g(y)) \in R'$ for all $x \in f(A)$. Thus, $f(A) \in PosRep(X', Y', R')$. Furthermore, the order relation among subsets is evidently preserved.

Exercise 1.37 Show that $PosRep$ is a functor by showing that composition of morphisms is preserved in order.

It is usually a good indication that a functor is present if our language includes the word "induces". This occurs with our first really nontrivial example of a functor:

the operation of taking duals. If we replace every vector space V with its dual V^T, then every linear map $\ell : V \to W$ induces a linear map $\ell^T : W^T \to V^T$. In fact, Proposition 1.6 and Theorem 1.5 establish that dualization defines a contravariant functor T : **Vec** $\to$ **Vec** called the *dualization functor*.

Exercise 1.38 Another example of a functor would seem to be the operation of taking quotients. Recall Theorem 1.3, which states that (under the appropriate conditions) a linear map $\ell : V \to W$ induces a linear map $V/U \to W/X$. Is this a functor **Vec** $\to$ **Vec** by our definition? If not, how can you modify the categories involved to make Theorem 1.3 define a functor?

Finally, we end with the notion of the *equivalence* of categories. Statements about a category can be transferred to any equivalent category.

Definition 1.18 An *equivalence of categories* is a functor $F : \mathbf{C} \to \mathbf{D}$ such that

1. Every object D in $\mathbf{D}$ (in the codomain of F) is isomorphic to an object $F(A)$ for some other object A in $\mathbf{C}$ (in the domain of F), and
2. For every pair of objects A, B in $\mathbf{C}$, the function $\mathrm{Hom}_{\mathbf{C}}(A, B) \to \mathrm{Hom}_{\mathbf{D}}(F(A), F(B))$ given by transforming each morphism $m : A \to B$ to a morphism $F(m) : F(A) \to F(B)$ is a bijection.

While it might be thought that we would want $F(A)$ to be *equal to* D in Definition 1.18, this is usually too stringent. Since morphisms in a category are only sensitive to isomorphism classes of objects (not the particular objects involved), we want the definition of equivalence to reflect that.

Example 1.30 The following are diagrams defining two separate preorders, P_1 on the set $\{a, b, c\}$ and P_2 on the set $\{d, e, f, g\}$.

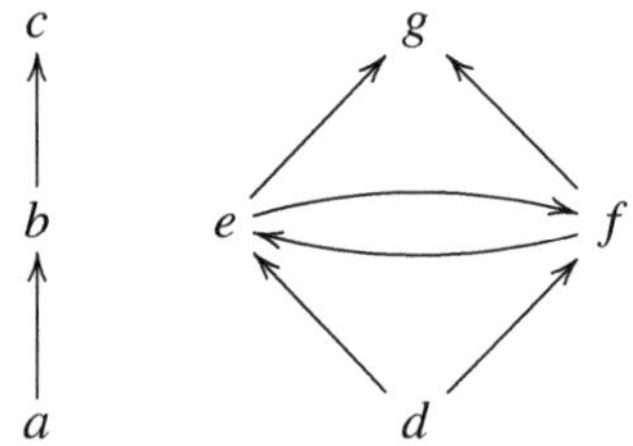

As usual, arrows point from smaller to larger elements, so $a \leq b$ for instance. Both of these preorders define categories $\mathbf{Pre}(P_1, \leq)$ and $\mathbf{Pre}(P_2, \leq)$ according to the construction in Definition 1.14. Even though these are not identical categories, since they have different numbers of objects, they are equivalent categories. To see this, define the functor $F : \mathbf{Pre}(P_1, \leq) \to \mathbf{Pre}(P_2, \leq)$ by its action on objects

$$F(a) := d, \;\; F(b) := e, \;\; \text{and } F(c) := g.$$

This defines a functor because F preserves the order relation. There is only one object not in the image of F, namely f. However, f is isomorphic to e, which is the image of b. Additionally, because there is at most one morphism between each pair of objects in both categories, the Hom set condition holds. We can compute the cardinalities of the Hom sets directly, namely

$$|\operatorname{Hom}_{\mathbf{Pre}(P_1,\leq)}(a,b)| = |\operatorname{Hom}_{\mathbf{Pre}(P_2,\leq)}(F(a),F(b))| = |\operatorname{Hom}_{\mathbf{Pre}(P_2,\leq)}(d,e)| = 1,$$

$$|\operatorname{Hom}_{\mathbf{Pre}(P_1,\leq)}(b,c)| = |\operatorname{Hom}_{\mathbf{Pre}(P_2,\leq)}(F(b),F(c))| = |\operatorname{Hom}_{\mathbf{Pre}(P_2,\leq)}(e,g)| = 1,$$

$$|\operatorname{Hom}_{\mathbf{Pre}(P_1,\leq)}(a,c)| = |\operatorname{Hom}_{\mathbf{Pre}(P_2,\leq)}(F(a),F(c))| = |\operatorname{Hom}_{\mathbf{Pre}(P_2,\leq)}(d,g)| = 1,$$

$$|\operatorname{Hom}_{\mathbf{Pre}(P_1,\leq)}(b,a)| = |\operatorname{Hom}_{\mathbf{Pre}(P_2,\leq)}(F(b),F(a))| = |\operatorname{Hom}_{\mathbf{Pre}(P_2,\leq)}(e,d)| = 0,$$

$$|\operatorname{Hom}_{\mathbf{Pre}(P_1,\leq)}(c,b)| = |\operatorname{Hom}_{\mathbf{Pre}(P_2,\leq)}(F(c),F(b))| = |\operatorname{Hom}_{\mathbf{Pre}(P_2,\leq)}(g,e)| = 0,$$

and

$$|\operatorname{Hom}_{\mathbf{Pre}(P_1,\leq)}(c,a)| = |\operatorname{Hom}_{\mathbf{Pre}(P_2,\leq)}(F(c),F(a))| = |\operatorname{Hom}_{\mathbf{Pre}(P_2,\leq)}(g,d)| = 0.$$

(There is nothing to check for self-morphisms $\operatorname{Hom}_{\mathbf{Pre}(P,\leq)}(a,a)$ for any preorder category.)

Exercise 1.39 Construct an equivalence functor $G : \mathbf{Pre}(P_2, \leq) \to \mathbf{Pre}(P_1, \leq)$ for Example 1.30.

1.5 Further Reading

Applications of linear algebra are almost too pervasive to mention (Golub and Loan 1996). Our formulation of the dimension theorem is similar to the presentation in Strang (2006). This formulation is motivated by the singular value decomposition and principal components analysis (PCA). PCA stands out as a particularly useful tool in many applications. Practical aspects of computing these decompositions is discussed in great detail in Isaacson and Keller (2012).

In contrast to linear algebra, the practical usefulness of category theory is not as widely appreciated. Recent expositions on the subject of applications of category theory are available in Fong and Spivak (2019) and Spivak (2014), which also have the benefit of providing a solid introduction to the theory as well.

References

Adámek J, Herrlich H, Strecker G (1990) Abstract and concrete categories. Wiley-Interscience
Fong B, Spivak DI (2019) An invitation to applied category theory: seven sketches in compositionality. Cambridge University Press
Golub GH, Loan CFV (1996) Matrix computations. Johns Hopkins University Press, pp 470–507
Isaacson E, Keller HB (2012) Analysis of numerical methods. Courier Corporation
Rydeheard D, Burstall R (1988) Computational category theory. Prentice-Hall
Spivak DI (2014) Category theory for the sciences. MIT Press
Strang G (2006) Linear algebra and its applications. Brooks and Cole

Chapter 2
Sequences and Chain Complexes

Our study of homological algebra begins in earnest in this chapter, by studying *sequences* of linear maps, one after the other

$$\cdots \to V^0 \xrightarrow{v^0} V^1 \xrightarrow{v^1} \cdots V^n \xrightarrow{v^n} \cdots$$

using superscripts—which are not exponents!—to identify the different spaces and maps involved. Most of this book is devoted to studying the structure of sequences of linear maps, and exploring some invariants that characterize them.

Single linear maps are classified by the dimensions of their domain, codomain, and image (or kernel). Finite dimensional linear maps admit a complete characterization in terms of these vector spaces, and as a result the subspace structure of a sequence of linear maps is key to characterizing sequences. Most of the machinery we need for such a characterization is most efficiently described using a decomposition of a vector space into a "sum" of other, smaller vector spaces.

In this chapter, we prove Theorem 2.1, which is the first of two theorems being analogous to Theorem 1.2 in Chap. 1, which gives a decomposition of each V^n space into a direct sum that accounts for each linear map v^n. (The other is Theorem 3.1, which we will prove in Chap. 3.) If the bases of each vector space V^n are carefully chosen to work with the linear maps, then we can characterize this sequence succinctly in a pictorial diagram called a *barcode diagram*.

Example 2.1 Consider the following sequence of linear maps.

$$\mathbb{R}^2 \xrightarrow{\begin{pmatrix} 1 & 1 \\ -1 & -1 \end{pmatrix}} \mathbb{R}^2 \xrightarrow{\begin{pmatrix} 1 & 2 \\ 0 & 1 \end{pmatrix}} \mathbb{R}^2 \xrightarrow{(1\ 0)} \mathbb{R}$$

M. Robinson, *Computational Homological Algebra*, Mathematical Engineering,
https://doi.org/10.1007/978-3-032-08634-1_2

(If not specified otherwise, we assume leading and trailing trivial vector spaces fill out the remainder of the sequence.) The kernel of the first matrix has dimension 1, and is spanned by $\begin{pmatrix} 1 \\ -1 \end{pmatrix}$. A convenient basis for the domain of this matrix is then

$$\left\{\begin{pmatrix} 1 \\ -1 \end{pmatrix}, \begin{pmatrix} 0 \\ 1 \end{pmatrix}\right\}.$$

Notice that the image of the second vector is

$$\begin{pmatrix} 1 & 1 \\ -1 & -1 \end{pmatrix}\begin{pmatrix} 0 \\ 1 \end{pmatrix} = \begin{pmatrix} 1 \\ -1 \end{pmatrix}.$$

If we apply the next matrix in the sequence, we obtain

$$\begin{pmatrix} 1 & 2 \\ 0 & 1 \end{pmatrix}\begin{pmatrix} 1 & 1 \\ -1 & -1 \end{pmatrix}\begin{pmatrix} 0 \\ 1 \end{pmatrix} = \begin{pmatrix} -1 \\ -1 \end{pmatrix}.$$

Multiplying by the last matrix in the sequence gives

$$(1\ 0)\begin{pmatrix} 1 & 2 \\ 0 & 1 \end{pmatrix}\begin{pmatrix} 1 & 1 \\ -1 & -1 \end{pmatrix}\begin{pmatrix} 0 \\ 1 \end{pmatrix} = (-1)\,.$$

This means that the original basis vector $\begin{pmatrix} 0 \\ 1 \end{pmatrix}$ "made it through" the entire sequence without getting sent to zero. It seems like a good choice of basis vector!

Returning to the second matrix in the sequence, notice that it has a trivial kernel. We can extend $\begin{pmatrix} 1 \\ -1 \end{pmatrix}$ into a basis of $\mathbb{R}^2$ (the second space) in almost any way we please. On the other hand, the last matrix in the sequence has a 1-dimensional kernel, and we already have one preferred basis vector selected, namely $\begin{pmatrix} -1 \\ -1 \end{pmatrix}$. Let us use our freedom in specifying the remaining basis vectors on the middle two spaces so that the remaining basis vector lies in the kernel of the third matrix. That is, let us select $\begin{pmatrix} 0 \\ 1 \end{pmatrix}$ for the other basis vector for the third space. Since the second matrix has trivial kernel, this vector has a unique preimage, namely $\begin{pmatrix} -2 \\ 1 \end{pmatrix}$.

It is convenient to express our findings graphically, using a *barcode diagram*.

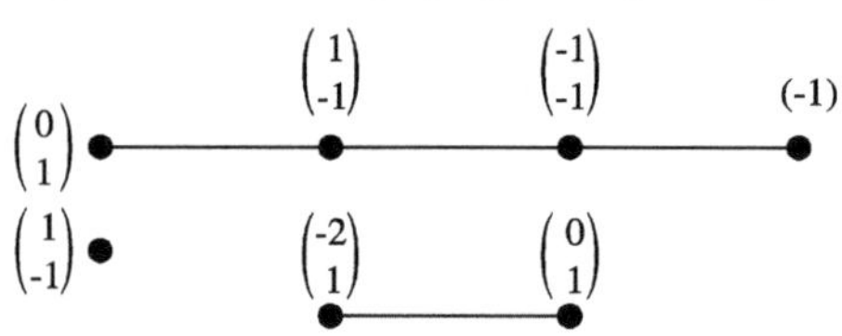

Each basis vector is marked with a heavy dot. You can see that there are two dots for the first three spaces, since each is two-dimensional. There is only one dot for the last space, since it is one-dimensional. Where a basis vector was "preserved" by the linear map listed, taking it from one space to the next, the dots are connected by a line segment. In the two instances where a basis vector was in the kernel, their corresponding bars do not continue to the next space.

In this chapter,

Section 2.1 Defines the concept of a *sequence of vector spaces.*
Section 2.2 Shows that sequences of vector spaces are characterized by their *barcode decomposition.*
Section 2.3 Relates neighboring terms of a sequence to one another.
Section 2.4 Turns our attention to a special class of sequences called *chain complexes.*
Section 2.5 Derives a summary of a chain complex, called *homology.*

2.1 Sequence Maps

Definition 2.1 A *sequence of vector spaces* (or simply a *sequence*) $(V^\bullet, v^\bullet)$ consists of an ordered pair: (1) an ordered list of vector spaces $\{\dots, V^{-1}, V^0, V^1, \dots\}$ and (2) and an ordered list of linear maps $v^n : V^n \to V^{n+1}$. In a sequence, the superscript n is called the *index* and each element V^n is called a *term* in the sequence. The $\bullet$ symbol[1] is used to refer to the collection of all indices.

We will often write a sequence using a diagram

$$\cdots \to V^0 \xrightarrow{v^0} V^1 \xrightarrow{v^1} \cdots V^n \xrightarrow{v^n} \cdots$$

If we wish for the indices to decrease rather than increase, as is common for *simplicial chain complexes* studied in Chaps. 4 and 5, we emphasize this fact by using subscripts instead of superscripts, like

$$\cdots \to V_n \xrightarrow{v_n} \cdots \xrightarrow{v_2} V_1 \xrightarrow{v_1} V_0 \xrightarrow{v_0} \cdots$$

using the convention that the index of a map matches the index of its domain.

Remark 2.1 Upon encountering a new sequence, the reader's first task should be to determine the appropriate indices for the terms in the sequence. Sometimes, like in the sequences above in Definition 2.1, the indices are given explicitly. But to avoid typographical clutter, the indices will not be shown on every sequence diagram. Instead, the indices may be specified in the text, implied by reference to another

[1] In LaTeX, you can typeset this symbol with `\bullet`.

sequence defined earlier, or left to the reader's discretion. With some practice, the reader will be able to give appropriate indices for every sequence in this book, even if they are not specified. For instance, it is usually a good guess that the left-most (or right-most) index is 0 or 1. Fortunately, many properties of sequences do not depend on the choice of indices.

By analogy with vector spaces, in which the structure of linear maps is paramount, the structure of maps between sequences is of particular interest. A reasonable definition of a map between sequences needs to transform each of the terms in the sequence by way of linear maps. We impose an additional compatibility condition between the linear maps in the sequence and the linear maps between sequence terms. Many interesting consequences follow from this compatibility condition!

Definition 2.2 A *sequence map* between two sequences of vector spaces $(V^\bullet, v^\bullet) \to (W^\bullet, w^\bullet)$ is a sequence of linear maps $m^n : V^n \to W^n$ so that

$$m^{n+1} \circ v^n = w^n \circ m^n$$

for every *index* n. The maps m^n are called *component maps* of the sequence map. If it is clear from context, we sometimes write $m^\bullet : V^\bullet \to W^\bullet$ for brevity. We will usually summarize this condition by a *commutative ladder diagram*

$$\begin{array}{ccccccccc}
\cdots & \longrightarrow & V^1 & \xrightarrow{v^1} & V^2 & \xrightarrow{v^2} & V^3 & \xrightarrow{v^3} & \cdots \\
 & & {\scriptstyle m^1}\big\downarrow & & {\scriptstyle m^2}\big\downarrow & & {\scriptstyle m^3}\big\downarrow & & \\
\cdots & \longrightarrow & W^1 & \xrightarrow[w^1]{} & W^2 & \xrightarrow[w^2]{} & W^3 & \xrightarrow[w^3]{} & \cdots
\end{array}$$

Recall that according to Definition 1.6 in Chap. 1, a commutative diagram is a diagram in which the composition of consecutive maps is independent of path; it only depends on the source and target.

Example 2.2 The diagram

$$\begin{array}{ccccccccccc}
\cdots & \longrightarrow & \mathbb{R} & \xrightarrow{\begin{pmatrix}1\\0\end{pmatrix}} & \mathbb{R}^2 & \xrightarrow{\begin{pmatrix}1&0\\0&1\end{pmatrix}} & \mathbb{R}^2 & \xrightarrow{\begin{pmatrix}1&1\end{pmatrix}} & \mathbb{R} & \longrightarrow & \cdots \\
 & & {\scriptstyle \begin{pmatrix}0\\1\end{pmatrix}}\Big\downarrow & & {\scriptstyle \begin{pmatrix}0&1\\1&0\end{pmatrix}}\Big\downarrow & & {\scriptstyle \begin{pmatrix}1&1\end{pmatrix}}\Big\downarrow & & \Big\downarrow & & \\
\cdots & \longrightarrow & \mathbb{R}^2 & \xrightarrow[\begin{pmatrix}1&0\\0&1\end{pmatrix}]{} & \mathbb{R}^2 & \xrightarrow[\begin{pmatrix}1&1\end{pmatrix}]{} & \mathbb{R} & \longrightarrow & 0 & \longrightarrow & \cdots
\end{array}$$

defines a sequence map. We merely need to check that each of the three squares commutes, by checking matrix products. For the leftmost square, we have that

$$\begin{pmatrix} 0 & 1 \\ 1 & 0 \end{pmatrix}\begin{pmatrix} 1 \\ 0 \end{pmatrix} = \begin{pmatrix} 0 \\ 1 \end{pmatrix} = \begin{pmatrix} 1 & 0 \\ 0 & 1 \end{pmatrix}\begin{pmatrix} 0 \\ 1 \end{pmatrix}.$$

For the middle square, we have that

$$\begin{pmatrix} 1 & 1 \end{pmatrix}\begin{pmatrix} 1 & 0 \\ 0 & 1 \end{pmatrix} = \begin{pmatrix} 1 & 1 \end{pmatrix} = \begin{pmatrix} 1 & 1 \end{pmatrix}\begin{pmatrix} 0 & 1 \\ 1 & 0 \end{pmatrix}.$$

Finally, the commutativity of the last square is trivial since the final term is the trivial space.

Exercise 2.1 Is the diagram below a sequence map?

$$\begin{array}{ccccccccc}
\mathbb{R} & \xrightarrow{\begin{pmatrix} 1 \\ 1 \\ 0 \end{pmatrix}} & \mathbb{R}^3 & \xrightarrow{\begin{pmatrix} 1 & -1 & 0 \\ 0 & 0 & 1 \end{pmatrix}} & \mathbb{R}^2 & \xrightarrow{\begin{pmatrix} 1 & 0 \\ 0 & 1 \end{pmatrix}} & \mathbb{R}^2 & \longrightarrow & 0 \\
{\scriptstyle \begin{pmatrix} 1 \\ 0 \end{pmatrix}}\downarrow & & \downarrow{\scriptstyle \begin{pmatrix} 0 & 1 & 0 \\ 0 & 0 & 1 \end{pmatrix}} & & \downarrow & & \downarrow & & \downarrow \\
\mathbb{R}^2 & \xrightarrow[\begin{pmatrix} 1 & 0 \\ 0 & 1 \end{pmatrix}]{} & \mathbb{R}^2 & \longrightarrow & 0 & \longrightarrow & 0 & \longrightarrow & 0.
\end{array}$$

Exercise 2.2 Is the diagram below a sequence map?

$$\begin{array}{ccccccccc}
0 & \longrightarrow & 0 & \longrightarrow & \mathbb{R} & \xrightarrow{(1)} & \mathbb{R} & \longrightarrow & 0 \\
 & & \downarrow & & \downarrow{\scriptstyle (0)} & & \downarrow{\scriptstyle \begin{pmatrix} 1 \\ 0 \end{pmatrix}} & & \\
0 & \longrightarrow & 0 & \longrightarrow & \mathbb{R} & \xrightarrow[\begin{pmatrix} 1 \\ 0 \end{pmatrix}]{} & \mathbb{R}^2 & \longrightarrow & 0
\end{array}$$

Lemma 2.1 *For every sequence of vector spaces, choosing the identity map at each index results in a sequence map, called the* identity sequence map. *Similarly, choosing the zero map at each index results in the* zero sequence map.

Proof Both of these statements involve ladder diagrams like the following

$$
\begin{array}{ccccccccc}
\cdots \longrightarrow & V^1 & \xrightarrow{v^1} & V^2 & \xrightarrow{v^2} & V^3 & \xrightarrow{v^3} & \cdots \\
 & \downarrow{\scriptstyle m^1} & & \downarrow{\scriptstyle m^2} & & \downarrow{\scriptstyle m^3} & & \\
\cdots \longrightarrow & V^1 & \xrightarrow[v^1]{} & V^2 & \xrightarrow[v^2]{} & V^3 & \xrightarrow[v^3]{} & \cdots
\end{array}
$$

If m^n is the identity map id_{V^n} for each n, then the first statement follows because

$$v^n \circ \mathrm{id}_{V^n} = v^n = \mathrm{id}_{V^{n+1}} \circ v^n.$$

Similarly, the second statement follows because composition of any linear map with the zero map is the zero map. □

Lemma 2.2 *Sequence maps can be composed by composing maps at each index. Specifically, if* $p^\bullet : (U^\bullet, u^\bullet) \to (V^\bullet, v^\bullet)$ *and* $q^\bullet : (V^\bullet, v^\bullet) \to (W^\bullet, w^\bullet)$ *are sequence maps, then the component maps of* $(q \circ p)^\bullet$ *are given by*

$$(q \circ p)^k = q^k \circ p^k.$$

Because composition of linear maps is associative, composition of sequence maps is also an associative operation.

We will therefore consider the notations $(q \circ p)^\bullet$ and $q^\bullet \circ p^\bullet$ to be interchangeable.

Proof Consider the diagram

$$
\begin{array}{ccccccccc}
\cdots \longrightarrow & U^1 & \xrightarrow{u^1} & U^2 & \xrightarrow{u^2} & U^3 & \xrightarrow{u^3} & \cdots \\
 & \downarrow{\scriptstyle p^1} & & \downarrow{\scriptstyle p^2} & & \downarrow{\scriptstyle p^3} & & \\
\cdots \longrightarrow & V^1 & \xrightarrow{v^1} & V^2 & \xrightarrow{v^2} & V_3 & \xrightarrow{v^3} & \cdots \\
 & \downarrow{\scriptstyle q^1} & & \downarrow{\scriptstyle q^2} & & \downarrow{\scriptstyle q^3} & & \\
\cdots \longrightarrow & W^1 & \xrightarrow[w^1]{} & W^2 & \xrightarrow[w^2]{} & W^3 & \xrightarrow[w^3]{} & \cdots
\end{array}
$$

noting that each square in the diagram commutes. For instance,

$$p^3 \circ u^2 = v^2 \circ p^2$$

and

$$q^3 \circ v^2 = w^2 \circ q^2.$$

Combining these two equations, we have that

$$\begin{aligned} w^2 \circ q^2 \circ p^2 &= q^3 \circ v^2 \circ p^2 \\ &= q^3 \circ p^3 \circ u^2, \end{aligned}$$

which means that the sequence of maps $(q^n \circ p^n)$ is a sequence map $(U^\bullet, u^\bullet) \to (W^\bullet, w^\bullet)$. □

Example 2.3 The composition of the sequence map

$$\begin{array}{ccccccccc} 0 & \longrightarrow & \mathbb{R}^2 & \xrightarrow{\begin{pmatrix} 1 & -1 \\ -1 & 1 \end{pmatrix}} & \mathbb{R}^2 & \xrightarrow{\begin{pmatrix} 1 & 1 \\ 1 & 1 \end{pmatrix}} & \mathbb{R}^2 & \longrightarrow & 0 \\ \downarrow & & \downarrow{\scriptstyle \mathrm{id}} & & \downarrow{\scriptstyle \mathrm{id}} & & \downarrow & & \downarrow \\ 0 & \longrightarrow & \mathbb{R}^2 & \xrightarrow[\begin{pmatrix} 1 & -1 \\ -1 & 1 \end{pmatrix}]{} & \mathbb{R}^2 & \longrightarrow & 0 & \longrightarrow & 0 \end{array}$$

and the sequence map

$$\begin{array}{ccccccccc} 0 & \longrightarrow & \mathbb{R}^2 & \xrightarrow{\begin{pmatrix} 1 & -1 \\ -1 & 1 \end{pmatrix}} & \mathbb{R}^2 & \longrightarrow & 0 & \longrightarrow & 0 \\ \downarrow & & {\scriptstyle \begin{pmatrix} 0 & 1 \\ 1 & 0 \end{pmatrix}}\downarrow & & \downarrow{\scriptstyle \begin{pmatrix} -1/2 & 1/2 \\ 1/2 & -1/2 \end{pmatrix}} & & \downarrow & & \downarrow \\ 0 & \longrightarrow & \mathbb{R}^2 & \xrightarrow[\begin{pmatrix} 1 & -1 \\ -1 & 1 \end{pmatrix}]{} & \mathbb{R}^2 & \longrightarrow & 0 & \longrightarrow & 0 \end{array}$$

is the sequence map

$$\begin{array}{ccccccccc} 0 & \longrightarrow & \mathbb{R}^2 & \xrightarrow{\begin{pmatrix} 1 & -1 \\ -1 & 1 \end{pmatrix}} & \mathbb{R}^2 & \xrightarrow{\begin{pmatrix} 1 & 1 \\ 1 & 1 \end{pmatrix}} & \mathbb{R}^2 & \longrightarrow & 0 \\ \downarrow & & {\scriptstyle \begin{pmatrix} 0 & 1 \\ 1 & 0 \end{pmatrix}}\downarrow & & \downarrow{\scriptstyle \begin{pmatrix} -1/2 & 1/2 \\ 1/2 & -1/2 \end{pmatrix}} & & \downarrow & & \downarrow \\ 0 & \longrightarrow & \mathbb{R}^2 & \xrightarrow[\begin{pmatrix} 1 & -1 \\ -1 & 1 \end{pmatrix}]{} & \mathbb{R}^2 & \longrightarrow & 0 & \longrightarrow & 0 \end{array}$$

Exercise 2.3 Show that composition of sequence maps is not commutative by constructing an example of two sequence maps that do not commute. Specifically, construct a pair of sequence maps $p^\bullet, q^\bullet : (V^\bullet, v^\bullet) \to (V^\bullet, v^\bullet)$ such that $p^\bullet \circ q^\bullet \neq q^\bullet \circ p^\bullet$.

Definition 2.3 We will denote the category **Seq**(**Vec**) (or **Seq** if the context is clear) of *sequences of vector spaces*, in which the sequences are objects and sequence maps are morphisms.

Corollary 2.1 *Lemmas 2.1 and 2.2 establish that* **Seq** *is a well defined category.*

Example 2.4 Given a finite subset $I \subseteq \mathbb{Z}$, there is a covariant functor $F_I : \mathbf{Seq} \to \mathbf{Vec}$ given by direct summing over the indices in I. Namely,

$$F_I((V^\bullet, v^\bullet)) := \bigoplus_{i \in I} V^i.$$

Notice that sequence maps get summed into linear maps by this transformation, though the $v^\bullet$ maps within the sequence are ignored.

Exercise 2.4 Prove or give a counterexample to the following statement: Every sequence map $m^\bullet : (V^\bullet, v^\bullet) \to (W^\bullet, w^\bullet)$ whose component maps $m^k : V^k \to W^k$ can be factored

$$m^k : V^k \xrightarrow{p^k} X^k \xrightarrow{q^k} W^k$$

factors into the composition of two sequence maps $p^\bullet : (V^\bullet, v^\bullet) \to (X^\bullet, x^\bullet)$ and $q^\bullet : (X^\bullet, x^\bullet) \to (W^\bullet, w^\bullet)$.

We will frequently form a new sequence by taking the term-by-term quotient of a sequence by a sequence of vector subspaces.

Definition 2.4 If $(V^\bullet, v^\bullet)$ and $(W^\bullet, w^\bullet)$ are two sequences in which $W^k \subseteq V^k$ and each w^k map is the restriction of v^k to W^k, then the *term-by-term sequence quotient* (or simply *quotient*) $(V^\bullet, v^\bullet)/(W^\bullet, w^\bullet)$ is given by

$$\cdots \longrightarrow V^k/W^k \xrightarrow{v'^k} V^{k+1}/W^{k+1} \xrightarrow{v'^{k+1}} \cdots$$

where $v'^\bullet$ are the maps induced by each $v^\bullet$ on the vector space quotients, term-by-term. We will often write the quotient $V^\bullet/W^\bullet$ when the maps are clear from context.

Exercise 2.5 Consider the following two sequences

$$\cdots \longrightarrow 0 \longrightarrow \mathbb{R} \longrightarrow \mathbb{R}^2 \longrightarrow \cdots \mathbb{R}^k \longrightarrow \cdots$$

$$\cdots \longrightarrow 0 \longrightarrow 0 \longrightarrow \mathbb{R} \longrightarrow \cdots \mathbb{R} \longrightarrow \cdots$$

where each linear map is of full rank.

1. Show that there is only one possible way to take the quotient of the two sequences.
2. Compute the quotient sequence.
3. Construct a sequence map, in which all components are of the largest rank possible, from one of the two sequences to the quotient sequence.

Exercise 2.6 Prove or give a counterexample to the following statement, which aims to generalize Definition 2.4: Every sequence map $m^\bullet : (V^\bullet, v^\bullet) \to (W^\bullet, w^\bullet)$ induces a sequence map $V^\bullet \to V^\bullet/W^\bullet$ on the term-by-term sequence of quotients.

Exercise 2.7 Is there any difference between the quotient of the following pair of sequences

$$0 \longrightarrow \mathbb{R} \xrightarrow{(1)} \mathbb{R} \longrightarrow 0$$

$$0 \longrightarrow \mathbb{R} \xrightarrow{(1)} \mathbb{R} \longrightarrow 0$$

versus the following pair?

$$0 \longrightarrow \mathbb{R} \xrightarrow{(1)} \mathbb{R} \longrightarrow 0$$

$$0 \longrightarrow \mathbb{R} \xrightarrow{(0)} \mathbb{R} \longrightarrow 0$$

Perhaps one of the most obvious things we can do with a sequence is to shift its index. For an integer n, we can reindex the sequence $(V^\bullet, v^\bullet)$ into a new one $(V^{\bullet-n}, v^{\bullet-n})$. This amounts to defining a *shift functor* σ^n : **Seq** $\to$ **Seq** by

$$\sigma^n(V^k) := V^{k-n}, \text{ and } \sigma^n(v^k) := v^{k-n}.$$

Because σ^1 arose historically from a particular construction for simplicial chain complexes, it is sometimes called *suspension* (see Definition 4.14 in Chap. 4 and Proposition 5.5 in Chap. 5).

Exercise 2.8 Show that the shift functors σ^n form an *abstract abelian group of transformations*, namely

1. Composition of shift functors is associative, $\sigma^m \circ (\sigma^n \circ \sigma^p) = (\sigma^m \circ \sigma^n) \circ \sigma^p$,
2. The functor σ^0 acts as an *identity element* $\sigma^0 \circ \sigma^n = \sigma^n = \sigma^n \circ \sigma^0$,
3. Each shift functor has a unique inverse, $\sigma^n \circ \sigma^{-n} = \sigma^0$, and
4. Composition of shift functors is commutative $\sigma^m \circ \sigma^n = \sigma^n \circ \sigma^m$.

Challenge: show that the group of shift functors is isomorphic to $(\mathbb{Z}, +)$, the group of integers under addition.

It is often convenient to avoid sequences with too many nontrivial terms. The following three subcategories of **Seq** will make their appearance occasionally.

Definition 2.5 The following are full subcategories of **Seq**, defined by which sequences are present as objects.

1. $\mathbf{Seq}^+$, *sequences bounded from below*. Each object is a sequence $(V^\bullet, v^\bullet)$ in which there is an integer N such that $V^n = 0$ for all $n < N$.
2. $\mathbf{Seq}^-$, *sequences bounded from above*. Each object is a sequence $(V^\bullet, v^\bullet)$ in which there is an integer N such that $V^n = 0$ for all $n > N$.
3. $\mathbf{Seq}^b$, *bounded sequences*. Each object is a sequence $(V^\bullet, v^\bullet)$ in which there are two integers $M < N$ such that $V^n = 0$ if $n < M$ or if $n > N$.

Example 2.5 Unlike the functor F_I defined in Example 2.4, in which the set I must be finite, the formula

$$F((V^\bullet, v^\bullet)) := \bigoplus_{n \in \mathbb{Z}} V^n$$

defines a functor $\mathbf{Seq}^b \to \mathbf{Vec}$ without any problems at all.

There is a notion of *isomorphic sequences*, just as there is a notion of isomorphic vector spaces, which comes from the definition of isomorphisms within the category **Seq**.

Definition 2.6 Two sequences of vector spaces $(V^\bullet, v^\bullet)$ and $(W^\bullet, w^\bullet)$ are *isomorphic* if there are two sequence maps $p^\bullet : (V^\bullet, v^\bullet) \to (W^\bullet, w^\bullet)$ and $q^\bullet : (W^\bullet, w^\bullet) \to (V^\bullet, v^\bullet)$ such that $p^\bullet \circ q^\bullet$ and $q^\bullet \circ p^\bullet$ are identity sequence maps.

Proposition 2.1 *If $p^\bullet : (V^\bullet, v^\bullet) \to (W^\bullet, w^\bullet)$ is a sequence map for two sequences of vector spaces $(V^\bullet, v^\bullet)$ and $(W^\bullet, w^\bullet)$ in which p^k is a vector space isomorphism for every k, then $p^\bullet$ is a sequence isomorphism.*

Proof Since each p^k is a vector space isomorphism, this means that each has an inverse $(p^k)^{-1} : W^k \to V^k$. We need to show two things: (1) that $(p^\bullet)^{-1}$ is a sequence map and that (2) the two compositions of the two sequence maps $p^\bullet \circ (p^\bullet)^{-1}$ and $(p^\bullet)^{-1} \circ p^\bullet$ are both identity sequence maps. Because $(p^k)^{-1}$ is the inverse of p^k, this implies (2) immediately. As for (1), we need to show that the diagram

$$\begin{array}{ccccccc}
\cdots & \longrightarrow & W^k & \xrightarrow{w^k} & W^{k+1} & \xrightarrow{w^{k+1}} & \cdots \\
 & & \Big\downarrow{\scriptstyle (p^k)^{-1}} & & \Big\downarrow{\scriptstyle (p^{k+1})^{-1}} & & \\
\cdots & \longrightarrow & V^k & \xrightarrow[v^k]{} & V^{k+1} & \xrightarrow[v^{k+1}]{} & \cdots
\end{array}$$

commutes for all k. We compute

$$\begin{aligned}
p^{k+1} \circ v^k \circ (p^k)^{-1} &= w^k \circ p^k \circ (p^k)^{-1} \\
p^{k+1} \circ v^k \circ (p^k)^{-1} &= w^k \\
v^k \circ (p^k)^{-1} &= (p^{k+1})^{-1} \circ w^k,
\end{aligned}$$

because the following diagram is commutative

$$\begin{array}{ccccccc}
\cdots & \longrightarrow & V^k & \xrightarrow{v^k} & V^{k+1} & \xrightarrow{v^{k+1}} & \cdots \\
 & & \Big\downarrow{\scriptstyle p^k} & & \Big\downarrow{\scriptstyle p^{k+1}} & & \\
\cdots & \longrightarrow & W^k & \xrightarrow[w^k]{} & W^{k+1} & \xrightarrow[w^{k+1}]{} & \cdots
\end{array}$$

□

Exercise 2.9 Construct an example of a sequence map $m^\bullet : (V^\bullet, v^\bullet) \to (V^\bullet, v^\bullet)$ that cannot be inverted.

Exercise 2.10 One way to compute the inverse of a bounded sequence map $m^\bullet$ is to use the functor $F : \mathbf{Seq}^b \to \mathbf{Vec}$ defined in Example 2.5. This results in a square matrix with a block diagonal structure, in which all of the blocks along the diagonal are invertible square matrices. Show that the inverse of this block diagonal matrix preserves this block structure. (Hint: use the commutative ladder diagram to help organize the proof.)

Exercise 2.11 Prove the converse to Proposition 2.1, namely that every sequence isomorphism $p^\bullet : (V^\bullet, v^\bullet) \to (W^\bullet, w^\bullet)$ has component maps p^k that are vector spaces isomorphisms. (Hint: does the diagram commutativity requirement impose any restrictions on invertibility of sequence maps beyond what is specified in Proposition 2.1?)

The dualization functor $T : \mathbf{Vec} \to \mathbf{Vec}$ lifts to a contravariant functor $\mathbf{Seq} \to \mathbf{Seq}$. Simply, a morphism in $\mathbf{Seq}$ (a sequence map) like

$$
\begin{array}{ccccccccc}
\cdots & \longrightarrow & V^1 & \xrightarrow{v^1} & V^2 & \xrightarrow{v^2} & V^3 & \xrightarrow{v^3} & \cdots \\
 & & \downarrow{\scriptstyle m^1} & & \downarrow{\scriptstyle m^2} & & \downarrow{\scriptstyle m^3} & & \\
\cdots & \longrightarrow & W^1 & \xrightarrow[w^1]{} & W^2 & \xrightarrow[w^2]{} & W^3 & \xrightarrow[w^3]{} & \cdots
\end{array}
$$

is transformed into another sequence map

$$
\begin{array}{ccccccccc}
\cdots & \longleftarrow & V_1 & \xleftarrow{(v^1)^T} & V_2 & \xleftarrow{(v^2)^T} & V_3 & \xleftarrow{(v^3)^T} & \cdots \\
 & & \uparrow{\scriptstyle (m^1)^T} & & \uparrow{\scriptstyle (m^2)^T} & & \uparrow{\scriptstyle (m^3)^T} & & \\
\cdots & \longleftarrow & W_1 & \xleftarrow[(w^1)^T]{} & W_2 & \xleftarrow[(w^2)^T]{} & W_3 & \xleftarrow[(w^3)^T]{} & \cdots
\end{array}
$$

in which all linear maps are reversed, and for which we have defined $V_k := (V^k)^T$ and $W_k := (W^k)^T$ in accordance with the convention that subscripts mean that the indices decrease with the linear maps.

Exercise 2.12 Under what conditions can we construct isomorphisms $g_k : V^k \to V_k$ for a sequence of vector spaces and their duals so that the following diagram commutes?

$$
\begin{array}{ccccccccc}
\cdots & \longrightarrow & V^1 & \xrightarrow{v^1} & V^2 & \xrightarrow{v^2} & V^3 & \xrightarrow{v^3} & \cdots \\
 & & \downarrow{\scriptstyle g_1} & & \downarrow{\scriptstyle g_2} & & \downarrow{\scriptstyle g_3} & & \\
\cdots & \longleftarrow & V_1 & \xleftarrow[(v^1)^T]{} & V_2 & \xleftarrow[(v^2)^T]{} & V_3 & \xleftarrow[(v^3)^T]{} & \cdots
\end{array}
$$

This is *not* a sequence map. (Hint: Consider using the proof of Theorem 1.5 of Chap. 1.)

2.2 The Barcode Decomposition

Just as a given linear map is best described by decomposing its domain and codomain into certain subspaces (via Theorem 1.2 of Chap. 1, for instance), it is best to describe a sequence by way of another, simpler, sequence that is isomorphic to it.

Like vector spaces, sequences have a notion of a *direct sum*, which allows more complicated sequences to be written in terms of simpler ones.

Definition 2.7 If $(V^\bullet, v^\bullet)$ and $(W^\bullet, w^\bullet)$ are two sequences of vector spaces, their *direct sum* $(V^\bullet, v^\bullet) \oplus (W^\bullet, w^\bullet)$ is given by the sequence

$$\cdots \longrightarrow V^{k-1} \oplus W^{k-1} \xrightarrow{v^{k-1} \oplus w^{k-1}} V^k \oplus W^k \xrightarrow{v^k \oplus w^k} V^{k+1} \oplus W^{k+1} \longrightarrow \cdots$$

Definition 2.8 A *bar* is a sequence of vector spaces in which there are two options, $\mathbb{R}$ and the trivial space 0, and occurrences of $\mathbb{R}$ must appear consecutively. The linear maps between spaces in a bar must all be of full rank. Briefly, a *bar* is a sequence of the form

$$\cdots \longrightarrow 0 \longrightarrow \mathbb{R} \longrightarrow \cdots \longrightarrow \mathbb{R} \longrightarrow 0 \longrightarrow \cdots$$

in which each map $\mathbb{R} \to \mathbb{R}$ above is an identity map.

The *length* of a bar is the number of occurrences of $\mathbb{R}$ in the bar, which might be infinite. The *birth* or *start* of a bar is the index of the term at which the first occurrence of $\mathbb{R}$ is found. Similarly the *death* or *end* of a bar is the index of the last occurrence of $\mathbb{R}$ in the bar. The birth or death of a bar is allowed to be infinite. We will occasionally write $\mathbb{R}^{(b,d)}$ (or $\mathbb{R}_{(b,d)}$ if indices decrease) to denote the bar with birth b and death d.

A *barcode* is a direct sum of bars.

Returning to the sequence in Example 2.1, the original sequence is isomorphic to the barcode that is the direct sum of the following bars

$$\mathbb{R} \longrightarrow \mathbb{R} \longrightarrow \mathbb{R} \longrightarrow \mathbb{R},$$

$$\mathbb{R} \longrightarrow 0 \longrightarrow 0 \longrightarrow 0, \text{ and}$$

$$0 \longrightarrow \mathbb{R} \longrightarrow \mathbb{R} \longrightarrow 0$$

which are listed in the same order as the barcode shown in that example.

Example 2.6 There is a covariant functor $G : \mathbf{Vec} \to \mathbf{Seq}$ defined by

$$G(V) := \bigoplus_{i=0}^{\dim V} \mathbb{R}^{(0,0)}.$$

Since each finite dimensional vector space is determined up to isomorphism by its dimension, this means that while $G(V)$ is a sequence

$$\cdots \longrightarrow 0 \longrightarrow 0 \longrightarrow \bigoplus_{i=0}^{\dim V} \mathbb{R} \longrightarrow 0 \longrightarrow 0 \longrightarrow \cdots$$

it is isomorphic to a sequence

$$\cdots \longrightarrow 0 \longrightarrow 0 \longrightarrow V \longrightarrow 0 \longrightarrow 0 \longrightarrow \cdots$$

in which V appears at index 0. Notice that every linear map $\ell : V \to W$ lifts to a sequence map

$$\begin{array}{ccccccccccc}
\cdots & \longrightarrow & 0 & \longrightarrow & 0 & \longrightarrow & V & \longrightarrow & 0 & \longrightarrow & 0 & \longrightarrow & \cdots \\
 & & \downarrow & & \downarrow & & \downarrow \ell & & \downarrow & & \downarrow & & \\
\cdots & \longrightarrow & 0 & \longrightarrow & 0 & \longrightarrow & W & \longrightarrow & 0 & \longrightarrow & 0 & \longrightarrow & \cdots
\end{array}$$

Exercise 2.13 Construct a functor $G^{(a,b)} : \mathbf{Vec} \to \mathbf{Seq}$ that turns a vector space into a bar with birth a and death b.

1. Prove that your construction is a functor.
2. Consider $I \subseteq \mathbb{Z}$. Characterize all possible behaviors of the compositions

$$G^{(a,b)} \circ F_I \text{ and } F_I \circ G^{(a,b)},$$

where F_I is defined as in Example 2.4.

Lemma 2.3 *Two barcodes are isomorphic if and only if they consist of the same bars.*

Proof Suppose that $(B^\bullet, b^\bullet)$ and $(C^\bullet, c^\bullet)$ are two barcodes. This means that

$$B^\bullet = \bigoplus_j B^{j,\bullet} \text{ and } C^\bullet = \bigoplus_k C^{k,\bullet}$$

where each $B^{j,\bullet}$ and each $C^{k,\bullet}$ is a bar.

If two barcodes consist of the same bars, this means that $B^{j,\bullet} = C^{\sigma(j),\bullet}$ for some permutation σ. To construct an isomorphism, we merely need to permute the components of each term. The isomorphism is shown in the following diagram

$$\begin{array}{ccccccc}
\cdots \longrightarrow & \bigoplus_j B^{j,m} & \xrightarrow{\oplus b^{j,m}} & \bigoplus_j B^{j,m+1} & \xrightarrow{\oplus b^{j,m+1}} & \cdots \\
& \downarrow & & \downarrow & & \\
\cdots \longrightarrow & \bigoplus_j C^{\sigma(j),m} & \xrightarrow{\oplus c^{\sigma(j),m}} & \bigoplus_j C^{\sigma(j),m+1} & \xrightarrow{\oplus c^{\sigma(j),m+1}} & \cdots \\
& \downarrow & & \downarrow & & \\
\cdots \longrightarrow & \bigoplus_k C^{k,m} & \xrightarrow{\oplus c^{k,m}} & \bigoplus_k C^{k,m+1} & \xrightarrow{\oplus c^{k,m+1}} & \cdots
\end{array}$$

in which the first down arrows are identity maps, and the second down arrows are permutations of direct summands.

Conversely, suppose that $(B^\bullet, b^\bullet)$ and $(C^\bullet, c^\bullet)$ are isomorphic and both are barcodes. Proposition 2.1 implies that there is a sequence of isomorphisms $i^\bullet$, making the diagram

$$\begin{array}{ccccccc}
\cdots \longrightarrow & \bigoplus_j B^{j,m} & \xrightarrow{\oplus b^{j,m}} & \bigoplus_j B^{j,m+1} & \xrightarrow{\oplus b^{j,m+1}} & \cdots \\
& \downarrow{\scriptstyle i^m} & & \downarrow{\scriptstyle i^{m+1}} & & \\
\cdots \longrightarrow & \bigoplus_k C^{k,m} & \xrightarrow{\oplus c^{k,m}} & \bigoplus_k C^{k,m+1} & \xrightarrow{\oplus c^{k,m+1}} & \cdots
\end{array}$$

commutative. Consider the bar $B^{j,\bullet}$. By definition, each term $B^{j,m}$ must be either 0 or $\mathbb{R}$. Since each i^m is injective, it must be that

$$i^m(B^{j,m}) \cong B^{j,m},$$

namely that the image of $B^{j,\bullet}$ is also a bar, and is isomorphic to $B^{j,\bullet}$. Since i^m is surjective,

$$C^{\bullet,m} = i^m(B^{\bullet,m}) = i^m\left(\bigoplus_j B^{j,m}\right) \cong \bigoplus_j B^{j,m}.$$

Put another way, $C^\bullet$ consists of exactly the same bars as $B^\bullet$, though perhaps taken in a different order. □

All sequences isomorphic to a specific sequence have a simple description in terms of their bases (Jacquard et al. 2023).

Theorem 2.1 (The barcode decomposition) *Every sequence* $(V^\bullet, v^\bullet) \in \mathbf{Seq}^+$ *of vector spaces that is bounded from below is isomorphic to a barcode* $\bigoplus_j B^j$.

Note that Lemma 2.3 establishes that this barcode $\bigoplus_j B^j$ is unique up to permutation of the bars, which means that it suffices to find *any* barcode isomorphic to $(V^\bullet, v^\bullet)$.

Proof We proceed using an induction-like argument on the term index k, showing how to find all of the bars that start at V^k. Once all of the bars starting at V^k are found, we proceed to V^{k+1} using the bars already found thus far as a starting point. Since all of the terms in our sequence are trivial spaces for negative indices, there are no bars started there. We can therefore start the process at V^0 without losing generality. Select an arbitrary nonzero vector e^0 from V^0. Consider the sequence of spaces

$$B^{0,k} = \begin{cases} \mathbb{R} & \text{if } k = 0, \\ \mathbb{R} & \text{if } \left(v^{k-1} \circ \cdots \circ v^0\right)(e^0) \neq 0 \\ 0 & \text{otherwise,} \end{cases}$$

with

$$b^{0,k} = \begin{cases} \mathrm{id}_{\mathbb{R}} & \text{if } \left(v^k \circ v^{k-1} \circ \cdots \circ v^0\right)(e^0) \neq 0 \\ 0 & \text{otherwise,} \end{cases}$$

which defines a bar $(B^{0,\bullet}, b^{0,\bullet})$ corresponding to the history of e^0 as it passes from one linear map in the sequence to the next.

We can form a sequence isomorphism term-by-term

$$\begin{array}{ccccccccccc} \cdots & \longrightarrow & 0 & \longrightarrow & 0 & \longrightarrow & B^{0,0} \oplus (V^0/B^{0,0}) & \xrightarrow{\;b^{0,0} \oplus v'^0\;} & B^{0,1} \oplus (V^1/B^{0,1}) & \xrightarrow{\;b^{0,1} \oplus v'^1\;} & \cdots \\ & & \downarrow & & \downarrow & & \downarrow & & \downarrow & & \\ \cdots & \longrightarrow & 0 & \longrightarrow & 0 & \longrightarrow & V^0 & \xrightarrow{\;v^0\;} & V^1 & \xrightarrow{\;v^1\;} & \cdots \end{array}$$

using Theorem 1.1 of Chap. 1 to establish that each vertical map is an isomorphism. Because the maps $b^{0,\bullet}$ are all of full rank, this ensures that the subspace condition of Theorem 1.3 of Chap. 1 is satisfied. This establishes the existence of the induced maps $v'^{\bullet}$.

Observe that the dimension of $V^0/B^{0,0}$ is one less than the dimension of V^0. Thus if we select another nonzero vector e^1 from V^0 that corresponds to a nonzero vector $[e^1]_{B^{0,0}}$ in the quotient, we can repeat the process. After a finite number of steps, we will have found a bar for each vector in a particular basis of V^0. This means that we have found an isomorphism between our original sequence and a sequence of the form

$$\cdots \longrightarrow 0 \longrightarrow 0 \longrightarrow \bigoplus_j B^{j,0} \longrightarrow \bigoplus_j B^{j,1} \oplus V'^1 \longrightarrow \cdots$$

in which each V'^k is the quotient $V^k / \bigoplus_j B^{j,k}$. This is the direct sum of a barcode and a sequence

$$\cdots \longrightarrow 0 \longrightarrow 0 \longrightarrow 0 \longrightarrow V'^1 \longrightarrow \cdots$$

This looks like our original sequence, but what was the first nontrivial space is now trivial. Thus, we can iterate the entire process $k + 1$ times to find all of the bars starting at index k. □

Remark 2.2 Bars are examples of *indecomposable submodules* or *interval submodules*, and many authors prefer to refer to them as such. Since that terminology seems needlessly verbose this book uses the briefer name "bar" exclusively.

While the proof of Theorem 2.1 is constructive and algorithmic, a faster algorithm is available if we do not need the details of the sequence isomorphism. Since we would like an algorithm suitable for implementation in software, let us restrict our attention to a sequence $(V^\bullet, v^\bullet)$ with finitely many nontrivial terms. Observe that every bar with finite length corresponds to a vector that eventually ends up in the kernel of a composition of linear maps in the sequence. For instance, consider the sequence $(V^\bullet, v^\bullet)$ and a vector $e \neq 0 \in V^0$ with $(v^1 \circ v^0)(e) = 0$. If we use e in the proof of Theorem 2.1, we obtain a commutative diagram like the following

$$\begin{array}{ccccccc}
\mathbb{R} & \xrightarrow{\text{id}} & \mathbb{R} & \longrightarrow & 0 & \longrightarrow & \cdots \\
\downarrow{\scriptstyle (e)} & & \downarrow & & \downarrow & & \\
V^0 & \xrightarrow[v^0]{} & V^1 & \xrightarrow[v^1]{} & V^2 & \xrightarrow[v^2]{} & \cdots
\end{array}$$

Such a diagram makes it clear that saying the bar corresponding to e ends at index 1 is equivalent to saying that $e \in \ker(v^1 \circ v^0)$. This suggests that we should rephrase the algorithm in terms of computing kernels.

Notice that the kernels of $v^k \circ \cdots \circ v^0$ eventually contain V^0 as k increases. This relies on the fact that the terms V^k are all trivial for sufficiently large k. This composition eventually becomes the zero map. The bars that end at $k - 1$ are therefore elements of the kernel of a composition of the form $v^k \circ \cdots$. All that remains to compute the barcode systematically is to keep track of which bars have started at which index.

Example 2.7 When a linear map is sparse—it has many zero entries—it may be easy to find its kernel by inspection. Algorithm 1 may be particularly easy to run for sequences of such matrices, since the computation of G_1 does not require any matrix operations. For instance, consider the following sequence

$$0 \longrightarrow \mathbb{R}^3 \xrightarrow{\begin{pmatrix} 0&0&0\\ 0&0&0\\ 0&0&1\\ 0&0&0\\ 0&1&0 \end{pmatrix}} \mathbb{R}^5 \xrightarrow{\begin{pmatrix} 0&0&0&0&1\\ 0&0&0&0&0\\ 0&0&0&1&0\\ 1&0&0&0&0\\ 0&1&0&0&0 \end{pmatrix}} \mathbb{R}^5 \xrightarrow{\begin{pmatrix} 0&0&0&0&1\\ 1&0&0&0&0\\ 0&0&0&0&0 \end{pmatrix}} \mathbb{R}^3 \longrightarrow 0.$$

Algorithm 1 Computing a barcode decomposition

Require: A sequence of vector spaces $(V^\bullet, v^\bullet)$ such that $V^k = 0$ for all $k < 0$ or for all $k > n$.
Ensure: A list of pairs (b_i, d_i) such that $(V^\bullet, v^\bullet)$ is isomorphic to the barcode $\bigoplus_i \mathbb{R}^{(b_i, d_i)}$

procedure BARCODE($(V^\bullet, v^\bullet)$)
  Initialize $B = \{\}$ ▷ The list of output pairs
  for $j = 0, 1, 2, \ldots, n$ **do** ▷ Candidate birth index for a bar
    for $k = j, j+1, \ldots, n$ **do** ▷ Candidate death index for a bar
      Compute $G_1 = \dim\ker(v^k \circ \cdots \circ v^j)$. ▷ The number of bars that end at k or earlier and start at j or earlier.
      Let G_2 be the number of elements (b_i, d_i) of B with $j \le d_i \le k$. ▷ The number of bars already computed that end within the interval under consideration.
      Append $G = G_1 - G_2$ copies of the pair (j, k) to B. ▷ The number of bars that start at j and end at k.
    end for
  end for
  return B.
end procedure

The kernel of each matrix is spanned by columns of zeros, since the columns with nonzero entries are linearly independent.

Let us assume that the first index is 0. Since that space is trivial, no bars start there. Therefore, let us find all the bars starting at index $j = 1$. Notice that in Algorithm 1, we loop until the last nontrivial space, which in this case has index 4. The algorithm runs as follows.

$j = 1,\ k = 1,\ B = \{\}$

$$v^1 = \begin{pmatrix} 0&0&0 \\ 0&0&0 \\ 0&0&1 \\ 0&0&0 \\ 0&1&0 \end{pmatrix}$$

so $G_1 = \dim\ker(v^1) = 1$. Since $B = \{\}$ there are no bars yet, so $G_2 = 0$. Thus, we have found $G = G_1 - G_2 = 1$ new bar. We therefore append $\{(1, 1)\}$ to B.
$j = 1,\ k = 2,\ B = \{(1, 1)\}$

$$v^2 \circ v^1 = \begin{pmatrix} 0&1&0 \\ 0&0&0 \\ 0&0&0 \\ 0&0&0 \\ 0&0&0 \end{pmatrix}$$

so $G_1 = \dim\ker(v^2 \circ v^1) = 2$. There is already one bar that has started at $j = 1$ and ended earlier than $k = 2$, so $G_2 = 1$. Thus there is $G = G_1 - G_2 = 1$ new bar to be appended, namely $(1, 2)$.
$j = 1,\ k = 3,\ B = \{(1, 1), (1, 2)\}$

$$v^3 \circ v^2 \circ v^1 = \begin{pmatrix} 0 & 0 & 0 \\ 0 & 1 & 0 \\ 0 & 0 & 0 \end{pmatrix}$$

so $G_1 = \dim\ker(v^3 \circ v^2 \circ v^1) = 2$. There are already two bars started at $j = 1$ that end earlier than $k = 3$, so $G_2 = 2$. This means that both of the dimensions of the kernel of $(v^3 \circ v^2 \circ v^1)$ correspond to existing bars. B need not be updated.
$j = 1,\ k = 4,\ B = \{(1, 1), (1, 2)\}$

$$v^4 \circ v^3 \circ v^2 \circ v^1 = \begin{pmatrix} 0 & 0 & 0 \end{pmatrix}$$

so $G_1 = \dim\ker(v^4 \circ v^3 \circ v^2 \circ v^1) = 3$. There are already two bars started at $j = 1$ that end earlier than $k = 4$, so $G_2 = 2$. Thus we append one copy of $(1, 4)$ to B.
$j = 2,\ k = 2,\ B = \{(1, 1), (1, 2), (1, 4)\}$

$$v^2 = \begin{pmatrix} 0 & 0 & 0 & 0 & 1 \\ 0 & 0 & 0 & 0 & 0 \\ 0 & 0 & 0 & 1 & 0 \\ 1 & 0 & 0 & 0 & 0 \\ 0 & 1 & 0 & 0 & 0 \end{pmatrix}$$

so $G_1 = \dim\ker(v^2) = 1$. There is one existing bar, namely $(1, 2)$, that ends at $j = k = 2$, so $G_2 = 1$. Thus no new bars need to be added to B.
$j = 2,\ k = 3,\ B = \{(1, 1), (1, 2), (1, 4)\}$

$$v^3 \circ v^2 = \begin{pmatrix} 0 & 1 & 0 & 0 & 0 \\ 0 & 0 & 0 & 0 & 1 \\ 0 & 0 & 0 & 0 & 0 \end{pmatrix}$$

so $G_1 = \dim\ker(v^3 \circ v^2) = 3$. Again, there is one bar, namely $(1, 2)$, that ends between $j = 2$ and $k = 3$, so $G_2 = 1$. Thus, we add $G = G_1 - G_2 = 2$ copies of $(2, 3)$ to B.
$j = 2,\ k = 4,\ B = \{(1, 1), (1, 2), (1, 4), (2, 3), (2, 3)\}$

$$v^4 \circ v^3 \circ v^2 = \begin{pmatrix} 0 & 0 & 0 & 0 & 0 \end{pmatrix}$$

so $G_1 = \dim\ker(v^4 \circ v^3 \circ v^2) = 5$. A total of four existing bars, $(1, 2)$, $(1, 4)$, $(2, 3)$, and $(2, 3)$, end between $j = 2$ and $k = 4$, so $G_2 = 4$. We therefore add one copy of $(2, 4)$ to B.
$j = 3,\ k = 3,\ B = \{(1, 1), (1, 2), (1, 4), (2, 3), (2, 3), (2, 4)\}$

$$v^3 = \begin{pmatrix} 0 & 0 & 0 & 0 & 1 \\ 1 & 0 & 0 & 0 & 0 \\ 0 & 0 & 0 & 0 & 0 \end{pmatrix}$$

so $G_1 = \dim\ker(v^3) = 3$. Two existing bars, (2, 3) and (2, 3), end at $j = k = 3$ so $G_2 = 2$. Thus we add one copy of (3, 3) to B.
$j = 3,\ k = 4,\ B = \{(1, 1), (1, 2), (1, 4), (2, 3), (2, 3), (2, 4), (3, 3)\}$

$$v^4 \circ v^3 = \begin{pmatrix} 0 & 0 & 0 & 0 & 0 \end{pmatrix}$$

so $G_1 = \dim\ker(v^4 \circ v^3) = 5$. There are five existing bars that end between $j = 3$ and $k = 4$, so no updates to B are needed.
$j = 4,\ k = 4,\ B = \{(1, 1), (1, 2), (1, 4), (2, 3), (2, 3), (2, 4), (3, 3)\}$ v^4 is the zero map, so $G_1 = 3$. At this final step, there are two bars, (1, 4) and (2, 4), ending at $j = k = 4$. Thus the final update to B adds one copy of (4, 4).

Exiting the loops, the list returned by the algorithm is

$$B = \{(1, 1), (1, 2), (1, 4), (2, 3), (2, 3), (2, 4), (3, 3), (4, 4)\}.$$

This means that the original sequence is isomorphic to the barcode

$$\mathbb{R}^{(1,1)} \oplus \mathbb{R}^{(1,2)} \oplus \mathbb{R}^{(1,4)} \oplus \mathbb{R}^{(2,3)} \oplus \mathbb{R}^{(2,3)} \oplus \mathbb{R}^{(2,4)} \oplus \mathbb{R}^{(3,3)} \oplus \mathbb{R}^{(4,4)}.$$

Exercise 2.14 Compute the barcode decomposition of the following sequence

$$0 \longrightarrow \mathbb{R}^2 \xrightarrow{\begin{pmatrix} 1 & 0 \end{pmatrix}} \mathbb{R} \xrightarrow{\begin{pmatrix} 0 \\ 1 \end{pmatrix}} \mathbb{R}^2 \xrightarrow{\begin{pmatrix} 1 & 0 \\ 0 & 0 \\ 0 & 0 \end{pmatrix}} \mathbb{R}^3 \longrightarrow 0.$$

Exercise 2.15 Compute the barcode decomposition of the following sequence

$$0 \longrightarrow \mathbb{R}^2 \xrightarrow{\begin{pmatrix} 0 & 0 \\ 1 & 0 \\ 0 & 1 \\ 0 & 0 \end{pmatrix}} \mathbb{R}^4 \xrightarrow{\begin{pmatrix} 0 & 0 & 0 & 1 \\ 1 & 0 & 0 & 0 \\ 0 & 1 & 0 & 0 \end{pmatrix}} \mathbb{R}^3 \xrightarrow{\begin{pmatrix} 1 & 0 & 0 \\ 0 & 1 & 0 \\ 0 & 0 & 1 \end{pmatrix}} \mathbb{R}^3 \longrightarrow 0.$$

Exercise 2.16 Compute the barcode decomposition of the following sequence

$$0 \longrightarrow \mathbb{R}^1 \xrightarrow{\begin{pmatrix}1\\0\end{pmatrix}} \mathbb{R}^2 \xrightarrow{\begin{pmatrix}2&4\\0&0\end{pmatrix}} \mathbb{R}^3 \xrightarrow{\begin{pmatrix}0&-4\\0&6\\0&-6\end{pmatrix}} \mathbb{R}^3 \xrightarrow{(0\ 3\ 3)} \mathbb{R} \longrightarrow 0.$$

Exercise 2.17 Compute the barcode decomposition of the following sequence

$$0 \longrightarrow \mathbb{R}^2 \xrightarrow{\begin{pmatrix}-9&-3\\9&2\\15&2\\0&2\end{pmatrix}} \mathbb{R}^4 \xrightarrow{\begin{pmatrix}-4&6&-6&-6\\-6&9&-9&-9\end{pmatrix}} \mathbb{R}^2 \xrightarrow{(9\ -6)} \mathbb{R} \xrightarrow{(3)} \mathbb{R} \longrightarrow 0.$$

Definition 2.9 The *Euler characteristic* of a bounded sequence $(V^\bullet, v^\bullet)$ is the sum

$$\chi(V^\bullet) := \sum_{k\in\mathbb{Z}} (-1)^k \dim V^k.$$

This sum is well defined because a bounded sequence only has finitely many nontrivial terms.

Lemma 2.4 *Suppose that $(V^\bullet, v^\bullet)$ and $(W^\bullet, w^\bullet)$ are two sequences, then*

$$\chi\left((V \oplus W)^\bullet\right) = \chi(V^\bullet) + \chi(W^\bullet).$$

Proof This is merely a calculation using the definition.

$$\begin{aligned}
\chi\left((V \oplus W)^\bullet\right) &= \sum_{k\in\mathbb{Z}} (-1)^k \dim(V \oplus W)^k \\
&= \sum_{k\in\mathbb{Z}} (-1)^k \dim\left(V^k \oplus W^k\right) \\
&= \sum_{k\in\mathbb{Z}} (-1)^k \left(\dim V^k + \dim W^k\right) \\
&= \sum_{k\in\mathbb{Z}} (-1)^k \dim V^k + \sum_{k\in\mathbb{Z}} (-1)^k \dim W^k \\
&= \chi(V^\bullet) + \chi(W^\bullet).
\end{aligned}$$

□

Lemma 2.5 *For a single bar,*

$$\chi\left(\mathbb{R}^{(a,b)}\right) = \begin{cases} (-1)^a & \text{if } b-a+1 \text{ is odd} \\ 0 & \text{if } b-a+1 \text{ is even} \end{cases}$$

noting that $b-a+1$ *is the length of the bar.*

Proof This is a calculation

$$\begin{aligned}\chi\left(\mathbb{R}^{(a,b)}\right) &= \sum_{a\le k\le b} (-1)^k \\ &= (-1)^a + (-1)^{a+1} + (-1)^{a+2} + \cdots + (-1)^b \\ &= (-1)^a - (-1)^a + (-1)^a - \cdots + (-1)^a(-1)^{b-a} \\ &= (-1)^a\left[1 - 1 + 1 - \cdots + (-1)^{b-a}\right].\end{aligned}$$

Notice that the quantity in the brackets is either 0 or 1 depending on the parity of the length of the bar. □

Corollary 2.2 *The Euler characteristic of a sequence is determined by its barcode. Hence, isomorphic sequences have the same Euler characteristic.*

Exercise 2.18 Compute the Euler characteristic of the following sequence

$$0 \longrightarrow \mathbb{R}^2 \xrightarrow{\begin{pmatrix} 0 & -3 \\ 1 & 0 \\ 0 & 1 \\ 0 & 0 \end{pmatrix}} \mathbb{R}^4 \xrightarrow{\begin{pmatrix} 0&0&2&1 \\ 1&0&0&0 \\ 0&1&0&0 \end{pmatrix}} \mathbb{R}^3 \xrightarrow{\begin{pmatrix} 1&0&0 \\ 0&1&0 \\ 4&5&1 \end{pmatrix}} \mathbb{R}^3 \longrightarrow 0.$$

Exercise 2.19 Exhibit two sequences with at least three terms that are not isomorphic but have the same Euler characteristic.

The barcode decomposition, while useful for understanding the structure of a single sequence, does not interact cleanly with sequence maps.

Example 2.8 Consider the sequence map given by the vertical maps in the diagram

$$\begin{array}{ccccccccc}
\cdots & \longrightarrow & 0 & \longrightarrow & \mathbb{R} & \xrightarrow{\mathrm{id}} & \mathbb{R} & \longrightarrow & \cdots \\
 & & \big\downarrow & & \big\downarrow \begin{pmatrix} 0 \\ 1 \end{pmatrix} & & \big\downarrow \mathrm{id} & & \\
\cdots & \longrightarrow & \mathbb{R} & \xrightarrow[\begin{pmatrix} 1 \\ 0 \end{pmatrix}]{} & \mathbb{R}^2 & \xrightarrow[(1\ 1)]{} & \mathbb{R} & \longrightarrow & \cdots
\end{array}$$

The top sequence is a bar already. Its image within the bottom sequence is also of the same form, namely

$$\cdots \longrightarrow 0 \longrightarrow \mathbb{R} \longrightarrow \mathbb{R} \longrightarrow \cdots$$

This image is incompatible with the barcode decomposition of the bottom sequence, which is given by the direct sum of the following two bars

$$\cdots \longrightarrow \mathbb{R} \longrightarrow \mathbb{R} \longrightarrow \mathbb{R} \longrightarrow \cdots$$

$$\cdots \longrightarrow 0 \longrightarrow \mathbb{R} \longrightarrow 0 \longrightarrow \cdots$$

This is not an accident: this shows that there is no way to align the barcode decompositions of these two sequences such that the sequence map takes bars in one sequence to bars in the other.

Exercise 2.20 Compute the sequence induced by taking term-by-term quotients of the bottom row by the images of the top row in the following diagram

$$\begin{array}{ccccccccc}
\cdots & \longrightarrow & 0 & \longrightarrow & \mathbb{R} & \xrightarrow{\mathrm{id}} & \mathbb{R} & \longrightarrow & \cdots \\
 & & \big\downarrow & & \big\downarrow \begin{pmatrix} 0 \\ 1 \end{pmatrix} & & \big\downarrow \mathrm{id} & & \\
\cdots & \longrightarrow & \mathbb{R} & \xrightarrow[\begin{pmatrix} 1 \\ 0 \end{pmatrix}]{} & \mathbb{R}^2 & \xrightarrow[(1\ 1)]{} & \mathbb{R} & \longrightarrow & \cdots
\end{array}$$

Explain whether the sequence you computed is induced by descending to the quotient as described in Theorem 1.3 in Chap. 1; if not explain where the failure lies.

Exercise 2.21 The following commutative diagram defines a sequence map $m^\bullet$

$$\begin{array}{ccccccccccc}
0 & \longrightarrow & \mathbb{R} & \xrightarrow{\begin{pmatrix}0\\0\\1\end{pmatrix}} & \mathbb{R}^3 & \xrightarrow{\begin{pmatrix}1&0&0\\0&0&0\end{pmatrix}} & \mathbb{R}^2 & \xrightarrow{\begin{pmatrix}0&0\end{pmatrix}} & \mathbb{R} & \longrightarrow & 0 \\
 & & \Big\downarrow{\scriptstyle (1)} & & \Big\downarrow{\scriptstyle (1\,0\,0)} & & \Big\downarrow{\scriptstyle \begin{pmatrix}0&0\\1&0\\0&1\end{pmatrix}} & & \Big\downarrow{\scriptstyle (0)} & & \\
0 & \longrightarrow & \mathbb{R} & \xrightarrow[(0)]{} & \mathbb{R} & \xrightarrow[\begin{pmatrix}0\\1\\0\end{pmatrix}]{} & \mathbb{R}^3 & \xrightarrow[(1\,0\,0)]{} & \mathbb{R} & \longrightarrow & 0
\end{array}$$

Unlike what happens in Example 2.8, this sequence map *does* induce a function taking bars of the top sequence to bars of the bottom sequence. Prove this fact using the following steps.

1. Verify that the diagram is commutative.
2. Compute the barcodes for the top and bottom sequences using Algorithm 1.
3. Compute the image of each bar from the top diagram, establishing that for this particular sequence map, there is a correspondence.
4. Reassemble the sequence map $m^\bullet$ from a direct sum of sequence maps $m_i^\bullet : \mathbb{R}^{(a_i,b_i)} \to \mathbb{R}^{(c_i,d_i)}$ from a bar in the top sequence to one in the bottom sequence. (Hint: it is not required that each of the $m_i^\bullet$ sequence maps have the same component maps.)

In Sect. 3.1 of Chap. 3, we will prove that for *chain complexes*—those sequences in which all bars have length 2 or less—individual bars in the domain map to bars in the codomain in a particular way.

2.3 Shifted Quotient Sequences

Consider an arbitrary sequence of linear maps

$$\cdots \to V^0 \xrightarrow{v^0} V^1 \xrightarrow{v^1} \cdots \longrightarrow V^n \xrightarrow{v^n} \cdots$$

What can we learn about this sequence? While the barcode is a useful and complete representation, there are other summary-like features that are also useful. Especially since sequence maps do not preserve bars, it is useful to find a simpler summary that does preserve bars. To this end, let us consider quotients of adjacent spaces in the sequence. Since the spaces may be different, we ought to consider the quotient of each space with the image of the map before it. We have a sequences of spaces (no arrows!)

$$\cdots \qquad V^1/\operatorname{image} v^0 \qquad V^2/\operatorname{image} v^1 \qquad \cdots \qquad V^n/\operatorname{image} v^{n-1} \qquad \cdots$$

But in fact a little more is true since we actually have a commutative diagram between the $V^\bullet$ sequence and a shift of it

$$\begin{array}{ccccccccc}
\cdots \longrightarrow & V^1 & \xrightarrow{v^1} & V^2 & \xrightarrow{v^2} \cdots \longrightarrow & V^n & \xrightarrow{v^n} & \cdots \\
& {\scriptstyle v^0}\uparrow & & \uparrow{\scriptstyle v^1} & & \uparrow{\scriptstyle v^{n-1}} & & \\
\cdots \longrightarrow & V^0 & \xrightarrow[v^0]{} & V^1 & \xrightarrow[v^1]{} \cdots \longrightarrow & V^{n-1} & \xrightarrow[v^{n-1}]{} & \cdots
\end{array}$$

so the v^k maps descend to the quotients

$$\cdots \longrightarrow V^1/\operatorname{image} v^0 \longrightarrow V^2/\operatorname{image} v^1 \longrightarrow \cdots \longrightarrow V^n/\operatorname{image} v^{n-1} \longrightarrow \cdots$$

However, since the image of each v^k map in this sequence appears in the quotient of the next term, all of these maps are zero maps.

The commutative ladder can be taken further by shifting different amounts. For instance, for a shift by a nonnegative integer k, let us define

$$v_k^n := v^{n+k} \circ \cdots \circ v^{n+1} \circ v^n.$$

Notice that $v_k^n : V^n \to V^{n+k}$, so that $v^n = v_1^n$. This notation also suggests that we take

$$v_0^n := \mathrm{id}_{V^n},$$

so we make that definition as well. The commutative ladder becomes

$$\begin{array}{ccccccccc}
\cdots \longrightarrow & V^1 & \xrightarrow{v^1} & V^2 & \xrightarrow{v^2} \cdots \longrightarrow & V^n & \xrightarrow{v^n} & \cdots \\
& {\scriptstyle v_k^{1-k}}\uparrow & & \uparrow{\scriptstyle v_k^{2-k}} & & \uparrow{\scriptstyle v_k^{n-k}} & & \\
\cdots \longrightarrow & V^{1-k} & \xrightarrow[v^{1-k}]{} & V^{2-k} & \xrightarrow[v^{2-k}]{} \cdots \longrightarrow & V^{n-k} & \xrightarrow[v^{n-k}]{} & \cdots
\end{array}$$

resulting in the *k-shifted quotient sequence* upon descending to quotients

$$\cdots \longrightarrow V^1/\,\text{image}\, v_k^{1-k} \xrightarrow{m^1} V^2/\,\text{image}\, v_k^{2-k} \xrightarrow{m^2} \cdots V^n/\,\text{image}\, v_k^{n-k} \xrightarrow{m^n} \cdots$$

This new sequence has the interesting property that the composition of k adjacent $m^\bullet$ maps is zero, namely

$$m^n \circ \cdots \circ m^{n-k+1} = 0$$

for all n.

Example 2.9 Consider the sequence

$$0 \longrightarrow \mathbb{R} \xrightarrow{\text{id}} \mathbb{R} \xrightarrow{\begin{pmatrix}1\\0\end{pmatrix}} \mathbb{R}^2 \xrightarrow{(0\ 1)} \mathbb{R} \longrightarrow 0$$

Shifting this sequence by two results in the following commutative ladder

$$\begin{array}{ccccccccccccc}
0 & \longrightarrow & \mathbb{R} & \xrightarrow{\text{id}} & \mathbb{R} & \xrightarrow{\begin{pmatrix}1\\0\end{pmatrix}} & \mathbb{R}^2 & \xrightarrow{(0\ 1)} & \mathbb{R} & \longrightarrow & 0 & \longrightarrow & 0 \\
\uparrow & & \uparrow & & \uparrow & & \uparrow{\scriptstyle \begin{pmatrix}1\\0\end{pmatrix}} & & \uparrow{\scriptstyle (0)} & & \uparrow & & \uparrow \\
0 & \longrightarrow & 0 & \longrightarrow & 0 & \longrightarrow & \mathbb{R} & \xrightarrow[\text{id}]{} & \mathbb{R} & \xrightarrow[\begin{pmatrix}1\\0\end{pmatrix}]{} & \mathbb{R}^2 & \xrightarrow[(0\ 1)]{} & \mathbb{R}
\end{array}$$

and descending to the quotients yields

$$0 \longrightarrow \mathbb{R} \xrightarrow{\text{id}} \mathbb{R} \xrightarrow{0} \mathbb{R} \xrightarrow{\text{id}} \mathbb{R} \longrightarrow 0 \longrightarrow 0$$

This new sequence is said to be *exact*, which means that it has the property that the kernel of each map is precisely the image of the previous one.

Example 2.10 Modifying the sequence in the previous example, consider the sequence of linear maps

$$0 \longrightarrow \mathbb{R} \xrightarrow{\text{id}} \mathbb{R} \xrightarrow{\begin{pmatrix}1\\0\end{pmatrix}} \mathbb{R}^2 \xrightarrow{(0\ 0)} \mathbb{R} \longrightarrow 0$$

Shifting this sequence by two results in the following commutative ladder

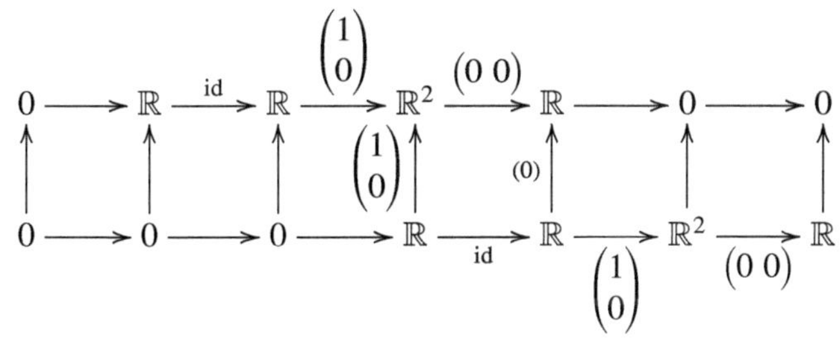

and descending to the quotients yields

$$0 \longrightarrow \mathbb{R} \xrightarrow{\text{id}} \mathbb{R} \xrightarrow{0} \boxed{\mathbb{R}} \xrightarrow{0} \mathbb{R} \longrightarrow 0 \longrightarrow 0$$

Unlike the result in the previous example, this new sequence is not exact since the image of the map ending at the boxed $\mathbb{R}$ is a proper subset of the kernel of the map starting at that index.

Proposition 2.2 *Sequence maps induce sequence maps on shifted quotient sequences.*

This means that we have defined a collection of *shifted quotient sequence functors* $Q_k : \mathbf{Seq} \to \mathbf{Seq}$.

Proof Suppose that we have a sequence map $m^\bullet : (V^\bullet, v^\bullet) \to (W^\bullet, w^\bullet)$

$$\begin{array}{ccccccccc}
\cdots \longrightarrow & V^1 & \xrightarrow{v^1} & V^2 & \xrightarrow{v^2} & V^3 & \xrightarrow{v^3} & \cdots \\
& \downarrow{\scriptstyle m^1} & & \downarrow{\scriptstyle m^2} & & \downarrow{\scriptstyle m^3} & & \\
\cdots \longrightarrow & W^1 & \xrightarrow[w^1]{} & W^2 & \xrightarrow[w^2]{} & W^3 & \xrightarrow[w^3]{} & \cdots
\end{array}$$

Using our construction of the quotient sequence shifted by an arbitrary k, we build the following diagram

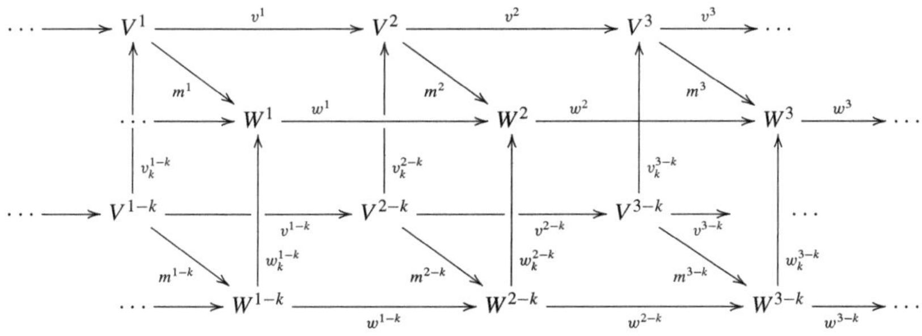

We need only show that each $m^n : V^n \to W^n$ descends to the quotients via Theorem 1.3 in Chap. 1. This follows from the fact that the diagram

$$\begin{array}{ccc} V^n & \xrightarrow{m^n} & W^n \\ {\scriptstyle v_k^{n-k}}\uparrow & & \uparrow{\scriptstyle w_k^{n-k}} \\ V^{n-k} & \xrightarrow{m^{n-k}} & W^{n-k} \end{array}$$

commutes because this means that $m^n(\text{image } v_k^{n-k}) \subseteq \text{image } w_k^{n-k}$. $\square$

Example 2.11 Consider the sequence map given by

$$\begin{array}{ccccccccc} 0 & \longrightarrow & 0 & \longrightarrow & \mathbb{R}^2 & \xrightarrow{\begin{pmatrix}1 & -1\end{pmatrix}} & \mathbb{R} & \longrightarrow & 0 \\ & & \downarrow & & \downarrow{\scriptstyle \begin{pmatrix}0 & 1\\ -2 & 4\\ -1 & 0\end{pmatrix}} & & \downarrow & & \\ 0 & \longrightarrow & \mathbb{R}^2 & \xrightarrow[\begin{pmatrix}-1 & -1\\ -10 & -2\\ -3 & 1\end{pmatrix}]{} & \mathbb{R}^3 & \longrightarrow & 0 & \longrightarrow & 0 \end{array}$$

If we wish to find the sequence map induced by Q_1, we first take Q_1 on each sequence.

Applying Q_1 to the top (domain) sequence is performed via the diagram

$$\begin{array}{ccccccccccc} 0 & \longrightarrow & 0 & \longrightarrow & \mathbb{R}^2 & \xrightarrow{\begin{pmatrix}1 & -1\end{pmatrix}} & \mathbb{R} & \longrightarrow & 0 & & \\ & & \uparrow & & \uparrow & & \uparrow{\scriptstyle \begin{pmatrix}1 & -1\end{pmatrix}} & & \uparrow & & \\ & & 0 & \longrightarrow & 0 & \longrightarrow & \mathbb{R}^2 & \xrightarrow[\begin{pmatrix}1 & -1\end{pmatrix}]{} & \mathbb{R} & \longrightarrow & 0 \end{array}$$

The resulting quotient sequence is

$$0 \longrightarrow 0 \longrightarrow \mathbb{R}^2 \longrightarrow 0 \longrightarrow 0$$

Applying Q_1 to the bottom sequence below (the codomain of the sequence map) is performed via the diagram

$$
\begin{array}{cccccccccccc}
0 & \longrightarrow & \mathbb{R}^2 & \xrightarrow{\begin{pmatrix} -1 & -1 \\ -10 & -2 \\ -3 & 1 \end{pmatrix}} & \mathbb{R}^3 & \longrightarrow & 0 & \longrightarrow & 0 & & \\
 & & \uparrow & & \uparrow {\scriptstyle \begin{pmatrix} -1 & -1 \\ -10 & -2 \\ -3 & 1 \end{pmatrix}} & & \uparrow & & \uparrow & & \\
 & & 0 & \longrightarrow & \mathbb{R}^2 & \xrightarrow[\begin{pmatrix} -1 & -1 \\ -10 & -2 \\ -3 & 1 \end{pmatrix}]{} & \mathbb{R}^3 & \longrightarrow & 0 & \longrightarrow & 0
\end{array}
$$

The resulting quotient sequence is

$$
0 \longrightarrow \mathbb{R}^2 \xrightarrow{\;0\;} \mathbb{R} \longrightarrow 0 \longrightarrow 0
$$

Now to find the induced maps, we use the quotient diagrams indexwise according to the diagram at the end of the proof of Proposition 2.2. We have the following three diagrams, one for each index,

$$
\begin{array}{ccc}
0 & \longrightarrow & \mathbb{R}^2 \\
\uparrow & & \uparrow \\
0 & \longrightarrow & 0
\end{array}
\qquad
\begin{array}{ccc}
\mathbb{R}^2 & \xrightarrow{\begin{pmatrix} 0 & 1 \\ -2 & 4 \\ -1 & 0 \end{pmatrix}} & \mathbb{R}^3 \\
\uparrow & & \uparrow {\scriptstyle \begin{pmatrix} -1 & -1 \\ -10 & -2 \\ -3 & 1 \end{pmatrix}} \\
0 & \longrightarrow & \mathbb{R}^2
\end{array}
\qquad
\begin{array}{ccc}
\mathbb{R} & \longrightarrow & 0 \\
\uparrow {\scriptstyle (1\ -1)} & & \uparrow \\
\mathbb{R}^2 & \longrightarrow & \mathbb{R}^3
\end{array}
$$

The only nontrivial induced map arises from the middle of these diagrams. It happens that both columns of the component map are in the image of the top map, which is to say that

$$
\begin{pmatrix} -1 & -1 \\ -10 & -2 \\ -3 & 1 \end{pmatrix} \begin{pmatrix} 1/4 \\ -1/4 \end{pmatrix} = \begin{pmatrix} 0 \\ -2 \\ -1 \end{pmatrix} \text{ and } \begin{pmatrix} -1 & -1 \\ -10 & -2 \\ -3 & 1 \end{pmatrix} \begin{pmatrix} -1/4 \\ -3/4 \end{pmatrix} = \begin{pmatrix} 1 \\ 4 \\ 0 \end{pmatrix}.
$$

This means that the induced map must be trivial. Therefore, the induced sequence map is given by

$$\begin{CD}
0 @>>> 0 @>>> \mathbb{R}^2 @>>> 0 @>>> 0 \\
@. @VVV @VV{0}V @VVV @. \\
0 @>>> \mathbb{R}^2 @>{0}>> \mathbb{R} @>>> 0 @>>> 0
\end{CD}$$

Proposition 2.3 *If $(V^\bullet, v^\bullet)$ is a sequence, then $Q_n\left((V^\bullet, v^\bullet)\right)$ has the same barcode decomposition as $(V^\bullet, v^\bullet)$, with the change that every bar longer than n is truncated to have length n, leaving its start unchanged.*

This proposition is the immediate consequence of two lemmas, which together explain the action of Q_n on single bars and on direct sums.

Lemma 2.6 *Recalling that $\mathbb{R}^{(a,b)}$ is a bar starting at a and ending at b,*

$$Q_n\left(\mathbb{R}^{(a,b)}\right) = \begin{cases} \mathbb{R}^{(a,b)} & \text{if } b \leq a+n-1 \\ \mathbb{R}^{(a,a+n-1)} & \text{otherwise.} \end{cases}$$

Proof If $b = a + n - 1$, then we have a diagram of the form

$$\begin{CD}
\cdots @>>> 0 @>>> \mathbb{R} @>>> \cdots (b-a+1) \text{ copies} \cdots @>>> \mathbb{R} @>>> 0 @>>> \cdots \\
@. @AAA @AAA @. @AAA @AAA @. \\
\cdots @>>> 0 @>>> 0 @>>> \cdots @>>> 0 @>>> \mathbb{R} @>>> \cdots
\end{CD}$$

which leaves the top sequence unchanged upon descending to the quotient. For $b < a + n - 1$, the bottom sequence is shifted further to the right so the quotient remains the same.

On the other hand if $b > a + n - 1$, the diagram is of the form

$$\begin{CD}
\cdots @>>> 0 @>>> \mathbb{R} @>>> \cdots @>>> \mathbb{R} @>>> 0 @>>> \cdots \\
@. @AAA @AAA @. @AAA @AAA @. \\
\cdots @>>> 0 @>>> 0 @>>> \cdots @>>> \mathbb{R} @>>> \mathbb{R} @>>> \cdots
\end{CD}$$

whose quotient sequence is the bar of the form

$$\cdots \longrightarrow 0 \longrightarrow \mathbb{R} \longrightarrow \cdots \longrightarrow \mathbb{R} \longrightarrow 0 \longrightarrow \cdots$$

of length n. □

Lemma 2.7 *Q_n preserves direct sums. That is, if $(V^\bullet, v^\bullet)$ and $(W^\bullet, w^\bullet)$ are two sequences, then*

$$Q_n\left((V^\bullet, v^\bullet) \oplus (W^\bullet, w^\bullet)\right) \cong Q_n\left((V^\bullet, v^\bullet)\right) \oplus Q_n\left((W^\bullet, w^\bullet)\right).$$

Proof Consider the following portion of the commutative diagram

$$
\begin{array}{ccccccc}
\cdots \longrightarrow & V^1 \oplus W^1 & \xrightarrow{v^1 \oplus w^1} & V^2 \oplus W^2 & \xrightarrow{v^2 \oplus w^2} & \cdots \\
& \uparrow{\scriptstyle v_n^{1-n} \oplus w_n^{1-n}} & & \uparrow{\scriptstyle v_n^{2-n} \oplus w_n^{2-n}} & & \\
\cdots \longrightarrow & V^{1-n} \oplus W^{1-n} & \xrightarrow[v^{1-n} \oplus w^{1-n}]{} & V^{2-n} \oplus W^{2-n} & \xrightarrow[v^{2-n} \oplus w^{2-n}]{} & \cdots
\end{array}
$$

The result follows from an investigation into the structure of each vertical map, which is of the form

$$v_n^{k-n} \oplus w_n^{k-n} : V^{k-n} \oplus W^{k-n} \to V^k \oplus W^k.$$

According to Theorem 1.2 in Chap. 1, we conclude

$$V^k \oplus W^k \cong \left((V^k \oplus W^k)/\operatorname{image}\left(v_n^{k-n} \oplus w_n^{k-n}\right)\right) \oplus \operatorname{image}\left(v_n^{k-n} \oplus w_n^{k-n}\right).$$

It follows from the definition of the direct sum that

$$\operatorname{image}\left(v_n^{k-n} \oplus w_n^{k-n}\right) \cong \left(\operatorname{image} v_n^{k-n}\right) \oplus \left(\operatorname{image} w_n^{k-n}\right).$$

This fact implies that

$$V^k \oplus W^k/\operatorname{image}\left(v_n^{k-n} \oplus w_n^{k-n}\right) \cong (V^k/\operatorname{image} v_n^{k-n}) \oplus (W^k/\operatorname{image} w_n^{k-n}).$$

This is just what we need, since it implies that Q_n preserves the direct sum structure of each term. Its action on the induced maps also preserves this structure. □

Exercise 2.22 Apply Q_2 to the following sequence

$$0 \to \mathbb{R} \xrightarrow{\begin{pmatrix} -1 \\ 1 \end{pmatrix}} \mathbb{R}^2 \xrightarrow{\begin{pmatrix} 0 & 0 \\ 0 & 0 \end{pmatrix}} \mathbb{R}^2 \xrightarrow{\begin{pmatrix} 4 & -6 \\ 4 & -5 \\ 3 & -4 \end{pmatrix}} \mathbb{R}^3 \xrightarrow{\begin{pmatrix} 2 & 3 & -6 \\ -3 & -5 & 10 \end{pmatrix}} \mathbb{R}^2 \xrightarrow{\begin{pmatrix} -4 & -2 \\ 5 & 3 \end{pmatrix}} \mathbb{R}^2 \to 0.$$

Hint: Compute the barcode decomposition first, and then apply Proposition 2.3 to the barcode.

Exercise 2.23 Apply Q_3 to the following sequence

$$0 \to \mathbb{R}^2 \xrightarrow{\begin{pmatrix} 0 & 0 \\ 2 & 0 \\ -5 & 1 \end{pmatrix}} \mathbb{R}^3 \xrightarrow{\begin{pmatrix} 21 & 10 & 6 \\ 23 & 10 & 8 \\ 32 & 16 & 8 \end{pmatrix}} \mathbb{R}^3 \xrightarrow{\begin{pmatrix} 0 & -4 & 4 \\ 0 & -3 & 3 \end{pmatrix}} \mathbb{R}^2 \xrightarrow{\begin{pmatrix} 1 & 0 \\ 6 & -6 \end{pmatrix}} \mathbb{R}^2 \to 0.$$

Exercise 2.24 Apply Q_2 to the following sequence

$$0 \longrightarrow \mathbb{R}^2 \xrightarrow{\begin{pmatrix} 0 & -1 \\ 4 & -1 \\ 2 & -2 \\ 0 & 1 \end{pmatrix}} \mathbb{R}^4 \xrightarrow{\begin{pmatrix} -20 & -4 & 8 & -8 \\ 20 & 4 & -8 & 8 \end{pmatrix}} \mathbb{R}^2 \longrightarrow 0.$$

Exercise 2.25 Apply Q_1 to the following sequence

$$0 \to \mathbb{R}^2 \xrightarrow{\begin{pmatrix} -3 & -3 \\ 3 & 3 \\ -2 & -2 \\ -2 & -2 \end{pmatrix}} \mathbb{R}^4 \xrightarrow{\begin{pmatrix} -19 & 6 & 15 & 21 \\ -10 & 6 & 12 & 12 \end{pmatrix}} \mathbb{R}^4 \xrightarrow{\begin{pmatrix} -6 & 9 \end{pmatrix}} \mathbb{R}^2 \to 0.$$

Proposition 2.4 *The image of* $Q_1 : \mathbf{Seq}^b \to \mathbf{Seq}^b$ *is a category that is equivalent to* **Vec**. *(Recall that* equivalence *for categories is given in Definition 1.18 in Chap. 1.)*

Proof Let us consider the functor F from $\mathbf{Seq}^b \to \mathbf{Vec}$ defined by

$$F\left((V^\bullet, v^\bullet)\right) := \bigoplus_{k \in \mathbb{Z}} V^k.$$

(See also Example 2.5.) Any vector space V can be realized as the image of the sequence

$$\cdots \longrightarrow 0 \longrightarrow V \longrightarrow 0 \longrightarrow \cdots$$

through F.

For any sequence map $m^\bullet : (V^\bullet, v^\bullet) \to (W^\bullet, w^\bullet)$, define

$$F(m^\bullet) := \bigoplus_{k \in \mathbb{Z}} m^k.$$

Notice that this is the functor $F_{\mathbb{Z}}$ defined in Example 2.4. Although F is clearly injective on the class of morphisms between any pair of bounded sequences (objects in $\mathbf{Seq}^b$), the linear maps $v^\bullet$ and $w^\bullet$ for those sequences are completely ignored by F.

Therefore, F will not establish an equivalence between $\mathbf{Seq}^b$ and $\mathbf{Vec}$. In particular, if we consider any collection of linear maps $p^k : V^k \to W^k$, then $F(p^\bullet)$ defines a linear map *even though* $p^\bullet$ may violate the commutativity conditions imposed by the linear maps $v^\bullet$ and $w^\bullet$. In the specific case of the image of Q_1, though, all of the $v^\bullet$ and $w^\bullet$ maps are zero maps. Commutativity is automatically assured for any such pair of sequences. □

Exercise 2.26 Prove that Q_n is *idempotent* for each n, which is to say that $Q_n \circ Q_n = Q_n$.

2.4 Chain Complexes and Exact Sequences

We should interpret Proposition 2.4 as stating that Q_1 strips away the sequence-specific structure from **Seq**, leaving only the structure of the vector spaces involved. Q_2 also strips away some of the structure of **Seq** but retains more than just the vector spaces involved. The rest of this chapter is largely devoted to the study of the image of Q_2.

Definition 2.10 A *chain complex* is a sequence of vector spaces $(C^\bullet, c^\bullet)$ such that $c^{m+1} \circ c^m = 0$ for all m. The category of chain complexes is denoted **Kom**, and is a full subcategory of **Seq**. Usually, we emphasize this fact by calling the morphisms of **Kom** *chain maps*, rather than sequence maps. As in the case of sequences, we also have $\mathbf{Kom}^+$, $\mathbf{Kom}^-$, and $\mathbf{Kom}^b$, categories of chain complexes bounded from below, from above, and on both ends, respectively.

Explicitly, suppose that $(C^\bullet, c^\bullet)$ and $(D^\bullet, d^\bullet)$ are two chain complexes. A collection of maps $m^k : C^k \to D^k$ such that the diagram

$$\begin{CD} C^k @>{c^k}>> C^{k+1} \\ @V{m^k}VV @VV{m^{k+1}}V \\ D^k @>>{d^k}> D^{k+1} \end{CD}$$

commutes for each k is called a *chain map* $m^\bullet : (C^\bullet, c^\bullet) \to (D^\bullet, d^\bullet)$. When no confusion can arise, we will usually write $m^\bullet : C^\bullet \to D^\bullet$. Each m^k is called a *component map* of the chain map.

Proposition 2.5 *Chain complexes are precisely those sequences whose barcodes only have bars of length* 0, 1, *or* 2. *Stated a little more succinctly, a sequence* $(V^\bullet, v^\bullet)$ *is a chain complex if and only if it is in the image of* Q_2.

This means that we should think of the shifted quotient Q_2 as a functor **Seq** $\to$ **Kom**.

Proof First, assume that we have an arbitrary sequence $(V^\bullet, v^\bullet)$. Its image through Q_2 is sequence isomorphic to the sequence $(V^\bullet / \operatorname{image} v_2^{\bullet-2}, m^\bullet)$, where $m^\bullet$ is the map induced by $v^\bullet$ on the quotients. The relevant diagram is

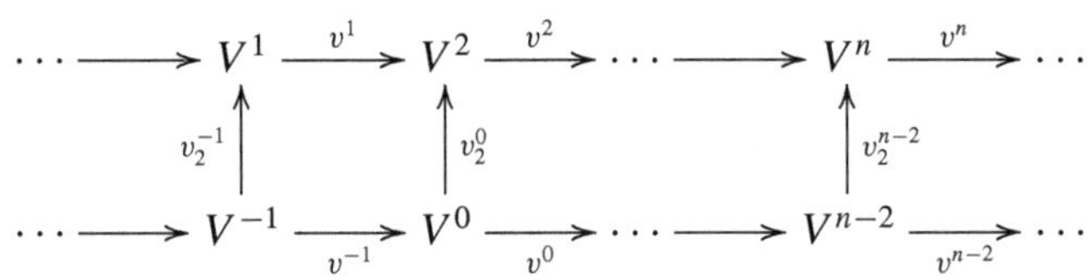

from which it follows that $m^{n+1} : V^{n+1} / \operatorname{image} v_2^{n-1} \to V^{n+2} / \operatorname{image} v_2^n$ is the map induced by v^n. Notice that

$$\begin{aligned}(m^{n+1} \circ m^n)(x) &\in \operatorname{image}(m^{n+1} \circ m^n)\\ &\in \operatorname{image} v_2^n,\end{aligned}$$

which means that the entire image of $m^{n+1} \circ m^n$ is contained in the zero coset of $V^{n+2} / \operatorname{image} v_2^n$. Thus, every sequence isomorphic to $(V^\bullet / \operatorname{image} v_2^{\bullet-2}, m^\bullet)$ is a chain complex.

Conversely, assume that $(V^\bullet, v^\bullet)$ is a chain complex, which is also a sequence in its own right. Observe that Q_2 leaves this sequence unchanged, namely that

$$Q_2\left((V^\bullet, v^\bullet)\right) = (V^\bullet, v^\bullet),$$

since $\operatorname{image} v_2^{\bullet-2} = 0$ by construction. □

Example 2.12 The following sequence is a chain complex, shown with its associated barcode

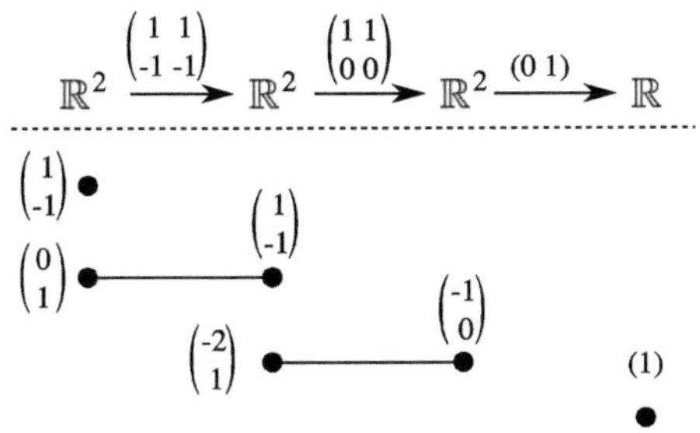

That this sequence is a chain complex can be easily verified by matrix multiplication,

$$\begin{pmatrix} 1 & 1 \\ 0 & 0 \end{pmatrix} \begin{pmatrix} 1 & 1 \\ -1 & -1 \end{pmatrix} = \begin{pmatrix} 0 & 0 \\ 0 & 0 \end{pmatrix}$$

and

$$(0\ 1) \begin{pmatrix} 1 & 1 \\ 0 & 0 \end{pmatrix} = (0\ 0)\,.$$

The barcode decomposition also reflects that it is a chain complex: the first and last bars listed have length 1, and the other two bars are of length 2.

Example 2.13 Consider the following sequence,

$$0 \longrightarrow \mathbb{R}^2 \xrightarrow{\begin{pmatrix} 0 & -2 \\ 0 & 2 \\ -1 & -7 \end{pmatrix}} \mathbb{R}^3 \xrightarrow{\begin{pmatrix} -4 & -3 & 0 \\ 6 & 4 & 0 \end{pmatrix}} \mathbb{R}^2 \xrightarrow{\begin{pmatrix} 0 & -2 \\ 2 & 3 \end{pmatrix}} \mathbb{R}^2 \longrightarrow 0$$

Because

$$\begin{pmatrix} -4 & -3 & 0 \\ 6 & 4 & 0 \end{pmatrix} \begin{pmatrix} 0 & -2 \\ 0 & 2 \\ -1 & -7 \end{pmatrix} = \begin{pmatrix} 0 & 2 \\ 0 & -4 \end{pmatrix}$$

is nonzero, this is not a chain complex. Additionally, note that

$$\begin{pmatrix} 0 & -2 \\ 2 & 3 \end{pmatrix} \begin{pmatrix} -4 & -3 & 0 \\ 6 & 4 & 0 \end{pmatrix} = \begin{pmatrix} -12 & -8 & 0 \\ 10 & 6 & 0 \end{pmatrix}$$

is also nonzero.

Proposition 2.5 asserts that if we apply Q_2 to the above sequence, this will result in a chain complex. To that end, construct the commutative diagram

$$
\begin{array}{ccccccccc}
\mathbb{R}^2 & \xrightarrow{\begin{pmatrix}0 & -2\\ 0 & 2\\ -1 & -7\end{pmatrix}} & \mathbb{R}^3 & \xrightarrow{\begin{pmatrix}-4 & -3 & 0\\ 6 & 4 & 0\end{pmatrix}} & \mathbb{R}^2 & \xrightarrow{\begin{pmatrix}0 & -2\\ 2 & 3\end{pmatrix}} & \mathbb{R}^2 & \longrightarrow & 0\\
\uparrow & & \uparrow & & \uparrow{\scriptstyle \begin{pmatrix}0 & 2\\ 0 & -4\end{pmatrix}} & & \uparrow{\scriptstyle \begin{pmatrix}-12 & -8 & 0\\ 10 & 6 & 0\end{pmatrix}} & & \uparrow\\
0 & \longrightarrow & 0 & \longrightarrow & \mathbb{R}^2 & \xrightarrow[\begin{pmatrix}0 & -2\\ 0 & 2\\ -1 & -7\end{pmatrix}]{} & \mathbb{R}^3 & \longrightarrow & \mathbb{R}^2
\end{array}
$$

The construction of Q_2 results in taking the quotients of the top row by the images of the bottom row. Since the product of the first two matrices on the top is precisely the matrix from the bottom at the third index, the induced maps on the quotients will multiply to zero. To find the induced maps, note that the orthogonal complement of the $(2\ \ -4)^T$ is $(4\ 2)^T$. This means that the only nontrivial induced map is

$$(4\ 2)\begin{pmatrix}-4 & -3 & 0\\ 6 & 4 & 0\end{pmatrix} = (-4\ -4\ 0)$$

Specifically, the quotient sequence is

$$\mathbb{R}^2 \xrightarrow{\begin{pmatrix}0 & -2\\ 0 & 2\\ -1 & -7\end{pmatrix}} \mathbb{R}^3 \xrightarrow{(-4\ -4\ 0)} \mathbb{R} \longrightarrow 0 \longrightarrow 0$$

This is easily verified to be a chain complex because

$$(-4\ -4\ 0)\begin{pmatrix}0 & -2\\ 0 & 2\\ -1 & -7\end{pmatrix} = (0\ 0)\,.$$

Exercise 2.27 Show that the following sequence is a chain complex

$$0 \longrightarrow \mathbb{R}^3 \xrightarrow{\begin{pmatrix}2 & -2 & 2\\ -4 & 8 & -10\end{pmatrix}} \mathbb{R}^2 \xrightarrow{(0\ 0)} \mathbb{R} \xrightarrow{(0)} \mathbb{R} \xrightarrow{(2)} \mathbb{R} \longrightarrow 0.$$

Exercise 2.28 Show that the following is not a chain complex

$$0 \to \mathbb{R}^2 \xrightarrow{\begin{pmatrix} 7 & 3 \\ -8 & -3 \\ 10 & 4 \end{pmatrix}} \mathbb{R}^3 \xrightarrow{\begin{pmatrix} 2 & 4 & 0 \\ -4 & -2 & 3 \end{pmatrix}} \mathbb{R}^2 \xrightarrow{(2\ 2)} \mathbb{R} \to 0$$

Exercise 2.29 Show that the following sequence is a chain complex

$$\mathbb{R} \xrightarrow{\begin{pmatrix} -1 \\ -1 \\ -2 \end{pmatrix}} \mathbb{R}^3 \xrightarrow{\begin{pmatrix} 5 & -11 & 3 \\ 15 & -29 & 7 \\ -13 & 27 & -7 \end{pmatrix}} \mathbb{R}^3 \xrightarrow{\begin{pmatrix} 0 & 0 & 0 \\ -14 & -4 & -10 \end{pmatrix}} \mathbb{R}^2 \xrightarrow{(-2\ 0)} \mathbb{R} \to 0.$$

Exercise 2.30 Show that the following sequence is a chain complex

$$0 \to \mathbb{R}^2 \xrightarrow{\begin{pmatrix} -1 & -1 \\ 2 & 2 \end{pmatrix}} \mathbb{R}^2 \xrightarrow{\begin{pmatrix} 4 & 2 \\ -8 & -4 \\ 0 & 0 \end{pmatrix}} \mathbb{R}^3 \xrightarrow{(-4\ -2\ -6)} \mathbb{R} \to 0.$$

Exercise 2.31 Suppose that $(C^\bullet, c^\bullet)$ is a chain complex in which only finitely many terms C^k are nontrivial vector spaces.

1. Show that it is possible to express the chain complex as a single matrix D for which $D^2 = 0$.
2. Demonstrate that this representation transforms chain maps into matrix multiplications.
3. Finally, show that this representation yields a covariant functor **Kom** $\to$ **Vec**.

It is sometimes useful to distinguish between a chain complex in which the indices increase $(C^\bullet, c^\bullet)$ and one in which the indices decrease $(D_\bullet, d_\bullet)$. Traditionally, chain complexes like $(D_\bullet, d_\bullet)$ with decreasing indices arise by way of a specific construction of interest to topologists, called the *simplicial chain complex*, which we'll discuss in Chap. 4. Vector space duality transforms the *simplicial chain complex* into a new sequence with increasing indices called the *simplicial cochain complex*, as we'll see in Sect. 5.6 of Chap. 5. As a result, topologists usually call a chain complex with increasing indices a *cochain complex* as a way to emphasize this situation, even if it does not arise by duality. Because the maps $d_\bullet$ in a *simplicial chain complex* describe how boundaries fit together, the $d_\bullet$ maps in a chain complex with decreasing indices are often called *boundary maps*. Dually, the $c^\bullet$ maps in a chain complex with increasing indices are sometimes called *coboundary maps*. Since we defer discussion of the *simplicial chain complex* until Chap. 4, we will not distinguish between chain and cochain complexes in this chapter.

Definition 2.11 A chain complex $(C^\bullet, c^\bullet)$ is said to be an *exact chain complex* or an *exact sequence* if $\text{image}\, c^k = \ker c^{k+1}$ for every k.

It is useful to characterize exactness by way of barcode decompositions.

Proposition 2.6 *A sequence is exact if and only if every nontrivial bar in its barcode decomposition has length* 2.

Proof Suppose that $\text{image}\, c^k = \ker c^{k+1}$ for every k. If $\mathbb{R}^{(k,m)}$ is a bar starting at index k, the image of this bar through c^k is in the kernel of the next map, c^{k+1}. That implies that $m \leq k+1$. The length of the bar must be 2 or less. On the other hand, if $m = k$, this means that the bar is of length 1 and so it must be in the kernel of c^k. Because $\text{image}\, c^{k-1} = \ker c^k$, this means that the bar cannot have started at index k, a contradiction.

Suppose that every bar in the barcode decomposition of $(C^\bullet, c^\bullet)$ has length 2. Then for every index k, C^k is a direct sum of bars that either start or end there. Without loss of generality, we need only consider sequences $(C'^\bullet, c'^\bullet)$ of the form

$$C' := \left(\bigoplus_{i=1}^{m} \mathbb{R}^{(k-1,k)}\right) \oplus \left(\bigoplus_{j=1}^{n} \mathbb{R}^{(k,k+1)}\right)$$

which consists of m bars that end at index k and n bars that start at index k. According to the definition of a bar, this implies that c'^{k-1} is an $(m+n) \times m$ matrix of the form

$$c'^{k-1} = \begin{pmatrix} 1 & 0 & \cdots & 0 \\ 0 & 1 & \cdots & 0 \\ & & \vdots & \\ 0 & 0 & \cdots & 1 \\ 0 & 0 & \cdots & 0 \\ 0 & 0 & \cdots & 0 \\ & & \vdots & \\ 0 & 0 & \cdots & 0 \end{pmatrix},$$

while c'^k is an $n \times (m+n)$ matrix of the form

$$c'^k = \begin{pmatrix} 0\,0\cdots 0\,1\,0\cdots 0 \\ 0\,0\cdots 0\,0\,1\cdots 0 \\ \vdots \qquad\quad \vdots \\ 0\,0\cdots 0\,0\,0\cdots 1 \end{pmatrix}.$$

From this, we can see that the kernel of c^k is m-dimensional and is precisely the image of c^{k-1}. □

Proposition 2.7 *The Euler characteristic for an exact sequence is* 0.

Proof The Euler characteristic of a length 2 bar is $\chi\left(\mathbb{R}^{(k,k+1)}\right) = 0$ by Lemma 2.5 since its length is even. Since all bars in an exact sequence have length 2, the sequence's Euler characteristic is a sum of 0 terms. □

The relationship between chain complexes and exact sequences can be summarized by saying that for a chain complex,

$$\text{image}\, c^k \subseteq \ker c^{k+1} \text{ for all } k,$$

but exactness requires equality of these subspaces. In an exact sequence, the c^k maps have the largest images possible for a chain complex.

Example 2.14 The following is an exact sequence

$$0 \longrightarrow \mathbb{R}^2 \xrightarrow{\begin{pmatrix} 1&0 \\ 0&0 \\ 0&1 \\ 0&0 \end{pmatrix}} \mathbb{R}^4 \xrightarrow{\begin{pmatrix} 0&1&0&0 \\ 0&0&0&1 \\ 0&1&0&1 \end{pmatrix}} \mathbb{R}^3 \xrightarrow{\begin{pmatrix} 1&1&-1 \end{pmatrix}} \mathbb{R} \longrightarrow 0.$$

It is a chain complex because

$$\begin{pmatrix} 0&1&0&0 \\ 0&0&0&1 \\ 0&1&0&1 \end{pmatrix} \begin{pmatrix} 1&0 \\ 0&0 \\ 0&1 \\ 0&0 \end{pmatrix} = \begin{pmatrix} 0&0 \\ 0&0 \\ 0&0 \end{pmatrix}$$

and

$$\begin{pmatrix} 1&1&-1 \end{pmatrix} \begin{pmatrix} 0&1&0&0 \\ 0&0&0&1 \\ 0&1&0&1 \end{pmatrix} = \begin{pmatrix} 0&0&0&0 \end{pmatrix}.$$

To verify exactness, we must compute kernels and images for each matrix.

1. The first map in the sequence is actually the zero map, so it has trivial image. The second map is

$$\begin{pmatrix} 1 & 0 \\ 0 & 0 \\ 0 & 1 \\ 0 & 0 \end{pmatrix},$$

which is an injective linear map, so it has trivial kernel.

2. The kernel of the third map consists of all vectors of the form

$$\begin{pmatrix} x \\ 0 \\ y \\ 0 \end{pmatrix}$$

all of which are in the image of the second map since

$$\begin{pmatrix} 1 & 0 \\ 0 & 0 \\ 0 & 1 \\ 0 & 0 \end{pmatrix} \begin{pmatrix} x \\ y \end{pmatrix} = \begin{pmatrix} x \\ 0 \\ y \\ 0 \end{pmatrix}.$$

3. The kernel of the fourth map consists of all vectors of the form

$$\begin{pmatrix} x \\ y \\ x + y \end{pmatrix},$$

which is in the image of the third map because

$$\begin{pmatrix} 0 & 1 & 0 & 0 \\ 0 & 0 & 0 & 1 \\ 0 & 1 & 0 & 1 \end{pmatrix} \begin{pmatrix} 0 \\ x \\ 0 \\ y \end{pmatrix} = \begin{pmatrix} x \\ y \\ x + y \end{pmatrix}.$$

4. Finally, the fifth map (a zero map) has a 1-dimensional kernel, which equals the image of the fourth map, since that map is surjective.

Exact sequences can encode injectivity and surjectivity of linear maps in a diagrammatic way. For instance, the assertion that the sequence

$$V \xrightarrow{\ell} W \longrightarrow 0$$

is exact means that image $\ell = W$, namely that ℓ is surjective. Dually, the assertion that the sequence

$$0 \longrightarrow V \xrightarrow{\ell} W$$

is exact means that ℓ is injective because $0 = \ker \ell$. Taken together, these two diagrams assert that ℓ is an isomorphism if and only if

$$0 \longrightarrow V \xrightarrow{\ell} W \longrightarrow 0$$

is exact.

Definition 2.12 A *short exact sequence* is an exact sequence which has at most three nontrivial terms, organized like the following diagram,

$$\cdots \longrightarrow 0 \longrightarrow A \xrightarrow{a} B \xrightarrow{b} C \xrightarrow{c} 0 \longrightarrow \cdots$$

Proposition 2.8 *Every short exact sequence of vector spaces*

$$0 \longrightarrow A \xrightarrow{a} B \xrightarrow{b} C \xrightarrow{c} 0$$

splits, *which means that (equivalently)*

$$A \cong B/C,$$

$$B \cong A \oplus C,$$

and

$$C \cong B/A.$$

Proof It suffices to establish the characterization of $B \cong A \oplus C$, as the other statements follow from Theorem 1.1 of Chap. 1. By our earlier discussion, exactness implies that a is injective and that c is surjective. Exactness also implies that

$$\ker b = \operatorname{image} a \cong A$$

and that

$$\operatorname{image} b = \ker c \cong C.$$

Thus Theorem 1.2 of Chap. 1 implies that

$$B \cong \ker b \oplus \operatorname{coimage} b \cong A \oplus \operatorname{image} b \cong A \oplus C. \qquad \square$$

Exact sequences generalize Theorem 1.2 of Chap. 1 (the dimension theorem) to sequences. If we assume that a sequence

$$0 \longrightarrow A \longrightarrow V \xrightarrow{\ell} W \longrightarrow B \longrightarrow 0.$$

is exact, this implies that

$$A \cong \ker \ell \text{ and } B \cong \operatorname{coker} \ell.$$

Exactness at the V term implies

$$V \cong A \oplus \operatorname{coimage} \ell \cong \ker \ell \oplus \operatorname{coimage} \ell,$$

which is precisely the conclusion of Theorem 1.2 of Chap. 1. Moreover, exactness at the next term implies

$$W \cong B \oplus \operatorname{image} \ell \cong \operatorname{coker} \ell \oplus \operatorname{image} \ell,$$

again in full agreement with Theorem 1.2 of Chap. 1.

Example 2.15 Consider the sequence

$$0 \longrightarrow \mathbb{R}^2 \xrightarrow{\begin{pmatrix} 0 & 1 \\ 4 & 1 \\ 2 & 2 \\ 4 & 2 \end{pmatrix}} \mathbb{R}^4 \xrightarrow{\begin{pmatrix} 3 & -3 & -6 & 6 \\ -9 & 3 & 12 & -9 \end{pmatrix}} \mathbb{R}^2 \longrightarrow 0$$

This is a chain complex because

$$\begin{pmatrix} 3 & -3 & -6 & 6 \\ -9 & 3 & 12 & -9 \end{pmatrix} \begin{pmatrix} 0 & 1 \\ 4 & 1 \\ 2 & 2 \\ 4 & 2 \end{pmatrix} = \begin{pmatrix} 0 & 0 \\ 0 & 0 \end{pmatrix}.$$

Moreover, both matrices are of full rank, which can be seen from the fact that

$$\begin{pmatrix} 0 & 1 \\ 4 & 1 \\ 2 & 2 \\ 4 & 2 \end{pmatrix}$$

has linearly independent columns and

$$\begin{pmatrix} 3 & -3 & -6 & 6 \\ -9 & 3 & 12 & -9 \end{pmatrix}$$

has linearly independent rows. This means that there are two bars that start on the left and end in the middle, but two additional bars start at the middle and end on the right. In other words, the barcode decomposition is

$$\mathbb{R} \longrightarrow \mathbb{R}$$

$$\mathbb{R} \longrightarrow \mathbb{R}$$

$$\mathbb{R} \longrightarrow \mathbb{R}$$

$$\mathbb{R} \longrightarrow \mathbb{R}$$

or more briefly

$$\left(\mathbb{R}^{(1,2)} \oplus \mathbb{R}^{(1,2)}\right) \oplus \left(\mathbb{R}^{(2,3)} \oplus \mathbb{R}^{(2,3)}\right),$$

which is in accordance with Proposition 2.8.

Proposition 2.8 is very specific to finite dimensional vector spaces, as it is false for sequences composed in other categories. These results usually will not generalize fully to the case of sequences of abelian groups. The ambitious reader should be on the lookout for every instance in which we use a split exact sequence of vector spaces without further comment. The barcode decomposition of a sequence of vector spaces is one such instance.

For instance, the following is an exact sequence of abelian groups

$$0 \longrightarrow \mathbb{Z} \xrightarrow{(2)} \mathbb{Z} \xrightarrow{p} \mathbb{Z}_2 \longrightarrow 0$$

in which (2) indicates multiplication by 2, $\mathbb{Z}_2$ is the group with two elements 0 and 1, and p returns 0 for each even number and 1 for each odd number. This sequence does not split, which means that the middle $\mathbb{Z}$ cannot be written as a direct sum. Since the focus of this book is on sequences of finite dimensional vector spaces, we will generally assume that issues of this sort do not arise, at least not until they become unavoidable in Chap. 6.

Remark 2.3 Alternative terminology is also used in the literature when considering bars of chain complexes. A length 1 bar $\mathbb{R}^{(k,k)}$ is sometimes called an "interval sphere in dimension k" and a length 2 bar $\mathbb{R}^{(k-1,k)}$ is called an "interval disk in dimension k."

Exercise 2.32 Prove the following statement or give a counterexample for it: The Euler characteristic of a chain complex can take any integer value. (Hint: consider Proposition 2.7.)

2.5 Homology of a Chain Complex

As an immediate consequence of Theorem 2.1, we have the following characterization of a chain complex. Every chain complex $(C^\bullet, c^\bullet)$ decomposes as the direct sum

$$(C^\bullet, c^\bullet) \cong (E^\bullet, e^\bullet) \oplus (H^\bullet, h^\bullet)$$

of an exact sequence $(E^\bullet, e^\bullet)$ (whose barcode consists of bars of length 2 only) and a sequence $(H^\bullet, h^\bullet)$ whose barcode consists of length 1 bars only.

Definition 2.13 The sequence $(H^\bullet, h^\bullet)$ is called the *homology*[2] of the chain complex $(C^\bullet, c^\bullet)$. Since all of the bars in $(H^\bullet, h^\bullet)$ are length 1, it follows that all of the maps h^k are zero maps, and therefore need not be listed. For the homology at index k, we write $H^k(C^\bullet, c^\bullet)$. We will usually write $H^\bullet(C^\bullet, c^\bullet)$ to refer to the entire homology sequence of a chain complex $(C^\bullet, c^\bullet)$. If the context is clear, $H^\bullet(C^\bullet)$ can be used instead even though the $c^\bullet$ maps are definitely important!

The dimensions

$$\beta^k = \dim H^k(C^\bullet, c^\bullet)$$

are called the *Betti numbers* of the chain complex. Intuitively, homology measures how far a chain complex is from being exact; it is exact when all its Betti numbers are zero.

Exercise 2.33 Use Theorem 2.1 to show that any direct sum of length 1 bars cannot be an exact sequence.

Example 2.16 Consider the chain complex $(C^\bullet, c^\bullet)$

$$0 \longrightarrow \mathbb{R} \longrightarrow 0 \longrightarrow \mathbb{R}^2 \xrightarrow{\begin{pmatrix} 0 & 0 \\ 0 & 0 \\ 3 & 1 \end{pmatrix}} \mathbb{R}^3 \longrightarrow 0.$$

Observe that the first nontrivial space corresponds to a bar of length 1 since it is mapped to zero immediately. The map $\mathbb{R}^2 \to \mathbb{R}^3$ has rank 1, yielding one length 2 bar, and three length 1 bars. In other words, the barcode is of the form

$$(C^\bullet, c^\bullet) \cong \mathbb{R}^{(1,1)} \oplus \mathbb{R}^{(3,3)} \oplus \mathbb{R}^{(3,4)} \oplus \mathbb{R}^{(4,4)} \oplus \mathbb{R}^{(4,4)},$$

[2] Readers already familiar with *simplicial chain complexes* and *simplicial cochain complexes* (defined in Chap. 4) take note: $H^k(C^\bullet, c^\bullet)$ is a covariant functor (Proposition 3.2 of Chap. 3) and is therefore *not cohomology*.

assuming the sequence starts with index 0. The homology of the chain complex is simply the direct sum of the length 1 bars, namely

$$H^\bullet(C^\bullet, c^\bullet) \cong \mathbb{R}^{(1,1)} \oplus \mathbb{R}^{(3,3)} \oplus \mathbb{R}^{(4,4)} \oplus \mathbb{R}^{(4,4)}.$$

We can also state this by saying that the homology $H^\bullet(C^\bullet, c^\bullet)$ is

$$H^k(C^\bullet, c^\bullet) = \begin{cases} \mathbb{R} & \text{if } k = 1, \\ \mathbb{R} & \text{if } k = 3, \\ \mathbb{R}^2 & \text{if } k = 4, \\ 0 & \text{otherwise.} \end{cases}$$

In other words, $\beta^1 = 1$, $\beta^3 = 1$, and $\beta^4 = 2$ are the only nonzero Betti numbers for this chain complex.

Exercise 2.34 Compute the homology of the following chain complex, assuming that the first space has index 0

$$0 \longrightarrow \mathbb{R}^2 \xrightarrow{\begin{pmatrix} -6 & 4 \end{pmatrix}} \mathbb{R} \xrightarrow{\begin{pmatrix} 0 \\ 0 \end{pmatrix}} \mathbb{R}^2 \xrightarrow{\begin{pmatrix} 0 & 0 \\ -3 & 1 \end{pmatrix}} \mathbb{R}^2 \longrightarrow 0.$$

Example 2.17 A preponderance of zero maps makes calculating homology easy, since length 1 bars are easy to spot. For instance, in the chain complex $(C^\bullet, c^\bullet)$

$$0 \longrightarrow \mathbb{R} \xrightarrow{\begin{pmatrix} 0 \\ 0 \\ 0 \end{pmatrix}} \mathbb{R}^3 \xrightarrow{\begin{pmatrix} 0 & 0 & 0 \\ 0 & 0 & 0 \end{pmatrix}} \mathbb{R}^2 \xrightarrow{\begin{pmatrix} 0 & 0 \end{pmatrix}} \mathbb{R} \longrightarrow 0,$$

every bar is of length 1. Hence the homology is simply

$$H_k(C^\bullet, c^\bullet) = \begin{cases} \mathbb{R} & \text{if } k = 1, \\ \mathbb{R}^3 & \text{if } k = 2, \\ \mathbb{R}^2 & \text{if } k = 3, \\ \mathbb{R} & \text{if } k = 4, \\ 0 & \text{otherwise.} \end{cases}$$

Exercise 2.35 Compute the homology of the following chain complex by computing its barcode decomposition.

$$0 \longrightarrow \mathbb{R}^2 \xrightarrow{\begin{pmatrix} 1 & 1 \\ -1 & 0 \\ 0 & -1 \end{pmatrix}} \mathbb{R}^3 \longrightarrow 0$$

Proposition 2.9 *For a chain complex* $(C^\bullet, c^\bullet)$*, its homology is given by*

$$H^k(C^\bullet, c^\bullet) = \ker c^k / \operatorname{image} c^{k-1}$$

for each k.

Corollary 2.5 implies that chain complexes can be characterized both by their bars being length 1 or 2 and equivalently by $c^k \circ c^{k-1} = 0$ for all k. This latter statement is equivalent to image $c^{k-1} \subseteq \ker c^k$ for all k, which means that the quotient formula in the statement is well defined. On the other hand, if a chain complex is exact it has no bars of length 1. Therefore, an exact chain complex has only trivial homology. On the other hand, Proposition 2.6 implies that for an exact sequence, image $c^{k-1} = \ker c^k$ for all k, agreeing with our proposed definition.

Proof Suppose that $(C^\bullet, c^\bullet)$ is isomorphic to the barcode

$$\left(\bigoplus_i \mathbb{R}^{(a_i,a_i)}\right) \oplus \left(\bigoplus_j \mathbb{R}^{(b_j,b_j+1)}\right)$$

in which the first summand contains all of the length 1 bars (the homology), and the second summand contains all the length 2 bars (an exact sequence). At index k in the sequence,

$$\ker c^k = \left(\bigoplus_{i\,:\,a_i=k} \mathbb{R}\right) \oplus \left(\bigoplus_{j\,:\,b_i+1=k} \mathbb{R}\right)$$

while

$$\operatorname{image} c^{k-1} = \bigoplus_{j\,:\,b_i=k-1} \mathbb{R},$$

but since $b_i + 1 = k$ is equivalent to $b_i = k - 1$, this means that

$$\ker c^k / \operatorname{image} c^{k-1} = \bigoplus_{i\,:\,a_i=k} \mathbb{R},$$

which is precisely the homology of $(C^\bullet, c^\bullet)$. □

The formula given in Proposition 2.9 for homology usually allows for a more direct calculation than the barcode decomposition and is often taken as the definition of homology. The formula also gives homology an interpretation in terms of the block structure of matrices as well, as the next example shows.

Example 2.18 Consider a rank r linear map $m : V \to V$ from an n-dimensional vector space to itself. Choosing an appropriate basis, we can write the matrix representation of m in the following block form (and upper triangular matrix)

$$m = \left(\begin{array}{c|c} 0 & * \\ \hline 0 & 0 \end{array}\right) \begin{array}{l} r\text{-th row} \\ \\ \end{array}$$
$$k\text{-th column}$$

in which the $*$ represents a block of possibly nonzero entries of size $r \times (n - k)$. The dimension of the kernel of m is at least k, while the dimension of the image is not more than r.

The sequence formed by iterating the map m

$$\cdots \xrightarrow{m} V \xrightarrow{m} V \xrightarrow{m} V \xrightarrow{m} \cdots$$

is a chain complex if $k \geq r$, because then the product of m with itself is zero. Expanding the block representation for m a little

$$m = \left(\begin{array}{c|c} 0 & * \\ \hline 0 & 0 \\ \hline 0 & 0 \end{array}\right) \begin{array}{l} r\text{-th row} \\ k\text{-th row} \\ \end{array}$$
$$k\text{-th column}$$

reveals that the difference between the image and the kernel is at least $k - r$. Thus, the homology of $(V^\bullet, m)$ is a feature of the block structure of m.

Example 2.19 Consider the chain complex $(C^\bullet, c^\bullet)$

$$0 \longrightarrow \mathbb{R} \xrightarrow{\begin{pmatrix} 1 \\ 1 \\ 1 \\ 0 \end{pmatrix}} \mathbb{R}^4 \xrightarrow{\begin{pmatrix} -5 & 1 & 4 & -3 \\ 1 & 0 & -1 & 1 \\ 1 & 0 & -1 & 1 \\ 0 & 1 & -1 & 2 \end{pmatrix}} \mathbb{R}^4 \xrightarrow{\begin{pmatrix} 1 & 2 & 3 & -1 \\ 0 & 1 & -1 & 0 \end{pmatrix}} \mathbb{R}^2 \longrightarrow 0.$$

Let us use Proposition 2.9 to compute the homology of this chain complex.

- $k = 1$ The kernel of c^1 is trivial (as is the image of c^0). Thus $H^1(C^\bullet) = 0$.

- $k = 2$ Since c^1 is injective, it is of full rank. Let us consider row reductions of c^2, starting with

$$\begin{pmatrix} -5 & 1 & 4 & -3 \\ 1 & 0 & -1 & 1 \\ 1 & 0 & -1 & 1 \\ 0 & 1 & -1 & 2 \end{pmatrix}.$$

Subtract the second and third row to obtain

$$\begin{pmatrix} -5 & 1 & 4 & -3 \\ 1 & 0 & -1 & 1 \\ 0 & 0 & 0 & 0 \\ 0 & 1 & -1 & 2 \end{pmatrix}.$$

Combining the first and second rows yields

$$\begin{pmatrix} -5 & 1 & 4 & -3 \\ 0 & 1 & -1 & 2 \\ 0 & 0 & 0 & 0 \\ 0 & 1 & -1 & 2 \end{pmatrix}.$$

Finally, subtracting the second and fourth row yields

$$\begin{pmatrix} -5 & 1 & 4 & -3 \\ 0 & 1 & -1 & 2 \\ 0 & 0 & 0 & 0 \\ 0 & 0 & 0 & 0 \end{pmatrix}.$$

Thus, c^2 has rank 2. Thus the quotient $H^2(C^\bullet) = \ker c^2 / \operatorname{image} c^1 \cong \mathbb{R}$ is of dimension 1.
- $k = 3$ From the previous step, we know that the rank of c^2 is 2. By inspection, the rank of c^3 is 2 also, implying that the kernel of c^3 is dimension 2. Thus $H^3(C^\bullet)$ is trivial.
- $k = 4$ Finally, since c^4 is a zero map, it has a kernel consisting of $\mathbb{R}^2$. This is the entire image of c^3, so $H^4(C^\bullet)$ is trivial.

The barcode decomposition of this chain complex is (by an application of Algorithm 1)

$$\mathbb{R}^{(1,2)} \oplus \mathbb{R}^{(2,2)} \oplus \mathbb{R}^{(2,3)} \oplus \mathbb{R}^{(2,3)} \oplus \mathbb{R}^{(3,4)} \oplus \mathbb{R}^{(3,4)},$$

which means that the only nonzero Betti number is $\beta^2 = 1$, in complete agreement with our findings using Proposition 2.9.

Exercise 2.36 Compute the homology of the following chain complex in two separate ways: (1) using Proposition 2.9 and (2) by computing its barcode decomposition.

$$0 \longrightarrow \mathbb{R}^3 \xrightarrow{\begin{pmatrix} 1 & -1 & 0 \\ -1 & 0 & 1 \\ 0 & 1 & -1 \end{pmatrix}} \mathbb{R}^3 \longrightarrow 0$$

Exercise 2.37 Compute the homology of the following chain complex in two separate ways: (1) using Proposition 2.9 and (2) by computing its barcode decomposition.

$$0 \longrightarrow \mathbb{R} \xrightarrow{\begin{pmatrix} 1 \\ 1 \\ 1 \end{pmatrix}} \mathbb{R}^3 \xrightarrow{\begin{pmatrix} 1 & -1 & 0 \\ -1 & 0 & 1 \\ 0 & 1 & -1 \end{pmatrix}} \mathbb{R}^3 \longrightarrow 0$$

Proposition 2.10 *The Euler characteristic is a homology invariant. Any two chain complexes with the same homology have the same Euler characteristic. Furthermore, for a chain complex* $(C^\bullet, c^\bullet)$,

$$\chi(C^\bullet) = \sum_{k\in\mathbb{Z}} (-1)^k \beta^k,$$

where $\beta^k = \dim H^k(C^\bullet, c^\bullet)$ *is the* k*-th Betti number.*

Proof Since the Euler characteristic of an exact sequence is zero, and every chain complex decomposes as the direct sum of its homology and an exact sequence (Lemma 2.4), the Euler characteristic depends only on the homology. □

Example 2.20 The Euler characteristic of a chain complex is one way to enrich integers into vector spaces. Given a positive integer m, one can associate a chain complex

$$0 \longrightarrow 0 \longrightarrow \mathbb{R}^m \longrightarrow 0$$

which has Euler characteristic m. By the same kind of construction, the chain complex

$$0 \longrightarrow \mathbb{R}^n \longrightarrow 0 \longrightarrow 0$$

has Euler characteristic $(-n)$. Because of Lemma 2.4, the direct sum of these two chain complexes

$$0 \longrightarrow \mathbb{R}^n \xrightarrow{0} \mathbb{R}^m \longrightarrow 0$$

has Euler characteristic $m - n$. Using this idea, any statement about integers implies an analogous statement about chain complexes.

Exercise 2.38 Prove or give a counterexample: If the Euler characteristic of a chain complex is 0, then the chain complex is an exact sequence. (Hint: this is the converse of Proposition 2.7).

If we use the Algorithm 1 (the barcode decomposition algorithm) in order to compute the homology of a chain complex, we do not compute the isomorphism between the chain complex and the barcode. This can be inconvenient, as it is often useful to have bases for the quotients involved. One can easily employ the kernel-image formula (Proposition 2.9) at each index to remedy this situation, and use the basis for the quotient computed using Algorithms 1 or 2 in Chap. 1. Another method uses orthogonal projections, as we now describe.

Consider a chain complex $(C^\bullet, c^\bullet)$. At index k, the kernel of $c^k : C^k \to C^{k+1}$ is a subspace of C^k. This means that there is an injective map $\kappa : \ker c^k \to C^k$, or an exact sequence

$$0 \longrightarrow \ker c^k \xrightarrow{\kappa} C^k \longrightarrow C^{k+1}.$$

On the other hand, since $(C^\bullet, c^\bullet)$ is a chain complex, the image of $c^{k-1} : C^{k-1} \to C^k$ is a subspace of $\ker c^k$. If we restrict the codomain of c^{k-1} to $\ker c^k$, we obtain a map $m : c^{k-1} \to \ker c^k$. We can write this as an exact sequence

$$0 \longrightarrow \operatorname{image} c^{k-1} \xrightarrow{m} \ker c^k.$$

The concatenation of these sequences to a longer one, namely

$$0 \longrightarrow \operatorname{image} c^{k-1} \xrightarrow{m} \ker c^k \xrightarrow{\kappa} C^k \longrightarrow C^{k+1},$$

is exact everywhere except at $\ker c^k$. At $\ker c^k$, this concatenated sequence has the same homology as our original chain complex at index k. Specifically, the homology

$$H^k(C^\bullet, c^\bullet) = \ker c^k / \operatorname{image} c^{k-1} \cong \ker \kappa / \operatorname{image} m = \ker c^k / \operatorname{image} m \cong \operatorname{coker} m$$

is the cokernel (in $\ker c^k$) of m. This can be expressed by the commutative diagram

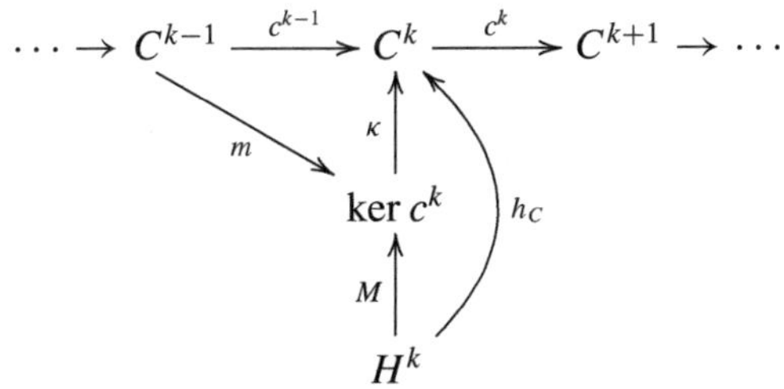

in which the columns of the matrix representation for h_C form a basis of $H^k(C^\bullet, c^\bullet)$. The matrix representations of the linear maps κ, m, M, and h_C are computed according to Algorithm 2.

Algorithm 2 Computing the homology of a chain complex

Require: Chain complex $(C^\bullet, c^\bullet)$ and an integer index k.

Ensure: A *homology basis matrix* h_C whose columns are elements of C^k that span the portion of the kernel of c^k that is outside the image of c^{k-1}

procedure HOMOLOGYPAIRWISE($(C^\bullet, c^\bullet)$)

  Compute a matrix κ whose columns are a basis for the kernel of c^k in C^k.

  Compute the induced map using the projection formula by solving the matrix linear system $\kappa \circ m = c^{k-1}$ for m,

$$\begin{aligned} \kappa \circ m &= c^{k-1} \\ \kappa^T \circ \kappa \circ m &= \kappa^T \circ c^{k-1} \\ m &= (\kappa^T \circ \kappa)^{-1} \circ \kappa^T \circ c^{k-1}. \end{aligned}$$

  Compute a matrix M whose columns are a basis for the cokernel of m.

  return $h_C := \kappa \circ M$.

end procedure

Example 2.21 Let us compute homology of the chain complex

$$0 \longrightarrow \mathbb{R}^2 \xrightarrow{\begin{pmatrix} -8 & 0 \\ -6 & 0 \end{pmatrix}} \mathbb{R}^2 \xrightarrow{\begin{pmatrix} -9 & 12 \\ -6 & 8 \end{pmatrix}} \mathbb{R}^2 \xrightarrow{\begin{pmatrix} 4 & -6 \\ -2 & 3 \end{pmatrix}} \mathbb{R}^2 \longrightarrow 0$$

using Algorithm 2.

Algorithm 2 works for individual indices separately. It will take four iterations of Algorithm 2 to compute the homology of this chain complex.

Index $k = 1$ Because the first arrow on the left corresponds to c^0 and is a 2×0 matrix, the homology at the first index is simply the kernel of c^1. The matrix $c^1 = \begin{pmatrix} -8 & 0 \\ -6 & 0 \end{pmatrix}$.

1. The map c^1 has a 1-dimensional kernel, spanned by the column vector

$$\kappa = \begin{pmatrix} 0 \\ 1 \end{pmatrix}.$$

2. Solving for m,

$$\begin{aligned} m &= (\kappa^T \circ \kappa)^{-1} \circ \kappa^T \circ c^0 \\ &= \left((1\ 0) \begin{pmatrix} 1 \\ 0 \end{pmatrix} \right)^{-1} (0\ 1) \begin{pmatrix} 0 \\ 0 \end{pmatrix} \\ &= (1)^{-1} (0) \\ &= 0. \end{aligned}$$

(As a crosscheck, notice that

$$\kappa \circ m = \begin{pmatrix} 0 \\ 1 \end{pmatrix} (0) = \begin{pmatrix} 0 \\ 0 \end{pmatrix} = c^0$$

as desired.)

3. The matrix M spanning the cokernel of m is

$$M = (1).$$

4. Therefore, the homology for index 1 is spanned by the column vector

$$h^1 = \kappa M = \begin{pmatrix} 0 \\ 1 \end{pmatrix}.$$

Index $k = 2$ The matrix $c^2 = \begin{pmatrix} -9 & 12 \\ -6 & 8 \end{pmatrix}$.

1. The c^2 has a 1-dimensional kernel, spanned by

$$\kappa = \begin{pmatrix} 4 \\ 3 \end{pmatrix}.$$

2. Solving for m,

$$m = (\kappa^T \circ \kappa)^{-1} \circ \kappa^T \circ c^1$$

$$
\begin{aligned}
&= \left((4\ 3) \begin{pmatrix} 4 \\ 3 \end{pmatrix} \right)^{-1} (4\ 3) \begin{pmatrix} -8 & 0 \\ -6 & 0 \end{pmatrix} \\
&= (25)^{-1} (-50\ 0) \\
&= \frac{1}{25} (-50\ 0) . \\
&= (-2\ 0) .
\end{aligned}
$$

(As a crosscheck, notice that

$$
\kappa \circ m = \begin{pmatrix} 4 \\ 3 \end{pmatrix} (-2\ 0) = \begin{pmatrix} -8 & 0 \\ -6 & 0 \end{pmatrix} = c^1
$$

as desired.)

3. The matrix M spanning the cokernel of m is trivial, since m is of full rank.
4. Thus the homology for index 2 is trivial.

Index $k = 3$. The matrix $c^3 = \begin{pmatrix} 4 & -6 \\ -2 & 3 \end{pmatrix}$.

1. The matrix c^3 has a 0-dimensional kernel, so κ is an empty matrix.
2. Therefore m is an empty matrix.
3. (This step is redundant!) The matrix M spanning the cokernel of m is

$$
M = \begin{pmatrix} 1 & 0 \\ 0 & 1 \end{pmatrix} .
$$

4. Thus the homology is trivial, since $h^3 = \kappa M$ and κ is trivial.

Index $k = 4$. The matrix c^4 is empty.

1. The c^4 has a 2-dimensional kernel, spanned by the columns of

$$
\kappa = \begin{pmatrix} 1 & 0 \\ 0 & 1 \end{pmatrix} .
$$

2. Solving for m,

$$
\begin{aligned}
m &= (\kappa^T \circ \kappa)^{-1} \circ \kappa^T \circ c^3 \\
&= \left(\begin{pmatrix} 1 & 0 \\ 0 & 1 \end{pmatrix} \begin{pmatrix} 1 & 0 \\ 0 & 1 \end{pmatrix} \right)^{-1} \begin{pmatrix} 1 & 0 \\ 0 & 1 \end{pmatrix} \begin{pmatrix} 4 & -6 \\ -2 & 3 \end{pmatrix} \\
&= \begin{pmatrix} 1 & 0 \\ 0 & 1 \end{pmatrix}^{-1} \begin{pmatrix} 4 & -6 \\ -2 & 3 \end{pmatrix} \\
&= \begin{pmatrix} 4 & -6 \\ -2 & 3 \end{pmatrix}
\end{aligned}
$$

(As a crosscheck, notice that

$$\kappa \circ m = \begin{pmatrix} 1 & 0 \\ 0 & 1 \end{pmatrix} \begin{pmatrix} 4 & -6 \\ -2 & 3 \end{pmatrix} = \begin{pmatrix} 4 & -6 \\ -2 & 3 \end{pmatrix} = c^3$$

as desired.)

3. Since m has a 1-dimensional kernel spanned by $\begin{pmatrix} 3 \\ 2 \end{pmatrix}$, the matrix M spanning the cokernel of m has many possibilities. One such possibility is

$$M = \begin{pmatrix} 1 \\ 0 \end{pmatrix}.$$

4. The homology for index 4 is 1-dimensional, spanned by

$$h^4 = \kappa \circ M = \begin{pmatrix} 1 \\ 0 \end{pmatrix}.$$

Exercise 2.39 Compute a basis for the homology of the following chain complex using Algorithm 2. Verify your answer against the barcode decomposition of the chain complex.

$$0 \longrightarrow \mathbb{R} \xrightarrow{\begin{pmatrix} 0 \\ 0 \end{pmatrix}} \mathbb{R}^2 \xrightarrow{\begin{pmatrix} 0 & 0 \end{pmatrix}} \mathbb{R} \longrightarrow 0$$

Exercise 2.40 Compute a basis for the homology of the following chain complex using Algorithm 2. Verify your answer against the barcode decomposition of the chain complex.

$$0 \longrightarrow \mathbb{R} \xrightarrow{\begin{pmatrix} 2 \\ 0 \end{pmatrix}} \mathbb{R}^2 \xrightarrow{\begin{pmatrix} 0 & 0 \end{pmatrix}} \mathbb{R} \longrightarrow 0$$

Exercise 2.41 Compute a basis for the homology of the following chain complex using Algorithm 2. Verify your answer against the barcode decomposition of the chain complex.

$$0 \to \mathbb{R}^4 \xrightarrow{\begin{pmatrix} 0 & 0 & 1 & 1 \\ 0 & 1 & 0 & -1 \\ 0 & -1 & -1 & 0 \\ 1 & 0 & 0 & 1 \\ -1 & 0 & 1 & 0 \\ 1 & 1 & 0 & 0 \end{pmatrix}} \mathbb{R}^6 \xrightarrow{\begin{pmatrix} -1 & -1 & -1 & 0 & 0 & 0 \\ 1 & 0 & 0 & -1 & -1 & 0 \\ 0 & 1 & 0 & 1 & 0 & -1 \\ 0 & 0 & -1 & 0 & -1 & -1 \end{pmatrix}} \mathbb{R}^4 \to 0$$

2.6 Further Reading

Our statement of Theorem 2.1 has a constructive, elementary proof. It is an expression of the famous *Krull-Schmidt theorem* in representation theory. This extends to a stronger statement that *every* sequence in **Seq** is isomorphic to a barcode, including those that are not bounded. This remains true even if the $v^\bullet$ maps do not all point in the same direction (from lower indices to higher ones). The interested reader can consult (Brüstle and Hanson 2025; Jacquard et al. 2023; Carlsson and de Silva 2010; Derksen and Weyman 2005; Gabriel 1972) for a proof of this more general result. Moreover, the statement of Theorem 2.1 still holds in certain settings where the terms in the sequence are infinite dimensional (Usher and Zhang 2016).

Depending on the context, what we call *sequences* are also known as *persistence modules*. They often arise from geometric data via *persistent homology* (Ghrist 2008; Zomorodian and Carlsson 2005; de Silva and Carlsson 2004). Sequences are also special cases of *quiver representations*. There is a large and detailed literature on quivers that the reader might enjoy. Many of the basics are covered in Oudot (2015). Statistical aspects of distributions of sequences are also interesting. Convenient representations for working with these distributions are known (Bubenik 2015; Mileyko et al. 2011; Cohen-Steiner et al. 2007).

The reader might ask what happens if instead of a sequence, we consider a larger grid of vector spaces and linear maps. Although Chap. 6 gives an elementary discussion of *grid diagrams*, a more advanced introduction is Botnan and Lesnick (2020).

References

Blanchette B, Brüstle T, Hanson EJ (2025) Exact structures for persistence modules. arXiv:2308.01790
Botnan MB, Lesnick M (2020) An introduction to multiparameter persistence. In: Representation of algebras and related structures. EMS Press, pp 77–150
Bubenik P (2015) Statistical topological data analysis using persistence landscapes. J Mach Learn Res 16(1):77–102
Carlsson G, de Silva V (2010) Zigzag persistence. Found Comput Math 10(4):367–405
Cohen-Steiner D, Edelsbrunner H, Harer J (2007) Stability of persistence diagrams. Discret Comput Geom 37(1):103–120
de Silva V, Carlsson G (2004) Topological estimation using witness complexes. In: Alexa M, Rusinkiewicz S (eds) Eurographics symposium on point-based graphics
Derksen H, Weyman J (2005) Quiver representations. Not AMS 52(2):200–206
Gabriel P (1972) Unzerlegbare darstellungen i. Manuscripta Mathematica 6:71–103
Ghrist R (2008) Barcodes: the persistent topology of data. Bull Am Math Soc 45(1):61
Jacquard E, Nanda V, Tillmann U (2023) The space of barcode bases for persistence modules. J Appl Comput Topol 7(1):1–30
Mileyko Y, Mukherjee S, Harer J (2011) Probability measures on the space of persistence diagrams. Inverse Probl 27(12):124007
Oudot SY (2015) Persistence theory: from quiver representations to data analysis, vol 209. American Mathematical Society
Usher M, Zhang J (2016) Persistent homology and Floer-Novikov theory. Geom & Topol 20(6):3333–3430
Zomorodian A, Carlsson G (2005) Computing persistent homology. Discret Comput Geom 33(2):249–274

Chapter 3
Chain Maps

This chapter is the fulcrum against which the force of linear algebra is magnified into a powerful tool for understanding chain maps. The Escolar-Hiraoka decomposition (Theorem 3.1) enables us to reason about chain maps in terms of their canonical decomposition into 10 different kinds of simpler chain maps. Not only does this dramatically simplify proofs of useful properties, it leads to Algorithm 1, which completely characterizes chain maps.

With the Escolar-Hiraoka decomposition in hand, we learn a few surprising facts about chain complexes of vector spaces. There are two—apparently different—equivalences between chain complexes, *quasi-isomorphism* and *chain homotopy equivalence*. While these two equivalences are quite different in general (as is discussed at length in Chap. 6), for chain complexes of vector spaces, they agree. In a roundabout way, these two equivalences are also related to exact sequences *of sequences*. If one starts with a three-term exact sequence (usually called a *short exact sequence*) of chain complexes (instead of vector spaces), one can derive a much longer exact sequence involving the homology of these three chain complexes, a fact called the Snake Lemma (Theorem 3.2).

In this chapter,

Section 3.1 Proves and algorithmically computes the Escolar-Hiraoka decomposition.
Section 3.2 Uses the Escolar-Hiraoka decomposition to compute the map on homology that is induced by a chain map.
Section 3.3 Touches upon chain complexes of sequences as a generalization of chain complexes of vector spaces (to be revisited in detail in Chap. 6).
Section 3.4 Explores the two main equivalences for chain complexes, *quasi-isomorphism* and *chain homotopy equivalence*
Section 3.5 Applies the machinery developed in this chapter to explore a useful duality between chain complexes.

M. Robinson, *Computational Homological Algebra*, Mathematical Engineering,
https://doi.org/10.1007/978-3-032-08634-1_3

3.1 The Escolar-Hiraoka Decomposition

We have already discussed how to construct a sequence as the direct sum of other sequences in Definition 2.7 in Chap. 2. Like decomposing linear maps between vector spaces, we can also decompose a sequence map into a direct sum of sequence maps, which works like block diagonal matrices.

Definition 3.1 Suppose that $m_1^\bullet : (V_1^\bullet, v_1^\bullet) \to (W_1^\bullet, w_1^\bullet)$ and $m_2^\bullet : (V_2^\bullet, v_2^\bullet) \to (W_2^\bullet, w_2^\bullet)$ are two sequence maps. Their *sequence map direct sum* is the sequence map $m^\bullet : (V_1^\bullet, v_1^\bullet) \oplus (V_2^\bullet, v_2^\bullet) \to (W_1^\bullet, w_1^\bullet) \oplus (W_2^\bullet, w_2^\bullet)$ with component maps $m_k : V_1^k \oplus V_2^k \to W_1^k \oplus W_2^k$ defined by $m^k = m_1^k \oplus m_2^k$.

Example 3.1 The direct sum of

$$
\begin{CD}
0 @>>> \mathbb{R} @>{\begin{pmatrix}0\end{pmatrix}}>> \mathbb{R} @>{\begin{pmatrix}1\end{pmatrix}}>> \mathbb{R} @>>> 0 \\
@. @V{\begin{pmatrix}1\end{pmatrix}}VV @V{\begin{pmatrix}0\\1\end{pmatrix}}VV @V{\begin{pmatrix}1\end{pmatrix}}VV @. \\
0 @>>> \mathbb{R} @>>{\begin{pmatrix}0\\0\end{pmatrix}}> \mathbb{R}^2 @>>{\begin{pmatrix}0 & 1\end{pmatrix}}> \mathbb{R} @>>> 0
\end{CD}
$$

and

$$
\begin{CD}
0 @>>> \mathbb{R}^2 @>>> 0 @>>> \mathbb{R} @>>> 0 \\
@. @V{\begin{pmatrix}1 & 0\\0 & 1\end{pmatrix}}VV @VVV @V{\begin{pmatrix}1\end{pmatrix}}VV @. \\
0 @>>> \mathbb{R}^2 @>>{\begin{pmatrix}0 & 0\end{pmatrix}}> \mathbb{R} @>>{\begin{pmatrix}1\end{pmatrix}}> \mathbb{R} @>>> 0
\end{CD}
$$

is

$$
\begin{CD}
0 @>>> \mathbb{R}\oplus\mathbb{R}^2 @>{\left(\begin{array}{c|cc}0 & 0 & 0\end{array}\right)}>> \mathbb{R} @>{\left(\begin{array}{c}1\\\hline 0\end{array}\right)}>> \mathbb{R}\oplus\mathbb{R} @>>> 0 \\
@. @V{\left(\begin{array}{c|cc}1 & 0 & 0\\\hline 0 & 1 & 0\\ 0 & 0 & 1\end{array}\right)}VV @V{\left(\begin{array}{c}0\\1\\\hline 0\end{array}\right)}VV @V{\left(\begin{array}{c|c}1 & 0\\\hline 0 & 1\end{array}\right)}VV @. \\
0 @>>> \mathbb{R}\oplus\mathbb{R}^2 @>>{\left(\begin{array}{c|cc}0 & 0 & 0\\ 0 & 0 & 0\\\hline 0 & 0 & 0\end{array}\right)}> \mathbb{R}^2\oplus\mathbb{R} @>>{\left(\begin{array}{cc|c}0 & 1 & 0\\\hline 0 & 0 & 1\end{array}\right)}> \mathbb{R}\oplus\mathbb{R} @>>> 0
\end{CD}
$$

Exercise 3.1 Check that the definition of the sequence map direct sum is well defined, by verifying that the diagram below commutes,

$$\begin{array}{ccc} V_1^k \oplus V_2^k & \xrightarrow{v_1^k \oplus v_2^k} & V_1^{k+1} \oplus V_2^{k+1} \\ {\scriptstyle m_1^k \oplus m_2^k}\downarrow & & \downarrow{\scriptstyle m_1^{k+1} \oplus m_2^{k+1}} \\ W_1^k \oplus W_2^k & \xrightarrow[w_1^k \oplus w_2^k]{} & W_1^{k+1} \oplus W_2^{k+1} \end{array}$$

Exercise 3.2 Construct the direct sum of

$$\begin{array}{ccccccccc} 0 & \longrightarrow & 0 & \longrightarrow & \mathbb{R}^2 & \xrightarrow{\begin{pmatrix}-2 & 2\end{pmatrix}} & \mathbb{R} & \longrightarrow & 0 \\ & & \downarrow & & \downarrow{\scriptstyle \begin{pmatrix}1 & 0\end{pmatrix}} & & \downarrow & & \\ 0 & \longrightarrow & \mathbb{R} & \xrightarrow[\begin{pmatrix}-1\end{pmatrix}]{} & \mathbb{R} & \longrightarrow & 0 & \longrightarrow & 0 \end{array}$$

and

$$\begin{array}{ccccccccc} 0 & \longrightarrow & \mathbb{R}^2 & \xrightarrow{\begin{pmatrix}1 & -1\end{pmatrix}} & \mathbb{R} & \xrightarrow{\begin{pmatrix}-4 \\ 0 \\ 2\end{pmatrix}} & \mathbb{R}^3 & \longrightarrow & 0 \\ & & \downarrow & & \downarrow & & \downarrow{\scriptstyle \begin{pmatrix}2 & 8 & 4\end{pmatrix}} & & \\ 0 & \longrightarrow & 0 & \longrightarrow & 0 & \longrightarrow & \mathbb{R} & \longrightarrow & 0 \end{array}$$

While we can sum several chain maps to form a chain map between "larger" chain complexes, we can also reverse the construction. Specifically, we can try to form a given chain map as the sum of simpler ones. While this idea works for sequence maps generally, the simpler sequence maps can themselves be arbitrarily complicated. The simplest *chain* maps are of limited complexity.

Theorem 3.1 (the Escolar-Hiraoka chain map decomposition, Escolar and Hiraoka (2016)) *Suppose that $m^\bullet : (C^\bullet, c^\bullet) \to (D^\bullet, d^\bullet)$ is a chain map. (That is, $m^\bullet$ is a sequence map and both sequences are chain complexes.) Then there is a barcode decomposition $\bigoplus_i P_i$ of $(C^\bullet, c^\bullet)$ and a barcode decomposition $\bigoplus_j Q_j$ of $(D^\bullet, d^\bullet)$ such that $m^\bullet$ decomposes as direct sum of chain maps $m_k^\bullet : P_k \to Q_k$ between bars in which each bar appears exactly once and each component of $m_k^\bullet$ is of full rank. (Zero length bars are permitted in the barcode decompositions.)*

Because the bars of both chain complexes appear exactly once in the decomposition, the canonicity of the barcode decomposition implies that this decomposition is canonical as well. We will refer to the decomposition of a chain map into chain maps between bars as its *Escolar-Hiraoka decomposition*, and each of the chain maps between bars will be called *bar chain maps*.

Proof Because of Theorem 2.1 in Chap. 2, the barcode decompositions $\bigoplus_i P_i$ and $\bigoplus_j Q_j$ are unique up to isomorphism. Since both sequences are chain complexes, no bar has length greater than 2. In order for a pair of bars to have a nontrivial sequence map $P_i \to Q_j$, the indices on which they have nontrivial terms must overlap. If this is not the case, for instance in the diagram

$$\begin{array}{ccccccccc} \cdots & \longrightarrow & \mathbb{R} & \longrightarrow & 0 & \longrightarrow & 0 & \longrightarrow & \cdots \\ & & \downarrow & & \downarrow & & \downarrow & & \\ \cdots & \longrightarrow & 0 & \longrightarrow & 0 & \longrightarrow & \mathbb{R} & \longrightarrow & \cdots \end{array}$$

the sequence map decomposes into a direct sum of sequence maps

$$\begin{array}{ccccccccc} \cdots & \longrightarrow & \mathbb{R} & \longrightarrow & 0 & \longrightarrow & 0 & \longrightarrow & \cdots \\ & & \downarrow & & \downarrow & & \downarrow & & \\ \cdots & \longrightarrow & 0 & \longrightarrow & 0 & \longrightarrow & 0 & \longrightarrow & \cdots \end{array}$$

and

$$\begin{array}{ccccccccc} \cdots & \longrightarrow & 0 & \longrightarrow & 0 & \longrightarrow & 0 & \longrightarrow & \cdots \\ & & \downarrow & & \downarrow & & \downarrow & & \\ \cdots & \longrightarrow & 0 & \longrightarrow & 0 & \longrightarrow & \mathbb{R} & \longrightarrow & \cdots \end{array}$$

Using this insight, we will not lose generality if we work with portions of the chain map that are not more than three terms long.

Let us first consider the case where the overlap is as large as possible, namely where we have a diagram like

$$\begin{array}{ccccccccccc} \cdots & \longrightarrow & 0 & \longrightarrow & C_1 & \longrightarrow & C_2 & \longrightarrow & 0 & \longrightarrow & \cdots \\ & & \downarrow & & \downarrow & & \downarrow & & \downarrow & & \\ \cdots & \longrightarrow & 0 & \longrightarrow & D_1 & \longrightarrow & D_2 & \longrightarrow & 0 & \longrightarrow & \cdots \end{array}$$

We can extract two three-term sequences from this diagram, namely

$$\begin{array}{ccccccccc} \cdots & \longrightarrow & 0 & \longrightarrow & C_1 & \longrightarrow & C_2 & & & & \\ & & & & & & \downarrow & & & & \\ & & & & & & D_2 & \longrightarrow & 0 & \longrightarrow & \cdots \end{array}$$

and

$$\begin{array}{ccccccccccc} \cdots & \longrightarrow & 0 & \longrightarrow & C_1 & & & & & & \\ & & & & \downarrow & & & & & & \\ & & & & D_1 & \longrightarrow & D_2 & \longrightarrow & 0 & \longrightarrow & \cdots \end{array}$$

If we compute the barcode decompositions for these two sequences independently, whatever we find for the $C_1 \to C_2$ and $D_1 \to D_2$ portions of the barcode decomposition will be bars in the decompositions of the original chain complexes. Due to the commutativity of the diagram, the barcode decompositions will be identical at the indices corresponding to C_1 and D_2, while they are unconstrained at the other two indices.[1] Thus, this construction sets up a chain map $m_k^\bullet$ from a bar P_i in $(C^\bullet, c^\bullet)$ and a bar Q_j in $(D^\bullet, d^\bullet)$. We therefore may decompose the chain map $m^\bullet$ into a direct sum of $m_k^\bullet$ and a chain map induced on the remainder of the two chain complexes, which is formed by term-by-term quotients with P_i and Q_j. Evidently, this prevents both P_i and Q_j from being reused in further sequence maps if either are nontrivial.

Now turning to the situation of partially overlapping bars, there are only two ways that this can occur in a chain complex, due to the length of the bars. The first

$$\begin{array}{ccccccccc} \cdots & \longrightarrow & C_1 & \longrightarrow & C_2 & \longrightarrow & 0 & \longrightarrow & \cdots \\ & & \downarrow & & \downarrow & & \downarrow & & \\ \cdots & \longrightarrow & 0 & \longrightarrow & D_2 & \longrightarrow & D_3 & \longrightarrow & \cdots \end{array}$$

violates commutativity if all of the spaces are nontrivial and the arrows represent full rank maps. The second

$$\begin{array}{ccccccccc} \cdots & \longrightarrow & 0 & \longrightarrow & C_2 & \longrightarrow & C_3 & \longrightarrow & \cdots \\ & & \downarrow & & \downarrow & & \downarrow & & \\ \cdots & \longrightarrow & D_1 & \longrightarrow & D_2 & \longrightarrow & 0 & \longrightarrow & \cdots \end{array}$$

is perfectly possible, and again sets up a correspondence between a bar in the domain chain complex and one in the codomain. This correspondence is independent of the

[1] Notice that we might not obtain the original bars if there is a nontrivial term to the left of D_1 or to the right of C_2, as this may apply an incompatible constraint on what the bars must be at D_1 or at C_2. This is what happened in Example 2.8 in Chap. 2.

other bars in the decompositions of both chain complexes. We may again decompose the chain map $m^\bullet$ as a direct sum of this smaller chain map and one induced by the quotient. □

Because of Example 2.8 in Chap. 2, a result like Theorem 3.1 cannot be true for sequence maps when the sequences have bars of length 3 or longer. Of the possible chain maps between bars of length 2 or less, commutativity rules out several diagrams. For instance the diagram

$$\begin{array}{ccc}\mathbb{R} & \xrightarrow{\text{id}} & \mathbb{R}\\ \downarrow & & \downarrow{\scriptstyle \text{id}}\\ 0 & \longrightarrow & \mathbb{R}\end{array}$$

does not commute. The remaining bar chain maps—with diagrams that commute—may occur in the Escolar-Hiraoka decomposition of a chain map.

Table 3.1 shows all the possible chain maps between bars, in which every arrow is of full rank. (Type 1 bar chain maps are trivial and will usually be ignored.) Because

Table 3.1 The Escolar-Hiraoka decomposition: all possible chain maps between bars

	Domain bar Length 0	Domain bar Length 1	Domain bar Length 2
Codomain bar length 0	Type 1: $\begin{array}{c}0\\ \downarrow\\ 0\end{array}$	Type 4: $\begin{array}{c}\mathbb{R}\\ \downarrow\\ 0\end{array}$	Type 7: $\begin{array}{ccc}\mathbb{R} & \longrightarrow & \mathbb{R}\\ \downarrow & & \downarrow\\ 0 & \longrightarrow & 0\end{array}$
Codomain bar length 1	Type 2: $\begin{array}{c}0\\ \downarrow\\ \mathbb{R}\end{array}$	Type 5: $\begin{array}{c}\mathbb{R}\\ \downarrow\\ \mathbb{R}\end{array}$	Type 8: $\begin{array}{ccc}\mathbb{R} & \longrightarrow & \mathbb{R}\\ \downarrow & & \downarrow\\ \mathbb{R} & \longrightarrow & 0\end{array}$
Codomain bar length 2	Type 3: $\begin{array}{ccc}0 & \longrightarrow & 0\\ \downarrow & & \downarrow\\ \mathbb{R} & \longrightarrow & \mathbb{R}\end{array}$	Type 6: $\begin{array}{ccc}0 & \longrightarrow & \mathbb{R}\\ \downarrow & & \downarrow\\ \mathbb{R} & \longrightarrow & \mathbb{R}\end{array}$	Type 9: $\begin{array}{ccc}\mathbb{R} & \longrightarrow & \mathbb{R}\\ \downarrow & & \downarrow\\ \mathbb{R} & \longrightarrow & \mathbb{R}\end{array}$
			Type 10: $\begin{array}{ccccc}0 & \longrightarrow & \mathbb{R} & \longrightarrow & \mathbb{R}\\ \downarrow & & \downarrow & & \downarrow\\ \mathbb{R} & \longrightarrow & \mathbb{R} & \longrightarrow & 0\end{array}$

we are working exclusively with finite dimensional vector spaces over $\mathbb{R}$, we lose no generality by requiring each arrow to be either an identity map $\mathbb{R} \to \mathbb{R}$ or a zero map. (This differs from the general case of chain complexes of arbitrary abelian groups, in which the maps must be specified.)

The tabulation of all possible chain maps between bars is the primary tool for studying chain complexes and chain maps in this book. The reader will frequently need to refer back to Table 3.1!

The limited number of possible chain maps between bars makes it possible to quickly assess properties of a chain map. Although Algorithm 1 for computing the canonical decomposition promised by Theorem 3.1 will be exhibited in a few pages, the decomposition already greatly simplifies manual calculations. Once the barcode decomposition is in hand, one refers to Table 3.1 to constrain the possible chain maps between bars. Any remaining ambiguity is then easy to resolve by a few checks.

Remark 3.1 Theorem 3.1 is a special case of what appears in the paper Escolar and Hiraoka (2016). Their main result is substantially more powerful, as it additionally constructs maps between each of the bar chain maps. For instance, a map from a Type 6 bar chain map to a Type 10 bar chain map is given by the diagonal maps in the diagram

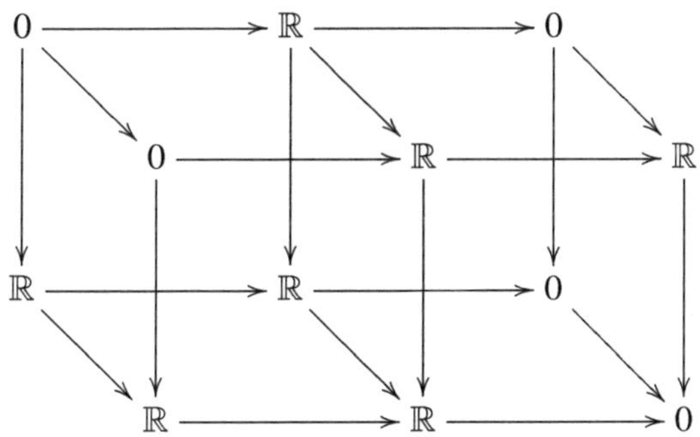

The collection of all possible maps between bar chain maps is organized into a labeled directed graph called the *Auslander-Reiten quiver*. It takes some work to obtain Table 3.1 from the appropriate Auslander-Reiten quiver (which appears in Escolar and Hiraoka (2016, Fig. 14)), because Table 3.1 does not distinguish between the same map in different indices but the Auslander-Reiten quiver does.

Example 3.2 The diagram below exhibits a chain map (columns) between two chain complexes (rows)

$$
\begin{array}{ccccccccc}
0 & \longrightarrow & \mathbb{R}^2 & \xrightarrow{\begin{pmatrix} 1 & -1 \\ -1 & 1 \end{pmatrix}} & \mathbb{R}^2 & \xrightarrow{\begin{pmatrix} 1 & 1 \\ 1 & 1 \end{pmatrix}} & \mathbb{R}^2 & \longrightarrow & 0 \\
 & & \big\downarrow \scriptstyle{\mathrm{id}} & & \big\downarrow \scriptstyle{\mathrm{id}} & & \big\downarrow & & \\
0 & \longrightarrow & \mathbb{R}^2 & \xrightarrow[\begin{pmatrix} 1 & -1 \\ -1 & 1 \end{pmatrix}]{} & \mathbb{R}^2 & \longrightarrow & 0 & \longrightarrow & 0
\end{array}
$$

We can verify this by showing that each square commutes.

Notice that both matrices on the top sequence have rank 1. The top sequence has barcode decomposition

$$\mathbb{R}^{(1,1)} \oplus \mathbb{R}^{(1,2)} \oplus \mathbb{R}^{(2,3)} \oplus \mathbb{R}^{(3,3)}.$$

Since the first component map is of full rank, this means that the first bar in the top sequence ($\mathbb{R}^{(1,1)}$) corresponds to a bar chain map of Type 5. For the same reason, reasoning from index 1, the second bar ($\mathbb{R}^{(1,2)}$) has to be part of either a bar chain map of Type 8 or 9. However, since component map in the middle (index 2) is also of full rank, this precludes Type 8. The second bar must correspond to a bar chain map of Type 9. Continuing with the third bar, again this must be of Type 8 or 9, but this time cannot be of Type 9 because the component map in index 3 is a zero map. The remaining bar ($\mathbb{R}^{(3,3)}$) must map to zero in the second sequence. It corresponds to a bar chain map of Type 4.

Summarizing the above, the decomposition that has been found is the sum of a Type 5 in index 1,

$$
\begin{array}{ccccccccc}
0 & \longrightarrow & \mathbb{R} & \longrightarrow & 0 & \longrightarrow & 0 & \longrightarrow & 0 \\
 & & \big\downarrow & & \big\downarrow & & \big\downarrow & & \\
0 & \longrightarrow & \mathbb{R} & \longrightarrow & 0 & \longrightarrow & 0 & \longrightarrow & 0
\end{array}
$$

a Type 9 in index 1,

$$
\begin{array}{ccccccccc}
0 & \longrightarrow & \mathbb{R} & \longrightarrow & \mathbb{R} & \longrightarrow & 0 & \longrightarrow & 0 \\
 & & \big\downarrow & & \big\downarrow & & \big\downarrow & & \\
0 & \longrightarrow & \mathbb{R} & \longrightarrow & \mathbb{R} & \longrightarrow & 0 & \longrightarrow & 0
\end{array}
$$

a Type 8 in index 2,

$$
\begin{array}{ccccccccc}
0 & \longrightarrow & 0 & \longrightarrow & \mathbb{R} & \longrightarrow & \mathbb{R} & \longrightarrow & 0 \\
 & & \big\downarrow & & \big\downarrow & & \big\downarrow & & \\
0 & \longrightarrow & 0 & \longrightarrow & \mathbb{R} & \longrightarrow & 0 & \longrightarrow & 0
\end{array}
$$

and a Type 4 in index 3,

$$\begin{array}{ccccccccc} 0 & \longrightarrow & 0 & \longrightarrow & 0 & \longrightarrow & \mathbb{R} & \longrightarrow & 0 \\ & & \downarrow & & \downarrow & & \downarrow & & \\ 0 & \longrightarrow & 0 & \longrightarrow & 0 & \longrightarrow & 0 & \longrightarrow & 0 \end{array}$$

Since the Escolar-Hiraoka decomposition is canonical, it yields unusually brief and complete theoretical results as well. For instance, it yields a complete characterization of isomorphisms between chain complexes. The proof of this fact merely consists of checking each entry in Table 3.1.

Corollary 3.1 *A chain map is an isomorphism between chain complexes if and only if its decomposition contains bar chain maps of Types* 1, 5, *or* 9 *only.*

Exercise 3.3 Given that chain maps are characterized by Theorem 3.1, describe the action of the functor Q_1 on chain maps. Specifically, describe the effect of Q_1 on each entry in Table 3.1, and then relate this back to an arbitrary chain map.

Equipped with Theorem 3.1 and Table 3.1, one can proceed systematically to decompose any chain map into a collection of chain maps between bars. The basic idea is to follow the barcode decomposition Algorithm 1 in Chap. 2 along the top chain complex, with the added step of tracing the bars discovered through the chain map. The process can be simplified if there is additional information. One famous application of this idea is the "five lemma", which shows how commutativity and exactness can constrain a component map of a chain map.

Proposition 3.1 (the five lemma for **Vec**) *Suppose that the following diagram defines a chain map between two exact sequences,*

$$\begin{array}{ccccccccc} A & \longrightarrow & B & \longrightarrow & C & \longrightarrow & D & \longrightarrow & E \\ \downarrow a & & \downarrow b & & \downarrow c & & \downarrow d & & \downarrow e \\ A' & \longrightarrow & B' & \longrightarrow & C' & \longrightarrow & D' & \longrightarrow & E'. \end{array}$$

If a, b, d, and e are isomorphisms, then c is also an isomorphism.

Proof Since the rows are exact, this means that their barcode decompositions consist of length 2 bars only. Since the diagram is assumed to be commutative, this means that it defines a chain map. The decomposition of the chain map using Theorem 3.1 will only have Types 3, 7, 9, or 10 in Table 3.1. But, since a, b, d, e are isomorphisms, this eliminates everything but Type 9 in the first two and last two indices (any bars starting at A or D). Let us consider a bar starting at B. Such a bar can only be in a bar chain map of Type 7, 9 or 10. But Types 7 and 10 require a or b to not be isomorphisms. This bar chain map must be of Type 9. Now consider a bar starting at C; the same argument limits this bar to be Type 9. Thus, the decomposition of this chain map only contains Type 9 bar chain maps. This must be a chain isomorphism by Corollary 3.1! We conclude that c must be an isomorphism. □

The proof of Proposition 3.1 exhibits a technique that will be used frequently: identifying which bar chain maps are possible in a given situation. For future reference, Table 3.2 compiles many of the important kinds of chain maps encountered in this chapter, and lists which bar chain maps could be possible (blank cell) or impossible "I". The rows are sorted lexicographically so that generally more restrictive classes of chain maps appear closer to the top of the table. The last column lists a reference to the proof in which this kind of chain map is classified. (Most of these are references to results proven later in this chapter. Do not be alarmed that some of chain maps are unfamiliar!) Interpret an entry "I" as definitive: no bar chain maps of that type can ever be present. For instance, a *quasi-isomorphism* cannot contain any bar chain maps of Type 2. Conversely, do *not* interpret a blank cell as meaning that a bar chain map *must* appear in that kind of chain map. For instance, there are chain isomorphisms that do not contain any bar chain maps of Type 5, even though that Type of bar chain map is allowable for chain isomorphisms.

Exercise 3.4 Another method for proving Proposition 3.1 involves a *diagram chase*, in which an element of one of the spaces is followed as it is mapped around the diagram. Under the hypotheses of Proposition 3.1,

1. Show that c is injective: suppose that you have $x \in C$ that is in the kernel of c. Using its images and (possible) preimages in the other nine spaces, conclude that x must actually be 0.

Table 3.2 Bar chain maps that can be present in a given kind of chain map: blank means *possible* and "I" means *impossible*

Chain map type	Bar chain map Type from Table 3.1										Reference
	1	2	3	4	5	6	7	8	9	10	
Map induced on the homology sequence by a quasi-isomorphism		I	I	I		I	I	I	I	I	Proposition 3.4
Chain isomorphism		I	I	I		I	I	I		I	Corollary 3.1
Long exact sequences of sequences		I		I	I	I		I		I	Lemma 3.1
Chain maps between exact sequences		I		I	I	I		I			Proposition 3.1
Quasi-isomorphism		I		I		I		I			Proposition 3.4
Induced maps on the homology sequence			I			I	I	I	I	I	Proposition 3.2
Null homotopic bar chain maps					I						Lemma 3.2

2. Show that c is surjective: suppose you have an arbitrary $y \in C'$. Again, using the images and (possible) preimages of y in the other nine spaces, show that there is a $z \in C$ satisfying the equation

$$c(z) = y.$$

Example 3.3 Consider the chain map given by the following diagram

$$\begin{array}{ccccccccccccc}
0 & \longrightarrow & \mathbb{R} & \xrightarrow{(0)} & \mathbb{R} & \xrightarrow{(0)} & \mathbb{R} & \xrightarrow{\begin{pmatrix}0\\1\\0\\0\end{pmatrix}} & \mathbb{R}^4 & \xrightarrow{\begin{pmatrix}1&0&0&0\\0&0&0&0\\0&0&1&0\\0&0&0&1\end{pmatrix}} & \mathbb{R}^4 & \longrightarrow & 0 \\
 & & \downarrow{\scriptstyle (1)} & & \downarrow{\scriptstyle (1)} & & \downarrow{\scriptstyle (0)} & & \downarrow{\scriptstyle \begin{pmatrix}0&0&0&0\\0&0&1&0\\0&0&0&1\end{pmatrix}} & & \downarrow{\scriptstyle \begin{pmatrix}0&1&0&0\\0&0&1&0\end{pmatrix}} & & \\
0 & \longrightarrow & \mathbb{R} & \xrightarrow[(0)]{} & \mathbb{R} & \xrightarrow[(0)]{} & \mathbb{R} & \xrightarrow[\begin{pmatrix}0\\0\\1\end{pmatrix}]{} & \mathbb{R}^3 & \xrightarrow[\begin{pmatrix}1&0&0\\0&1&0\end{pmatrix}]{} & \mathbb{R}^2 & \longrightarrow & 0.
\end{array}$$

First of all, that this is a chain map can be verified by inspection: composition of consecutive maps in both rows yield zero maps.

To decompose the diagram into chain maps between bars, we follow the construction in the proof of Theorem 3.1. This instructs us to iteratively remove the chain maps shown in Table 3.1 as direct summands. To start the process, notice that the first bar in the top sequence starting at the first nontrivial space is of length 1 and maps to a bar of length 1 in the bottom sequence. Thus the first summand in the decomposition of the chain map is

$$\begin{array}{ccccccccccccc}
0 & \longrightarrow & \mathbb{R} & \longrightarrow & 0 & \longrightarrow & 0 & \longrightarrow & 0 & \longrightarrow & 0 & \longrightarrow & 0 \\
 & & \downarrow{\scriptstyle (1)} & & \downarrow & & \downarrow & & \downarrow & & \downarrow & & \\
0 & \longrightarrow & \mathbb{R} & \longrightarrow & 0 & \longrightarrow & 0 & \longrightarrow & 0 & \longrightarrow & 0 & \longrightarrow & 0.
\end{array}$$

If we quotient this from the chain map diagram, we obtain the simpler chain map diagram

$$
\begin{array}{ccccccccccccc}
0 & \longrightarrow & 0 & \longrightarrow & \mathbb{R} & \xrightarrow{(0)} & \mathbb{R} & \xrightarrow{\begin{pmatrix}0\\1\\0\\0\end{pmatrix}} & \mathbb{R}^4 & \xrightarrow{\begin{pmatrix}1&0&0&0\\0&0&0&0\\0&0&1&0\\0&0&0&1\end{pmatrix}} & \mathbb{R}^4 & \longrightarrow & 0\\
 & & \downarrow & & \downarrow{\scriptstyle (1)} & & \downarrow{\scriptstyle (0)} & & \downarrow{\scriptstyle \begin{pmatrix}0&0&0&0\\0&0&1&0\\0&0&0&1\end{pmatrix}} & & \downarrow{\scriptstyle \begin{pmatrix}0&1&0&0\\0&0&1&0\end{pmatrix}} & & \\
0 & \longrightarrow & 0 & \longrightarrow & \mathbb{R} & \xrightarrow[(0)]{} & \mathbb{R} & \xrightarrow[\begin{pmatrix}0\\0\\1\end{pmatrix}]{} & \mathbb{R}^3 & \xrightarrow[\begin{pmatrix}1&0&0\\0&1&0\end{pmatrix}]{} & \mathbb{R}^2 & \longrightarrow & 0
\end{array}
$$

from which our work will proceed. Again, the first available bar in the top chain complex is of length 1, corresponding to a chain map

$$
\begin{array}{ccccccccccccc}
0 & \longrightarrow & 0 & \longrightarrow & \mathbb{R} & \longrightarrow & 0 & \longrightarrow & 0 & \longrightarrow & 0 & \longrightarrow & 0\\
 & & \downarrow & & \downarrow{\scriptstyle (1)} & & \downarrow & & \downarrow & & \downarrow & & \\
0 & \longrightarrow & 0 & \longrightarrow & \mathbb{R} & \longrightarrow & 0 & \longrightarrow & 0 & \longrightarrow & 0 & \longrightarrow & 0
\end{array}
$$

leaving us with the diagram

$$
\begin{array}{ccccccccccccc}
0 & \longrightarrow & 0 & \longrightarrow & 0 & \longrightarrow & \mathbb{R} & \xrightarrow{\begin{pmatrix}0\\1\\0\\0\end{pmatrix}} & \mathbb{R}^4 & \xrightarrow{\begin{pmatrix}1&0&0&0\\0&0&0&0\\0&0&1&0\\0&0&0&1\end{pmatrix}} & \mathbb{R}^4 & \longrightarrow & 0\\
 & & \downarrow & & \downarrow & & \downarrow{\scriptstyle (0)} & & \downarrow{\scriptstyle \begin{pmatrix}0&0&0&0\\0&0&1&0\\0&0&0&1\end{pmatrix}} & & \downarrow{\scriptstyle \begin{pmatrix}0&1&0&0\\0&0&1&0\end{pmatrix}} & & \\
0 & \longrightarrow & 0 & \longrightarrow & 0 & \longrightarrow & \mathbb{R} & \xrightarrow[\begin{pmatrix}0\\0\\1\end{pmatrix}]{} & \mathbb{R}^3 & \xrightarrow[\begin{pmatrix}1&0&0\\0&1&0\end{pmatrix}]{} & \mathbb{R}^2 & \longrightarrow & 0.
\end{array}
$$

The next bar in the top chain complex is of length 2. Notice that since

$$
\begin{pmatrix}0&0&0&0\\0&0&1&0\\0&0&0&1\end{pmatrix}\begin{pmatrix}0\\1\\0\\0\end{pmatrix} = \begin{pmatrix}0\\0\\0\end{pmatrix},
$$

this length 2 bar is in the kernel of the chain map. Thus it corresponds to a direct summand

$$\begin{array}{ccccccccccccc}
0 & \longrightarrow & 0 & \longrightarrow & 0 & \longrightarrow & \mathbb{R} & \longrightarrow & \mathbb{R} & \xrightarrow{\qquad\qquad\qquad} & 0 & \longrightarrow & 0\\
 & & \downarrow & & \downarrow & & \downarrow{\scriptstyle (0)} & & \downarrow{\scriptstyle (0)} & & \downarrow & & \\
0 & \longrightarrow & 0 & \longrightarrow & 0 & \longrightarrow & 0 & \longrightarrow & 0 & \xrightarrow{\qquad\qquad\qquad} & & \longrightarrow & 0.
\end{array}$$

When we take the quotient with this chain map, nothing in the lower chain complex will be changed. This leaves us with

$$\begin{array}{ccccccccccccc}
0 & \longrightarrow & 0 & \longrightarrow & 0 & \longrightarrow & 0 & \longrightarrow & \mathbb{R}^3 & \xrightarrow{\begin{pmatrix}1&0&0\\0&0&0\\0&1&0\\0&0&1\end{pmatrix}} & \mathbb{R}^4 & \longrightarrow & 0\\
 & & \downarrow & & \downarrow & & \downarrow & & \downarrow{\scriptstyle \begin{pmatrix}0&0&0\\0&1&0\\0&0&1\end{pmatrix}} & & \downarrow{\scriptstyle \begin{pmatrix}0&1&0&0\\0&0&1&0\end{pmatrix}} & & \\
0 & \longrightarrow & 0 & \longrightarrow & 0 & \longrightarrow & \mathbb{R} & \xrightarrow[\begin{pmatrix}0\\0\\1\end{pmatrix}]{} & \mathbb{R}^3 & \xrightarrow[\begin{pmatrix}1&0&0\\0&1&0\end{pmatrix}]{} & \mathbb{R}^2 & \longrightarrow & 0.
\end{array}$$

as the remaining portion of the chain map to decompose. Starting at the next nontrivial space in the top chain complex, there are three bars of length 2. The bar corresponding to the first component of $\mathbb{R}^3$ in the top sequence maps to zero along the chain map, and so has the form of

$$\begin{array}{ccc}
\mathbb{R} & \longrightarrow & \mathbb{R}\\
\downarrow & & \downarrow\\
0 & \longrightarrow & ?
\end{array}$$

where the question mark needs to be determined. According to Table 3.1, there is only one possibility for this bar chain map. Namely, the space marked with a question mark must be trivial. Therefore, in context, the bar chain map must be of the form

$$\begin{array}{ccccccccccccc}
0 & \longrightarrow & 0 & \longrightarrow & 0 & \longrightarrow & 0 & \longrightarrow & \mathbb{R} & \xrightarrow{\quad (1) \quad} & \mathbb{R} & \longrightarrow & 0\\
 & & \downarrow & & \downarrow & & \downarrow & & \downarrow{\scriptstyle (0)} & & \downarrow{\scriptstyle (0)} & & \\
0 & \longrightarrow & 0 & \longrightarrow & 0 & \longrightarrow & 0 & \longrightarrow & 0 & \xrightarrow{\qquad\qquad\qquad} & 0 & \longrightarrow & 0.
\end{array}$$

This means that the remaining chain map to decompose is

$$\begin{array}{ccccccccccccc}
0 & \rightarrow & 0 & \rightarrow & 0 & \rightarrow & 0 & \rightarrow & \mathbb{R}^2 & \xrightarrow{\begin{pmatrix}0&0\\1&0\\0&1\end{pmatrix}} & \mathbb{R}^3 & \rightarrow & 0 \\
 & & \downarrow & & \downarrow & & \downarrow & & \downarrow{\scriptsize\begin{pmatrix}0&0\\1&0\\0&1\end{pmatrix}} & & \downarrow{\scriptsize\begin{pmatrix}1&0&0\\0&1&0\end{pmatrix}} & & \\
0 & \rightarrow & 0 & \rightarrow & 0 & \rightarrow & \mathbb{R} & \xrightarrow[\begin{pmatrix}0\\0\\1\end{pmatrix}]{} & \mathbb{R}^3 & \xrightarrow[\begin{pmatrix}1&0&0\\0&1&0\end{pmatrix}]{} & \mathbb{R}^2 & \rightarrow & 0.
\end{array}$$

Consider the next bar of length 2 in the top sequence, namely the one corresponding to the basis vector $\begin{pmatrix}1\\0\end{pmatrix}$ in the $\mathbb{R}^2$ space. This basis vector maps to the vector $\begin{pmatrix}0\\1\\0\end{pmatrix}$ in $\mathbb{R}^3$ of the lower sequence. This vector is not in the image of the previous map in the sequence, and is not in the kernel of the next map of the sequence. Thus, we conclude that the bar we are considering is part of a chain map

$$\begin{array}{ccccccccccccc}
0 & \rightarrow & 0 & \rightarrow & 0 & \rightarrow & 0 & \rightarrow & \mathbb{R} & \xrightarrow{(1)} & \mathbb{R} & \rightarrow & 0 \\
 & & \downarrow & & \downarrow & & \downarrow & & \downarrow{\scriptsize(1)} & & \downarrow{\scriptsize(1)} & & \\
0 & \rightarrow & 0 & \rightarrow & 0 & \rightarrow & 0 & \rightarrow & \mathbb{R} & \xrightarrow[(1)]{} & \mathbb{R} & \rightarrow & 0.
\end{array}$$

Thus we are left with the task of decomposing

$$\begin{array}{ccccccccccccc}
0 & \rightarrow & 0 & \rightarrow & 0 & \rightarrow & 0 & \rightarrow & \mathbb{R} & \xrightarrow{\begin{pmatrix}0\\1\end{pmatrix}} & \mathbb{R}^2 & \rightarrow & 0 \\
 & & \downarrow & & \downarrow & & \downarrow & & \downarrow{\scriptsize\begin{pmatrix}0\\1\end{pmatrix}} & & \downarrow{\scriptsize(1\ 0)} & & \\
0 & \rightarrow & 0 & \rightarrow & 0 & \rightarrow & \mathbb{R} & \xrightarrow[\begin{pmatrix}0\\1\end{pmatrix}]{} & \mathbb{R}^2 & \xrightarrow[(1\ 0)]{} & \mathbb{R} & \rightarrow & 0.
\end{array}$$

The remaining length 2 bar in the top chain complex has a nontrivial image in the bottom chain complex, and is visibly of the form of the only length 3 chain map in Table 3.1. Therefore, there is a direct summand

$$
\begin{array}{ccccccccccccc}
0 & \longrightarrow & 0 & \longrightarrow & 0 & \longrightarrow & 0 & \longrightarrow & \mathbb{R} & \xrightarrow{\quad (1) \quad} & \mathbb{R} & \longrightarrow & 0 \\
 & & \downarrow & & \downarrow & & \downarrow & & \downarrow{\scriptstyle (1)} & & \downarrow & & \\
0 & \longrightarrow & 0 & \longrightarrow & 0 & \longrightarrow & \mathbb{R} & \xrightarrow[(1)]{} & \mathbb{R} & \longrightarrow & 0 & \longrightarrow & 0.
\end{array}
$$

This leaves the final direct summand

$$
\begin{array}{ccccccccccccc}
0 & \longrightarrow & 0 & \longrightarrow & 0 & \longrightarrow & 0 & \longrightarrow & 0 & \longrightarrow & \mathbb{R} & \longrightarrow & 0 \\
 & & \downarrow & & \downarrow & & \downarrow & & \downarrow & & \downarrow{\scriptstyle (1)} & & \\
0 & \longrightarrow & 0 & \longrightarrow & 0 & \longrightarrow & 0 & \longrightarrow & \mathbb{R} & \xrightarrow[\quad (1) \quad]{} & \mathbb{R} & \longrightarrow & 0,
\end{array}
$$

which completes the decomposition.

The process from the above example can be formalized into Algorithm 1. Identification of the bar chain maps in Table 3.1 requires some logic that is sensitive to the fact that the entries in the first column do not correspond to bars in the domain chain complex.

Exercise 3.5 Using Algorithm 1, decompose the following chain map into a direct sum of chain maps between bars shown in Table 3.1.

$$
\begin{array}{ccccccccccccccc}
0 & \longrightarrow & \mathbb{R}^2 & \xrightarrow{(0\ 0)} & \mathbb{R} & \xrightarrow{(0)} & \mathbb{R} & \xrightarrow{\begin{pmatrix}0\\0\\1\end{pmatrix}} & \mathbb{R}^3 & \xrightarrow{\begin{pmatrix}1&0&0\\0&0&0\end{pmatrix}} & \mathbb{R}^2 & \xrightarrow{\begin{pmatrix}0&0\\0&1\end{pmatrix}} & \mathbb{R}^2 & \longrightarrow & 0 \\
 & & \downarrow{\scriptstyle \begin{pmatrix}1&0\\0&0\end{pmatrix}} & & \downarrow{\scriptstyle \begin{pmatrix}0\\1\end{pmatrix}} & & \downarrow{\scriptstyle (1)} & & \downarrow{\scriptstyle \begin{pmatrix}0&0&1\\0&0&0\end{pmatrix}} & & \downarrow{\scriptstyle (0\ 0)} & & \downarrow{\scriptstyle (1\ 0)} & & \\
0 & \longrightarrow & \mathbb{R}^2 & \xrightarrow[\begin{pmatrix}0&1\\0&0\end{pmatrix}]{} & \mathbb{R}^2 & \xrightarrow[(0\ 0)]{} & \mathbb{R} & \xrightarrow[\begin{pmatrix}1\\0\end{pmatrix}]{} & \mathbb{R}^2 & \xrightarrow[(0\ 0)]{} & \mathbb{R} & \xrightarrow[(1)]{} & \mathbb{R} & \longrightarrow & 0
\end{array}
$$

Exercise 3.6 Using Algorithm 1, decompose the following chain map into a direct sum of chain maps between bars shown in Table 3.1.

Algorithm 1 Decomposing a chain map

Require: A chain map $m^\bullet : C^\bullet \to D^\bullet$ from one bounded chain complex $(C^\bullet, c^\bullet)$ to another $(D^\bullet, d^\bullet)$.
Ensure: A list of pairs (s_j, t_j) where s_j represents the starting index and t_j represents the Type of a bar chain map $b_j^\bullet$ such that $m^\bullet$ is their direct sum. ▷ The Types refer to entries in Table 3.1

procedure DECOMPOSECHAINMAP($\{m^\bullet\}$,$(C^\bullet, c^\bullet)$,$(D^\bullet, d^\bullet)$)
  Initialize $j = 0$.
  while do
    if there are no nontrivial spaces found among $C^\bullet$ **then**
      Compute barcode decomposition of $D^\bullet$ using Algorithm 1 in Chap. 2.
      for each bar $\delta_k = (\sigma_k, \tau_k)$ found in the barcode decomposition of $D^\bullet$ **do**
        Append the start of each bar found σ_k to the list $s_\bullet$.
        if $\tau_k - \sigma_k = 0$ **then** Append Type 2 to the list $t_\bullet$. ▷ bar of length 1
        if $\tau_k - \sigma_k = 1$ **then** Append Type 3 to the list $t_\bullet$. ▷ bar of length 2
      end for
      return the lists $s_\bullet$ and $t_\bullet$.
    else ▷ There are nontrivial spaces in $C^\bullet$
      Let i be the index of the first nontrivial space C^i in the domain chain complex.
      ▷ Define $v \in C^i$ according to the following cases:
      if $\ker c^i$ is nontrivial **then** let v be a nontrivial element of $\ker c^i$.
      elseif $(m^i)^{-1}(\text{image}\, d^{i-1})$ is nontrivial **then** let v be a nontrivial element of $(m^i)^{-1}(\text{image}\, d^{i-1})$ ▷ this step helps to identify Types 6 and 10
      else let

$$v := \begin{pmatrix} 1 & 0 & \cdots & 0 \end{pmatrix}^T .$$

      ▷ Classify the bar chain map in Table 3.1 that contains v:
      if $c^i(v) = 0$ **then** ▷ domain bar is of length 1
        if $m^i(v) = 0$ **then** $b_j^\bullet$ is of Type 4,
        elseif $m^i(v)$ is not in the image of d^{i-1} **then** $b_j^\bullet$ is of Type 5,
        else $b_j^\bullet$ is of Type 6. ▷ $m^i(v)$ is in the image of d^{i-1}
      else ▷ domain bar is of length 2
        if $m^i(v) = 0$ and $(m^{i+1} \circ c^i)(v) = 0$ **then** $b_j^\bullet$ is of Type 7,
        elseif $m^i(v) \neq 0$ and $(m^{i+1} \circ c^i)(v) \neq 0$, **then** $b_j^\bullet$ is of Type 9,
        elseif $m^i(v)$ is not in the image of d^{i-1}, **then** $b_j^\bullet$ is of Type 8,
        else $b_j^\bullet$ is of Type 10.
      end if
      Record that $s_j = i$, and record that t_j is the Type just found.
      Construct the bar chain map $b_j^\bullet$ using the starting index i and type t_j.
      Replace the chain map $m^\bullet : C^\bullet \to D^\bullet$ with the one induced by descending to the quotient of $m^\bullet$ and $b_j^\bullet$. ▷ Use Algorithm 4 in Chap. 1 or the projection formula on all implicated maps.
      Increment j.
    end if
  end while
end procedure

$$
\begin{array}{ccccccccccccc}
\mathbb{R} & \longrightarrow & 0 & \longrightarrow & \mathbb{R} & \xrightarrow{\begin{pmatrix}0\\0\\1\\0\end{pmatrix}} & \mathbb{R}^4 & \xrightarrow{\begin{pmatrix}0&0&0&0\\0&0&0&1\end{pmatrix}} & \mathbb{R}^2 & \xrightarrow{(1\ 0)} & \mathbb{R} & \longrightarrow & 0\\
\Big\downarrow{\scriptstyle (0)} & & \Big\downarrow & & \Big\downarrow{\scriptstyle (1)} & & \Big\downarrow{\scriptstyle (0\ 0\ 0\ 1)} & & \Big\downarrow{\scriptstyle \begin{pmatrix}1&0\\0&1\\0&0\\0&0\end{pmatrix}} & & \Big\downarrow{\scriptstyle (0)} & & \\
\mathbb{R} & \longrightarrow & 0 & \longrightarrow & \mathbb{R} & \xrightarrow[(0)]{} & \mathbb{R} & \xrightarrow[\begin{pmatrix}0\\1\\0\\0\end{pmatrix}]{} & \mathbb{R}^4 & \xrightarrow[(0\ 0\ 1\ 0)]{} & \mathbb{R} & \longrightarrow & 0
\end{array}
$$

Exercise 3.7 Using Algorithm 1, decompose the following chain map into a direct sum of bar sequence maps shown in Table 3.1.

$$
\begin{array}{ccccccccccc}
0 & \longrightarrow & \mathbb{R} & \xrightarrow{\begin{pmatrix}0\\0\\0\\1\end{pmatrix}} & \mathbb{R}^4 & \xrightarrow{\begin{pmatrix}0&0&1&0\\0&0&0&0\end{pmatrix}} & \mathbb{R}^2 & \xrightarrow{(0\ 1)} & \mathbb{R} & \longrightarrow & 0\\
 & & \Big\downarrow{\scriptstyle \begin{pmatrix}0\\0\\1\end{pmatrix}} & & \Big\downarrow{\scriptstyle \begin{pmatrix}0&0&0&0\\1&0&0&0\\0&1&0&0\\0&0&1&0\\0&0&0&0\\0&0&0&1\end{pmatrix}} & & \Big\downarrow{\scriptstyle \begin{pmatrix}1&0\\0&0\end{pmatrix}} & & \Big\downarrow & & \\
0 & \longrightarrow & \mathbb{R}^3 & \xrightarrow[\begin{pmatrix}0&0&0\\0&0&0\\1&0&0\\0&0&0\\0&1&0\\0&0&1\end{pmatrix}]{} & \mathbb{R}^6 & \xrightarrow[\begin{pmatrix}0&0&0&1&0&0\\0&0&0&0&0&0\end{pmatrix}]{} & \mathbb{R}^2 & \longrightarrow & 0 & \longrightarrow & 0
\end{array}
$$

3.2 The Induced Map on the Homology Sequence

The decomposition of a chain map into its bar chain maps is a complete characterization and aligns well with the use of homology as a summary of a chain complex. That is to say, every chain map *induces* maps on the homology sequence. Because of how homology picks out the length 1 bars from a chain complex, the induced map on the homology sequence is easy to obtain. In other words, *homology is a summary of a chain complex* in a rather precise sense. Specifically, homology is a functor that transforms chain complexes into vector spaces and transforms chain maps into linear maps.

Proposition 3.2 *Homology is a covariant functor* $H^\bullet : \mathbf{Kom}^b \to \mathbf{Vec}$.

We will usually write either $(m_*)^\bullet$ or $H^\bullet(m^\bullet)$ for the linear map $H^\bullet(C^\bullet, c^\bullet) \to H^\bullet(D^\bullet, d^\bullet)$ induced on the homology sequence by a chain map $m^\bullet : (C^\bullet, c^\bullet) \to (D^\bullet, d^\bullet)$.

Proof H^k is a function $H^k : Ob(\mathbf{Kom}) \to Ob(\mathbf{Vec})$ on objects, since it converts a chain complex into a vector space according to the formula

$$H^k(C^\bullet, c^\bullet) = \ker c^k / \operatorname{image} c^{k-1}.$$

Considering all of the spaces in the homology merely results in a larger space

$$H^\bullet(C^\bullet, c^\bullet) = \bigoplus_{k \in \mathbb{Z}} H^k(C^\bullet, c^\bullet),$$

which is a finite dimensional vector space if $(C^\bullet, c^\bullet)$ is a bounded chain complex in $\mathbf{Kom}^b$.

Suppose we have a chain map $m^\bullet : (C^\bullet, c^\bullet) \to (D^\bullet, d^\bullet)$. By Theorem 2.1 in Chap. 2 we can decompose the domain and codomain chain complexes into $m^\bullet : (E^\bullet, e^\bullet) \oplus H^\bullet(C^\bullet) \to (E'^\bullet, e'^\bullet) \oplus H^\bullet(D^\bullet)$ where $(E^\bullet, e^\bullet)$ and $(E'^\bullet, e'^\bullet)$ are exact sequences. Using Theorem 3.1, we can conclude that $m^\bullet$ decomposes into a direct sum of chain maps that respect these two decompositions. Namely,

$$m^\bullet = m_E^\bullet \oplus (m_*)^\bullet$$

where $(m_*)^\bullet : H^\bullet(C^\bullet) \to H^\bullet(D^\bullet)$. Consulting Table 3.1, this decomposition asserts that $(m_E)^\bullet$ will be a direct sum of bar chain maps of Types 1, 3, 7, 9, and 10, while $(m_*)^\bullet$ can only contain Types 1, 2, 4, and 5. Each of Types 1, 2, 4, and 5 are defined by at most one nontrivial linear map, acting on bars of length 0 or 1. Any bar chain maps of Type 6 will induce bar chain maps of Type 4 on the homology sequence, since the length 2 bar in the codomain of a Type 6 map will be removed when forming the homology sequence. Similarly, any bar chain maps of Type 8 will induce bar chain maps of Type 2 on the homology sequence. Types 3, 7, 9, and 10 induce bar chain maps of Type 1 on the homology sequence. Thus, by decomposing $(m_*)^\bullet$ into bar chain maps, we have also decomposed it in a way that respects the direct sum over indices in the homology sequence $H^\bullet(C^\bullet)$. Thus, $(m_*)^\bullet$ is a linear map, or in other words a morphism in **Vec**.

Finally, we need to establish that homology respects composition of chain maps. Let us assume that we have two chain maps $p^\bullet : (C^\bullet, c^\bullet) \to (D^\bullet, d^\bullet)$ and $q^\bullet : (D^\bullet, d^\bullet) \to (E^\bullet, e^\bullet)$. Their composition $q^\bullet \circ p^\bullet$ is also a well defined chain map. What we need to establish is that for each index k,

$$((q^\bullet \circ p^\bullet)_*)^k = (q_*)^k \circ (p_*)^k.$$

The right side of this equation is a composition of linear maps, while the left side is a composition of sequence maps. Theorem 3.1 asserts that we can assume each of the chain complexes decomposes according to its barcode decomposition, and that these bars are preserved when we decompose the columns of the diagram

$$\begin{array}{ccccccccc}
\cdots \longrightarrow & C^{k-1} & \xrightarrow{c^{k-1}} & C^k & \xrightarrow{c^k} & C^{k+1} & \xrightarrow{c^{k+1}} & \cdots \\
& \downarrow{\scriptstyle p^{k-1}} & & \downarrow{\scriptstyle p^k} & & \downarrow{\scriptstyle p^{k+1}} & & \\
\cdots \longrightarrow & D^{k-1} & \xrightarrow{d^{k-1}} & D^k & \xrightarrow{d^k} & D^{k+1} & \xrightarrow{d^{k+1}} & \cdots \\
& \downarrow{\scriptstyle q^{k-1}} & & \downarrow{\scriptstyle q^k} & & \downarrow{\scriptstyle q^{k+1}} & & \\
\cdots \longrightarrow & E^{k-1} & \xrightarrow{e^{k-1}} & E^k & \xrightarrow{e^k} & E^{k+1} & \xrightarrow{e^{k+1}} & \cdots
\end{array}$$

into bars as well. As for the homology sequence, we have a diagram induced by the chain maps $p^\bullet$ and $q^\bullet$

$$\begin{array}{ccccc}
\cdots & H^{k-1}(C^\bullet) & H^k(C^\bullet) & H^{k+1}(C^\bullet) & \cdots \\
& \downarrow{\scriptstyle (p_*)^{k-1}} & \downarrow{\scriptstyle (p_*)^k} & \downarrow{\scriptstyle (p_*)^{k+1}} & \\
\cdots & H^{k-1}(D^\bullet) & H^k(D^\bullet) & H^{k+1}(D^\bullet) & \cdots \\
& \downarrow{\scriptstyle (q_*)^{k-1}} & \downarrow{\scriptstyle (q_*)^k} & \downarrow{\scriptstyle (q_*)^{k+1}} & \\
\cdots & H^{k-1}(E^\bullet) & H^k(E^\bullet) & H^{k+1}(E^\bullet) & \cdots
\end{array}$$

Notice in particular that there are no horizontal arrows in the diagram! We also have the diagram induced by their composition $(q^\bullet \circ p^\bullet)_*$.

$$\begin{array}{ccccc}
\cdots & H^{k-1}(C^\bullet) & H^k(C^\bullet) & H^{k+1}(C^\bullet) & \cdots \\
& \downarrow{\scriptstyle ((q^\bullet \circ p^\bullet)_*)^{k-1}} & \downarrow{\scriptstyle ((q^\bullet \circ p^\bullet)_*)^k} & \downarrow{\scriptstyle ((q^\bullet \circ p^\bullet)_*)^{k+1}} & \\
\cdots & H^{k-1}(E^\bullet) & H^k(E^\bullet) & H^{k+1}(E^\bullet) & \cdots
\end{array}$$

Observe that this latter diagram decomposes into bar chain maps of Types 1, 2, 4, and 5, none of which require horizontal arrows. According to the construction of each of the induced maps $(p_*)^\bullet$, $(q_*)^\bullet$ and $((q^\bullet \circ p^\bullet)_*)^\bullet$, each nontrivial bar chain map in the above diagrams arose either as a bar chain map of the same Type in $p^\bullet$, $q^\bullet$, or $(q^\bullet \circ p^\bullet)$ or as a transformation of a bar chain map of Type 6 or 8. Each bar chain map of Type 5 in $(q^\bullet \circ p^\bullet)$ at index k is in one-to-one correspondence with one in $p^\bullet$ *and* one in $q^\bullet$ all at the same index k. This relationship is preserved in the homology sequence. Any bar chain map of Type 2 in $(q^\bullet \circ p^\bullet)$ at index k corresponds to a bar of length 1 or 2 in the sequence

$$C^k \xrightarrow{p^k} D^k \xrightarrow{q^k} E^k$$

that ends at E^k. Again, this relationship is preserved in the homology sequences. A similar relationship exists between each bar chain map of Type 4 in $(q^\bullet \circ p^\bullet)$ and a bar of length 1 or 2 starting at C^k in the above sequence. Any bar chain maps of Type 8 in $(q^\bullet \circ p^\bullet)$ will similarly relate to bars that end in the $E^\bullet$ sequence and do not start a length 1 bar in $C^\bullet$. By duality, bar chain maps of Type 6 in $(q^\bullet \circ p^\bullet)$ correspond to bars that do not end at $E^\bullet$ in a length 1 bar. □

Remark 3.2 Proposition 3.2 is not a special case of the shifted quotient functor Q_1 (Proposition 2.4 in Chap. 2), since the shifted quotient Q_1 truncates all bars to length 1 rather than selecting just the length 1 bars.

Example 3.4 Consider the chain map given by the diagram

$$\begin{CD}
\mathbb{R} @>{\begin{pmatrix}1\\1\\0\end{pmatrix}}>> \mathbb{R}^3 @>{\begin{pmatrix}1&-1&0\\0&0&1\end{pmatrix}}>> \mathbb{R}^2 @>{\begin{pmatrix}1&0\\0&1\end{pmatrix}}>> \mathbb{R}^2 @>>> 0\\
@V{\begin{pmatrix}1\\0\end{pmatrix}}VV @VV{\begin{pmatrix}0&1&0\\0&0&1\end{pmatrix}}V @VVV @VVV @VVV\\
\mathbb{R}^2 @>>{\begin{pmatrix}1&0\\0&1\end{pmatrix}}> \mathbb{R}^2 @>>> 0 @>>> 0 @>>> 0.
\end{CD}$$

Since the bottom row is an exact chain complex, the induced map on the homology sequence is the zero map at all indices.

Example 3.5 Consider the chain map given by the diagram

$$\begin{CD}
0 @>>> \mathbb{R}^2 @>{\begin{pmatrix}1&0\\0&1\end{pmatrix}}>> \mathbb{R}^2 @>>> 0 @>>> 0 @>>> 0\\
@. @VV{\begin{pmatrix}1&0\\0&1\end{pmatrix}}V @VV{(1\ 0)}V @VVV @VVV\\
0 @>>> \mathbb{R}^2 @>>{(1\ 0)}> \mathbb{R} @>>{(0)}> \mathbb{R} @>>{(1)}> \mathbb{R} @>>> 0.
\end{CD}$$

Let us first decompose this chain map using Algorithm 1. The algorithm starts with

$$v = \begin{pmatrix} 1 \\ 0 \end{pmatrix}$$

in the upper left corner of the diagram. Following this vector through the first map along the upper chain complex, and then through the second component map, we compute

$$\begin{pmatrix} 1 & 0 \end{pmatrix} \begin{pmatrix} 1 & 0 \\ 0 & 1 \end{pmatrix} \begin{pmatrix} 1 \\ 0 \end{pmatrix} = \begin{pmatrix} 1 \end{pmatrix} \neq 0.$$

Since the diagram commutes, we don't have to look further: this vector v is part of a bar chain map of Type 9.

Descending to the quotient, the chain map becomes

$$\begin{array}{ccccccccccc}
0 & \longrightarrow & \mathbb{R} & \xrightarrow{(1)} & \mathbb{R} & \longrightarrow & 0 & \longrightarrow & 0 & \longrightarrow & 0 \\
 & & \downarrow{\scriptstyle (1)} & & \downarrow{\scriptstyle (0)} & & \downarrow & & \downarrow & & \\
0 & \longrightarrow & \mathbb{R} & \xrightarrow[(0)]{} & 0 & \longrightarrow & \mathbb{R} & \xrightarrow[(1)]{} & \mathbb{R} & \longrightarrow & 0.
\end{array}$$

Inspecting this diagram, it is immediately clear that it is the direct sum of a bar chain map of Type 8 (in the first index) and Type 3 (in the third index). None of Types 9, 8, or 3 result in any nontrivial induced map on the homology sequence. The components of this induced map must be all zero maps.

Example 3.6 Consider the chain map given by the diagram

$$\begin{array}{ccccccccc}
0 & \longrightarrow & \mathbb{R}^2 & \xrightarrow{\begin{pmatrix} 1 & 0 \\ 0 & 0 \end{pmatrix}} & \mathbb{R}^2 & \xrightarrow{\begin{pmatrix} 0 & 0 \end{pmatrix}} & \mathbb{R} & \longrightarrow & 0 \\
 & & \downarrow & & \downarrow{\scriptstyle \begin{pmatrix} 0 & 1 \\ 0 & 0 \end{pmatrix}} & & \downarrow{\scriptstyle \begin{pmatrix} 1 \\ 0 \end{pmatrix}} & & \\
0 & \longrightarrow & 0 & \longrightarrow & \mathbb{R}^2 & \xrightarrow[\begin{pmatrix} 0 & 0 \\ 0 & 1 \end{pmatrix}]{} & \mathbb{R}^2 & \longrightarrow & 0.
\end{array}$$

Let us decompose this chain map using Algorithm 1, starting with the vector

$$v_1 = \begin{pmatrix} 1 \\ 0 \end{pmatrix}$$

in the first index of the domain chain complex. This vector has image 0 in the codomain chain complex, and has a nontrivial image in the next index of the domain chain complex. Consulting Table 3.1, the only way this can happen is with a bar chain map of Type 7. Thus, we can descend to the quotient to obtain the reduced chain map

$$\begin{array}{ccccccccc} 0 & \longrightarrow & \mathbb{R} & \xrightarrow{(0)} & \mathbb{R} & \xrightarrow{(0)} & \mathbb{R} & \longrightarrow & 0 \\ & & \downarrow & & \downarrow\begin{pmatrix}1\\0\end{pmatrix} & & \downarrow\begin{pmatrix}1\\0\end{pmatrix} & & \\ 0 & \longrightarrow & 0 & \longrightarrow & \mathbb{R}^2 & \xrightarrow[\begin{pmatrix}0&0\\0&1\end{pmatrix}]{} & \mathbb{R}^2 & \longrightarrow & 0. \end{array}$$

The next nontrivial vector is $v_2 = 1$, still in the first index of the domain chain complex. Since the next map in the domain chain complex is a zero map, this means that we have found a bar chain map of Type 4 in index 1. Descending to the quotient to remove this bar chain map, we obtain

$$\begin{array}{ccccccccc} 0 & \longrightarrow & 0 & \longrightarrow & \mathbb{R} & \xrightarrow{(0)} & \mathbb{R} & \longrightarrow & 0 \\ & & \downarrow & & \downarrow\begin{pmatrix}1\\0\end{pmatrix} & & \downarrow\begin{pmatrix}1\\0\end{pmatrix} & & \\ 0 & \longrightarrow & 0 & \longrightarrow & \mathbb{R}^2 & \xrightarrow[\begin{pmatrix}0&0\\0&1\end{pmatrix}]{} & \mathbb{R}^2 & \longrightarrow & 0. \end{array}$$

The next nontrivial space in the domain is in index 2, with a nontrivial vector $v_3 = 1$. This vector has image 0 in the next domain space, and has image $\begin{pmatrix}1\\0\end{pmatrix}$ in the codomain at index 2. *That* vector is in the kernel of the next codomain map. We have found a bar chain map of Type 5 in index 2. Removing this bar chain map, we have

$$\begin{array}{ccccccccc} 0 & \longrightarrow & 0 & \longrightarrow & 0 & \longrightarrow & \mathbb{R} & \longrightarrow & 0 \\ & & \downarrow & & \downarrow & & \downarrow\begin{pmatrix}1\\0\end{pmatrix} & & \\ 0 & \longrightarrow & 0 & \longrightarrow & \mathbb{R} & \xrightarrow[\begin{pmatrix}0\\1\end{pmatrix}]{} & \mathbb{R}^2 & \longrightarrow & 0. \end{array}$$

The next nontrivial space is in index 3, and the component of the chain map at that index has a nontrivial image $\begin{pmatrix}1\\0\end{pmatrix}$ in the codomain. The codomain map from the previous index does not have this vector in its image. There is a bar chain map of Type 5 at index 3, leaving us with

$$
\begin{array}{ccccccccc}
0 & \longrightarrow & 0 & \longrightarrow & 0 & \longrightarrow & 0 & \longrightarrow & 0 \\
 & & \downarrow & & \downarrow & & \downarrow & & \\
0 & \longrightarrow & 0 & \longrightarrow & \mathbb{R} & \xrightarrow[(1)]{} & \mathbb{R} & \longrightarrow & 0,
\end{array}
$$

which is a bar chain map of Type 3 in index 2.

If we select only the bar chain maps that correspond to induced maps on the homology sequence, we have a Type 4 in index 1, a Type 5 in index 2, and a Type 5 in index 3. As a diagram, this can be shown

$$
\begin{array}{ccccc}
0 & \mathbb{R} & \mathbb{R} & \mathbb{R} & 0 \\
\downarrow & \downarrow & \downarrow \mathrm{id} & \downarrow \mathrm{id} & \downarrow \\
0 & 0 & \mathbb{R} & \mathbb{R} & 0
\end{array}
$$

in which the first (leftmost) index is zero. Put another way, if the original chain map is named $m^\bullet$

$$
\begin{aligned}
(m_*)^1 &= 0_{\mathbb{R},0} : \mathbb{R} \to 0, \\
(m_*)^2 &= \mathrm{id}_{\mathbb{R}} : \mathbb{R} \to \mathbb{R}, \\
(m_*)^3 &= \mathrm{id}_{\mathbb{R}} : \mathbb{R} \to \mathbb{R}.
\end{aligned}
$$

Exercise 3.8 Consider the following chain map,

$$
\begin{array}{ccccc}
\mathbb{R}^2 & \xrightarrow{\begin{pmatrix} 0 & 1 \\ -1 & 0 \\ 1 & 1 \end{pmatrix}} & \mathbb{R}^3 & \xrightarrow{\begin{pmatrix} -1 & 1 & 1 \\ 0 & 0 & 0 \end{pmatrix}} & \mathbb{R}^2 \\
\begin{pmatrix} 0 & 1 \\ 1 & 0 \end{pmatrix}\Big\downarrow & & \begin{pmatrix} 0 & 1 & 0 \\ 1 & 0 & 0 \end{pmatrix}\Big\downarrow & & \Big\downarrow\begin{pmatrix} 0 & 1 \\ 0 & 0 \end{pmatrix} \\
\mathbb{R}^2 & \xrightarrow[\begin{pmatrix} 0 & -1 \\ 1 & 0 \end{pmatrix}]{} & \mathbb{R}^2 & \xrightarrow[\begin{pmatrix} 0 & 0 \\ 0 & 0 \end{pmatrix}]{} & \mathbb{R}^2
\end{array}
$$

1. Show that this is a chain map. You must verify that both rows are chain complexes, and that the diagram commutes.
2. Show that both rows are actually exact.
3. Explain what the induced map on the homology sequence must be as a result of parts (1) and (2).

Exercise 3.9 Show that the diagram below is not a chain map

$$
\begin{array}{ccccccccc}
0 & \longrightarrow & \mathbb{R}^2 & \xrightarrow{\begin{pmatrix} 1 & -1 \\ -1 & 1 \end{pmatrix}} & \mathbb{R}^2 & \longrightarrow & 0 & \longrightarrow & 0 \\
 & & \big\downarrow \mathrm{id} & & \big\downarrow \mathrm{id} & & \big\downarrow & & \\
0 & \longrightarrow & \mathbb{R}^2 & \xrightarrow[\begin{pmatrix} 1 & -1 \\ -1 & 1 \end{pmatrix}]{} & \mathbb{R}^2 & \xrightarrow[\begin{pmatrix} 1 & 1 \\ 1 & 1 \end{pmatrix}]{} & \mathbb{R}^2 & \longrightarrow & 0
\end{array}
$$

What happens—specifically—if you attempt to compute the induced map on the homology sequence in the diagram above? Exhibit a homology vector in the top row that does not have a well defined image in the homology of the bottom row. (Hint: you will not be able to apply the chain map decomposition Theorem 3.1.)

Exercise 3.10 Compute the map induced on the homology sequence by the chain map

$$
\begin{array}{ccccccccc}
0 & \longrightarrow & 0 & \longrightarrow & \mathbb{R} & \xrightarrow{(0)} & \mathbb{R} & \longrightarrow & 0 \\
 & & \big\downarrow & & \big\downarrow (0) & & \big\downarrow \begin{pmatrix} 1 \\ 0 \end{pmatrix} & & \\
0 & \longrightarrow & 0 & \longrightarrow & \mathbb{R} & \xrightarrow[\begin{pmatrix} 1 \\ 0 \end{pmatrix}]{} & \mathbb{R}^2 & \longrightarrow & 0
\end{array}
$$

1. Run Algorithm 1 on the chain map to obtain its decomposition into bar chain maps.
2. Determine the homology sequences of both chain complexes using your answer to (1).
3. Translate the bar chain maps into the induced maps for each nontrivial homology space.

Exercise 3.11 Compute the map induced on the homology sequence by the chain map

$$
\begin{array}{ccccccccccccc}
0 & \longrightarrow & \mathbb{R}^3 & \xrightarrow{\begin{pmatrix}1&0&0\\0&0&1\end{pmatrix}} & \mathbb{R}^2 & \xrightarrow{\begin{pmatrix}0&0\\0&0\end{pmatrix}} & \mathbb{R}^2 & \xrightarrow{\begin{pmatrix}1&0\\0&0\end{pmatrix}} & \mathbb{R}^2 & \xrightarrow{(0\ 1)} & \mathbb{R} & \longrightarrow & 0 \\
 & & \downarrow{\scriptsize\begin{pmatrix}1&0&0\\0&0&0\\0&0&1\end{pmatrix}} & & \downarrow{\scriptsize\begin{pmatrix}0&0\\0&0\end{pmatrix}} & & \downarrow{\scriptsize\begin{pmatrix}0&0\\1&0\\0&0\\0&1\end{pmatrix}} & & \downarrow{\scriptsize\begin{pmatrix}0&0\\0&1\end{pmatrix}} & & \downarrow & & \\
0 & \longrightarrow & \mathbb{R}^3 & \xrightarrow[\begin{pmatrix}0&0&0\\0&1&0\end{pmatrix}]{} & \mathbb{R}^2 & \xrightarrow[\begin{pmatrix}0&0\\1&0\\0&0\\0&0\end{pmatrix}]{} & \mathbb{R}^4 & \xrightarrow[\begin{pmatrix}1&0&0&0\\0&0&1&0\end{pmatrix}]{} & \mathbb{R}^2 & \longrightarrow & 0 & \longrightarrow & 0
\end{array}
$$

1. Run Algorithm 1 on the chain map to obtain its decomposition into bar chain maps.
2. Determine the homology sequences of both chain complexes using your answer to (1).
3. Translate the bar chain maps into the induced maps for each nontrivial homology space.

Exercise 3.12 Compute the map induced on the homology sequence by the chain map

$$
\begin{array}{ccccccccccccccc}
0 & \longrightarrow & \mathbb{R} & \xrightarrow{(0)} & \mathbb{R} & \xrightarrow{\begin{pmatrix}1\\0\end{pmatrix}} & \mathbb{R}^2 & \xrightarrow{\begin{pmatrix}0&0\\0&1\end{pmatrix}} & \mathbb{R}^2 & \xrightarrow{\begin{pmatrix}0&0\\1&0\end{pmatrix}} & \mathbb{R}^2 & \xrightarrow{(1\ 0)} & \mathbb{R} & \longrightarrow & 0 \\
 & & \downarrow{\scriptsize(1)} & & \downarrow{\scriptsize(1)} & & \downarrow{\scriptsize\begin{pmatrix}0&0\\1&0\\0&1\end{pmatrix}} & & \downarrow{\scriptsize\begin{pmatrix}0&0\\1&0\end{pmatrix}} & & \downarrow{\scriptsize(1\ 0)} & & \downarrow & & \\
0 & \longrightarrow & \mathbb{R} & \xrightarrow[(0)]{} & \mathbb{R} & \xrightarrow[\begin{pmatrix}0\\1\\0\end{pmatrix}]{} & \mathbb{R}^3 & \xrightarrow[\begin{pmatrix}0&0&0\\1&0&0\end{pmatrix}]{} & \mathbb{R}^2 & \xrightarrow[(1\ 0)]{} & \mathbb{R} & \longrightarrow & 0 & \longrightarrow & 0
\end{array}
$$

1. Run Algorithm 1 on the chain map to obtain its decomposition into bar chain maps.

2. Determine the homology sequences of both chain complexes using your answer to (1).
3. Translate the bar chain maps into the induced maps for each nontrivial homology space.

Although one can compute the induced map on the homology sequence by selecting bar chain maps of Types 1, 2, 4, and 5 out of a chain map and by converting Types 6 and 8 to Types 4 and 2 (respectively), there is a more direct approach, in which one does not need to first compute barcode decompositions. This direct approach is summarized by the commutative diagram in which the induced map $(m_*)^\bullet$ is computed using the projection formula.

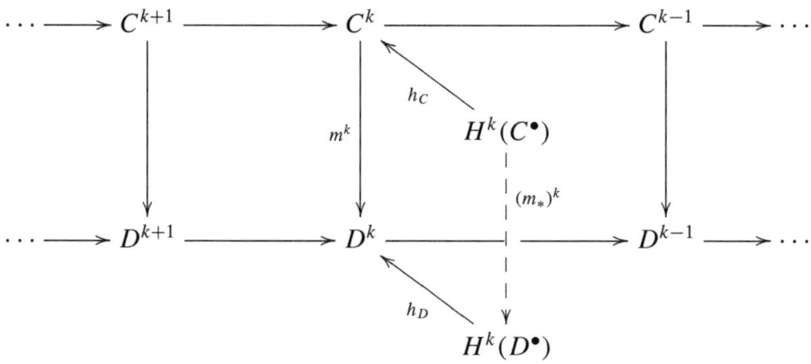

Algorithm 2 computes this via the projection formula

$$\begin{aligned} h_D \circ (m_*)^k &= m^k \circ h_C \\ h_D^T \circ h_D \circ (m_*)^k &= h_D^T \circ m^k \circ h_C \\ (m_*)^k &= (h_D^T \circ h_D)^{-1} \circ h_D^T \circ m^k \circ h_C. \end{aligned}$$

This is well defined because the h_D computed by Algorithm 2 in Chap. 2 is of full rank, making $h_D^T \circ h_D$ invertible.

Example 3.7 Let us use Algorithm 2 to redo Example 3.6. Again, the chain map is given by

$$\begin{array}{ccccccccc} 0 & \longrightarrow & \mathbb{R}^2 & \xrightarrow{\begin{pmatrix} 1 & 0 \\ 0 & 0 \end{pmatrix}} & \mathbb{R}^2 & \xrightarrow{\begin{pmatrix} 0 & 0 \end{pmatrix}} & \mathbb{R} & \longrightarrow & 0 \\ & & \downarrow & & \downarrow \begin{pmatrix} 0 & 1 \\ 0 & 0 \end{pmatrix} & & \downarrow \begin{pmatrix} 1 \\ 0 \end{pmatrix} & & \\ 0 & \longrightarrow & 0 & \longrightarrow & \mathbb{R}^2 & \xrightarrow[\begin{pmatrix} 0 & 0 \\ 0 & 1 \end{pmatrix}]{} & \mathbb{R}^2 & \longrightarrow & 0. \end{array}$$

Algorithm 2 Computing the induced map on the homology sequence

Require: A chain map $m^\bullet : C^\bullet \to D^\bullet$ from one bounded chain complex $(C^\bullet, c^\bullet)$ to another $(D^\bullet, d^\bullet)$ and an integer index k.
Ensure: The linear map $(m_*)^k : H^k(C^\bullet) \to H^k(D^\bullet)$

1: **procedure** HomologyInducedMap($m^\bullet$,k)
2: Construct the homology basis matrix h_C using Algorithm 2 in Chapter 2.
3: Construct the homology basis matrix h_D using Algorithm 2 in Chapter 2.
4: **return** the induced map,

$$(m_*)^k = (h_D^T \circ h_D)^{-1} \circ h_D^T \circ m^k \circ h_C.$$

5: **end procedure**

Algorithm 2 iterates once for each index k.

Index $k = 1$ Applying Algorithm 2:

1. We construct the homology basis matrix h_C^1 using Algorithm 2 in Chap. 2.
   a. We compute the matrix κ for the kernel of c^1

$$\kappa = \begin{pmatrix} 0 \\ 1 \end{pmatrix}.$$

   b. We compute the induced map μ using the projection formula

$$\mu = (\kappa^T \circ \kappa)^{-1} \circ \kappa^T \circ c^0 = (1)^{-1} \begin{pmatrix} 0 & 1 \end{pmatrix} \begin{pmatrix} 1 & 0 \\ 0 & 0 \end{pmatrix} = \begin{pmatrix} 0 & 0 \end{pmatrix}.$$

   c. Compute the cokernel of μ, spanned by columns of

$$M = \begin{pmatrix} 1 \end{pmatrix}.$$

   d. The homology basis matrix is

$$h_C^1 = \kappa \circ M = \begin{pmatrix} 0 \\ 1 \end{pmatrix}.$$

2. We could construct the homology basis matrix h_D^1 using Algorithm 2 in Chap. 2, but a quick inspection reveals that $h_D^1 = 0$.
3. The induced map

$$(m_*)^1 = ((h_D^1)^T \circ h_D^1)^{-1} \circ (h_D^1)^T \circ m^1 \circ h_C^1$$

   is therefore a zero map. Since h_C^1 has 1 column, $(m_*)^1$ is really a map $\mathbb{R} \to 0$, which agrees with Example 3.6 in a Type 4 bar chain map in index 1.

Index $k = 2$ Applying Algorithm 2:

1. We construct the homology basis matrix h_C^2 using Algorithm 2 in Chap. 2.

   a. We compute the matrix κ for the kernel of c^2

   $$\kappa = \begin{pmatrix} 1 & 0 \\ 0 & 1 \end{pmatrix}.$$

   b. We compute the induced map μ using the projection formula

   $$\mu = (\kappa^T \circ \kappa)^{-1} \circ \kappa^T \circ c^1 = \begin{pmatrix} 1 & 0 \\ 0 & 0 \end{pmatrix}.$$

   c. Compute the cokernel of μ, spanned by columns of

   $$M = \begin{pmatrix} 0 \\ 1 \end{pmatrix}.$$

   d. The homology basis matrix is

   $$h_C^2 = \kappa \circ M = \begin{pmatrix} 0 \\ 1 \end{pmatrix}.$$

2. We construct the homology basis matrix h_D^2 using Algorithm 2 in Chap. 2.

   a. We compute the matrix κ for the kernel of d^2

   $$\kappa = \begin{pmatrix} 1 \\ 0 \end{pmatrix}.$$

   b. We compute the induced map μ using the projection formula

   $$\mu = (\kappa^T \circ \kappa)^{-1} \circ \kappa^T \circ d^1 = \begin{pmatrix} 1 & 0 \end{pmatrix} \begin{pmatrix} 0 \\ 0 \end{pmatrix} = 0.$$

   Notice that since κ is a (2×1) matrix, μ is a $(1 \times 2)(2 \times 0) = (1 \times 0)$ matrix.

   c. Compute the cokernel of μ, spanned by columns of

   $$M = (1).$$

   Again, notice that since μ is a (1×0) matrix, M can only have one row.

   d. The homology basis matrix is

   $$h_D^2 = \kappa \circ M = \begin{pmatrix} 1 \\ 0 \end{pmatrix}.$$

3. The induced map is calculated as

$$(m_*)^2 = ((h_D^2)^T \circ h_D^2)^{-1} \circ (h_D^2)^T \circ m^2 \circ h_C^2 = (1)\begin{pmatrix}1 & 0\end{pmatrix}\begin{pmatrix}0 & 1\\ 0 & 0\end{pmatrix}\begin{pmatrix}0\\ 1\end{pmatrix} = (1).$$

This also means that there is a bar chain map of Type 5 in index 2, a finding which agrees with Example 3.6.

Index $k = 3$ Applying Algorithm 2:

1. We construct the homology basis matrix h_C^3 using Algorithm 2 in Chap. 2.
   a. We compute the matrix κ for the kernel of c^3

   $$\kappa = \begin{pmatrix}1\end{pmatrix}.$$

   b. We compute the induced map μ using the projection formula

   $$\mu = (\kappa^T \circ \kappa)^{-1} \circ \kappa^T \circ c^2 = \begin{pmatrix}0 & 0\end{pmatrix}.$$

   c. Compute the cokernel of μ, spanned by columns of

   $$M = \begin{pmatrix}1\end{pmatrix}.$$

   d. The homology basis matrix is

   $$h_C^3 = \kappa \circ M = \begin{pmatrix}1\end{pmatrix}.$$

2. We construct the homology basis matrix h_D^3 using Algorithm 2 in Chap. 2.
   a. We compute the matrix κ for the kernel of d^3

   $$\kappa = \begin{pmatrix}1 & 0\\ 0 & 1\end{pmatrix}.$$

   b. We compute the induced map μ using the projection formula

   $$\mu = (\kappa^T \circ \kappa)^{-1} \circ \kappa^T \circ d^2 = \begin{pmatrix}0 & 0\\ 0 & 1\end{pmatrix}.$$

   c. Compute the cokernel of μ, spanned by columns of

   $$M = \begin{pmatrix}1\\ 0\end{pmatrix}.$$

   d. The homology basis matrix is

$$h_D^3 = \kappa \circ M = \begin{pmatrix} 1 & 0 \\ 0 & 1 \end{pmatrix} \begin{pmatrix} 1 \\ 0 \end{pmatrix} = \begin{pmatrix} 1 \\ 0 \end{pmatrix}.$$

3. The induced map is calculated as

$$(m_*)^3 = ((h_D^3)^T \circ h_D^3)^{-1} \circ (h_D^3)^T \circ m^3 \circ h_C^3 = (1) \begin{pmatrix} 1 & 0 \end{pmatrix} \begin{pmatrix} 1 \\ 0 \end{pmatrix} (1) = (1).$$

Once again, this agrees with the finding of a Type 5 bar chain map at index 3 as in Example 3.6

Exercise 3.13 Compute the induced map on the homology sequence for the chain map shown in Exercise 3.10 using Algorithm 2.

Exercise 3.14 Compute the induced map on the homology sequence for the chain map shown in Exercise 3.11 using Algorithm 2.

Exercise 3.15 Compute the induced map on the homology sequence for the chain map shown in Exercise 3.12 using Algorithm 2.

3.3 The Snake Lemma

This chapter has explored the structure of sequences built from stringing linear maps one after the other. Since we have the idea of a sequence map, it is very natural to ask what happens if we build sequences *of sequences* in exactly the same way. The algebraic structure of sequences of sequences of vector spaces is substantially more intricate than that of sequences of vector spaces. Because arbitrary sequence maps do not respect the barcode structure of sequences while chain maps do, it is wise to limit the discussion to chain complexes of sequences. While we will defer an in-depth theoretical classification of such structures until Chap. 6, the idea is useful enough in Chap. 5 to warrant a preliminary exploration here.

Definition 3.2 A sequence of sequence maps

$$\cdots \longrightarrow (A^\bullet, a^\bullet) \xrightarrow{p^\bullet} (B^\bullet, b^\bullet) \xrightarrow{q^\bullet} (C^\bullet, c^\bullet) \xrightarrow{r^\bullet} \cdots$$

is called *exact* if each row in the diagram

$$
\begin{array}{ccccccccc}
 & & \vdots & & \vdots & & \vdots & & \\
 & & \downarrow & & \downarrow & & \downarrow & & \\
\cdots & \longrightarrow & A^{k-1} & \xrightarrow{p^{k-1}} & B^{k-1} & \xrightarrow{q^{k-1}} & C^{k-1} & \xrightarrow{r^{k-1}} & \cdots \\
 & & \downarrow{\scriptstyle a^{k-1}} & & \downarrow{\scriptstyle b^{k-1}} & & \downarrow{\scriptstyle c^{k-1}} & & \\
\cdots & \longrightarrow & A^{k} & \xrightarrow{p^{k}} & B^{k} & \xrightarrow{q^{k}} & C^{k} & \xrightarrow{r^{k}} & \cdots \\
 & & \downarrow{\scriptstyle a^{k}} & & \downarrow{\scriptstyle b^{k}} & & \downarrow{\scriptstyle c^{k}} & & \\
\cdots & \longrightarrow & A^{k+1} & \xrightarrow{p^{k+1}} & B^{k+1} & \xrightarrow{q^{k+1}} & C^{k+1} & \xrightarrow{r^{k+1}} & \cdots \\
 & & \downarrow{\scriptstyle a^{k+1}} & & \downarrow{\scriptstyle b^{k+1}} & & \downarrow{\scriptstyle c^{k+1}} & & \\
 & & \vdots & & \vdots & & \vdots & &
\end{array}
$$

is an exact sequence of vector spaces. (Notice that the above diagram is written so that the sequence maps are horizontal.)

Exercise 3.16 Construct the terms of an exact sequence that starts with $\mathbb{R}^{(0,1)}$, namely construct the $A^{\bullet,0}$, $A^{\bullet,1}$, ... so that

$$
\begin{array}{ccccccccc}
 & & \vdots & & \vdots & & \vdots & & \\
 & & \downarrow & & \downarrow & & \downarrow & & \\
0 & \longrightarrow & 0 & \longrightarrow & A^{-1,0} & \longrightarrow & A^{-1,1} & \longrightarrow & \cdots \\
 & & \downarrow & & \downarrow & & \downarrow & & \\
0 & \longrightarrow & \mathbb{R} & \longrightarrow & A^{0,0} & \longrightarrow & A^{0,1} & \longrightarrow & \cdots \\
 & & \downarrow{\scriptstyle \mathrm{id}} & & \downarrow & & \downarrow & & \\
0 & \longrightarrow & \mathbb{R} & \longrightarrow & A^{1,0} & \longrightarrow & A^{1,1} & \longrightarrow & \cdots \\
 & & \downarrow & & \downarrow & & \downarrow & & \\
0 & \longrightarrow & 0 & \longrightarrow & A^{2,0} & \longrightarrow & A^{2,1} & \longrightarrow & \cdots \\
 & & \downarrow & & \downarrow & & \downarrow & & \\
 & & \vdots & & \vdots & & \vdots & &
\end{array}
$$

has exact rows. There are many possible answers.

Exact sequences of sequences are fundamentally different from exact sequences of vector spaces. Unlike short exact sequences of vector spaces, which always split according to Proposition 2.8 in Chap. 2, short exact sequences of sequences may not split.

Example 3.8 The following diagram shows a short exact sequence of three sequences,

$$
\begin{array}{ccccccccc}
 & & \vdots & & \vdots & & \vdots & & \\
 & & \downarrow & & \downarrow & & \downarrow & & \\
0 & \longrightarrow & 0 & \longrightarrow & 0 & \longrightarrow & 0 & \longrightarrow & 0 \\
 & & \downarrow & & \downarrow & & \downarrow & & \\
0 & \longrightarrow & 0 & \longrightarrow & \mathbb{R} & \xrightarrow{\text{id}} & \mathbb{R} & \longrightarrow & 0 \\
 & & \downarrow & & \downarrow{\scriptstyle \text{id}} & & \downarrow & & \\
0 & \longrightarrow & \mathbb{R} & \xrightarrow{\text{id}} & \mathbb{R} & \longrightarrow & 0 & \longrightarrow & 0 \\
 & & \downarrow & & \downarrow & & \downarrow & & \\
0 & \longrightarrow & 0 & \longrightarrow & 0 & \longrightarrow & 0 & \longrightarrow & 0 \\
 & & \downarrow & & \downarrow & & \downarrow & & \\
 & & \vdots & & \vdots & & \vdots & &
\end{array}
$$

in which the sequence maps are shown horizontally. The barcode decomposition of each row has either no nontrivial bars or exactly one bar of length 2. Each row is indeed exact. If this sequence were to be split, this would mean that the sequence in the middle column would be the direct sum of the sequences on either side of it. However, the direct sum of those two sequences

$$\cdots \longrightarrow 0 \longrightarrow \mathbb{R} \longrightarrow 0 \longrightarrow 0 \longrightarrow \cdots$$

$$\oplus$$

$$\cdots \longrightarrow 0 \longrightarrow 0 \longrightarrow \mathbb{R} \longrightarrow 0 \longrightarrow \cdots$$

is isomorphic to the sequence

$$\cdots \longrightarrow 0 \longrightarrow \mathbb{R} \xrightarrow{(0)} \mathbb{R} \longrightarrow 0 \longrightarrow \cdots$$

This sequence cannot be isomorphic to the sequence in the middle column because the middle map is a zero map instead of an identity map. This is therefore a short exact sequence of sequences that does not split.

It is helpful to think of a grid-like diagram like the one in the above example as a sequence of sequences. Within this diagram is hidden a *long exact sequence.*

Lemma 3.1 (the snake lemma for sequences) *Suppose that*

$$0 \longrightarrow (C^\bullet, c^\bullet) \xrightarrow{\gamma^\bullet} (D^\bullet, d^\bullet) \xrightarrow{\delta^\bullet} (E^\bullet, e^\bullet) \longrightarrow 0$$

is an exact sequence of sequences. There exists a sequence map $\varepsilon^\bullet : (E^\bullet, e^\bullet) \to (C^{\bullet+1}, c^{\bullet+1})$ such that the sequence

$$\begin{array}{ccccccc} \cdots \longrightarrow (C^{\bullet-k}, c^{\bullet-k}) & \xrightarrow{\gamma^{\bullet-k}} & (D'^{\bullet-k}, d'^{\bullet-k}) & \xrightarrow{\delta^{\bullet-k}} & (E^{\bullet-k}, e^{\bullet-k}) & & \\ & & & & \Big\downarrow \varepsilon^{\bullet-k} & & \\ & & & & (C^{\bullet-k+1}, c^{\bullet-k+1}) & \xrightarrow{\gamma^{\bullet-k+1}} & \cdots \end{array}$$

of sequence maps is exact for each $k \in \mathbb{Z}$, where the sequence $(D'^\bullet, d'^\bullet)$ is a restriction of the original $(D^\bullet, d^\bullet)$ to a subspace: $D'^k \subseteq D^k$ and

$$\begin{array}{ccc} D'^k & \longrightarrow & D^k \\ & \searrow^{d'^k} & \Big\downarrow d^k \\ & & E^k \end{array}$$

commutes for all $k \in \mathbb{Z}$.

We usually call this new exact sequence the *long exact sequence* associated to the short exact sequence of sequences. The sequence map $\varepsilon^\bullet$ is called the *connecting map.*

Proof The hypothesis of the lemma posits that the diagram

$$\begin{array}{ccccccccc} & & \vdots & & \vdots & & \vdots & & \\ & & \Big\downarrow & & \Big\downarrow & & \Big\downarrow & & \\ 0 & \longrightarrow & C^k & \xrightarrow{\gamma^k} & D^k & \xrightarrow{\delta^k} & E^k & \longrightarrow & 0 \\ & & \Big\downarrow c^k & & \Big\downarrow d^k & & \Big\downarrow e^k & & \\ 0 & \longrightarrow & C^{k+1} & \xrightarrow[\gamma^{k+1}]{} & D^{k+1} & \xrightarrow[\delta^{k+1}]{} & E^{k+1} & \longrightarrow & 0 \\ & & \Big\downarrow & & \Big\downarrow & & \Big\downarrow & & \\ & & \vdots & & \vdots & & \vdots & & \end{array}$$

has exact rows. Although we had been thinking of the horizontal maps as being the components of a sequence map, let us change perspective: think of the rows as chain complexes! Under this new perspective, the vertical maps become components of a chain map. Apply Theorem 3.1 to decompose this chain map into bar chain maps. Since the rows are exact, the only bar chain maps appearing in the decomposition are Types 3, 7, 9, and 10. Under the decomposition, the space E^k is spanned by bar chain maps of Types 7, 9, and 10. Explicitly, the chain map decomposition induces a decomposition of the vector space

$$E^k = \left(\bigoplus_{i=1}^{I} u_i\right) \oplus \left(\bigoplus_{j=1}^{J} v_j\right) \oplus \left(\bigoplus_{m=1}^{M} w_m\right)$$

where u_i corresponds to a bar of Type 7, v_j corresponds to a bar of Type 9, and w_m corresponds to a bar of Type 10. Similarly, the chain map decomposition induces a decomposition of

$$C^{k+1} = \left(\bigoplus_{n=1}^{N} x_n\right) \oplus \left(\bigoplus_{p=1}^{P} y_p\right) \oplus \left(\bigoplus_{m=1}^{M} w_m\right)$$

where x_n corresponds to a bar of Type 3, y_p corresponds to a bar of Type 9, and w_m corresponds to a bar of Type 10. Since the bar chain maps of Type 9 are of length 2, the v_j and y_p basis vectors are unrelated. It is crucial to notice that the situation is different for bar chain maps of Type 10: they have length 3, which means that the w_m appearing in the decompositions for E^k and C^{k+1} are *identical*. Define the linear map $\varepsilon^k : E^k \to C^{k+1}$ according to its action on a vector z from the above basis

$$\varepsilon^k(z) := \begin{cases} w_m & \text{if } z = w_m \\ 0 & \text{otherwise,} \end{cases}$$

and extend by linearity. In this way, ε^k uses the bar chain maps of Type 10 to *connect* the sequence $E^\bullet$ to the sequence $C^{\bullet+1}$.

Using this definition of $\varepsilon^\bullet$, the sequence

$$\cdots \longrightarrow D^k \xrightarrow{\delta^k} E^k \xrightarrow{\varepsilon^k} C^{k+1} \xrightarrow{\gamma^{k+1}} D^{k+1} \xrightarrow{\delta^{k+1}} \cdots$$

of vector spaces is not quite exact, because each of the original Type 10 bar chain maps has turned into a bar of length 4 in this sequence. These length 4 bars start at D^k and end at D^{k+1}. In essence, we have a diagram like

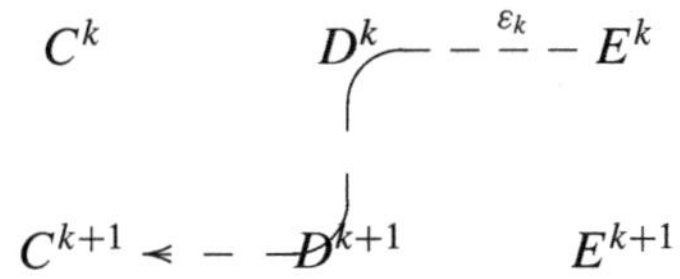

To obtain the long exact sequence we desire, we need to descend to the quotient of D^k with respect to the subspace $\bigoplus_{m=1}^{M} w_m$ (spanned by the Type 10 bar chain maps) at each index k. A given w_m will appear both in D^k and in D^{k+1}, since both appear in the support of any given Type 10 bar chain map. If we call the quotient $D'^{\bullet}$, we essentially allow the connecting map to avoid the $D'^{\bullet}$ sequence altogether, suggested by the diagram

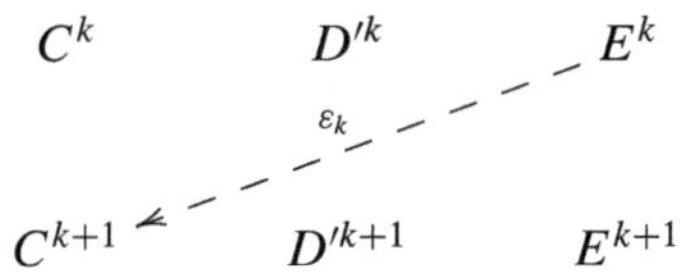

This truncates the length 4 bars to length 2 by shortening *both* ends by 1 index. □

Example 3.9 Consider the following exact sequence of sequences, given by the diagram

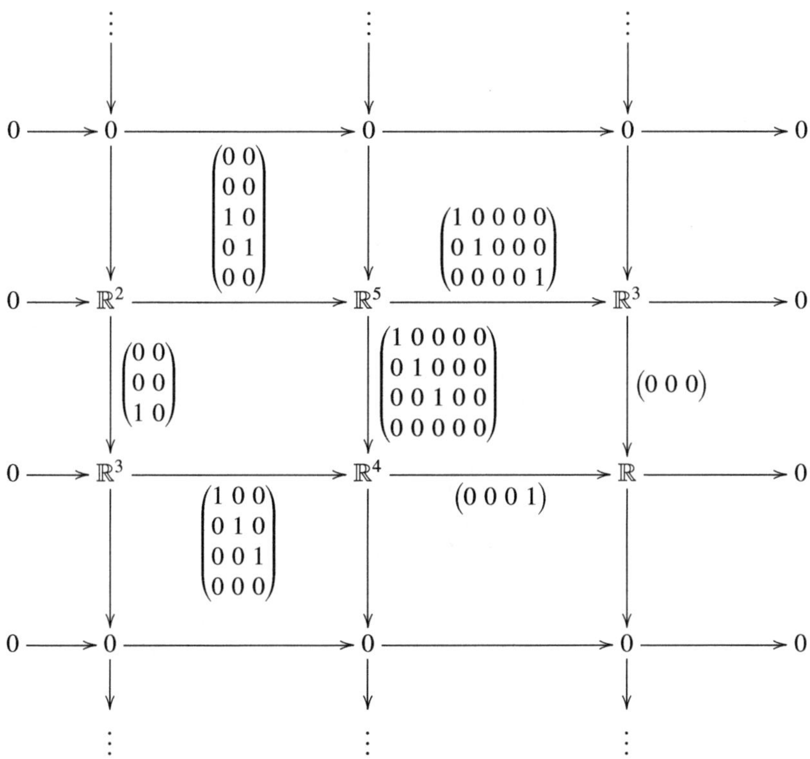

Let us verify that the rows of this diagram are both exact. Notice that all of the boundary matrices in both nontrivial rows are of full rank. In the first nontrivial row, the third and fourth rows of the first nontrivial matrix are nonzero. These fall into the third and fourth columns of the next matrix, which are the only zero columns. Thus,

$$\text{image}\begin{pmatrix} 0 & 0 \\ 0 & 0 \\ 1 & 0 \\ 0 & 1 \\ 0 & 0 \end{pmatrix} = \ker\begin{pmatrix} 1 & 0 & 0 & 0 & 0 \\ 0 & 1 & 0 & 0 & 0 \\ 0 & 0 & 0 & 0 & 1 \end{pmatrix}.$$

A similar argument establishes that the bottom row is also exact.

Given that the rows are exact, the diagram satisfies the hypotheses of Lemma 3.1. Let us compute the connecting sequence map, following the construction from the proof of Lemma 3.1. To this end, we need to decompose that last nontrivial space $\mathbb{R}^3$ in the first nontrivial row based on the bar chain maps of Types 7, 9, and 10. Since the component of chain map starting at that space is a zero map $\mathbb{R}^3 \to \mathbb{R}$, there are no bar chain maps of Type 9 involved in the decomposition.

Since the boundary map into $\mathbb{R}^3$ is surjective, let us examine the preimages of a basis

$$z_1 = \begin{pmatrix} 1 \\ 0 \\ 0 \end{pmatrix}, \; z_2 = \begin{pmatrix} 0 \\ 1 \\ 0 \end{pmatrix}, \text{ and } z_3 = \begin{pmatrix} 0 \\ 0 \\ 1 \end{pmatrix}$$

of $\mathbb{R}^3$ in $\mathbb{R}^5$. These preimages are

$$w_1 = \begin{pmatrix} 1 \\ 0 \\ 0 \\ 0 \\ 0 \end{pmatrix}, \; w_2 = \begin{pmatrix} 0 \\ 1 \\ 0 \\ 0 \\ 0 \end{pmatrix}, \text{ and } u_1 = \begin{pmatrix} 0 \\ 0 \\ 0 \\ 0 \\ 1 \end{pmatrix},$$

respectively. Following these along the middle component map into $\mathbb{R}^4$, the first two, w_1 and w_2, have nontrivial images,

$$w_1' = \begin{pmatrix} 1 \\ 0 \\ 0 \\ 0 \end{pmatrix} \text{ and } w_2' = \begin{pmatrix} 0 \\ 1 \\ 0 \\ 0 \end{pmatrix},$$

while u_1 has a trivial image. Thus, the vector u_1 corresponds to a bar chain map of Type 7. Since we already showed that the rows are exact, the vectors w_1 and w_2 cannot correspond to anything but bar chain maps of Type 10. If we trace these back through the diagram

$$
\begin{array}{ccc}
 & w_i & \longrightarrow & z_i \\
 & \big\downarrow & & \\
z_i' & \longrightarrow \; w_i' & &
\end{array}
$$

for $i = 1, 2$, we have

$$z_1' = \begin{pmatrix} 1 \\ 0 \\ 0 \end{pmatrix} \text{ and } z_2' = \begin{pmatrix} 0 \\ 1 \\ 0 \end{pmatrix}.$$

Therefore, the connecting map $\mathbb{R}^3 \to \mathbb{R}^3$ is given by the matrix

$$\begin{pmatrix} 1 & 0 & 0 \\ 0 & 1 & 0 \\ 0 & 0 & 0 \end{pmatrix}.$$

We can extend the sequence map using this connecting map

$$
\begin{array}{ccccccccccc}
 & & \vdots & & \vdots & & \vdots & & \vdots & & \\
 & & \downarrow & & \downarrow & & \downarrow & & \downarrow & & \\
0 & \longrightarrow & \mathbb{R}^2 & \xrightarrow{\begin{pmatrix} 0 & 0 \\ 0 & 0 \\ 1 & 0 \\ 0 & 1 \\ 0 & 0 \end{pmatrix}} & \mathbb{R}^5 & \xrightarrow{\begin{pmatrix} 1 & 0 & 0 & 0 & 0 \\ 0 & 1 & 0 & 0 & 0 \\ 0 & 0 & 0 & 0 & 1 \end{pmatrix}} & \mathbb{R}^3 & \xrightarrow{\begin{pmatrix} 1 & 0 & 0 \\ 0 & 1 & 0 \\ 0 & 0 & 0 \end{pmatrix}} & \mathbb{R}^3 & \xrightarrow{\begin{pmatrix} 1 & 0 & 0 \\ 0 & 1 & 0 \\ 0 & 0 & 1 \\ 0 & 0 & 0 \end{pmatrix}} & \cdots \\
 & & \Big\downarrow {\scriptstyle \begin{pmatrix} 0 & 0 \\ 0 & 0 \\ 1 & 0 \end{pmatrix}} & & \Big\downarrow {\scriptstyle \begin{pmatrix} 1 & 0 & 0 & 0 & 0 \\ 0 & 1 & 0 & 0 & 0 \\ 0 & 0 & 1 & 0 & 0 \\ 0 & 0 & 0 & 0 & 0 \end{pmatrix}} & & \Big\downarrow {\scriptstyle (0\ 0\ 0)} & & \Big\downarrow & & \\
0 & \longrightarrow & \mathbb{R}^3 & \xrightarrow[\begin{pmatrix} 1 & 0 & 0 \\ 0 & 1 & 0 \\ 0 & 0 & 1 \\ 0 & 0 & 0 \end{pmatrix}]{} & \mathbb{R}^4 & \xrightarrow[(0\ 0\ 0\ 1)]{} & \mathbb{R} & \longrightarrow & 0 & \longrightarrow & \cdots \\
 & & \downarrow & & \downarrow & & \downarrow & & \downarrow & & \\
 & & \vdots & & \vdots & & \vdots & & \vdots & &
\end{array}
$$

The rows in this sequence map are no longer exact, since the bar w_i starting at index 2 is of length 4, and is essentially the result of "straightening out" the Type 10 bar chain map into a diagram

$$w_i \longrightarrow z_i \longrightarrow z_i' \longrightarrow w_i'.$$

To remedy the lack of exactness, the proof of Lemma 3.1 instructs us to perform a quotient to truncate the bars coming from bar chain maps of Type 10. In our particular diagram, this only impacts the top row at indices 2 and 5. This results in the sequence map

$$
\begin{array}{ccccccccccc}
 & & \vdots & & \vdots & & \vdots & & \vdots & & \\
 & & \downarrow & & \downarrow & & \downarrow & & \downarrow & & \\
0 & \longrightarrow & \mathbb{R}^2 & \xrightarrow{\begin{pmatrix}1&0\\0&1\\0&0\end{pmatrix}} & \mathbb{R}^3 & \xrightarrow{\begin{pmatrix}0&0&0\\0&0&0\\0&0&1\end{pmatrix}} & \mathbb{R}^3 & \xrightarrow{\begin{pmatrix}1&0&0\\0&1&0\\0&0&0\end{pmatrix}} & \mathbb{R}^3 & \xrightarrow{\begin{pmatrix}0&0&1\\0&0&0\end{pmatrix}} & \cdots \\
 & & \Big\downarrow{\scriptsize\begin{pmatrix}0&0\\0&0\\1&0\end{pmatrix}} & & \Big\downarrow{\scriptsize\begin{pmatrix}0&0&0\\0&0&0\\1&0&0\\0&0&0\end{pmatrix}} & & \Big\downarrow{\scriptsize\begin{pmatrix}0&0&0\end{pmatrix}} & & \Big\downarrow & & \\
0 & \longrightarrow & \mathbb{R}^3 & \xrightarrow[\begin{pmatrix}1&0&0\\0&1&0\\0&0&1\\0&0&0\end{pmatrix}]{} & \mathbb{R}^4 & \xrightarrow[\begin{pmatrix}0&0&0&1\end{pmatrix}]{} & \mathbb{R} & \longrightarrow & 0 & \longrightarrow & \cdots \\
 & & \downarrow & & \downarrow & & \downarrow & & \downarrow & & \\
 & & \vdots & & \vdots & & \vdots & & \vdots & &
\end{array}
$$

in which the rows are exact.

Exercise 3.17 Construct the connecting sequence map for the following exact sequence of sequences

$$
\begin{array}{ccccccccc}
 & & \vdots & & \vdots & & \vdots & & \\
 & & \downarrow & & \downarrow & & \downarrow & & \\
0 & \longrightarrow & \mathbb{R}^2 & \xrightarrow{\begin{pmatrix}1&0\\0&1\\0&0\end{pmatrix}} & \mathbb{R}^3 & \xrightarrow{\begin{pmatrix}0&0&1\end{pmatrix}} & \mathbb{R} & \longrightarrow & 0 \\
 & & \Big\downarrow{\scriptsize\begin{pmatrix}1&0\\0&0\\0&0\end{pmatrix}} & & \Big\downarrow{\scriptsize\begin{pmatrix}1&0&0\\0&0&1\\0&0&0\\0&0&0\\0&0&0\end{pmatrix}} & & \Big\downarrow{\scriptsize\begin{pmatrix}0\\0\end{pmatrix}} & & \\
0 & \longrightarrow & \mathbb{R}^3 & \xrightarrow[\begin{pmatrix}1&0&0\\0&1&0\\0&0&1\\0&0&0\\0&0&0\end{pmatrix}]{} & \mathbb{R}^5 & \xrightarrow[\begin{pmatrix}0&0&0&1&0\\0&0&0&0&1\end{pmatrix}]{} & \mathbb{R}^2 & \longrightarrow & 0 \\
 & & \downarrow & & \downarrow & & \downarrow & & \\
 & & \vdots & & \vdots & & \vdots & &
\end{array}
$$

Exercise 3.18 Construct the connecting sequence map for the following exact sequence of sequences

$$\begin{array}{ccccccccc}
 & & \vdots & & \vdots & & \vdots & & \\
 & & \downarrow & & \downarrow & & \downarrow & & \\
0 & \longrightarrow & \mathbb{R} & \xrightarrow{\begin{pmatrix} 0 \\ 1 \\ 0 \\ -2 \\ 2 \end{pmatrix}} & \mathbb{R}^5 & \xrightarrow{\begin{pmatrix} -6 & -4 & 5 & -7 & -5 \\ -64 & -60 & 32 & -52 & -22 \\ -68 & -68 & 35 & -55 & -21 \\ 4 & 0 & -3 & 5 & 5 \end{pmatrix}} & \mathbb{R}^4 & \longrightarrow & 0 \\
 & & \Big\downarrow \begin{pmatrix} -2 \\ 3 \\ -1 \end{pmatrix} & & \Big\downarrow \begin{pmatrix} 90 & 69 & -52.5 & 82.5 & 52.5 \\ 27 & 21 & -15 & 24 & 15 \\ 78 & 57 & -45 & 72 & 48 \\ 45 & 33 & -28.5 & 43.5 & 28.5 \end{pmatrix} & & \Big\downarrow \begin{pmatrix} -12 & -9 & 9 & -15 \end{pmatrix} & & \\
0 & \longrightarrow & \mathbb{R}^3 & \longrightarrow & \mathbb{R}^4 & \xrightarrow[\begin{pmatrix} 2 & -10 & 2 & -2 \end{pmatrix}]{} & \mathbb{R} & \longrightarrow & 0 \\
 & & \Big\downarrow & \begin{pmatrix} 8 & 8 & -1 \\ -3 & 0 & 3 \\ -3 & 3 & 6 \\ 20 & 11 & -10 \end{pmatrix} & \Big\downarrow & & \Big\downarrow & & \\
 & & \vdots & & \vdots & & \vdots & &
\end{array}$$

If we consider an exact sequence of chain complexes, then the induced maps on the homology sequences become worthy of study.

Proposition 3.3 *Suppose that*

$$0 \longrightarrow (C^\bullet, c^\bullet) \xrightarrow{\gamma^\bullet} (D^\bullet, d^\bullet) \xrightarrow{\delta^\bullet} (E^\bullet, e^\bullet) \longrightarrow 0$$

is an exact sequence of chain complexes. Then the induced maps

$$H^\bullet(C^\bullet, c^\bullet) \xrightarrow{(\gamma_*)^\bullet} H^\bullet(D^\bullet, d^\bullet) \xrightarrow{(\delta_*)^\bullet} H^\bullet(E^\bullet, e^\bullet)$$

satisfy

$$\text{image}(\gamma_*)^k = \ker(\delta_*)^k$$

for each $k \in \mathbb{Z}$.

Proof As in the proof of Lemma 3.1, we focus our attention on the diagram

$$\begin{array}{ccccccccc} & & \vdots & & \vdots & & \vdots & & \\ & & \downarrow & & \downarrow & & \downarrow & & \\ 0 & \longrightarrow & C^k & \xrightarrow{\gamma^k} & D^k & \xrightarrow{\delta^k} & E^k & \longrightarrow & 0 \\ & & \downarrow{\scriptstyle c^k} & & \downarrow{\scriptstyle d^k} & & \downarrow{\scriptstyle e^k} & & \\ 0 & \longrightarrow & C^{k+1} & \xrightarrow[\gamma^{k+1}]{} & D^{k+1} & \xrightarrow[\delta^{k+1}]{} & E^{k+1} & \longrightarrow & 0 \\ & & \downarrow & & \downarrow & & \downarrow & & \\ & & \vdots & & \vdots & & \vdots & & \end{array}$$

which has exact rows. The exactness ensures that if we interpret this diagram as a chain map, its decomposition will only contain bar chain maps of Types 3, 7, 9, and 10. The Type 9 maps do not impact the homology of any of the chain complexes under discussion, and so may be safely ignored. A bar chain map of Type 7 in this diagram is of the form

$$\begin{array}{ccccccccc} 0 & \longrightarrow & \mathbb{R} & \xrightarrow{\gamma^k} & \mathbb{R} & \xrightarrow{\delta^k} & 0 & \longrightarrow & 0 \\ & & \downarrow & & \downarrow & & \downarrow & & \\ 0 & \longrightarrow & 0 & \xrightarrow[\gamma^{k+1}]{} & 0 & \xrightarrow[\delta^{k+1}]{} & 0 & \longrightarrow & 0 \end{array}$$

or

$$\begin{array}{ccccccccc} 0 & \longrightarrow & 0 & \xrightarrow{\gamma^k} & \mathbb{R} & \xrightarrow{\delta^k} & \mathbb{R} & \longrightarrow & 0 \\ & & \downarrow & & \downarrow & & \downarrow & & \\ 0 & \longrightarrow & 0 & \xrightarrow[\gamma^{k+1}]{} & 0 & \xrightarrow[\delta^{k+1}]{} & 0 & \longrightarrow & 0 \end{array}$$

and so corresponds to a nontrivial basis element x of $H^k(C^\bullet)$ or $H^k(D^\bullet)$ being mapped to $H^k(D^\bullet)$ or $H^k(E^\bullet)$ respectively. Upon descending to the quotient (to compute homology sequences), a bar chain map of Type 7 is unchanged—becoming a bar of length 2 in the sequence of homology spaces. For instance, this means that if $x \in H^k(C^\bullet)$, it will be in both the image of $(\gamma_*)^k$ (since it is in the image of γ^k) and the kernel of $(\delta_*)^k$ (since it is in the kernel of δ^k). On the other hand, if $x \in H^k(D^\bullet)$, then it is not in the kernel of $(\delta_*)^k$ and is also not in the image of $(\gamma_*)^k$. The situation is the same for bar chain maps of Type 3, but with $k+1$ replacing k.

It remains to examine the situation of a Type 10 bar chain map, which has the form

$$\begin{array}{ccccccccc}
0 & \longrightarrow & 0 & \xrightarrow{\gamma^k} & \mathbb{R} & \xrightarrow{\delta^k} & \mathbb{R} & \longrightarrow & 0 \\
 & & \downarrow & & \downarrow{\scriptstyle d^k} & & \downarrow & & \\
0 & \longrightarrow & \mathbb{R} & \xrightarrow[\gamma^{k+1}]{} & \mathbb{R} & \xrightarrow[\delta^{k+1}]{} & 0 & \longrightarrow & 0
\end{array}$$

This kind of bar chain map corresponds only to trivial elements in $H^k(D^\bullet)$ or $H^{k+1}(D^\bullet)$, and so impacts neither the kernel of $(\delta_*)^k$ nor the image of $(\gamma_*)^k$. □

Exercise 3.19 Prove or give a counterexample to the following statement: If

$$0 \longrightarrow (C^\bullet, c^\bullet) \xrightarrow{\gamma^\bullet} (D^\bullet, d^\bullet) \xrightarrow{\delta^\bullet} (E^\bullet, e^\bullet) \longrightarrow 0$$

is a short exact sequence of chain complexes, then it is always the case that the following sequence of induced maps

$$0 \longrightarrow H^k(C^\bullet, c^\bullet) \xrightarrow{(\gamma_*)^k} H^k(D^\bullet, d^\bullet) \xrightarrow{(\delta_*)^k} H^k(E^\bullet, e^\bullet) \longrightarrow 0$$

is an exact sequence of vector spaces.

Theorem 3.2 (the snake lemma for chain complexes) *Suppose that*

$$0 \longrightarrow (C^\bullet, c^\bullet) \xrightarrow{\gamma^\bullet} (D^\bullet, d^\bullet) \xrightarrow{\delta^\bullet} (E^\bullet, e^\bullet) \longrightarrow 0$$

is an exact sequence of chain complexes. Then the connecting map $\varepsilon^\bullet : (E^\bullet, e^\bullet) \to (C^{\bullet+1}, c^{\bullet+1})$ *induces a long exact sequence*

$$\cdots \to H^k(C^\bullet, c^\bullet) \longrightarrow H^k(D^\bullet, d^\bullet) \longrightarrow H^k(E^\bullet, e^\bullet) \xrightarrow{(\varepsilon_*)^k} H^{k+1}(C^\bullet, c^\bullet) \to \cdots$$

of homology vector spaces.

Notice that $H^k(D^\bullet, d^\bullet)$ appears in the long exact sequence, not $H^k(D'^\bullet, d'^\bullet)$; the proof establishes that these sequences have the same homology.

Proof From Lemma 3.1, we can conclude that there is a chain map whose components are given by the vertical maps in the diagram

$$
\begin{array}{ccccccccccc}
\cdots \to E^{n-k-1} & \xrightarrow{\varepsilon^{n-k-1}} & C^{n-k} & \xrightarrow{\gamma^{n-k}} & D'^{n-k} & \xrightarrow{\delta^{n-k}} & E^{n-k} & \xrightarrow{\varepsilon^{n-k}} & C^{n-k+1} & \xrightarrow{\gamma^{n-k+1}} & \cdots \\
\downarrow e^{n-k-1} & & \downarrow c^{n-k} & & \downarrow d'^{n-k} & & \downarrow e^{n-k} & & \downarrow c^{n-k+1} & & \\
\cdots \to E^{n-k} & \xrightarrow[\varepsilon^{n-k}]{} & C^{n+1-k} & \xrightarrow[\gamma^{n+1-k}]{} & D'^{n+1-k} & \xrightarrow[\delta^{n+1-k}]{} & E^{n+1-k} & \xrightarrow[\varepsilon^{n+1-k}]{} & C^{n-k+2} & \xrightarrow[\gamma^{n-k+2}]{} & \cdots
\end{array}
$$

in which the rows are exact for each integer n and k. From the proof of Lemma 3.1, the only nontrivial bar chain maps that can occur in the decomposition of this chain map are of Types 3, 7, 9, and 10. As in the proof of Proposition 3.3, bar chain maps of Type 9 have no impact on homology, and may be ignored. Furthermore, the construction of the connecting sequence map transforms each bar chain map of Type 10 that starts at C^{n+1-k} in the bottom row in the original short exact sequence into one of Type 3 (starting at E^{n-k} in the bottom row) and one of Type 7 (starting at E^{n-k} in the top row) in the above diagram. Thus the above diagram decomposes into bar chain maps of Types 3, 7, and 9. This means that the sequence of induced maps

$$
\cdots \to H^k(C^\bullet, c^\bullet) \longrightarrow H^k(D'^\bullet, d'^\bullet) \longrightarrow H^k(E^\bullet, e^\bullet) \xrightarrow{(\varepsilon_*)^k} H^{k+1}(C^\bullet, c^\bullet) \to \cdots
$$

is exact without further work.

To obtain the final statement, we need to establish that $H^k(D^\bullet, d^\bullet) \cong H^k(D'^\bullet, d'^\bullet)$. Notice that the only difference between $D'^\bullet$ and $D^\bullet$ occurs when there was a Type 10 bar chain map in the original short exact sequence. We constructed $D'^\bullet$ by deleting length 2 bars from $D^\bullet$, which does not change the homology. □

From the proof of Theorem 3.2, we have a clear interpretation of Type 10 maps that is also useful in calculations.

Corollary 3.2 *The connecting map* $(\varepsilon_*)^k : H^k(E^\bullet, e^\bullet) \to H^{k+1}(C^\bullet, c^\bullet)$ *in Theorem 3.2 is nontrivial if and only if the short exact sequence*

$$
\begin{array}{ccccccccc}
0 & \longrightarrow & C^k & \longrightarrow & D^k & \longrightarrow & E^k & \longrightarrow & 0 \\
 & & \downarrow & & \downarrow & & \downarrow & & \\
0 & \longrightarrow & C^{k+1} & \longrightarrow & D^{k+1} & \longrightarrow & E^{k+1} & \longrightarrow & 0
\end{array}
$$

contains a bar chain map of Type 10.

Exercise 3.20 Given the following short exact sequence of chain complexes, in which the sequence maps are horizontal in the diagram

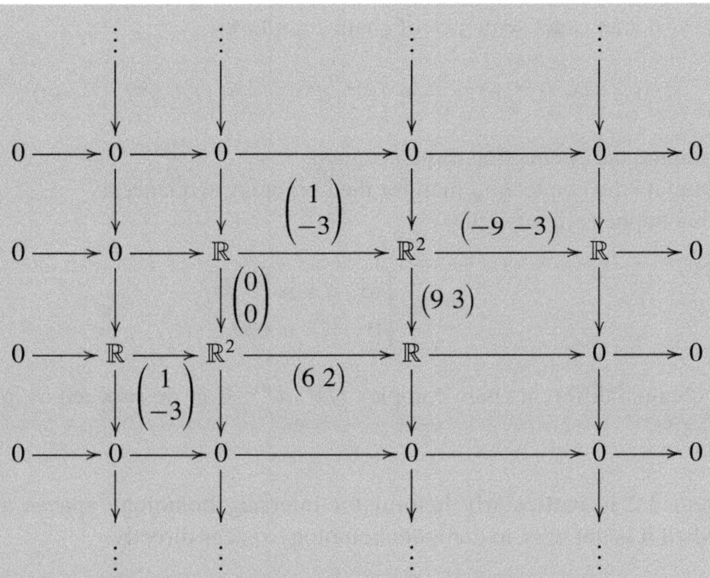

compute all of the nontrivial connecting maps.

Exercise 3.21 Suppose that

$$C^k = \begin{cases} \mathbb{R} & \text{if } k \text{ is even,} \\ 0 & \text{if } k \text{ is odd,} \end{cases}$$

$$E^k = \begin{cases} 0 & \text{if } k \text{ is even,} \\ \mathbb{R} & \text{if } k \text{ is odd,} \end{cases}$$

and

$$D^k = \mathbb{R}$$

for all k. Under these definitions, we can form chain complexes $(C^\bullet, c^\bullet)$ and $(E^\bullet, e^\bullet)$ in exactly one way. Define

$$d^k = \begin{cases} 0 & \text{if } k \text{ is even,} \\ \text{id} & \text{if } k \text{ is odd,} \end{cases}$$

so that $(D^\bullet, d^\bullet)$ is also a chain complex.

1. Construct an exact sequence of chain complexes

$$0 \longrightarrow (C^\bullet, c^\bullet) \longrightarrow (D^\bullet, d^\bullet) \longrightarrow (E^\bullet, e^\bullet) \longrightarrow 0.$$

2. Compute the connecting sequence map.
3. Compute the connecting map for the homology sequences.
4. What happens if we define

$$d'^k = \begin{cases} \text{id} & \text{if } k \text{ is even,} \\ 0 & \text{if } k \text{ is odd,} \end{cases}$$

to obtain a different chain complex $(D^\bullet, d'^\bullet)$? Can we proceed as in parts (1)–(3)?

Theorem 3.2 is particularly helpful for inferring homology spaces at certain indices when it is not easy to compute homology spaces directly.

Example 3.10 Consider the following short exact sequence of chain complexes, in which the sequence maps are horizontal in the diagram

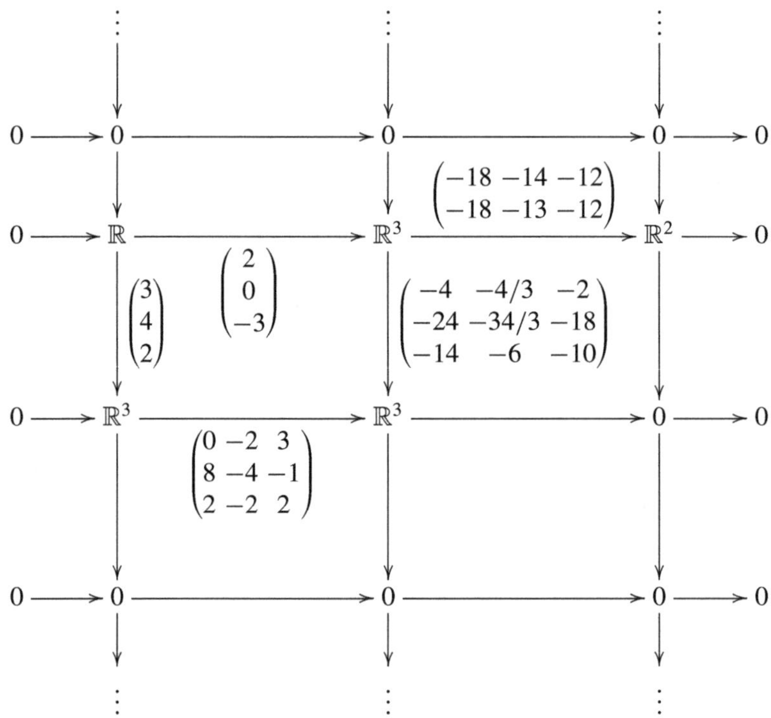

Although the numerical values of the component maps look pretty unpleasant, a little calculation reveals that the matrix

$$\begin{pmatrix} -4 & -4/3 & -2 \\ -24 & -34/3 & -18 \\ -14 & -6 & -10 \end{pmatrix}$$

is of full rank. Therefore, the middle column is an exact sequence, which implies it has trivial homology. Since the rows are all exact as well, Theorem 3.2 implies that the right column is quasi-isomorphic to the left column after a shift. On the other hand, the right column's homology is evident simply because it has exactly one nontrivial term. Hence, the homology of the left column must also be nontrivial in exactly one index. This fact is somewhat obscured from the outset, though one can recognize that a bar of length 2 begins in the first nontrivial index for the left column (the second row).

Exercise 3.22 Suppose that the following is a short exact sequence of chain complexes

$$0 \longrightarrow (C^\bullet, c^\bullet) \longrightarrow (D^\bullet, d^\bullet) \longrightarrow (E^\bullet, e^\bullet) \longrightarrow 0.$$

If you know that

1. $H^2(C^\bullet, c^\bullet) \cong \mathbb{R}^2$ and $H^1(D^\bullet, d^\bullet) \cong \mathbb{R}^3$ are the only nontrivial homology spaces appearing in $H^\bullet(C^\bullet, c^\bullet)$ and $H^\bullet(D^\bullet, d^\bullet)$, compute $H^1(E^\bullet)$.
2. Compute $H^k(D^\bullet)$ if $H^k(C^\bullet) \cong \mathbb{R}$ for every k, but $(D^\bullet, d^\bullet)$ is exact.
3. What can you conclude if $H^k(E^\bullet, e^\bullet) \cong 0$ for every even k?

Exercise 3.23 Prove Theorem 3.2 directly by way of a diagram chase on

$$\begin{array}{ccccccccc}
 & & \vdots & & \vdots & & \vdots & & \\
 & & \downarrow{\scriptstyle c^{k-1}} & & \downarrow{\scriptstyle d^{k-1}} & & \downarrow{\scriptstyle e^{k-1}} & & \\
0 & \longrightarrow & C^k & \xrightarrow{\gamma^k} & D^k & \xrightarrow{\delta^k} & E^k & \longrightarrow & 0 \\
 & & \downarrow{\scriptstyle c^{k}} & & \downarrow{\scriptstyle d^{k}} & & \downarrow{\scriptstyle e^{k}} & & \\
0 & \longrightarrow & C^{k+1} & \xrightarrow{\gamma^{k+1}} & D^{k+1} & \xrightarrow{\delta^{k+1}} & E^{k+1} & \longrightarrow & 0 \\
 & & \downarrow{\scriptstyle c^{k+1}} & & \downarrow{\scriptstyle d^{k+1}} & & \downarrow{\scriptstyle e^{k+1}} & & \\
 & & \vdots & & \vdots & & \vdots & &
\end{array}$$

Specifically, start with an element x of E^k in the kernel of e^k and show that the connecting map takes it to an element of the kernel of c^{k+1}. Then show that every element y of E^k in the image of e^{k-1} maps to an element of the image of c^k. (You will be recounting the construction of the connecting map!)

Exercise 3.24 Here is another form of Theorem 3.2. Consider the commutative diagram of vector spaces

$$\begin{array}{ccccccc} & & A & \longrightarrow & B & \longrightarrow & C & \longrightarrow & 0 \\ & & \downarrow{\scriptstyle a} & & \downarrow{\scriptstyle b} & & \downarrow{\scriptstyle c} & & \\ 0 & \longrightarrow & D & \longrightarrow & E & \longrightarrow & F & & \end{array}$$

in which the rows are exact. Show that this diagram induces an exact sequence

$$\ker a \longrightarrow \ker b \longrightarrow \ker c \longrightarrow \operatorname{coker} a \longrightarrow \operatorname{coker} b \longrightarrow \operatorname{coker} c$$

using a diagram chase. Challenge: prove this same statement, but where A, B, C, D, E, and F are chain complexes.

3.4 Quasi-Isomorphisms

Since every chain complex decomposes into an exact sequence and a homology sequence, and chain maps induce maps on the homology sequence, it is useful to explore the classes of chain complexes that have the same homology. We already know that chain complexes that are isomorphic as sequences have the same homology. In this section, we introduce two other equivalence relations on chain complexes that preserve homology and explore some of their elementary properties. These three equivalences are listed in Table 3.3. All chain complexes in an equivalence class for each of the three relations listed in Table 3.3 have the same homology, but the size of

Table 3.3 Types of equivalences between chain complexes $(C^\bullet, c^\bullet)$ and $(D^\bullet, d^\bullet)$

Assume there exist chain maps $f^\bullet : (C^\bullet, c^\bullet) \to (D^\bullet, d^\bullet)$ and $g^\bullet : (D^\bullet, d^\bullet) \to (C^\bullet, c^\bullet)$		
Chain isomorphism (Definition 2.1 in Chap. 2)	Homotopy equivalence (Definition 3.6)	Quasi-isomorphism (Definition 3.3)
$f^\bullet \circ g^\bullet = \mathrm{id}_{(D^\bullet, d^\bullet)}$ and $g^\bullet \circ f^\bullet = \mathrm{id}_{(C^\bullet, c^\bullet)}$	$f^\bullet \circ g^\bullet \simeq \mathrm{id}_{(D^\bullet, d^\bullet)}$ and $g^\bullet \circ f^\bullet \simeq \mathrm{id}_{(C^\bullet, c^\bullet)}$	$f^\bullet_* : H^\bullet(C^\bullet, c^\bullet) \to H^\bullet(D^\bullet, d^\bullet)$ is an isomorphism (Equivalently, $g^\bullet_*$ is an isomorphism)

the equivalence class increases moving from left to right in the table. The relations are nested, in the sense that every pair of chain complexes that are chain isomorphic are also *homotopy equivalent*. Every pair of *homotopy equivalent* chain complexes are also *quasi-isomorphic*. Chain isomorphism is the finest equivalence relation (smallest equivalence class) and *quasi-isomorphism* is the coarsest[2] equivalence relation (largest equivalence class).

Definition 3.3 A *quasi-isomorphism* is a chain map whose induced map on the homology sequence is an isomorphism. Two chain complexes are said to be *quasi-isomorphic* if there is a quasi-isomorphism between them.

Proposition 3.4 *Isomorphic chain complexes are also quasi-isomorphic.*

Proof Corollary 3.1 states that isomorphisms between chain complexes decompose into bar chain map Types 1, 5, and 9. On the other hand, the proof of Proposition 3.2 implies that induced maps on the homology sequence decompose into Types 1, 2, 4, and 5. This means that isomorphisms between chain complexes induce bar chain maps of Type 1 or 5 on the homology sequence, both of which are vector space isomorphisms. □

The converse of Proposition 3.4 is not true because quasi-isomorphisms do not necessarily have inverses. That is, if $q^\bullet : (C^\bullet, c^\bullet) \to (D^\bullet, d^\bullet)$ is a quasi-isomorphism, there may not exist a chain map $r^\bullet : (D^\bullet, d^\bullet) \to (C^\bullet, c^\bullet)$ with $r^\bullet \circ q^\bullet = \mathrm{id}_{(C^\bullet, c^\bullet)}$ and $q^\bullet \circ r^\bullet = \mathrm{id}_{(D^\bullet, d^\bullet)}$.

Example 3.11 Any bar chain map of Type 7 is a quasi-isomorphism trivially, since both the domain and codomain have trivial homology. Such a bar chain map is not an isomorphism, and there is no chain map that is its inverse.

Example 3.12 The following are two chain complexes that are evidently not isomorphic, because the spaces are of different dimensions:

$$(C^\bullet, c^\bullet) := 0 \longrightarrow \mathbb{R}^4 \xrightarrow{\begin{pmatrix} 1 & 0 & 0 & -1 \\ -1 & 1 & 0 & 0 \\ 0 & -1 & 1 & 0 \\ 0 & 0 & -1 & 1 \end{pmatrix}} \mathbb{R}^4 \longrightarrow 0.$$

and

$$(D^\bullet, d^\bullet) := 0 \longrightarrow \mathbb{R}^3 \xrightarrow{\begin{pmatrix} 1 & 0 & -1 \\ -1 & 1 & 0 \\ 0 & -1 & 1 \end{pmatrix}} \mathbb{R}^3 \longrightarrow 0$$

[2] Although we show in Theorem 3.4 that *homotopy equivalence* and *quasi-isomorphism* are identical for chain complexes of vector spaces, we will continue to distinguish them because they are differ under more general settings, as will be discussed in Chap. 6.

These two chain complexes are quasi-isomorphic, though, since the following is a chain map

$$
\begin{array}{ccccccc}
0 & \longrightarrow & \mathbb{R}^4 & \xrightarrow{\begin{pmatrix} 1 & 0 & 0 & -1 \\ -1 & 1 & 0 & 0 \\ 0 & -1 & 1 & 0 \\ 0 & 0 & -1 & 1 \end{pmatrix}} & \mathbb{R}^4 & \longrightarrow & 0 \\
 & & \Big\downarrow \begin{pmatrix} 1 & 0 & 0 & 0 \\ 0 & 1 & 0 & 0 \\ 0 & 0 & 0 & 1 \end{pmatrix} & & \Big\downarrow \begin{pmatrix} 1 & 0 & 0 & 0 \\ 0 & 1 & 0 & 0 \\ 0 & 0 & 1 & 1 \end{pmatrix} & & \\
0 & \longrightarrow & \mathbb{R}^3 & \xrightarrow[\begin{pmatrix} 1 & 0 & -1 \\ -1 & 1 & 0 \\ 0 & -1 & 1 \end{pmatrix}]{} & \mathbb{R}^3 & \longrightarrow & 0
\end{array}
$$

that induces isomorphisms on the homology sequence. To see this, first let us verify that the diagram commutes,

$$
\begin{pmatrix} 1 & 0 & 0 & 0 \\ 0 & 1 & 0 & 0 \\ 0 & 0 & 1 & 1 \end{pmatrix}
\begin{pmatrix} 1 & 0 & 0 & -1 \\ -1 & 1 & 0 & 0 \\ 0 & -1 & 1 & 0 \\ 0 & 0 & -1 & 1 \end{pmatrix}
=
\begin{pmatrix} 1 & 0 & 0 & -1 \\ -1 & 1 & 0 & 0 \\ 0 & -1 & 0 & 1 \end{pmatrix},
$$

and

$$
\begin{pmatrix} 1 & 0 & -1 \\ -1 & 1 & 0 \\ 0 & -1 & 1 \end{pmatrix}
\begin{pmatrix} 1 & 0 & 0 & 0 \\ 0 & 1 & 0 & 0 \\ 0 & 0 & 0 & 1 \end{pmatrix}
=
\begin{pmatrix} 1 & 0 & 0 & -1 \\ -1 & 1 & 0 & 0 \\ 0 & -1 & 0 & 1 \end{pmatrix}.
$$

Now as for the homology,

$$
H^1(D^\bullet) = \ker d^1 / \operatorname{image} d^0 = \ker d^1 = \operatorname{span}\left\{ \begin{pmatrix} 1 \\ 1 \\ 1 \\ 1 \end{pmatrix} \right\},
$$

$$
H^2(D^\bullet) = \ker d^2 / \operatorname{image} d^1 = \operatorname{coker} d^1 = \operatorname{span}\left\{ \begin{pmatrix} 1 \\ 1 \\ 1 \\ 1 \end{pmatrix} \right\},
$$

and

$$H^1(C^\bullet) = \ker c^1 / \operatorname{image} c^0 = \ker c^1 = \operatorname{span}\left\{\begin{pmatrix}1\\1\\1\end{pmatrix}\right\},$$

$$H^2(C^\bullet) = \ker c^2 / \operatorname{image} c^1 = \operatorname{coker} c^1 = \operatorname{span}\left\{\begin{pmatrix}1\\1\\1\end{pmatrix}\right\}.$$

Observe that since

$$\begin{pmatrix}1&0&0&0\\0&1&0&0\\0&0&0&1\end{pmatrix}\begin{pmatrix}1\\1\\1\\1\end{pmatrix} = \begin{pmatrix}1\\1\\1\end{pmatrix},$$

this means that the chain map induces an isomorphism on H^1.

For H^2,

$$\begin{aligned}\begin{pmatrix}1&0&0&0\\0&1&0&0\\0&0&1&1\end{pmatrix}\begin{pmatrix}1\\1\\1\\1\end{pmatrix} &= \begin{pmatrix}1\\1\\2\end{pmatrix}\\ &= \frac{4}{3}\begin{pmatrix}1\\1\\1\end{pmatrix} + \begin{pmatrix}1&0&-1\\-1&1&0\\0&-1&1\end{pmatrix}\begin{pmatrix}0\\-1/3\\1/3\end{pmatrix}\\ &\in \frac{4}{3}\begin{pmatrix}1\\1\\1\end{pmatrix} + \operatorname{image} d^1,\end{aligned}$$

meaning that the chain map induces an isomorphism on H^2 as well. All of the other homology spaces are trivial so there is nothing else to prove.

Proposition 3.5 *Quasi-isomorphisms are an equivalence relation on chain complexes. Specifically,*

Reflexivity The identity chain map $\mathrm{id}_{(C^\bullet,c^\bullet)} : (C^\bullet, c^\bullet) \to (C^\bullet, c^\bullet)$ *for any chain complex is a quasi-isomorphism,*

Symmetry If there is a quasi-isomorphism $q_1^\bullet : (C^\bullet, c^\bullet) \to (D^\bullet, d^\bullet)$*, then there is also a quasi-isomorphism* $q_2^\bullet : (D^\bullet, d^\bullet) \to (C^\bullet, c^\bullet)$*, and*

Transitivity If there is a quasi-isomorphism $q_1^\bullet : (C^\bullet, c^\bullet) \to (D^\bullet, d^\bullet)$ *and a quasi-isomorphism* $q_2^\bullet : (D^\bullet, d^\bullet) \to (E^\bullet, e^\bullet)$*, then their composition* $q_2^\bullet \circ q_1^\bullet : (C^\bullet, c^\bullet) \to (E^\bullet, e^\bullet)$ *is also a quasi-isomorphism.*

Proof We proceed constructively. Since every quasi-isomorphism is also a chain map we can treat it via its decomposition into bar chain maps.

Reflexivity The identity chain map $\mathrm{id}_{(C^\bullet, c^\bullet)} : (C^\bullet, c^\bullet) \to (C^\bullet, c^\bullet)$ for any chain complex is an isomorphism. Proposition 3.4 implies that it is also a quasi-isomorphism.

Symmetry Since $q_1^\bullet$ is a quasi-isomorphism, this means that $(C^\bullet, c^\bullet)$ and $(D^\bullet, d^\bullet)$ have isomorphic homology sequences. In order to construct a quasi-isomorphism $q_2^\bullet : (D^\bullet, d^\bullet) \to (C^\bullet, c^\bullet)$, it is not necessary to construct an inverse to $q_1^\bullet$. (An inverse to $q_1^\bullet$ might not exist!) Suppose that we decompose our two chain complexes into their homology and exact parts so that

$$(C^\bullet, c^\bullet) \cong H^\bullet(C^\bullet, c^\bullet) \oplus (E^\bullet, e^\bullet)$$

with $(E^\bullet, e^\bullet)$ being exact, and

$$(D^\bullet, d^\bullet) \cong H^\bullet(D^\bullet, d^\bullet) \oplus (F^\bullet, f^\bullet)$$

with $(F^\bullet, f^\bullet)$ being exact. The fact that $q_1^\bullet$ is a quasi-isomorphism means that

$$(q_1)_*^\bullet : H^\bullet(C^\bullet, c^\bullet) \to H^\bullet(D^\bullet, d^\bullet)$$

is an isomorphism. In other words, there exists an inverse

$$((q_1)_*^\bullet)^{-1} : H^\bullet(D^\bullet, d^\bullet) \to H^\bullet(C^\bullet, c^\bullet).$$

We can therefore construct a chain map componentwise: for each k, let

$$q_2^k := (q_1^k)^{-1} \oplus 0_{F^k, E^k} : \left(H^k(D^\bullet, d^\bullet) \oplus (F^k, f^k)\right) \to \left(H^k(C^\bullet, c^\bullet) \oplus (E^k, e^k)\right).$$

Although this definition essentially ignores the exact parts of the two chain complexes, it induces the isomorphism $(q_1)_*^\bullet$ on the homology sequences. Therefore, $q_2^\bullet$ is a quasi-isomorphism.

Transitivity Notice that $q_2^\bullet \circ q_1^\bullet$ is a chain map, and that both $q_1^\bullet$ and $q_2^\bullet$ induce an isomorphism on the homology sequence. Their composition must also induce an isomorphism on the homology sequence. □

Theorem 3.3 (quasi-isomorphism is equivalent to isomorphic homology for vector spaces) *Two chain complexes are quasi-isomorphic if and only if they have isomorphic homology.*

Together with Proposition 3.5, this theorem means that quasi-isomorphism equivalence classes consist of those chain complexes that all have the same homology.

Proof The definition states that quasi-isomorphic chain complexes have isomorphic homology. To prove this theorem we need to show the converse: if two chain complexes have isomorphic homology, then there is a chain map that induces an isomorphism on the homology sequence.

Suppose that we have two chain complexes $(C^\bullet, c^\bullet)$ and $(D^\bullet, d^\bullet)$ with

$$H^k(C^\bullet) \cong H^k(D^\bullet)$$

for all $k \in \mathbb{Z}$. Without loss of generality, we may use Theorem 2.1 in Chap. 2 (the barcode decomposition) and assume that both chain complexes are actually barcodes. The fact that the homologies agree means that there is a bijection ϕ between the length 1 bars in $(C^\bullet, c^\bullet)$ and the length 1 bars in $(D^\bullet, d^\bullet)$. We can then proceed to construct the required quasi-isomorphism, assembling a chain map as a direct sum of bar chain maps of Types 3, 5, 7 only. The isomorphism on the homology sequence is carried by the Type 5 bar chain maps, matching up the length 1 bars according to the bijection ϕ. The bar chain maps of Types 3 and 7 mean that we do not bother to match the length 2 bars from one complex to the other. □

Although the proof of Theorem 3.3 is constructive, it is quite non-canonical! That is, there may be several chain maps that induce the same isomorphism on the homology sequence, each with very different decompositions into bar chain maps. For instance, the next example shows a different quasi-isomorphism for the chain complexes shown in Example 3.12.

Example 3.13 Let us use the construction from the proof of Theorem 3.3 for the chain complexes shown in Example 3.12. Applying Algorithm 1 in Chap. 2 to the two chain complexes, we have that

$$(C^\bullet, c^\bullet) \cong \mathbb{R}^{(1,1)} \oplus \mathbb{R}^{(1,2)} \oplus \mathbb{R}^{(1,2)} \oplus \mathbb{R}^{(1,2)} \oplus \mathbb{R}^{(2,2)}.$$

and

$$(D^\bullet, d^\bullet) \cong \mathbb{R}^{(1,1)} \oplus \mathbb{R}^{(1,2)} \oplus \mathbb{R}^{(1,2)} \oplus \mathbb{R}^{(2,2)},$$

The construction requests us to use

- One Type 5 bar chain map starting at index 1,
- Two Type 3 bar chain maps starting at index 1,
- Three Type 7 bar chain maps starting at index 1, and
- One Type 5 bar chain map starting at index 2.

Assembling these pieces, we obtain

$$\begin{array}{ccccccc}
0 & \longrightarrow & \mathbb{R}^4 & \xrightarrow{\begin{pmatrix} 0&1&0&0\\ 0&0&1&0\\ 0&0&0&1\\ 0&0&0&0 \end{pmatrix}} & \mathbb{R}^4 & \longrightarrow & 0 \\
 & & \Big\downarrow \begin{pmatrix} 1&0&0&0\\ 0&0&0&0\\ 0&0&0&0 \end{pmatrix} & & \Big\downarrow \begin{pmatrix} 0&0&0&0\\ 0&0&0&0\\ 0&0&0&1 \end{pmatrix} & & \\
0 & \longrightarrow & \mathbb{R}^3 & \xrightarrow[\begin{pmatrix} 0&1&0\\ 0&0&1\\ 0&0&0 \end{pmatrix}]{} & \mathbb{R}^3 & \longrightarrow & 0,
\end{array}$$

which is—up to sequence isomorphism—a chain map that induces an isomorphism on the homology sequence. The two sequences in this chain map are isomorphic (but are not equal) to the original $(C^\bullet, c^\bullet)$ and $(D^\bullet, d^\bullet)$ chain complexes. The ranks of the component maps of this chain map are both 1. In contrast, the component ranks for the chain map shown in Example 3.12 are both 3.

Exercise 3.25 Are the following two chain complexes quasi-isomorphic?

$$0 \to \mathbb{R}^2 \xrightarrow{\begin{pmatrix} 6 & 6 \\ -9 & 0 \\ 3 & 2 \\ 6 & 4 \end{pmatrix}} \mathbb{R}^4 \xrightarrow{\begin{pmatrix} -12 & 4 & -20 & 28 \end{pmatrix}} \mathbb{R} \xrightarrow{(0)} \mathbb{R} \to 0$$

and

$$0 \to \mathbb{R} \xrightarrow{\begin{pmatrix} -1 \\ 2 \\ 0 \end{pmatrix}} \mathbb{R}^3 \xrightarrow{\begin{pmatrix} 6 & 3 & 0 \\ 12 & 6 & 0 \end{pmatrix}} \mathbb{R}^2 \xrightarrow{\begin{pmatrix} 6 & -3 \\ 0 & 0 \\ -2 & 1 \end{pmatrix}} \mathbb{R}^3 \xrightarrow{\begin{pmatrix} -2 & 4 & -6 \end{pmatrix}} \mathbb{R} \to 0$$

If not, at which index do the chain complexes have different homology? If they are quasi-isomorphic, exhibit two distinct chain maps that induce an isomorphism on the homology sequence.

Exercise 3.26 Are the following two chain complexes quasi-isomorphic?

$$0 \longrightarrow \mathbb{R}^2 \xrightarrow{\begin{pmatrix} 6 & 0 \end{pmatrix}} \mathbb{R} \longrightarrow 0$$

and

$$0 \longrightarrow \mathbb{R} \xrightarrow{(2)} \mathbb{R} \xrightarrow{\begin{pmatrix} 0 \\ 0 \\ 0 \\ 0 \end{pmatrix}} \mathbb{R}^4 \xrightarrow{\begin{pmatrix} -8 & -32 & 34 & -18 \\ 6 & 16 & -18 & 10 \\ -3 & -6 & 7 & -4 \end{pmatrix}} \mathbb{R}^3 \longrightarrow 0$$

If not, at which index do the chain complexes have different homology? If they are quasi-isomorphic, exhibit two distinct chain maps that induce an isomorphism on the homology sequence.

Exercise 3.27 Are the following two chain complexes quasi-isomorphic?

$$0 \to \mathbb{R}^2 \xrightarrow{\begin{pmatrix} 6 & 6 \\ -9 & 0 \\ 3 & 2 \\ 6 & 4 \end{pmatrix}} \mathbb{R}^4 \xrightarrow{\begin{pmatrix} -12 & 4 & -20 & 28 \end{pmatrix}} \mathbb{R} \xrightarrow{(0)} \mathbb{R} \to 0$$

and

$$0 \to \mathbb{R}^2 \xrightarrow{\begin{pmatrix} -4 & -2 \\ -5 & -4 \\ 7 & 5 \end{pmatrix}} \mathbb{R}^3 \xrightarrow{\begin{pmatrix} -3 & -6 & -6 \\ -2 & -4 & -4 \end{pmatrix}} \mathbb{R}^2 \xrightarrow{(0\ 0)} \mathbb{R} \to 0$$

If not, at which index do the chain complexes have different homology? If they are quasi-isomorphic, exhibit two distinct chain maps that induce an isomorphism on the homology sequence.

One important source of quasi-isomorphisms are *homotopy equivalences*. In order to define a homotopy equivalence, we need to develop the idea of a *chain homotopy* between chain maps, which is a particular kind of relation between chain maps.

Definition 3.4 A *chain homotopy* (or simply a *homotopy*) between two chain maps $f^\bullet, g^\bullet : (C^\bullet, c^\bullet) \to (D^\bullet, d^\bullet)$ consists of a set of maps $h^k : C^k \to D^{k-1}$ such that

$$h^{k+1} \circ c^k + d^{k-1} \circ h^k = f^k - g^k$$

for each k. We say that the chain maps $f^\bullet$ and $g^\bullet$ are *chain homotopic* if there is a chain homotopy h between them, and will write $f^\bullet \simeq_h g^\bullet$ (or simply $f^\bullet \simeq g^\bullet$) to express this fact.[3]

Although a chain homotopy can be summarized by the following diagram,

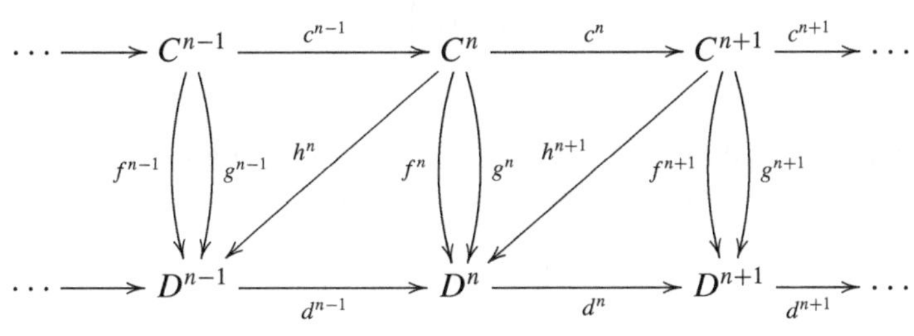

such a diagram may not be commutative.

[3] We will not write $h^\bullet$ for a chain homotopy to avoid mistaking it for a sequence map.

Example 3.14 Consider again the two chain complexes $(C^\bullet, c^\bullet)$ and $(D^\bullet, d^\bullet)$ from Example 3.12 and the quasi-isomorphism between them. The following is a different chain map that also induces isomorphisms on the homology sequence.

$$\begin{array}{ccccccc}
0 & \longrightarrow & \mathbb{R}^4 & \xrightarrow{\begin{pmatrix} 1 & 0 & 0 & -1 \\ -1 & 1 & 0 & 0 \\ 0 & -1 & 1 & 0 \\ 0 & 0 & -1 & 1 \end{pmatrix}} & \mathbb{R}^4 & \longrightarrow & 0 \\
 & & {\begin{pmatrix} 1&0&0&0 \\ 0&0&1&0 \\ 0&0&0&1 \end{pmatrix}}\Big\downarrow & & \Big\downarrow{\begin{pmatrix} 1&0&0&0 \\ 0&1&1&0 \\ 0&0&0&1 \end{pmatrix}} & & \\
0 & \longrightarrow & \mathbb{R}^3 & \xrightarrow[\begin{pmatrix} 1 & 0 & -1 \\ -1 & 1 & 0 \\ 0 & -1 & 1 \end{pmatrix}]{} & \mathbb{R}^3 & \longrightarrow & 0.
\end{array}$$

Let us show that these two chain maps are chain homotopic. If we call the chain map above $f^\bullet$ and the chain map from Example 3.12 $g^\bullet$, the difference $f^\bullet - g^\bullet$ can be expressed as a *non-commuting* diagram

$$\begin{array}{ccccccc}
0 & \longrightarrow & \mathbb{R}^4 & \xrightarrow{\begin{pmatrix} 1 & 0 & 0 & -1 \\ -1 & 1 & 0 & 0 \\ 0 & -1 & 1 & 0 \\ 0 & 0 & -1 & 1 \end{pmatrix}} & \mathbb{R}^4 & \longrightarrow & 0 \\
 & & {\begin{pmatrix} 0&0&0&0 \\ 0&-1&1&0 \\ 0&0&0&0 \end{pmatrix}}\Big\downarrow & \overset{h^2}{\swarrow} & \Big\downarrow{\begin{pmatrix} 0&0&0&0 \\ 0&0&1&0 \\ 0&0&-1&0 \end{pmatrix}} & & \\
0 & \longrightarrow & \mathbb{R}^3 & \xrightarrow[\begin{pmatrix} 1 & 0 & -1 \\ -1 & 1 & 0 \\ 0 & -1 & 1 \end{pmatrix}]{} & \mathbb{R}^3 & \longrightarrow & 0,
\end{array}$$

where we have inserted a map $h^2 : C^2 \to D^1$ for the homotopy. For this to be a chain homotopy, the following two equations must be satisfied

$$h^2 \circ c^1 + d^0 \circ h^1 = f^1 - g^1$$

$$h^2 \circ \begin{pmatrix} 1 & 0 & 0 & -1 \\ -1 & 1 & 0 & 0 \\ 0 & -1 & 1 & 0 \\ 0 & 0 & -1 & 1 \end{pmatrix} = \begin{pmatrix} 0 & 0 & 0 & 0 \\ 0 & -1 & 1 & 0 \\ 0 & 0 & 0 & 0 \end{pmatrix},$$

and

$$h^3 \circ c^2 + d^1 \circ h^2 = f^2 - g^2$$
$$\begin{pmatrix} 1 & 0 & -1 \\ -1 & 1 & 0 \\ 0 & -1 & 1 \end{pmatrix} \circ h^2 = \begin{pmatrix} 0 & 0 & 0 & 0 \\ 0 & 0 & 1 & 0 \\ 0 & 0 & -1 & 0 \end{pmatrix},$$

where we have made use of the fact that h^1 and h^3 must be zero maps. Notice that h^2 selects the third row from c^1 and the second column from d^1. A choice of matrix for h^2 that performs both of these can be found by inspection, namely

$$h^2 = \begin{pmatrix} 0 & 0 & 0 & 0 \\ 0 & 0 & 1 & 0 \\ 0 & 0 & 0 & 0 \end{pmatrix}.$$

Therefore, we can conclude that $f^\bullet \simeq_h g^\bullet$.

The fact that both chain maps in the previous example are quasi-isomorphic is not an accident, as Proposition 3.6 below states.

Proposition 3.6 *If two chain maps are homotopic, then they induce the same map on the homology sequence.*

Proof Let $f^\bullet, g^\bullet : (C^\bullet, c^\bullet) \to (D^\bullet, d^\bullet)$ be two chain maps and h a chain homotopy between them. The defining equation of a chain homotopy can be rearranged to yield

$$f^k = g^k + h^{k+1} \circ c^k + d^{k-1} \circ h^k$$

for each k. If we have a nontrivial $x \in H^k(C^\bullet)$, this corresponds to a length 1 bar in $(C^\bullet, c^\bullet)$. Thus $c^k(x) = 0$, so the equation is

$$f^k(x) = g^k(x) + (d^{k-1} \circ h^k)(x).$$

The latter term on the right is clearly an element of image d^{k-1}. It is trivial when we descend to the quotient

$$H^k(D^\bullet) = \ker d^k / \operatorname{image} d^{k-1}.$$

Thus $f^k(x)$ and $g^k(x)$ correspond to the same coset in $H^k(D^\bullet)$. □

Another way to express that two chain maps are homotopic is to say that their difference is *null homotopic*.

Definition 3.5 A chain map $f^\bullet$ is said to be *null homotopic* or a *null homotopy* if $f^\bullet \simeq 0^\bullet$.

Lemma 3.2 *The only bar chain map that is not null homotopic is Type 5.*

As a consequence, any zero chain map is automatically null homotopic. Readers familiar with abelian groups should take note: Lemma 3.2 is *not true* for chain complexes of abelian groups; see Exercise 3.28.

Proof We proceed case-by-case from Table 3.1.

Type 1 Since Type 1 bar chain maps are completely trivial, the chain homotopy required is also trivial. There is nothing else to show.

Types 2, 3, 4, and 7 Since the components of Type 2, 3, 4, and 7 bar chain maps are zero maps, the chain homotopy simply consists of zero maps.

Type 5 If Type 5 bar chain maps were to be null homotopic, we would have a diagram like

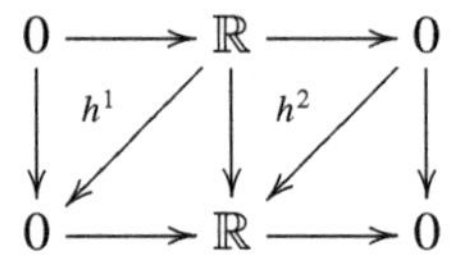

where

$$h^2 \cdot 0 + 0 \cdot h^1 = 1,$$

but since $\mathbb{R}$ has no zero divisors, there are no values of h^1 or h^2 that can satisfy this equation.

Type 6 Consider the following diagram

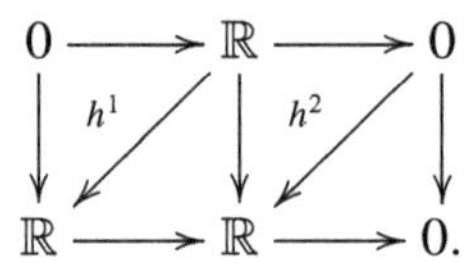

In this diagram, h^2 must be a zero map, but selecting $h^1 = (1)$ results in a chain homotopy.

Type 8 Consider the following diagram

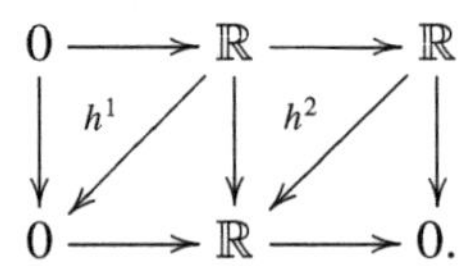

In this diagram, h^1 must be a zero map, but selecting $h^2 = (1)$ results in a chain homotopy.

Type 9 Consider the following diagram

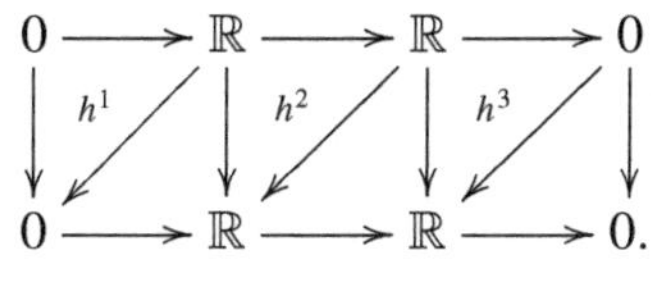

As before, h^1 and h^3 must be zero maps, but selecting $h^2 = (1)$ results in a chain homotopy.

Type 10 Consider the following diagram

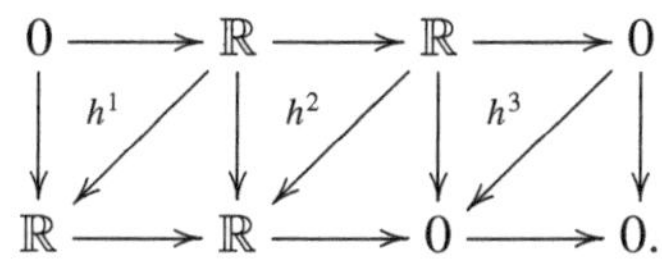

In the diagram, h^3 must be a zero map. There are two chain homotopies that are easily checked:

1. $h^1 = (1)$, $h^2 = (0)$, and
2. $h^1 = (0)$, $h^2 = (1)$.

Either of these suffices to show that Type 10 bar chain maps are null homotopic. □

Exercise 3.28 For readers familiar with abelian groups, the following is an example of a chain map

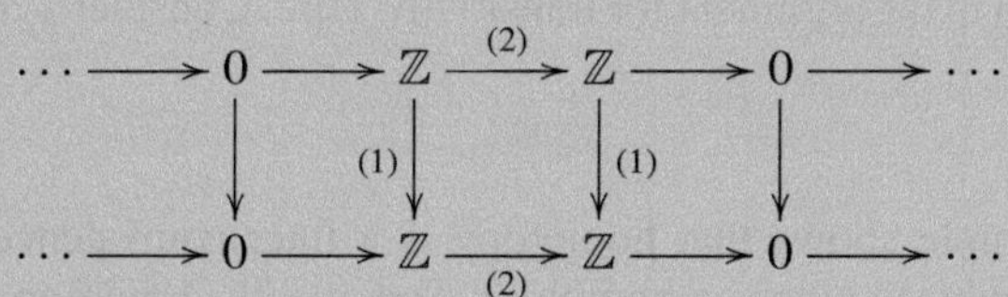

that is not null homotopic. Prove this fact.

Proposition 3.7 *Homotopy is an equivalence relation between chain maps.*

Proof Assume that $f^\bullet, g^\bullet, h^\bullet : (C^\bullet, c^\bullet) \to (D^\bullet, d^\bullet)$ are three chain maps between two chain complexes. We establish each axiom for an equivalence relation in turn.

- Reflexivity Observe that $f^\bullet - f^\bullet$ is the zero chain map, which is evidently null homotopic. Therefore $f^\bullet \simeq f^\bullet$ for any chain map.

- Symmetry Suppose that $f^\bullet \simeq g^\bullet$, which is equivalent to the statement that $f^\bullet - g^\bullet$ is null homotopic. Since $g^\bullet - f^\bullet$ has the same chain map decomposition as $f^\bullet - g^\bullet$ (they differ by a change in basis only), it follows that $g^\bullet - f^\bullet$ is null homotopic as well. Therefore, $g^\bullet \simeq f^\bullet$.
- Transitivity Suppose that $f^\bullet \simeq_p g^\bullet$ and $g^\bullet \simeq_q h^\bullet$, which implies that $f^\bullet - g^\bullet$ and $g^\bullet - h^\bullet$ are null homotopic. If the chain homotopies are p and q respectively, this means that

$$p^{k+1} \circ c^k + d^{k-1} \circ p^k = f^k - g^k$$

and

$$q^{k+1} \circ c^k + d^{k-1} \circ q^k = g^k - h^k$$

for all k. Because

$$\begin{aligned} f^k - h^k &= f^k - h^k + 0^k \\ &= f^k - h^k - g^k + g^k \\ &= (f^k - g^k) + (g^k - h^k) \\ &= (p^{k+1} \circ c^k + d^{k-1} \circ p^k) + (q^{k+1} \circ c^k + d^{k-1} \circ q^k) \\ &= (p^{k+1} + q^{k+1}) \circ c^k + d^{k-1} \circ (p^k + q^k), \end{aligned}$$

which is to say that $p + q$ is a homotopy between $f^\bullet$ and $h^\bullet$. Therefore $f^\bullet \simeq_{(p+q)} h^\bullet$. □

Exercise 3.29 Modify the proof of Proposition 3.7 to use the Escolar-Hiraoka decomposition in the argument for transitivity. Hint: Consider Type 5 bar chain maps.

The next example shows that homotopy is a finer equivalence between chain maps than their induced maps on homology. Specifically, homotopy can distinguish between two quasi-isomorphisms even though they both induce the same map on homology.

Example 3.15 Consider the chain map $f^\bullet$ given by the diagram

$$\begin{array}{ccccc} 0 & \longrightarrow & \mathbb{R} & \longrightarrow & 0 \\ \downarrow & & \downarrow{\scriptstyle (2)} & & \downarrow \\ 0 & \longrightarrow & \mathbb{R} & \longrightarrow & 0 \end{array}$$

and the chain map $g^\bullet$ given by the similar-looking diagram

$$\begin{array}{ccccc} 0 & \longrightarrow & \mathbb{R} & \longrightarrow & 0 \\ \downarrow & & \downarrow{\scriptstyle (1)} & & \downarrow \\ 0 & \longrightarrow & \mathbb{R} & \longrightarrow & 0 \end{array}$$

Both of these chain maps are quasi-isomorphisms. Furthermore the barcode decompositions of all the sequences involved contain no length 2 bars so the induced maps on the homology sequence are given by the same diagrams. However, the difference $f^\bullet - g^\bullet$ between these two chain maps is just $g^\bullet$ again, and this contains a Type 5 chain map. The difference $f^\bullet - g^\bullet$ is therefore not null homotopic. There is no chain homotopy between $f^\bullet$ and $g^\bullet$.

Although chain homotopies are equivalences between chain maps—not between chain complexes—we can use this equivalence to define *homotopy equivalence* between chain complexes. The intuition is that a homotopy equivalence is a chain map that has an inverse up to homotopy, even if it does not have an actual inverse.

Definition 3.6 Two chain complexes $(C^\bullet, c^\bullet)$ and $(D^\bullet, d^\bullet)$ are said to be *homotopy equivalent* if there are two chain maps $f^\bullet : (C^\bullet, c^\bullet) \to (D^\bullet, d^\bullet)$ and $g^\bullet : (D^\bullet, d^\bullet) \to (C^\bullet, c^\bullet)$ such that $f^\bullet \circ g^\bullet \simeq \mathrm{id}_{(D^\bullet, d^\bullet)}$ and $g^\bullet \circ f^\bullet \simeq \mathrm{id}_{(C^\bullet, c^\bullet)}$. We call $f^\bullet$ (or $g^\bullet$) a *homotopy equivalence* in this situation.

Lemma 3.3 *Every exact sequence of vector spaces is homotopy equivalent to the zero sequence.*

This lemma does not hold for arbitrary abelian groups, as shown in Exercise 3.30.

Proof The zero chain map is the only chain map from an arbitrary chain complex to the zero sequence, and is also the only chain map from the zero sequence to an arbitrary chain complex. Therefore, what we need to establish is that the zero map from an exact sequence to itself is chain homotopic to the identity chain map. Namely, we need to show that the identity chain map for an exact sequence is null homotopic. Since the identity map for an exact sequence decomposes into bar chain maps of Types 1 and 9 only, and both of these are null homotopic by Lemma 3.2, this means that the identity map is null homotopic. □

Proposition 3.8 *Homotopy equivalences are coarser than chain isomorphisms and finer than quasi-isomorphisms. Specifically,*

1. *Homotopy equivalence is an equivalence relation,*
2. *Isomorphic chain complexes are also homotopy equivalent, and*
3. *Homotopy equivalent chain complexes are quasi-isomorphic.*

This establishes the placement of homotopy equivalence in the middle column of Table 3.3.

Proof We establish each statement in turn.

1. To establish that homotopy equivalence is an equivalence relation, we need to show

   Reflexivity The identity chain map is a homotopy equivalence. Explicitly, $\text{id}^\bullet \circ \text{id}^\bullet - \text{id}^\bullet$ is a null homotopy because it is a zero chain map.

   Symmetry Suppose that a chain complex $(C^\bullet, c^\bullet)$ is homotopy equivalent to another $(D^\bullet, d^\bullet)$ by way of the chain maps $f^\bullet : (C^\bullet, c^\bullet) \to (D^\bullet, d^\bullet)$ and $g^\bullet : (D^\bullet, d^\bullet) \to (C^\bullet, c^\bullet)$. Since the definition of homotopy equivalent is itself symmetric, this means that $(D^\bullet, d^\bullet)$ is homotopy equivalent to $(C^\bullet, c^\bullet)$ merely by switching the roles of $f^\bullet$ and $g^\bullet$.

   Transitivity Suppose that $f^\bullet : (C^\bullet, c^\bullet) \to (D^\bullet, d^\bullet)$ and $g^\bullet : (D^\bullet, d^\bullet) \to (C^\bullet, c^\bullet)$ form a homotopy equivalence and that $\phi^\bullet : (D^\bullet, d^\bullet) \to (E^\bullet, e^\bullet)$ and $\gamma^\bullet : (E^\bullet, e^\bullet) \to (D^\bullet, d^\bullet)$ also form a homotopy equivalence. Explicitly, we assume that there are four homotopies h_1, h_2, h_3, and h_4 such that

$$\begin{aligned} g^\bullet \circ f^\bullet &\simeq_{h_1} \text{id}_{C^\bullet}, \\ f^\bullet \circ g^\bullet &\simeq_{h_2} \text{id}_{D^\bullet}, \\ \gamma^\bullet \circ \phi^\bullet &\simeq_{h_3} \text{id}_{D^\bullet}, \text{ and} \\ \phi^\bullet \circ \gamma^\bullet &\simeq_{h_4} \text{id}_{E^\bullet} . \end{aligned}$$

   To show transitivity, we must show that $(\phi^\bullet \circ f^\bullet)$ and $(g^\bullet \circ \gamma^\bullet)$ form a homotopy equivalence. We do this explicitly, by constructing homotopies

$$\begin{aligned} \text{id}_{C^\bullet} &\simeq (g^\bullet \circ \gamma^\bullet) \circ (\phi^\bullet \circ f^\bullet) = g^\bullet \circ (\gamma^\bullet \circ \phi^\bullet) \circ f^\bullet, \text{ and} \\ \text{id}_{E^\bullet} &\simeq (\phi^\bullet \circ f^\bullet) \circ (g^\bullet \circ \gamma^\bullet) = \phi^\bullet \circ (f^\bullet \circ g^\bullet) \circ \gamma^\bullet. \end{aligned}$$

   Let us start with the first of these, and begin by using the homotopies h_1 and h_3. For each index k, we have that

$$\begin{aligned} \text{id}_{C^k} - g^k \circ f^k &= h_1^{k+1} \circ c^k + c^{k-1} \circ h_1^k \\ \text{id}_{C^k} &= g^k \circ f^k + h_1^{k+1} \circ c^k + c^{k-1} \circ h_1^k \end{aligned}$$

   Similarly, we have that

$$\text{id}_{D^k} = \gamma^k \circ \phi^k + h_3^{k+1} \circ d^k + d^{k-1} \circ h_3^k.$$

   Combining these facts, we have that

$$\begin{aligned} \text{id}_{C^k} &= g^k \circ \text{id}_{D^k} \circ f^k + h_1^{k+1} \circ c^k + c^{k-1} \circ h_1^k \\ &= g^k \circ \left(\gamma^k \circ \phi^k + h_3^{k+1} \circ d^k + d^{k-1} \circ h_3^k\right) \circ f^k + h_1^{k+1} \circ c^k + c^{k-1} \circ h_1^k \\ &= g^k \circ \left(\gamma^k \circ \phi^k\right) \circ f^k + g^k \circ h_3^{k+1} \circ d^k \circ f^k + g^k \circ d^{k-1} \circ h_3^k \circ f^k + \\ &\quad h_1^{k+1} \circ c^k + c^{k-1} \circ h_1^k. \end{aligned}$$

Making use of the fact that the diagrams for the chain maps $f^\bullet$ and $g^\bullet$ are commutative, we have that

$$d^k \circ f^k = f^{k+1} \circ c^k \text{ and } g^k \circ d^{k-1} = c^{k-1} \circ g^{k-1}.$$

Substituting these we obtain,

$$\begin{aligned}
\mathrm{id}_{C^k} &= g^k \circ \left(\gamma^k \circ \phi^k\right) \circ f^k + g^k \circ h_3^{k+1} \circ f^{k+1} \circ c^k + c^{k-1} \circ g^{k-1} \circ h_3^k \circ f^k + \\
&\quad h_1^{k+1} \circ c^k + c^{k-1} \circ h_1^k \\
&= g^k \circ \left(\gamma^k \circ \phi^k\right) \circ f^k + \left(g^k \circ h_3^{k+1} \circ f^{k+1} + h_1^{k+1}\right) \circ c^k + \\
&\quad c^{k-1} \circ \left(g^{k-1} \circ h_3^k \circ f^k + h_1^k\right),
\end{aligned}$$

which means that if we define

$$h^k := g^{k-1} \circ h_3^k \circ f^k + h_1^k,$$

we have constructed a homotopy

$$\mathrm{id}_{C^\bullet} \simeq_h (g^\bullet \circ \gamma^\bullet) \circ (\phi^\bullet \circ f^\bullet).$$

Making the necessary substitutions (exchanging g and ϕ, exchanging f and γ, using h_2 in place of h_1, and using h_4 in place of h_3), we can obtain another homotopy h'

$$\mathrm{id}_{E^\bullet} \simeq_{h'} (\phi^\bullet \circ f^\bullet) \circ (g^\bullet \circ \gamma^\bullet).$$

2. Start with a chain isomorphism, in which we have two chain maps $f^\bullet$ and $g^\bullet$ that are inverses of each other. Then, the difference $f^\bullet \circ g^\bullet - \mathrm{id}^\bullet$ is null homotopic because it is a zero chain map. Similarly, $g^\bullet \circ f^\bullet - \mathrm{id}^\bullet$ is null homotopic, because it is a zero chain map.
3. Suppose that a chain complex $(C^\bullet, c^\bullet)$ is homotopy equivalent to another $(D^\bullet, d^\bullet)$ by way of the chain maps $f^\bullet : (C^\bullet, c^\bullet) \to (D^\bullet, d^\bullet)$ and $g^\bullet : (D^\bullet, d^\bullet) \to (C^\bullet, c^\bullet)$. This means that

$$\begin{aligned}
g^\bullet \circ f^\bullet &\simeq \mathrm{id}_{C^\bullet} \text{ and} \\
f^\bullet \circ g^\bullet &\simeq \mathrm{id}_{D^\bullet}.
\end{aligned}$$

Proposition 3.6 means that the map $(g^\bullet \circ f^\bullet)_*$ induced on the homology sequence is the same as the one induced by the identity map, and a similar statement is true for $(f^\bullet \circ g^\bullet)_*$. Since homology is a covariant functor (Proposition 3.2), we have that

$$(g^\bullet \circ f^\bullet)_* = (g_*)^\bullet \circ (f_*)^\bullet = \mathrm{id}_{H^\bullet(C^\bullet)} \text{ and}$$
$$(f^\bullet \circ g^\bullet)_* = (f_*)^\bullet \circ (g_*)^\bullet = \mathrm{id}_{H^\bullet(D^\bullet)}\,.$$

Thus $(f_*)^k$ and $(g_*)^k$ are isomorphisms for each index k, which means that $f^\bullet$ and $g^\bullet$ are quasi-isomorphisms. □

Homotopy equivalences respect direct sums, which is often convenient in calculations.

Lemma 3.4 *If* $(B^\bullet, b^\bullet)$ *is homotopy equivalent to* $(C^\bullet, c^\bullet)$*, then* $(A^\bullet, a^\bullet) \oplus (B^\bullet, b^\bullet)$ *is homotopy equivalent to* $(A^\bullet, a^\bullet) \oplus (C^\bullet, c^\bullet)$.

Proof Suppose that the assumed homotopy equivalence is given by the chain maps $f^\bullet : (B^\bullet, b^\bullet) \to (C^\bullet, c^\bullet)$ and $g^\bullet : (C^\bullet, c^\bullet) \to (B^\bullet, b^\bullet)$ and the chain homotopies

$$g^\bullet \circ f^\bullet \simeq_h \mathrm{id}_{B^\bullet} \text{ and } f^\bullet \circ g^\bullet \simeq_{h'} \mathrm{id}_{C^\bullet}\,.$$

The result follows because $\mathrm{id}_{A^\bullet} \oplus f^\bullet$ and $\mathrm{id}_{A^\bullet} \oplus g^\bullet$ form a homotopy equivalence. Specifically,

$$(\mathrm{id}_{A^\bullet} \oplus g^\bullet) \circ (\mathrm{id}_{A^\bullet} \oplus f^\bullet) - \mathrm{id}_{A^\bullet \oplus B^\bullet} = 0_{A^\bullet, A^\bullet} \oplus (g^\bullet \circ f^\bullet - \mathrm{id}_{B^\bullet}).$$

This is null homotopic through the homotopy defined by $0_{A^k, A^{k-1}} \oplus h^k$ for each index k. The same reasoning establishes that $(\mathrm{id}_{A^\bullet} \oplus f^\bullet) \circ (\mathrm{id}_{A^\bullet} \oplus g^\bullet)$ is homotopic to $id_{C^\bullet}$. □

Exercise 3.30 Continuing Exercise 3.28 for readers familiar with abelian groups, the following is an exact sequence of abelian groups

$$0 \longrightarrow \mathbb{Z} \xrightarrow{(2)} \mathbb{Z} \xrightarrow{p} \mathbb{Z}_2 \longrightarrow 0,$$

in which p computes the parity (evenness or oddness) of an integer.

1. Prove that this sequence is indeed exact.
2. Does this sequence have a barcode decomposition? Why or why not?
3. Show that the zero sequence map is a quasi-isomorphism.
4. Show that the identity sequence map

$$\begin{array}{ccccccccc}
0 & \longrightarrow & \mathbb{Z} & \xrightarrow{(2)} & \mathbb{Z} & \xrightarrow{p} & \mathbb{Z}_2 & \longrightarrow & 0 \\
 & & \downarrow{\scriptstyle (1)} & & \downarrow{\scriptstyle (1)} & & \downarrow{\scriptstyle (1)} & & \\
 & \longrightarrow & \mathbb{Z} & \xrightarrow[(2)]{} & \mathbb{Z} & \xrightarrow[p]{} & \mathbb{Z}_2 & \longrightarrow & 0
\end{array}$$

is *not* null homotopic.

Use your answers to (1)–(4) to conclude that this sequence is an example of one which is quasi-isomorphic to the zero sequence but not homotopy equivalent to the zero sequence.

Exercise 3.30 not withstanding, the converse of Proposition 3.8 is true for chain complexes of vector spaces.

Theorem 3.4 (quasi-isomorphism is equivalent to homotopy equivalence for vector spaces) *Two chain complexes of vector spaces are quasi-isomorphic if and only if they are homotopy equivalent. Specifically, if* $q^\bullet : (C^\bullet, c^\bullet) \to (D^\bullet, d^\bullet)$ *is a quasi-isomorphism, then the two chain complexes are homotopy equivalent.*

Before proving this Theorem, we note that it is *not* true for chain complexes of other things (not vector spaces). We address this more general setting in Chap. 6. It is often easier to construct homotopy equivalences than quasi-isomorphisms. For instance, in Chap. 5, we will construct homotopy equivalences for chain complexes that come from simplicial complexes. At least for chain complexes of vector spaces, we lose no generality by this approach.

Proof We need only establish the converse of Proposition 3.8. Begin by decomposing the two chain complexes into their homology and exact sequences, noting that $q^\bullet$ induces a chain isomorphism on the homology sequences by assumption. By Lemma 3.3, both exact sequences are homotopy equivalent to the zero sequence.[4] They are therefore homotopy equivalent to each other by the transitivity established in Proposition 3.8. With the homology sequences being chain isomorphic, and the exact sequences being homotopy equivalent, Lemma 3.4 establishes that the original sequences are homotopy equivalent. □

Exercise 3.31 The following two chain complexes

$$0 \longrightarrow \mathbb{R}^2 \xrightarrow{\quad (3\ 6) \quad} \mathbb{R} \longrightarrow 0$$

and

$$0 \longrightarrow \mathbb{R}^2 \xrightarrow{\quad (2\ 1) \quad} \mathbb{R} \longrightarrow 0$$

are quasi-isomorphic. Construct a homotopy equivalence between them.

[4] Lemma 3.3 holds for chain complexes of vector spaces, but is not true for general chain complexes since it relies upon the details of the bar chain map decomposition. It is at this point that the proof fails in general!.

Exercise 3.32 The following two chain complexes

$$0 \longrightarrow \mathbb{R} \xrightarrow{(0)} \mathbb{R} \xrightarrow{\begin{pmatrix} -4 \\ -3 \\ 0 \\ 2 \end{pmatrix}} \mathbb{R}^4 \xrightarrow{\begin{pmatrix} 2 & -4 & 0 & -2 \\ 2 & 0 & -2 & 4 \end{pmatrix}} \mathbb{R}^2 \longrightarrow 0$$

and

$$0 \to \mathbb{R}^3 \xrightarrow{\begin{pmatrix} 24 & -12 & 16 \\ -13 & 6 & -8 \end{pmatrix}} \mathbb{R}^2 \xrightarrow{(0\ 0)} \mathbb{R} \xrightarrow{(0)} \mathbb{R} \xrightarrow{(-3)} \mathbb{R} \to 0$$

are quasi-isomorphic. Construct a homotopy equivalence between them.

Exercise 3.33 The following two chain complexes

$$0 \longrightarrow \mathbb{R}^3 \xrightarrow{\begin{pmatrix} 3 & -4 & -4 \\ 0 & 0 & 0 \\ 0 & 0 & 0 \\ 0 & 1 & 1 \end{pmatrix}} \mathbb{R}^4 \xrightarrow{\begin{pmatrix} 0 & 3 & 3 & 0 \\ 0 & 2 & 3 & 0 \end{pmatrix}} \mathbb{R}^2 \longrightarrow 0$$

and

$$0 \longrightarrow \mathbb{R} \xrightarrow{\begin{pmatrix} 0 \\ 3 \\ -1 \end{pmatrix}} \mathbb{R}^3 \xrightarrow{(-4\ 4\ 12)} \mathbb{R} \longrightarrow 0$$

are quasi-isomorphic. Construct a homotopy equivalence between them.

Exercise 3.34 The following two chain complexes

$$0 \longrightarrow \mathbb{R}^3 \xrightarrow{\begin{pmatrix} 3 & -4 & -4 \\ 0 & 0 & 0 \\ 0 & 0 & 0 \\ 0 & 1 & 1 \end{pmatrix}} \mathbb{R}^4 \xrightarrow{\begin{pmatrix} 0 & 3 & 3 & 0 \\ 0 & 2 & 3 & 0 \end{pmatrix}} \mathbb{R}^2 \longrightarrow 0$$

and

$$0 \longrightarrow \mathbb{R}^2 \xrightarrow{\begin{pmatrix} 0 & 0 \\ 0 & 0 \end{pmatrix}} \mathbb{R}^2 \xrightarrow{\begin{pmatrix} 1 & -3 \\ 0 & 3 \\ -1 & 1 \\ 0 & 3 \end{pmatrix}} \mathbb{R}^4 \xrightarrow{\begin{pmatrix} -3 & 13 & -3 & -15 \\ 0 & -4 & 0 & 4 \end{pmatrix}} \mathbb{R}^2 \longrightarrow 0$$

are not quasi-isomorphic. Prove that there is no homotopy equivalence between them. Challenge: show this directly without computing the homology or barcodes of the chain complexes.

Remark 3.3 The reader should be aware that there are several different sign conventions for chain homotopies in the literature. Another popular sign convention is

$$h^{k+1} \circ c^k + (-1)^k d^{k-1} \circ h^k = f^k - g^k.$$

3.5 The Hodge Theorem

Recall that dualization T : **Vec** $\to$ **Vec**, which converts each vector space V into the space V^T of linear functions $V \to \mathbb{R}$, is a contravariant functor (see Sect. 1.3 in Chap. 1, and Proposition 1.6 in Chap. 1). In this section, we show that the dualization functor can simply be applied to every map in a chain complex to obtain another chain complex. In a similar way, chain maps are taken to new chain maps. This section explores this new perspective by defining the *Laplacian operators*, whose kernels determine the homology of a chain complex according to the *Hodge theorem*. The Laplacian operators also have other uses, as will be seen in Sect. 5.6 of Chap. 5.

Example 3.16 Consider the chain complex

$$0 \longrightarrow \mathbb{R}^2 \xrightarrow{\begin{pmatrix} 6 & -9 \\ -2 & 3 \end{pmatrix}} \mathbb{R}^2 \xrightarrow{\begin{pmatrix} 0 & 0 \end{pmatrix}} \mathbb{R} \xrightarrow{\begin{pmatrix} 1 \\ 0 \end{pmatrix}} \mathbb{R}^2 \xrightarrow{\begin{pmatrix} 0 & 1 \end{pmatrix}} \mathbb{R} \longrightarrow 0$$

This chain complex has the barcode decomposition

$$\begin{array}{ccccccccc}
\mathbb{R} & & & & & & & & \\
\mathbb{R} & \longrightarrow & \mathbb{R} & & & & & & \\
 & & \mathbb{R} & & & & & & \\
 & & & & \mathbb{R} & \longrightarrow & \mathbb{R} & & \\
 & & & & & & \mathbb{R} & \longrightarrow & \mathbb{R}
\end{array}$$

Dualizing each map in the chain complex is easy—just take transposes. The resulting sequence is

$$0 \longleftarrow \mathbb{R}^2 \xleftarrow{\begin{pmatrix} 6 & -2 \\ -9 & 3 \end{pmatrix}} \mathbb{R}^2 \xleftarrow{\begin{pmatrix} 0 \\ 0 \end{pmatrix}} \mathbb{R} \xleftarrow{(1\ 0)} \mathbb{R}^2 \xleftarrow{\begin{pmatrix} 0 \\ 1 \end{pmatrix}} \mathbb{R} \longleftarrow 0$$

which is easily seen to be a chain complex by multiplying consecutive matrices. The barcode decomposition of this new chain complex is

$$\begin{array}{ccccccccc}
\mathbb{R} & & & & & & & & \\
\mathbb{R} & \longleftarrow & \mathbb{R} & & & & & & \\
 & & \mathbb{R} & & & & & & \\
 & & & & \mathbb{R} & \longleftarrow & \mathbb{R} & & \\
 & & & & & & \mathbb{R} & \longleftarrow & \mathbb{R}
\end{array}$$

which (aside from the arrow reversal) is the same as the original decomposition.

The process of dualizing all maps in a chain complex reverses the direction of the indices. When we dualize each map of

$$\cdots \longrightarrow C^1 \xrightarrow{c^1} C^2 \xrightarrow{c^2} C^3 \xrightarrow{c^3} \cdots$$

we obtain

$$\cdots \longleftarrow (C^1)^T \xleftarrow{(c^1)^T} (C^2)^T \xleftarrow{(c^2)^T} (C^3)^T \xleftarrow{(c^3)^T} \cdots.$$

There is a shift in the indices of the maps, because the map with domain $C_k = (C^k)^T$ is not $(c^k)^T$ but rather $(c^{k-1})^T$. Let us define $c_k := (c^{k-1})^T$, which means that we must write $(C_\bullet, c_\bullet) = ((C^\bullet)^T, (c^{\bullet-1})^T)$ for the dual of $(C^\bullet, c^\bullet)$.

In Example 3.16, the dual chain complex and the original chain complex have the same barcode decomposition. This is not an accident, as Lemma 3.5 below will show. In order to see this, it is easiest to work with the barcode decomposition directly. Any chain complex $(C^\bullet, c^\bullet)$ has a barcode decomposition into length 1 and length 2 bars. This means we can select a basis

$$e^{k,1}, e^{k,2}, \ldots, e^{k,n_k}$$

for C^k such that $c^k(e^{k,j})$ is either zero or a basis element $e^{k+1,m}$ of C^{k+1} (the index m depends on j). As discussed in Sect. 1.3 in Chap. 1, corresponding to each basis of C^k is a dual basis of $C_k = (C^k)^T$, consisting of elements

$$e_{k,1}, e_{k,2}, \ldots, e_{k,n_k}.$$

Accordingly, let us define a specific isomorphism $g_k : C^k \to C_k$ by extending the formula

$$g_k(e^{k,j}) := e_{k,j}, \tag{3.1}$$

by linearity. We will call these g_k the *dualizing isomorphisms* for $(C^\bullet, c^\bullet)$. These isomorphisms are not canonical. This means that their definition depends on $(C^\bullet, c^\bullet)$, the specifics of its barcode decomposition, and the choice of basis.

Lemma 3.5 *The dual of a chain complex is a chain complex itself. Specifically, if $(C^\bullet, c^\bullet)$ is a chain complex, then $(C_\bullet, c_\bullet) = ((C^\bullet)^T, (c^{\bullet-1})^T)$ is also a chain complex. Both chain complexes have exactly the same barcode decomposition (accounting for the difference in index direction). Any bar of length ℓ that starts at index k in $(C^\bullet, c^\bullet)$ corresponds to a bar of length ℓ that ends at index k in $(C_\bullet, c_\bullet)$.*

Proof Without loss of generality, suppose that $(C^\bullet, c^\bullet)$ is a barcode. According to Theorem 1.5 in Chap. 1, if V is a finite dimensional vector space then there is an isomorphism $g : V \to V^T$ from V to its dual vector space V^T. This means that if we let $C_k := (C^k)^T$ and $c_k := (c^{k-1})^T$, we can define a sequence of isomorphisms $g_k : C^k \to C_k$. The lemma follows by using the resulting *non-commuting* diagram

$$\begin{array}{ccccccccc}
\cdots & \longrightarrow & C^1 & \xrightarrow{c^1} & C^2 & \xrightarrow{c^2} & C^3 & \xrightarrow{c^3} & \cdots \\
 & & \downarrow g_1 & & \downarrow g_2 & & \downarrow g_3 & & \\
\cdots & \longleftarrow & C_1 & \xleftarrow{c_2} & C_2 & \xleftarrow{c_3} & C_3 & \xleftarrow{c_4} & \cdots
\end{array}$$

to establish that the barcodes of $(C^\bullet, c^\bullet)$ and $(C_\bullet, c_\bullet)$ are identical.

Because we assumed that $(C^\bullet, c^\bullet)$ is a barcode, each bar corresponds to a basis element from

$$e^{k,1}, e^{k,2}, \ldots, e^{k,n_k}$$

such that $c^k(e^{k,j})$ is either zero or a basis element $e^{k+1,m}$ of C^{k+1}. Use the dualizing isomorphisms $g_k : C^k \to C_k$ given by

$$g_k(e^{k,j}) := e_{k,j},$$

where $e_{k,j}$ is the dual basis element for $e^{k,j}$.

Suppose that $e^{k,j} \in C^k$ corresponds to the start of a length 2 bar. It is a basis element whose image in C^{k+1} is $e^{k+1,m}$. This means that

$$\begin{aligned}
(c_{k+1} \circ g_{k+1} \circ c^k)(e^{k,j}) &= (c_{k+1} \circ g_{k+1}) \left(c^k(e^{k,j})\right) \\
&= (c_{k+1} \circ g_{k+1}) \left(e^{k+1,m}\right) \\
&= c_{k+1}(e_{k+1,m}) \\
&= (c^k)^T(e_{k+1,m}) \\
&= e_{k,j} \\
&= g_k(e^{k,j}).
\end{aligned}$$

Therefore, the corresponding bar in $(C_\bullet, c_\bullet)$ is also of length 2 but it *ends* at $e_{k,j}$ rather than starting there.

Now suppose that $e^{k,j} \in C^k$ corresponds to a length 1 bar. This means that it has a trivial image in C^{k+1}. The above calculation can be repeated to show that

$$\begin{aligned}
(c_{k+1} \circ g_{k+1} \circ c^k)(e^{k,j}) &= (c_{k+1} \circ g_{k+1}) \left(c^k(e^{k,j})\right) \\
&= (c_{k+1} \circ g_{k+1}) \left(e^{k+1,m}\right) \\
&= 0.
\end{aligned}$$

This means that $e_{k,j}$ also corresponds to a length 1 bar in $(C_\bullet, c_\bullet)$ at index k. □

Remark 3.4 As the reader might suspect, Lemma 3.5 fails to hold (and most of the rest of this section fails with it) if the chain complexes are built from modules or rings instead of vector spaces! See de Silva et al. (2011) for many more interesting details.

We can lift the dualization functor T : **Vec** → **Vec** for vector spaces to a new functor for chain complexes by simply applying it to each map in turn.

Example 3.17 Consider the following chain map

$$\begin{array}{ccccccccccc}
0 & \longrightarrow & \mathbb{R} & \xrightarrow{\begin{pmatrix}2\\0\end{pmatrix}} & \mathbb{R}^2 & \xrightarrow{\begin{pmatrix}0&4\\0&6\end{pmatrix}} & \mathbb{R}^2 & \xrightarrow{\begin{pmatrix}3&-2\\-3&2\end{pmatrix}} & \mathbb{R}^2 & \longrightarrow & 0 \\
 & & \Big\downarrow \begin{pmatrix}-2\\-1\end{pmatrix} & & \Big\downarrow \begin{pmatrix}-2&-4\\-2&4\\0&2\end{pmatrix} & & \Big\downarrow \begin{pmatrix}-2&-2/3\\2&-8/3\\2&4/3\end{pmatrix} & & \Big\downarrow \begin{pmatrix}-1&-1\end{pmatrix} & & \\
0 & \longrightarrow & \mathbb{R}^2 & \xrightarrow[\begin{pmatrix}4&-4\\5&-6\\0&0\end{pmatrix}]{} & \mathbb{R}^3 & \xrightarrow[\begin{pmatrix}0&0&-6\\0&0&-4\\0&0&8\end{pmatrix}]{} & \mathbb{R}^3 & \xrightarrow[\begin{pmatrix}18&3&15\end{pmatrix}]{} & \mathbb{R} & \longrightarrow & 0
\end{array}$$

Let us call the domain (the top row) $(C^\bullet, c^\bullet)$ and the codomain (the bottom row) $(D^\bullet, d^\bullet)$, so that we may call the chain map $m^\bullet : (C^\bullet, c^\bullet) \to (D^\bullet, d^\bullet)$. If we dualize every map in the diagram we obtain

$$
\begin{array}{ccccccccccc}
0 & \longleftarrow & \mathbb{R} & \xleftarrow{\begin{pmatrix}2 & 0\end{pmatrix}} & \mathbb{R}^2 & \xleftarrow{\begin{pmatrix}0 & 0\\ 4 & 6\end{pmatrix}} & \mathbb{R}^2 & \xleftarrow{\begin{pmatrix}3 & -3\\ -2 & 2\end{pmatrix}} & \mathbb{R}^2 & \longleftarrow & 0\\
 & & \Big\uparrow \begin{pmatrix}-2 & -1\end{pmatrix} & & \Big\uparrow \begin{pmatrix}-2 & -2 & 0\\ -4 & 4 & 2\end{pmatrix} & & \Big\uparrow \begin{pmatrix}-2 & 2 & 2\\ -2/3 & -8/3 & 4/3\end{pmatrix} & & \Big\uparrow \begin{pmatrix}-1\\ -1\end{pmatrix} & & \\
0 & \longleftarrow & \mathbb{R}^2 & \xleftarrow[\begin{pmatrix}4 & 5 & 0\\ -4 & -6 & 0\end{pmatrix}]{} & \mathbb{R}^3 & \xleftarrow[\begin{pmatrix}0 & 0 & 0\\ 0 & 0 & 0\\ -6 & -4 & 8\end{pmatrix}]{} & \mathbb{R}^3 & \xleftarrow[\begin{pmatrix}18\\ 3\\ 15\end{pmatrix}]{} & \mathbb{R} & \longleftarrow & 0
\end{array}
$$

which is still a chain map! Since all of the maps have reversed direction, this means that the new chain map $m_\bullet = (m^\bullet)^T$ has reversed direction as well. In particular, it is a chain map $m_\bullet : (D_\bullet, d_\bullet) \to (C_\bullet, c_\bullet)$.

Proposition 3.9 *Dualization defines a contravariant functor T : **Kom** $\to$ **Kom** given by*

$$T(C^\bullet, c^\bullet) := ((C^\bullet)^T, (c^{\bullet -1})^T) = (C_\bullet, c_\bullet).$$

Proof Apply the functor T : **Vec** $\to$ **Vec** to every linear map in a typical chain map $m^\bullet : (V^\bullet, v^\bullet) \to (W^\bullet, w^\bullet)$ given by

$$
\begin{array}{ccccccccc}
\cdots & \longrightarrow & V^1 & \xrightarrow{v^1} & V^2 & \xrightarrow{v^2} & V^3 & \xrightarrow{v^3} & \cdots\\
 & & \Big\downarrow m^1 & & \Big\downarrow m^2 & & \Big\downarrow m^3 & & \\
\cdots & \longrightarrow & W^1 & \xrightarrow[w^1]{} & W^2 & \xrightarrow[w^2]{} & W^3 & \xrightarrow[w^3]{} & \cdots
\end{array}
$$

This process results in another chain map

$$
\begin{array}{ccccccccc}
\cdots & \longleftarrow & (V^1)^T & \xleftarrow{(v^1)^T} & (V^2)^T & \xleftarrow{(v^2)^T} & (V^3)^T & \xleftarrow{(v^3)^T} & \cdots\\
 & & \Big\uparrow (m^1)^T & & \Big\uparrow (m^2)^T & & \Big\uparrow (m^3)^T & & \\
\cdots & \longleftarrow & (W^1)^T & \xleftarrow[(w^1)^T]{} & (W^2)^T & \xleftarrow[(w^2)^T]{} & (W^3)^T & \xleftarrow[(w^3)^T]{} & \cdots
\end{array}
$$

in which all linear maps are reversed. To see that this is a chain map, we reason as follows, starting from the commutativity of the original chain map

$$
\begin{aligned}
w^k \circ m^k &= m^{k+1} \circ v^k \\
(w^k \circ m^k)^T &= (m^{k+1} \circ v^k)^T \\
(m^k)^T \circ (w^k)^T &= (v^k)^T \circ (m^{k+1})^T,
\end{aligned}
$$

which establishes commutativity as desired.

This new chain map is of the form $(m_\bullet) : (W_\bullet, w_\bullet) \to (V_\bullet, v_\bullet)$,

$$
\begin{array}{ccccccccc}
\cdots & \longleftarrow & V_1 & \xleftarrow{v_1} & V_2 & \xleftarrow{v_2} & V_3 & \xleftarrow{v_3} & \cdots \\
 & & \uparrow{\scriptstyle m_1} & & \uparrow{\scriptstyle m_2} & & \uparrow{\scriptstyle m_2} & & \\
\cdots & \longleftarrow & W_1 & \xleftarrow[w_2]{} & W_2 & \xleftarrow[w_3]{} & W_3 & \xleftarrow[w_4]{} & \cdots
\end{array}
$$

if we define $m_k := (m^k)^T$ in accordance with the convention that subscripts mean that the indices decrease with the linear maps. □

It is useful to have a distinct concept of the homology of the dual chain complex. This is usually called the *cohomology* of the chain complex.

Definition 3.7 For a chain complex $(C^\bullet, c^\bullet)$, its *cohomology in index k* is given by[5]

$$H_k(C^\bullet, c^\bullet) := \left(H^k \circ T\right)(C^\bullet, c^\bullet).$$

Since the cohomology H_k is the composition of a covariant functor H^k and a contravariant functor T, it follows immediately that H_k is a contravariant functor H_k : **Kom** $\to$ **Vec** for each index k.

Proposition 3.10 *If $(C^\bullet, c^\bullet)$ is a chain complex, then*

$$H_k(C^\bullet, c^\bullet) = \left(H^k \circ T\right)(C^\bullet, c^\bullet) \cong \left(H^k(C^\bullet, c^\bullet)\right)^T$$

for all k.

Proof This is an immediate consequence of Lemma 3.5, because $T(C^\bullet, c^\bullet) = ((C^\bullet)^T, (c^{\bullet-1})^T)$ has the same number of length 1 bars in each index as $(C^\bullet, c^\bullet)$. □

Lemma 3.5 established that the barcode decompositions of $(C^\bullet, c^\bullet)$ and its dual $(C_\bullet, c_\bullet)$ are identical. The proof involved a non-commuting diagram that related each C^k to C_k. We could as well have written the diagram "the other way," which would have used dualizing isomorphisms $C_k \to C^k$; the proof would have been unchanged. If we combine both of these diagrams, we obtain a larger non-commuting diagram,

[5] Caution! The index direction differs from that in Chaps. 4—5. This is a merely notational difference.

$$\begin{array}{ccccccccc}
\cdots & \longrightarrow & C^1 & \xrightarrow{c^1} & C^2 & \xrightarrow{c^2} & C^3 & \xrightarrow{c^3} & \cdots \\
 & & \Big\downarrow{\scriptstyle g_1} & & \Big\downarrow{\scriptstyle g_2} & & \Big\downarrow{\scriptstyle g_3} & & \\
\cdots & \longleftarrow & C_1 & \xleftarrow{c_2} & C_2 & \xleftarrow{c_3} & C_3 & \xleftarrow{c_4} & \cdots \\
 & & \Big\downarrow{\scriptstyle g_1^{-1}} & & \Big\downarrow{\scriptstyle g_2^{-1}} & & \Big\downarrow{\scriptstyle g_3^{-1}} & & \\
\cdots & \longrightarrow & C^1 & \xrightarrow{c^1} & C^2 & \xrightarrow{c^2} & C^3 & \xrightarrow{c^3} & \cdots
\end{array}$$

in which there are many distinct paths from C^2 in the top row to C^2 in the bottom row. The three shortest such paths are

$$\begin{array}{c}
C^2 \\
\Big\downarrow{\scriptstyle g_2} \\
C_2 \\
\Big\downarrow{\scriptstyle g_2^{-1}} \\
C^2
\end{array}
\qquad
\begin{array}{ccc}
 & & C^2 \\
 & & \Big\downarrow{\scriptstyle g_2} \\
C_1 & \xleftarrow{c_2} & C_2 \\
\Big\downarrow{\scriptstyle g_1^{-1}} & & \\
C^1 & \xrightarrow{c^1} & C^2
\end{array}
\qquad
\begin{array}{ccc}
C^2 & \xrightarrow{c^2} & C^3 \\
 & & \Big\downarrow{\scriptstyle g_3} \\
C_2 & \xleftarrow{c_3} & C_3 \\
\Big\downarrow{\scriptstyle g_2^{-1}} & & \\
C^2 & &
\end{array}$$

Longer paths are possible by composing consecutive horizontal maps, but any such path will compose to the zero map if $(C^\bullet, c^\bullet)$ is assumed to be a chain complex. Therefore, every nontrivial map $C^2 \to C^2$ formed by composing maps along a path from the top row to the bottom row will be one of these three.

Since the diagram does not commute, these define three (potentially) different linear maps $C^2 \to C^2$. Evidently the leftmost of these is simply the identity map $\mathrm{id}_{C^2} : C^2 \to C^2$. The *Laplacian operator* L^2 measures the difference between the other two maps, and is given by[6]

$$L^2 := g_2^{-1} \circ c_3 \circ g_3 \circ c^2 - c^1 \circ g_1^{-1} \circ c_2 \circ g_2. \tag{3.2}$$

Notice that any element of homology $H^2(C^\bullet, c^\bullet)$ will necessarily lie in the kernel of L^2 because it corresponds to a bar of length 1. Each of the two paths represented in the definition of L^2 requires traversing a path of length 2 in $(C^\bullet, c^\bullet)$. Specifically, any element of homology $H^2(C^\bullet, c^\bullet)$ lies in the kernel of c^2 and also lies outside the image of c^1. Since c_2 is the dual of c^1, whenever a vector lies outside the image of c^1, it lies in the kernel of c_2 according to Theorem 1.6 in Chap. 1.

On the other hand, if b corresponds to a bar of length 2 that starts at C^2, then the first term of $L^2(b)$ as given in (3.2) will be nonzero. However, the second term of $L^2(b)$ will be zero, since b lies outside the image of c^1. Hence $L^2(b) \neq 0$. Conversely, if $b' \in C^2$ is the end of a bar of length 2 ending at C^2, this means that $c^2(b') = 0$ but

[6] The reader should be aware that various other sign conventions are in wide use. Always be sure to check the signs!.

that b' is in the image of c^1. Therefore, $L^2(b') \neq 0$. We can therefore conclude that the kernel of L^2 is precisely the homology $H^2(C^\bullet, c^\bullet)$.

We can generalize the definition of L^2 to arbitrary indices to obtain a map $L^k : C^k \to C^k$. Theorem 3.5 (in a formulation inspired by Lim (2020)) below asserts that the kernel of L^k agrees with the homology at index k.

Definition 3.8 Suppose that $(C^\bullet, c^\bullet)$ is a chain complex and $(C_\bullet, c_\bullet)$ is its dual chain complex. Select a basis for $(C^\bullet, c^\bullet)$ corresponding to its barcode decomposition, and select the dual basis for $(C_\bullet, c_\bullet)$ so that $g_k : C^k \to C_k$ is the dualizing isomorphism for each index k. For each index k, the *Hodge k-Laplacian operator* (or simply the *k-Laplacian*) is a linear map $L^k : C^k \to C^k$ given by

$$L^k := g_k^{-1} \circ c_{k+1} \circ g_{k+1} \circ c^k - c^{k-1} \circ g_{k-1}^{-1} \circ c_k \circ g_k.$$

Elements of $\ker L^k$ are called *harmonic elements* of the chain complex.

Theorem 3.5 (the Hodge theorem) *If* $(C^\bullet, c^\bullet)$ *is a chain complex, then for all k*

$$H^k(C^\bullet, c^\bullet) \cong \ker L^k.$$

In short, the harmonic elements of a chain complex form a basis for its homology.

Proof An element of $\ker L^k$ is an element $v \in C^k$ such that

$$(g_k^{-1} \circ c_{k+1} \circ g_{k+1} \circ c^k)(v) = (c^{k-1} \circ g_{k-1}^{-1} \circ c_k \circ g_k)(v). \tag{3.3}$$

Suppose that v is at the start of a length 2 bar in $(C^\bullet, c^\bullet)$. This implies that v has a nontrivial image in C^{k+1}. Then the calculation in the proof of Lemma 3.5 establishes that

$$(g_k^{-1} \circ c_{k+1} \circ g_{k+1} \circ c^k)(v) = v,$$

because g_k and g_{k+1} are dualizing isomorphisms. Combining this fact with (3.3) implies that v is in the image of c^{k-1}. Therefore, v is at the end of a bar starting at C^{k-1}. This means that v is actually in the middle of a length 3 bar, which cannot occur with a chain complex.

Conversely, if we assume that v is at the end of a length 2 bar in $(C^\bullet, c^\bullet)$, this means that

$$v = (c^{k-1} \circ g_{k-1}^{-1} \circ c_k \circ g_k)(v),$$

by a calculation similar to the one in the proof of Lemma 3.5. This of course means that $c^k(v) \neq 0$ because (3.3) must hold. This again is a contradiction.

Taking these two facts together, any v corresponding to a length 2 bar cannot correspond to any element of $\ker L^k$. On the other hand, if v is a length 1 bar in C^k, then both sides of (3.3) must vanish. This means that $v \in \ker L^k$. Since we are only considering chain complexes, there is nothing else to prove. □

Example 3.18 Consider the chain complex

$$0 \longrightarrow \mathbb{R}^2 \xrightarrow{\begin{pmatrix}-3 & -6\\ 0 & 0\\ 4 & 8\end{pmatrix}} \mathbb{R}^3 \xrightarrow{\begin{pmatrix}8 & -12 & 6\\ 4 & -6 & 3\end{pmatrix}} \mathbb{R}^2 \longrightarrow 0$$

This chain complex has barcode decomposition

$$\begin{array}{ccccc}\mathbb{R} & & & & \\ \mathbb{R} & \longrightarrow & \mathbb{R} & & \\ & & \mathbb{R} & & \\ & & \mathbb{R} & \longrightarrow & \mathbb{R} \\ & & & & \mathbb{R}\end{array}$$

In particular, if we assume that the leftmost index is 0, then the Betti numbers (dimensions of the homology spaces) are

$$\beta_1 = 1, \ \beta_2 = 1, \ \text{and } \beta_3 = 1.$$

If we assert that the dual chain complex is obtained simply by transposing, we are simply using identity matrices for the dualizing isomorphisms. The formula for the Laplacians becomes a difference between matrix products, which is straightforward to compute. if we consider the first index on the left to be 0, then

$$\begin{aligned} L^1 &= \begin{pmatrix}-3 & 0 & 4\\ -6 & 0 & 8\end{pmatrix}\begin{pmatrix}-3 & -6\\ 0 & 0\\ 4 & 8\end{pmatrix} - \begin{pmatrix}0\\ 0\end{pmatrix}(0\ 0) \\ &= \begin{pmatrix}25 & 50\\ 50 & 100\end{pmatrix}, \end{aligned}$$

$$\begin{aligned} L^2 &= \begin{pmatrix}8 & 4\\ -12 & -6\\ 6 & 3\end{pmatrix}\begin{pmatrix}8 & -12 & 6\\ 4 & -6 & 3\end{pmatrix} - \begin{pmatrix}-3 & -6\\ 0 & 0\\ 4 & 8\end{pmatrix}\begin{pmatrix}-3 & 0 & 4\\ -6 & 0 & 8\end{pmatrix} \\ &= \begin{pmatrix}35 & -120 & 120\\ -120 & 180 & -90\\ 120 & -90 & -35\end{pmatrix}, \end{aligned}$$

and

$$L^3 = (0\ 0)\begin{pmatrix}0\\0\end{pmatrix} - \begin{pmatrix}8 & -12 & 6\\ 4 & -6 & 3\end{pmatrix}\begin{pmatrix}8 & 4\\ -12 & -6\\ 6 & 3\end{pmatrix}$$
$$= \begin{pmatrix}-244 & -122\\ -122 & -61\end{pmatrix}.$$

Notice that each of the Laplacians has a symmetric matrix.

It is straightforward to compute the kernel of each map, namely

$$\ker L^1 = \operatorname{span}\left\{\begin{pmatrix}-2\\1\end{pmatrix}\right\},$$

$$\ker L^2 = \operatorname{span}\left\{\begin{pmatrix}24\\25\\18\end{pmatrix}\right\},$$

and

$$\ker L^3 = \operatorname{span}\left\{\begin{pmatrix}-1\\2\end{pmatrix}\right\},$$

which agree with the Betti number calculations obtained from the barcode decomposition.

Exercise 3.35 Construct all of the nontrivial Laplacians from the chain complex

$$0 \longrightarrow \mathbb{R} \xrightarrow{\begin{pmatrix}2\\0\end{pmatrix}} \mathbb{R}^2 \xrightarrow{\begin{pmatrix}0 & 4\\0 & 6\end{pmatrix}} \mathbb{R}^2 \xrightarrow{\begin{pmatrix}3 & -2\\-3 & 2\end{pmatrix}} \mathbb{R}^2 \longrightarrow 0$$

Exercise 3.36 Compute the Laplacians for a chain complex that is nontrivial at just two indices. Work through all of the cases that can occur if these indices are consecutive.

Proposition 3.11 *The Laplacians of a chain complex* $(C^\bullet, c^\bullet)$ *have symmetric matrices when dual bases are used for the dual chain complex* $(C_\bullet, c_\bullet)$.

Proof It suffices to assume that $(C^\bullet, c^\bullet)$ is a barcode. If we use the dual basis in $(C_\bullet, c_\bullet)$, then the dualizing isomorphisms g_k are represented by identity matrices. In this situation, L^k has the matrix representation

$$L^k = c_{k+1} \circ c^k - c^{k-1} \circ c_k.$$

Dualizing is simply transposing matrices in this basis so that

$$\begin{aligned}(L^k)^T &= (c_{k+1} \circ c^k - c^{k-1} \circ c_k)^T \\ &= (c_{k+1} \circ c^k)^T - (c^{k-1} \circ c_k)^T \\ &= (c^k)^T \circ (c_{k+1})^T - (c_k)^T \circ (c^{k-1})^T \\ &= c_{k+1} \circ c^k - c^{k-1} \circ c_k = L^k,\end{aligned}$$

which establishes that the matrix for L^k is symmetric. □

Because Laplacians have symmetric matrices, they have real eigenvalues. If only the Betti numbers of a chain complex are desired, it is often more convenient to obtain them by counting the number of zero eigenvalues of the Laplacians. Moreover, the eigenvectors of a symmetric matrix can be obtained quickly by variations on the *Lanczos method* Golub and Loan (1996). If the complete barcode decomposition of a chain complex is not needed, it is therefore often quicker to work entirely with a Laplacian at a specific index.

Example 3.19 Consider the chain complex

$$0 \longrightarrow \mathbb{R} \xrightarrow{\begin{pmatrix}-1\\0\end{pmatrix}} \mathbb{R}^2 \xrightarrow{\begin{pmatrix}0&-1\\0&-3\\0&0\end{pmatrix}} \mathbb{R}^3 \xrightarrow{\begin{pmatrix}0&0&2\\0&0&-3\end{pmatrix}} \mathbb{R}^2 \longrightarrow 0,$$

which has the barcode decomposition

$$\begin{array}{cccc}
\mathbb{R} \longrightarrow & \mathbb{R} & & \\
 & \mathbb{R} \longrightarrow & \mathbb{R} & \\
 & & \mathbb{R} & \\
 & & \mathbb{R} \longrightarrow & \mathbb{R} \\
 & & & \mathbb{R}
\end{array}$$

The Laplacians for this chain complex are given by a direct application of the definition as,

$$L^1 = (-1\ 0)\begin{pmatrix}-1\\0\end{pmatrix} = (1),$$

$$L^2 = \begin{pmatrix} 0 & 0 & 0 \\ -1 & -3 & 0 \end{pmatrix} \begin{pmatrix} 0 & -1 \\ 0 & -3 \\ 0 & 0 \end{pmatrix} - \begin{pmatrix} -1 \\ 0 \end{pmatrix} \begin{pmatrix} -1 & 0 \end{pmatrix} = \begin{pmatrix} -1 & 0 \\ 0 & 10 \end{pmatrix},$$

$$L^3 = \begin{pmatrix} 0 & 0 \\ 0 & 0 \\ 2 & -3 \end{pmatrix} \begin{pmatrix} 0 & 0 & 2 \\ 0 & 0 & -3 \end{pmatrix} - \begin{pmatrix} 0 & -1 \\ 0 & -3 \\ 0 & 0 \end{pmatrix} \begin{pmatrix} 0 & 0 & 0 \\ -1 & -3 & 0 \end{pmatrix} = \begin{pmatrix} -1 & -3 & 0 \\ -3 & -9 & 0 \\ 0 & 0 & 13 \end{pmatrix},$$

and

$$L^4 = \begin{pmatrix} 0 \\ 0 \end{pmatrix} \begin{pmatrix} 0 & 0 \end{pmatrix} - \begin{pmatrix} 0 & 0 & 2 \\ 0 & 0 & -3 \end{pmatrix} \begin{pmatrix} 0 & 0 \\ 0 & 0 \\ 2 & -3 \end{pmatrix} = \begin{pmatrix} -4 & 6 \\ 6 & -9 \end{pmatrix}.$$

The eigenvalues of each Laplacian can be computed using their respective characteristic polynomials, though in this case L^1 and L^2 can be computed by inspection:

- The eigenvalue of L^1 is 1, and
- The eigenvalues of L^2 are -1 and 10.

This implies that the homologies $H^1(C^\bullet, c^\bullet)$ and $H^2(C^\bullet, c^\bullet)$ are both trivial, a fact which can be confirmed by the non-existence of length 1 bars in those indices in the barcode decomposition.

Recall that $\det \ell$ is the determinant of a linear map ℓ. The characteristic equation of L^3 is

$$\begin{aligned} 0 &= \det \begin{pmatrix} -1-\lambda & -3 & 0 \\ -3 & -9-\lambda & 0 \\ 0 & 0 & 13-\lambda \end{pmatrix} \\ &= (13-\lambda) \det \begin{pmatrix} -1-\lambda & -3 \\ -3 & -9-\lambda \end{pmatrix} \\ &= (13-\lambda)\left((-1-\lambda)(-9-\lambda) - 9\right) \\ &= (13-\lambda)\left(9 + 10\lambda + \lambda^2 - 9\right) \\ &= (13-\lambda)(10+\lambda)\,\lambda, \end{aligned}$$

which implies that the eigenvalues are $\lambda = -10, 0, 13$. The single zero eigenvalue is consistent with the Betti number $\beta_3 = 1$.

The eigenvalues of L^4 are obtained by its characteristic equation

$$\begin{aligned} 0 &= \det \begin{pmatrix} -4-\lambda & 6 \\ 6 & -9-\lambda \end{pmatrix} \\ &= (-4-\lambda)(-9-\lambda) - 36 \\ &= 36 + 13\lambda + \lambda^2 - 36 \\ &= (13+\lambda)\lambda, \end{aligned}$$

implying that the eigenvalues are $\lambda = 0, -13$. Once again, this is consistent with $\beta_4 = 1$.

Exercise 3.37 Consider the following chain complex

$$0 \longrightarrow \mathbb{R} \xrightarrow{(0)} \mathbb{R} \xrightarrow{\begin{pmatrix}-2\\-1\end{pmatrix}} \mathbb{R}^2 \xrightarrow{\begin{pmatrix}0&0\\0&0\end{pmatrix}} \mathbb{R}^2 \longrightarrow 0.$$

1. Compute the Laplacian matrices for the chain complex.
2. Diagonalize each Laplacian matrix.
3. Verify that the eigenvectors corresponding to zero eigenvalues form a basis for each homology space.

Exercise 3.38 Consider the following chain complex

$$0 \longrightarrow \mathbb{R}^2 \xrightarrow{\begin{pmatrix}0&0\\-2&3\\0&0\end{pmatrix}} \mathbb{R}^3 \xrightarrow{\begin{pmatrix}0&0&0\\-4&0&3\end{pmatrix}} \mathbb{R}^2 \longrightarrow 0.$$

1. Compute the Laplacian matrices for the chain complex.
2. Diagonalize each Laplacian matrix.
3. Verify that the eigenvectors corresponding to zero eigenvalues form a basis for each homology space.

Exercise 3.39 Consider the chain map

$$\begin{array}{ccccccccccc}
0 & \longrightarrow & \mathbb{R}^2 & \xrightarrow{\begin{pmatrix}-2&-2\\-1&-4\end{pmatrix}} & \mathbb{R}^2 & \xrightarrow{\begin{pmatrix}0&0\\0&0\end{pmatrix}} & \mathbb{R}^2 & \xrightarrow{(3\ -12)} & \mathbb{R} & \longrightarrow & 0 \\
\big\downarrow & & \big\downarrow{\scriptstyle\begin{pmatrix}12&6\\12&6\\6&0\end{pmatrix}} & & \big\downarrow{\scriptstyle\begin{pmatrix}0&0\\-2&-2\end{pmatrix}} & & \big\downarrow{\scriptstyle\begin{pmatrix}5&-16\\-4&14\end{pmatrix}} & & \big\downarrow{\scriptstyle(1)} & & \\
\mathbb{R} & \xrightarrow[\begin{pmatrix}-3\\-2\\1\end{pmatrix}]{} & \mathbb{R}^3 & \xrightarrow[\begin{pmatrix}0&0&0\\-7&9&-3\end{pmatrix}]{} & \mathbb{R}^2 & \xrightarrow[\begin{pmatrix}-2&0\\1&0\end{pmatrix}]{} & \mathbb{R}^2 & \xrightarrow[(-1\ -2)]{} & \mathbb{R} & \longrightarrow & 0
\end{array}$$

Compute the induced map on homology of this chain map using Laplacians. Use the following procedure:

1. Construct the Laplacian matrices for both chain complexes.
2. Compute the kernel of each Laplacian matrix for the domain chain complex.
3. Transform each kernel element you found in (2) by the component maps.
4. Verify that each element you found in (3) is in the kernel of its corresponding Laplacian in the codomain chain complex.

3.6 Further Reading

For a pair of sequences $(V^\bullet, v^\bullet)$ and $(W^\bullet, w^\bullet)$, one can ask about the structure of the collection of all sequence maps $m^\bullet : (V^\bullet, v^\bullet) \to (W^\bullet, w^\bullet)$. Many authors have studied the conditions under which these sequence maps exist, and if they do exist, which are the "smallest" in some sense. One starts by constructing a "matching" between the bars of $V^\bullet$ and the bars of $W^\bullet$. This can often be performed algorithmically. The interested reader is referred to Giusti and Lee (2023), Yoon et al. (2023), Bauer and Lesnick (2014).

Conversely, if one wants to *infer* a sequence map by matching the bars of the domain with those in the codomain, this can be done. The problem is under determined. There are many possibilities for the sequence map. Details of one possible way to do this efficiently are described in Gonzalez-Diaz et al. (2023).

While this chapter discussed chain maps, sequence maps between sequences whose bars have length 2 or less, Escolar and Hiraoka (2016) establishes that similar results hold for longer bars. Specifically, under the right conditions, there is a decomposition of a sequence map into a direct sum of finitely many different kinds of simpler, *indecomposable* sequence maps, even though the bars of the two sequences are not preserved. The conditions require that each indecomposable (as a whole) is of length 4 or less, which is certainly satisfied for chain maps. On the other hand, for sequences whose indecomposables have five nontrivial terms, there is an infinite collection of indecomposable sequence maps. The interested reader is encouraged to read Escolar and Hiraoka (2016) for full details, including a complete characterization of the indecomposable diagrams and the Auslander-Reiten quiver that shows all of the relationships between them, of which Table 3.1 in this chapter is a special case.

Much of the proof of Theorem 3.1 generalizes rather beautifully to sequences with non-integer indices (Jacquard et al. 2023). The overlappingness that shows up in the proof of Theorem 3.1 plays an important role for general sequences of vector spaces. If there is no overlapping, there is a canonical decomposition; otherwise there may be several distinct choices. In the chain map case, this overlapping condition degenerates, which is why we get quite the zoo of bar chain maps.

The Hodge Laplacians discussed in Sect. 3.5 play an interesting role in signal processing, since they extend a notion of Fourier analysis in signal processing (Aksoy and Young 2025; Lim 2020; Krim et al. 2016; Krim and Hamza 2015). They also provide a concrete way to discover "hidden" coordinates of a data set (Scoccola and Perea 2023; Scoccola et al. 2023; Perea and Carlsson 2014).

References

Aksoy SG, Young SJ (2025) Re-imagining spectral graph theory. https://arxiv.org/abs/2504.10624

Bauer U, Lesnick M (2014) Induced matchings of barcodes and the algebraic stability of persistence. In: Proceedings of the thirtieth annual symposium on computational geometry, pp 355–364

de Silva V, Morozov D, Vejdemo-Johansson M (2011) Dualities in persistent (co)homology. Inverse Probl. 27(12):124003. https://doi.org/10.1088/0266-5611/27/12/124003

Escolar EG, Hiraoka Y (2016) Persistence modules on commutative ladders of finite type. Discret Comput Geom 55(1):100–157

Giusti C, Lee D (2023) Signatures, Lipschitz-free spaces, and paths of persistence diagrams. SIAM J Appl Algebr Geom 7(4):828–866

Golub GH, Loan CFV (1996) Matrix computations. Johns Hopkins University Press, pp 470–507

Gonzalez-Diaz R, Soriano-Trigueros M, Torras-Casas A (2023) Partial matchings induced by morphisms between persistence modules. Comput Geom 112:101985

Jacquard E, Nanda V, Tillmann U (2023) The space of barcode bases for persistence modules. J Appl Comput Topol 7(1):1–30

Krim H, Gentimis T, Chintakunta H (2016) Discovering the whole by the coarse: a topological paradigm for data analysis. IEEE Signal Proc Mag 33(2):95–104

Krim H, Hamza AB (2015) Geometric methods in signal and image analysis. Cambridge University Press

Lim LH (2020) Hodge Laplacians on graphs. SIAM Rev. 62(3):685–715

Perea JA, Carlsson G (2014) A Klein-bottle-based dictionary for texture representation. Int J Comput Vis 107:75–97

Scoccola L, Gakhar H, Bush J, Schonsheck N, Rask T, Zhou L, Perea JA (2023) Toroidal coordinates: decorrelating circular coordinates with lattice reduction. https://arxiv.org/abs/2212.07201

Scoccola L, Perea JA (2023) Approximate and discrete Euclidean vector bundles. In: Forum of mathematics, Sigma. Cambridge University Press, vol 11, p e20

Yoon HR, Ghrist R, Giusti C (2023) Persistent extensions and analogous bars: data-induced relations between persistence barcodes. J Appl Comput Topol 7(3):571–617

Chapter 4
Abstract Simplicial Complexes

Chain complexes were not conceived on their own, but were discovered in the course of exploring various abstract models of space. An *abstract simplicial complex* is a representation of a space that is both structurally rich yet requires few axioms to describe. Given an abstract simplicial complex, one can construct its *simplicial chain complex* and compute its *simplicial homology*, which are algebraic summaries of its structure.

To build some intuition through an example, consider a collection of 10 points in the plane, shown in Fig. 4.1a. Each point is disconnected from the rest, so there are 10 connected "pieces". It is natural to connect some of the points with line segments, which reduces the number of connected pieces. For instance, in Fig. 4.1b, we have added two line segments. These two segments effectively reduce the number of connected pieces by 2, suggesting the formula

$$\text{\# pieces} = \text{\# points} - \text{\# line segments}.$$

This formula is not really correct, though. Consider Fig. 4.1c, where we have added a total of 5 line segments. The formula would suggest that the number of connected pieces is 5 even though the correct number is 6. What has happened is that the triangle has 3 points and 3 line segments, which the formula says counts as 0 pieces! To restore order, we need to intentionally add the triangle back, yielding a new formula

$$\text{\# pieces} = \text{\# points} - \text{\# line segments} + \text{\# triangles}.$$

Thus in Fig. 4.1d, the revised formula produces the correct total number of pieces: 6. Of course, the reader may well reason that if we have ill-placed triangles, we may again end up with the wrong number of pieces. This is correct! For instance, a tetrahedron has 4 points, 6 line segments, and 4 triangular faces. Our formula counts the number of pieces for this one tetrahedron as 2. We need to deduct an extra tetrahedron to get the correct answer.

M. Robinson, *Computational Homological Algebra*, Mathematical Engineering,
https://doi.org/10.1007/978-3-032-08634-1_4

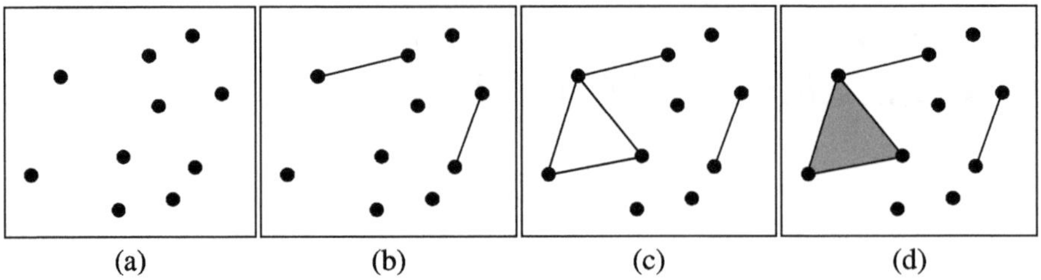

Fig. 4.1 Progressively adding simplices spanned by a set of points: **a** a collection of 10 disconnected points, **b** two edges added to the points, **c** three more edges added, and finally **d** a triangle added

The formula we are constructing is an alternating sum called the *Euler characteristic* of an abstract simplicial complex

$$\chi := \sum_{k=0}^{\infty}(-1)^k \text{ (\#}k\text{-dimensional simplices)} ,$$

in which a point is a 0-dimensional simplex, a line segment is a 1-dimensional simplex, a triangle is a 2-dimensional simplex, and so on. We already defined the Euler characteristic earlier in Sect. 2.2 of Chap. 2 for sequences of vector spaces, and this is no accident. The Euler characteristic for simplices can be derived from the one for sequences of vector spaces, and this motivates the idea of the *simplicial chain complex*. Moreover, since we have already proven that the Euler characteristic is sensitive only to the homology of a chain complex, it stands to reason that the homology associated to the simplicial chain complex will carry interesting information about the simplicial complex itself.

In this chapter,
Section 4.1 Defines *abstract simplicial complexes*, which model "abstract spaces". Section 4.2 Explores some key combinatorial properties of abstract simplicial complexes that are useful technical tools in what follows. Section 4.3 Defines a few ways to construct abstract simplicial complexes from other combinatorial objects. Section 4.4 Defines the *category of abstract simplicial complexes*. Section 4.5 Defines *simplicial chain complexes*, showing how they generalize the Euler characteristic discussed above.

4.1 Abstract Simplicial Complexes

When trying to develop a notion of connectedness between different sets of items, a natural place to start is that of a *graph*. We declare that the items form the *vertices* of a *graph*.

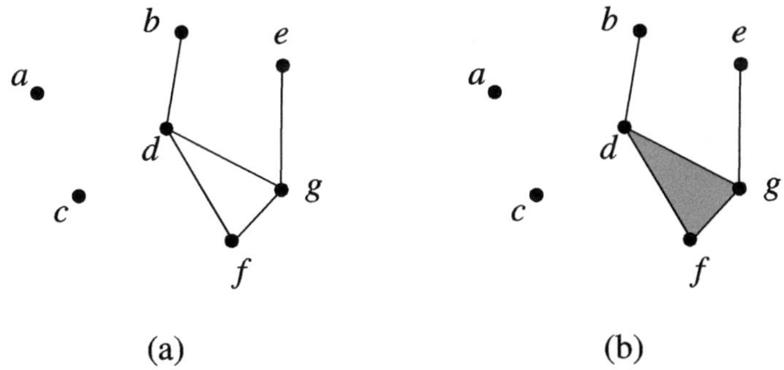

Fig. 4.2 **a** A graph on 7 vertices (which is also an abstract simplicial complex with only 0 and 1 dimensional simplices), **b** an abstract simplicial complex on the same 7 vertices with a 2 dimensional simplex. See Example 4.1

Definition 4.1 A *graph* $G = (V, E)$ consists of a set V of *vertices* and a set E of *edges*, in which each element of E must be a two element subset of V. We usually say that an edge $e = \{v, w\}$ is an edge *between* or *connecting* vertices v and w. The *degree* of v is the number of edges that contain v.

Example 4.1 Figure 4.2a shows a graph on the seven vertices $V = \{a, b, c, d, e, f, g\}$. The edges in this graph are $E = \{\{b, d\}, \{d, f\}, \{d, g\}, \{e, g\}, \{f, g\}\}$. Notice that while all pairs of the edges $\{d, f\}$, $\{d, g\}$, and $\{f, g\}$ have nonempty intersections, there is no vertex in common among all three. This is an example of a *loop* in the graph.

The edges in a graph are fundamentally limited to modeling pairwise relations between vertices. This constraint is unnatural but easily relaxed. The most convenient generalization for algebraic purposes is that of an *abstract simplicial complex*, which allows the edges to consist of larger subsets of vertices.

Definition 4.2 An *abstract simplicial complex* (or *simplicial complex*) (X, V_X) is a collection X of subsets of a set V_X satisfying two properties:

Coverage $\bigcup X = V_X$, and
Closure If σ is a subset of V_X that is in X and τ is some subset of σ, then τ is also in X.

Often when no confusion can arise, we will refer to X as a simplicial complex rather than writing (X, V_X). We call each element of V_X a *vertex*, and each element of X a *simplex*. The *dimension* of a simplex $\sigma \in X$ is defined to be one less than its cardinality. The maximum dimension of any simplex in X is called the *dimension of X*. If σ is a simplex of dimension k, we often call σ a *k-simplex*. If σ and τ are simplices of X for which $\sigma \subseteq \tau$ holds, we say that σ is a *face* of τ.

We use square brackets to indicate the vertices included in a given simplex. For instance, if $\sigma = [v_0, v_1, \ldots, v_k]$ is a simplex, we say that the dimension of σ is k.

The coverage axiom for a simplicial complex ensures that every vertex is contained in at least one simplex. This is useful because the definition of an abstract simplicial complex distinguishes between a vertex $v \in V_X$ and a 0-dimensional simplex $[v] \in X$. Moreover, every simplicial complex contains the empty set $\emptyset$, and the empty set has dimension -1.

Example 4.2 Every graph $G = (V, E)$ can be rendered as a 1-dimensional simplicial complex with vertex set V, by taking each vertex v as a simplex $[v]$ and also each edge $\{v_1, v_2\} \in E$ as a simplex $[v_1, v_2]$. Every 1-dimensional simplicial complex can also be converted to a graph simply by identifying the 1-dimensional simplices as edges. Using this interpretation, the graph shown in Fig. 4.2a, and whose edges are listed in Example 4.1, becomes the simplicial complex

$$\{\emptyset, [a], [b], [c], [d], [e], [f], [g], [b, d], [d, f], [d, g], [e, g], [f, g]\}.$$

With the added structure of a simplicial complex, we can distinguish between the situation of the loop in the graph, in which edges $[d, f]$, $[d, g]$, and $[f, g]$ have no common vertex across all three, and a situation where there is an additional simplex $[d, f, g]$, since

$$\{\emptyset, [a], [b], [c], [d], [e], [f], [g], [b, d], [d, f], [d, g], [e, g], [f, g], [d, f, g]\}$$

is also a simplicial complex. This larger complex is represented pictorially in Fig. 4.2b.

Example 4.3 Assuming that $V = \{a, b, c\}$ is a set of vertices, if a simplicial complex contains $[a, b, c]$, then it must contain all subsets of this, namely each of the following

$$\emptyset, [a], [b], [c], [a, b], [a, c], \text{ and } [b, c],$$

as shown schematically in Fig. 4.3.

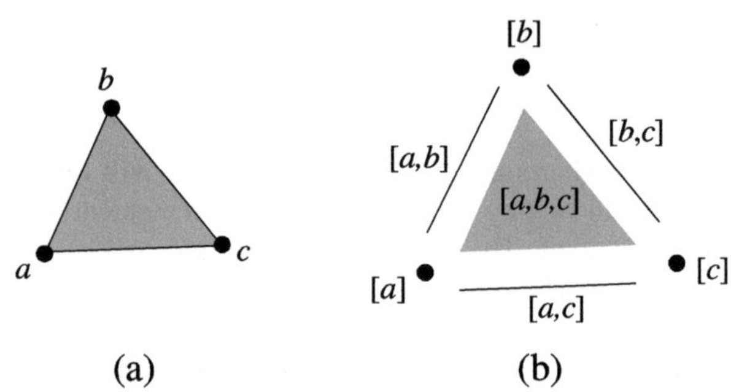

Fig. 4.3 **a** An abstract simplicial complex consisting of one 2-dimensional simplex, and **b** each of its nonempty simplices. See Example 4.3

If we expand the vertex set to $V' = \{a, b, c, d\}$,

$$\{\emptyset, [a], [b], [c], [a, b], [a, c], [b, c], [a, b, c]\}$$

is not a valid simplicial complex because it violates the coverage axiom: $[d]$ is not present.

On the other hand,

$$\{\emptyset, [b], [c], [d], [a, c], [b, c], [a, b, c]\}$$

is not a valid simplicial complex because it violates closure. It is missing $[a]$ and $[a, b]$.

Exercise 4.1 Suppose that $k > 0$ is a positive integer. Prove that the process of taking a simplicial complex X and removing all simplices of dimension k or greater yields a new simplicial complex.

Exercise 4.2 Assuming that $V = \{a, b, c, d, e\}$ is a set of vertices, list all of the simplices that must be present in any simplicial complex that contains $[a, b, c, d]$, $[b, c, d, e]$, $[a, c, d, e]$, and $[a, b, c, e]$.

Exercise 4.3 Assuming that $V = \{a, b, c, d\}$ is a set of vertices, list all of the simplices that must be present in any simplicial complex that contains $[a]$, $[a, b]$, $[a, b, c]$, and $[a, b, c, d]$.

Listing all of the simplices in anything but the smallest simplicial complexes quickly becomes tedious. Fortunately, it is usually not necessary to list all simplices.

Definition 4.3 The simplicial complex *generated by* a set Y of subsets of V is the smallest simplicial complex X that contains Y. Equivalently, the simplicial complex generated by Y is the intersection of all simplicial complexes that contain Y. We call Y the *generating set* of X.

Example 4.4 The simplicial complex generated by $[a, b, c]$ and $[b, c, d, e]$ is shown in Fig. 4.4. Notice in particular that $[b, c]$ is a simplex in this complex but that $[a, b, c, d]$ is not.

Exercise 4.4 Assuming that $V = \{a, b, c, d, e, f, g\}$ is a set of vertices, list all of the simplices generated by $[a, b, c, d]$, $[a, e, f]$, $[b, f]$, and $[g]$.

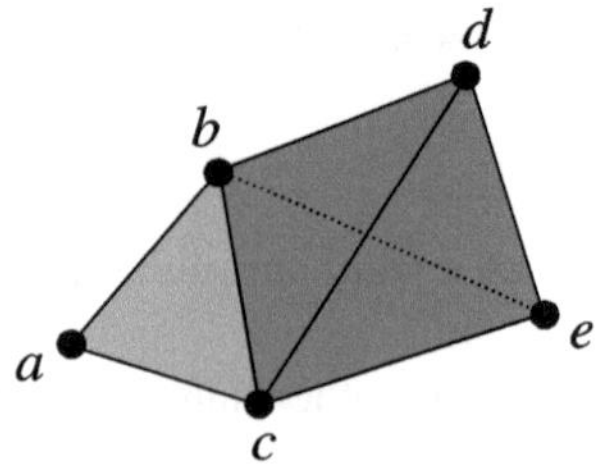

Fig. 4.4 The abstract simplicial complex generated by $[a, b, c]$ and $[b, c, d, e]$ in Example 4.4

Exercise 4.5 Assuming that $V = \{2, 3, \dots\}$ is a set of vertices, list all of the simplices generated by $[n, 2n, 3n, 4n, \dots]$ for each $n \in V$.

Proposition 4.1 *Minimal (in the sense of inclusion) generating sets are unique.*

Proof Suppose that X is a simplicial complex generated by a set $G \subseteq X$ of simplices. Furthermore, assume that G is minimal, which is to say that if F is any other set of simplices that generates X, then $G \subseteq F$. Let us assume that there is a different minimal generating set G'. Evidently, since G' is a generating set, $G \subseteq G'$ by the minimality of G. But since G' is also minimal, $G' \subseteq G$. Thus $G = G'$. □

Example 4.5 Let $X = \{a, b, c, d\}$ and $Y = \{A, B, C, D, E\}$. We can represent a relation $R \subseteq X \times Y$ as Boolean matrix with one row for each element of X and one column for each element of Y. A 1 in a given entry in the matrix means that ordered pair is an element of R, while a 0 means that ordered pair is not an element of R. For instance, if we order rows and columns alphabetically, the matrix

$$r := \begin{pmatrix} 1 & 0 & 1 & 1 & 1 \\ 1 & 1 & 0 & 0 & 1 \\ 0 & 1 & 1 & 1 & 0 \\ 0 & 0 & 1 & 0 & 0 \end{pmatrix}$$

defines a relation R for which (a, A) is a member but (d, A) is not. The matrix r can be used to construct a simplicial complex by reading off each column and selecting the rows that contain a 1. The resulting simplicial complex is generated by the simplices $[a, b]$, $[b, c]$, and $[a, c, d]$, as shown in Fig. 4.5a. These defining simplices correspond to the first three columns. Notice that the fourth column implies the existence of a simplex $[a, c]$ which is already present because it is a face of $[a, c, d]$. Similarly, the last column is a duplicate of the first, and therefore contributes no new simplices.

Definition 4.4 Suppose that $R \subseteq X \times Y$ is a binary relation between two finite sets X and Y, such that for every $x \in X$, there is at least one $y \in Y$ with $(x, y) \in R \subseteq X \times Y$. The *Dowker complex* $D(X, Y, R)$ is the abstract simplicial complex constructed according to the following recipe: each subset $\sigma \subseteq X$ is a simplex of $D(X, Y, R)$ provided there is a $y \in Y$ such that $(x, y) \in R$ for each $x \in \sigma$.

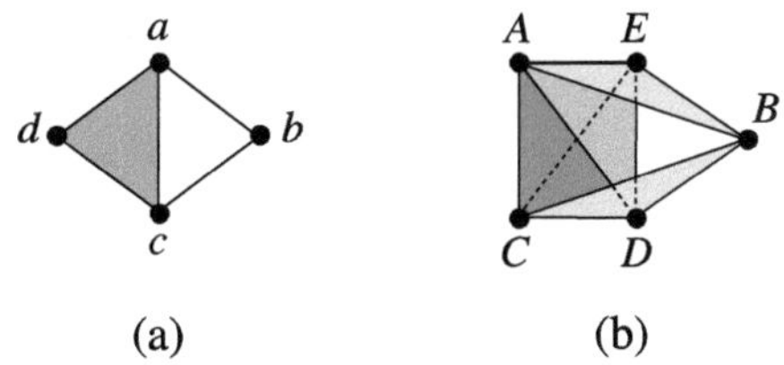

Fig. 4.5 The two Dowker simplicial complexes formed from the relation R defined in Example 4.5: **a** the complex $D(X, Y, R)$, **b** the complex $D(Y, X, R^T)$

The simplicial complex defined in Example 4.5 is the Dowker complex $D(X, Y, R)$ for the relation R.

Proposition 4.2 *The Dowker complex $D(X, Y, R)$ is a well defined abstract simplicial complex, which is to say that Definition 4.4 implies that the closure axiom holds.*

Proof Suppose that σ is a simplex of $D(X, Y, R)$ and that $\tau \subset \sigma$. Since σ is a simplex of $D(X, Y, R)$, this means that there is a $y \in Y$ such that $(x, y) \in R$ for each $x \in \sigma$. This same y works for every x in τ since $\tau \subset \sigma$, so τ is a simplex of $D(X, Y, R)$ as well. □

Example 4.6 Notice that the relation R in Example 4.5 is not symmetric. Indeed, symmetry makes little sense since X and Y are different sets! However, the transpose r^T of r, given by the matrix

$$r^T = \begin{pmatrix} 1 & 1 & 0 & 0 \\ 0 & 1 & 1 & 0 \\ 1 & 0 & 1 & 1 \\ 1 & 0 & 1 & 0 \\ 1 & 1 & 0 & 0 \end{pmatrix},$$

defines a relation $R^T \subseteq Y \times X$. We can therefore construct the Dowker complex $D(Y, X, R^T)$, which is generated by $[A, C, D, E]$, $[A, B, E]$, and $[B, C, D]$, as shown in Fig. 4.5b.

Evidently the simplicial complexes $D(X, Y, R)$ and $D(Y, X, R^T)$ are not the same, even if we relabel vertices, but they are related. A famous theorem of Dowker (1952) asserts that they have the same *simplicial homology* $H_\bullet(D(X, Y, R)) \cong H_\bullet(D(Y, X, R^T))$. Simplicial homology will be defined later in Sect. 5.1 of Chap. 5.

4.2 Combinatorial Properties of Simplicial Complexes

Several useful subsets of a simplicial complex help characterize its structure.

Definition 4.5 If σ is a simplex of an abstract simplicial complex X, the *star of* σ consists of the set

$$\operatorname{star} \sigma := \{\tau \in X : \sigma \subseteq \tau\}$$

of simplices that contain σ as a subset. Dually, *closure of* σ consists of the set

$$\operatorname{cl}\sigma := \{\tau \in X : \tau \subseteq \sigma\}$$

of simplices contained by σ. Finally, the *link of* σ consists of the set

$$\operatorname{lk}\sigma := \{\tau \in X : \sigma \cap \tau = \emptyset, \text{ and } \sigma \cup \tau \in X\}$$

of simplices that are disjoint from σ but whose union with σ is a larger simplex (see also Munkres 1984, p. 371).

Example 4.7 Consider the simplicial complex generated by $[a, b, c]$, $[c, d]$, $[a, e]$, $[e, f]$, $[c, f]$, $[g]$ shown in Fig. 4.6. Figure 4.7 shows the star, closure, and link of $[e]$ and $[e, f]$ in this simplicial complex.

- Two simplices properly contain e, so $\operatorname{star}[e] = \{[e], [a, e], [e, f]\}$.
- Since no simplices properly contain $[e, f]$, $\operatorname{star}[e, f] = \{[e, f]\}$.
- The subsets of $[e]$ include $\emptyset$, so $\operatorname{cl}[e] = \{\emptyset, [e]\}$.
- The subsets of $[e, f]$ are $\operatorname{cl}[e, f] = \{\emptyset, [e], [f], [e, f]\}$.
- Exactly two simplices properly contain e, namely $[a, e]$ and $[e, f]$. The only candidates for the link of $[e]$ involve vertices a and f, so $\operatorname{lk}[e] = \{[a], [f]\}$.
- Since no simplices properly contain $[e, f]$, the link $\operatorname{lk}[e, f]$ must be empty.

Exercise 4.6 Consider the simplicial complex generated by $[a, b, c, d]$, $[b, c, e]$, $[a, c, e]$, $[a, b, e]$.

1. Compute $\operatorname{star}[a]$,
2. Compute $\operatorname{cl}[a]$,
3. Compute $\operatorname{lk}[a]$,
4. Compute $\operatorname{star}[a, c, e]$,
5. Compute $\operatorname{cl}[a, c, e]$,

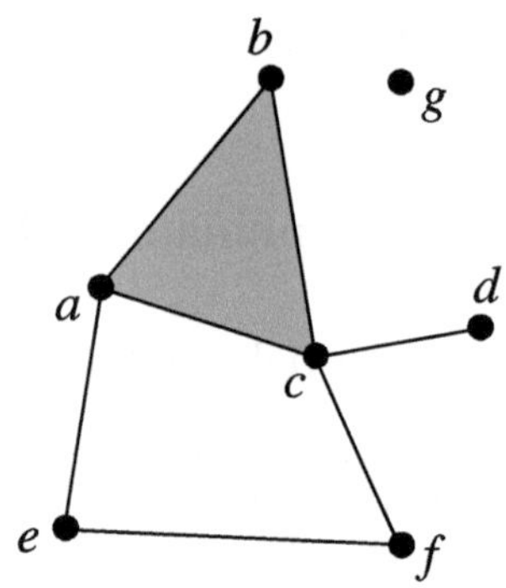

Fig. 4.6 The simplicial complex in Examples 4.7, 4.8, and 4.9

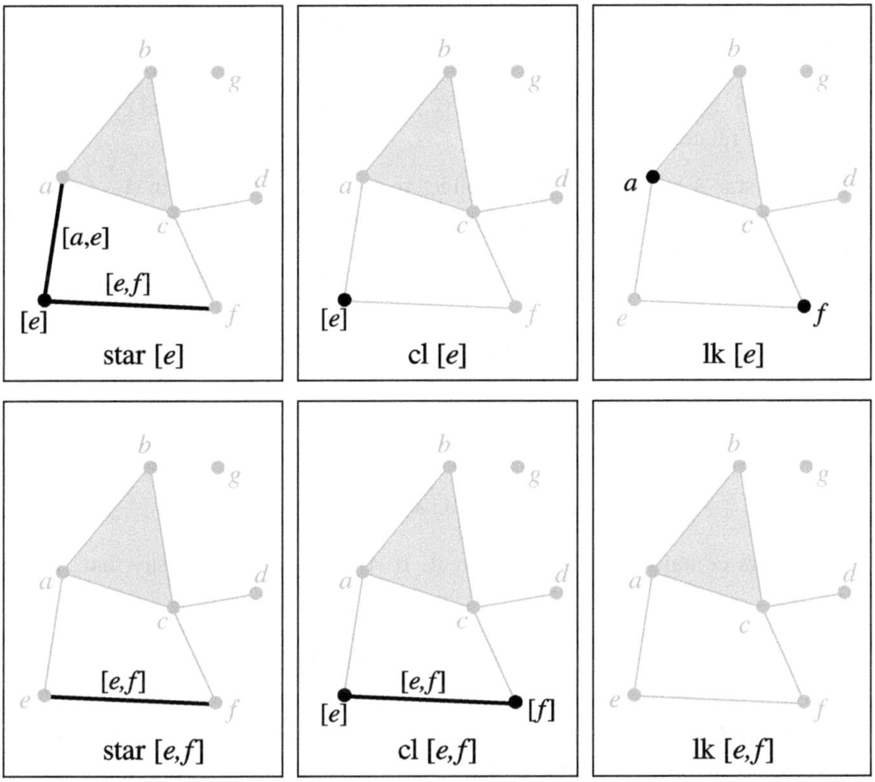

Fig. 4.7 The star, closure, and link of $[e]$ and $[e, f]$ in the simplicial complex shown in Fig. 4.6 appearing in Examples 4.7, 4.8, and 4.9. See also Definition 4.5 for details

6. Compute $\mathrm{lk}[a, c, e]$,
7. Compute $\mathrm{star}[a, d]$,
8. Compute $\mathrm{cl}[a, d]$, and
9. Compute $\mathrm{lk}[a, d]$.

Exercise 4.7 Consider the simplicial complex generated by $[a, b, c]$, $[a, b, e]$, $[a, c, d]$, $[a, d, e]$, $[b, f]$, $[d, f]$.

1. Compute $\mathrm{star}[a]$,
2. Compute $\mathrm{star}[f]$,
3. Compute $\mathrm{cl}[b, e]$,
4. Compute $\mathrm{lk}[a]$, and
5. Compute $\mathrm{lk}[f]$.

The operations of star, closure, and link for a simplex can be generalized to sets of simplices.

Definition 4.6 Starting with a subset $A \subseteq X$ of an abstract simplicial complex, the following are related subsets:

1. The *star* star A is the set of all simplices that contain a simplex in A.
2. The *closure* cl A is the set of all simplices that are contained by a simplex in A. If $A = \text{cl } A$, then we say that A is a *closed set*.
3. The *link* lk A is the set of all simplices in cl star A whose vertex sets are disjoint from A, or
$$\text{lk } A := (\text{cl star } A) \backslash (\text{star } A \cup \text{cl } A).$$
4. The *interior* int A is the union
$$\text{int } A := \bigcup_{\sigma \in A \,:\, \text{star}\{\sigma\} \subseteq A} \text{star}\{\sigma\}$$
of all stars contained entirely within A. If $A = \text{int } A$, then we say that A is an *open set*.
5. The *frontier*[1] is
$$\text{fr } A := \text{cl } A \cap \text{cl}(X \backslash A).$$

Exercise 4.8 Show that if σ is a simplex, then the two definitions of the star of σ agree. In other words show that
$$\text{star}\{\sigma\} = \text{star } \sigma.$$
Additionally, show that
$$\text{int}\{\text{star } \sigma\} = \text{star } \sigma.$$

Exercise 4.9 Show that for any subset A of simplices in a simplicial complex,
$$\text{int } A = \text{cl } A \backslash \text{ fr } A.$$
Hint: There are several ways to approach this problem. The most algebraic is to use the definition of fr A to restate the right side of the equation in terms of closed sets.

[1] The frontier is often called the *boundary*, but then this term can be confused with other senses of the word "boundary".

Exercise 4.10 Show that the frontier of a single simplex σ in a simplicial complex X is equal to all of the proper subsets of σ.

Exercise 4.11 (*For readers who know topology*) Show that the set of stars of all subsets of an abstract simplicial constitutes a topology on the set of all simplices.

Definition 4.7 A *path* in a simplicial complex X is an ordered sequence of vertices $\{v_0, v_1, \ldots, v_n\}$ such that every pair of consecutive vertices v_i, v_{i+1} is a subset of some simplex of X. Such a path is said to *start* at v_0 and *end* at v_n. A simplicial complex is said to be *path connected* if for every pair of vertices a, b in X, there is a path starting at a and ending at b.

Example 4.8 Consider the simplicial complex generated by $[a, b, c]$, $[c, d]$, $[a, e]$, $[e, f]$, $[c, f]$ shown in Fig. 4.6. The path e, f, c, a, b is shown in Fig. 4.8. This path begins at e and ends at b and is a valid path because the simplicial complex contains all of the following 1-dimensional simplices: $[e, f]$, $[c, f]$, $[a, c]$, and $[a, b]$.

Example 4.9 In the simplicial complex generated by $[a, b, c]$, $[c, d]$, $[a, e]$, $[e, f]$, $[c, f]$ shown in Fig. 4.6, there is no path from vertex a to vertex g. Therefore, this simplicial complex is not path connected.

Exercise 4.12 In the simplicial complex generated by $[a, b, c]$, $[c, d]$, $[a, e]$, $[e, f]$, $[c, f]$ shown in Fig. 4.6, enumerate all paths from b to d.

Proposition 4.3 *Existence of a path connecting a pair of vertices is an equivalence relation. The equivalence classes of this relation are called the* path connected components *or simply the* path components *of the abstract simplicial complex.*

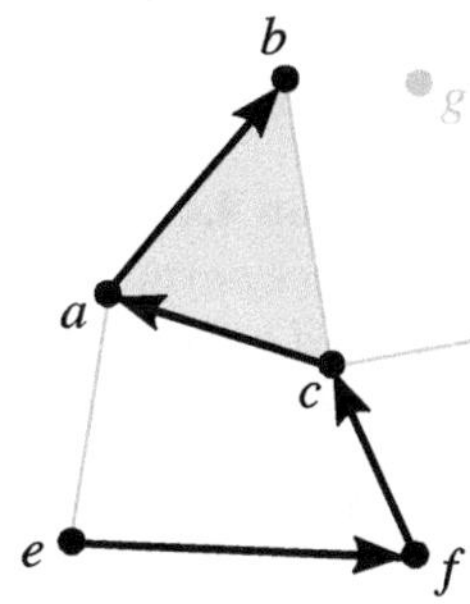

Fig. 4.8 The path e, f, c, a, b in the simplicial complex shown in Fig. 4.6, discussed in Example 4.8

Proof Proceed constructively,

Reflexivity Because of the coverage axiom, each vertex v is contained in at least one simplex. Therefore, there is a path from each vertex v to itself, namely the path of length 1, $\{v\}$.

Symmetry Suppose that $\{v_0, v_1, \ldots, v_n\}$ is a path from v_0 to v_n. Taking the vertices in the opposite order yields a path $\{v_n, v_{n-1}, \ldots, v_0\}$ starting at v_n and ending at v_0.

Transitivity If $\{v_0, v_1, \ldots, v_n\}$ is a path from v_0 to v_n and $\{v_n, w_1, \ldots, w_m\}$ is a path from v_n to w_m, then their concatenation $\{v_0, v_1, \ldots, v_n, w_1, \ldots, w_m\}$ is a path from v_0 to w_m. □

Exercise 4.13 Consider the abstract simplicial complex generated by $X = \{[a, b, c], [a, d], [b, d]\}$.

1. Sketch a diagram of X.
2. How many path components does X have? Circle each component in your diagram.
3. List the simplices of the largest simplicial complex Y that arises when you delete the vertex a from X.

4.3 Simplicial Complexes and Partial Orders

Recall the definition of a *partially ordered set* from Definition 1.13 in Chap. 1. Specifically, a partially ordered set consists of an ordered pair $(P, \leq_P)$, a set P and an order relation $\leq_P$ on elements of P such that the following three axioms hold:

1. (Reflexivity) $x \leq_P x$ for all $x \in P$,
2. (Transitivity) If $x \leq_P y$ and $y \leq_P z$, then $x \leq_P z$, and
3. (Antisymmetry) $x \leq_P y$ and $y \leq_P x$ implies $x = y$.

Notice that a partial order does not require that either $x \leq_P y$ or $y \leq_P x$ for an arbitrary pair of elements $x, y \in P$. But if at least one of these is always true, then we call $\leq_P$ a *total order*.

Definition 4.8 The *face partial order* for an abstract simplicial complex X has the simplices of X as its elements, and $\sigma \leq \tau$ whenever $\sigma \subseteq \tau$.

Proposition 4.4 *The star of a simplex σ consists of all simplices above (containing) it in the face partial order. The closure of a simplex consists of all the simplices below (contained by) it in the face partial order. The intersection $\sigma \cap \tau$ of two simplices σ and τ is the maximal simplex in the face partial order that is below both σ and τ.*

Proof The first two statements simply rephrase the definition of star and closure from Sect. 4.2.

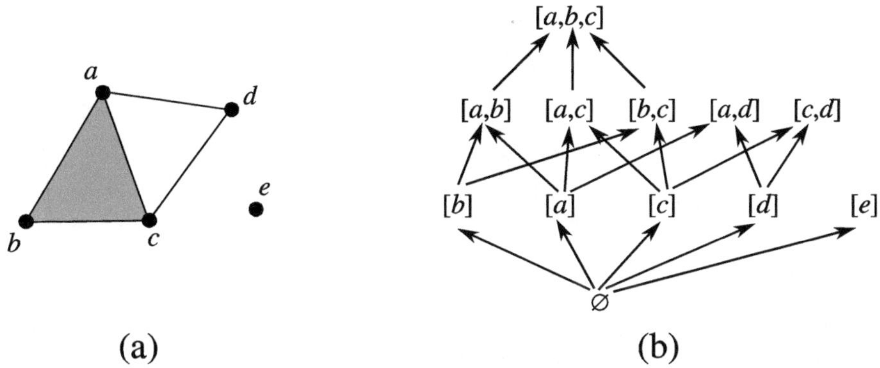

Fig. 4.9 **a** The simplicial complex generated by $[a, b, c]$, $[c, d]$, $[a, d]$, $[e]$, and **b** its face partial order. See Example 4.10

To prove the last statement, suppose that there is another simplex η below both σ and τ. This means that $\eta \subseteq \sigma$ and $\eta \subseteq \tau$. But then, if $v \in \eta$, we must conclude that $v \in \sigma$ and $v \in \tau$. Thus, $v \in (\sigma \cap \tau)$, which establishes that $\eta \leq (\sigma \cap \tau)$. If $\eta \geq (\sigma \cap \tau)$ in the face partial order, this means that $(\sigma \cap \tau) \subseteq \eta$. We must conclude that $(\sigma \cap \tau) = \eta$. □

Example 4.10 Figure 4.9 shows the simplicial complex generated by $[a, b, c]$, $[c, d]$, $[a, d]$, $[e]$, and its face partial order. Notice that there is no path connecting e to any other vertex. This is visible in the partial order since $[e]$ is a maximal element, which means that there is no simplex containing it.

Exercise 4.14 Compute the face partial order for the simplicial complex generated by $[a, b, c, d]$, $[a, e, f]$, $[b, f]$, and $[g]$.

Exercise 4.15 Compute the face partial order for the simplicial complex generated by $[a]$, $[a, b]$, $[a, b, c]$, and $[a, b, c, d]$.

There is a distinction between *path connected* and a more general kind of *connected* in the theory of abstract topological spaces. This distinction does not occur for abstract simplicial complexes, as the next exercise asks the reader to show.

Exercise 4.16 Show that a simplicial complex is path connected if and only if whenever it is partitioned into two pieces, the closure of one intersects the other. (This latter notion is that of *connectedness* for general topological spaces.)

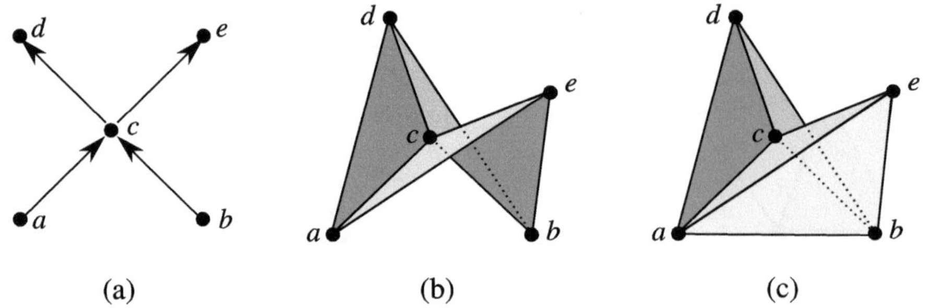

Fig. 4.10 Two simplicial complexes formed from a partial order: **a** the partial order, **b** the order complex and **c** the nerve complex. See Examples 4.11 and 4.12

Every simplicial complex implies a partial order on its simplices, but the reverse construction is also possible. There are several different simplicial complexes that can be created from a partial order. Two of the most commonly used are the (1) *order complex* and (2) the *nerve complex*.

Definition 4.9 Suppose that $(P, \leq)$ is a partially ordered set. The *order complex* is the simplicial complex $O(P, \leq)$ whose vertex set is P, and each simplex is a subset $[p_0, p_1, \ldots, p_k]$ that is *totally ordered*, namely

$$p_0 < p_1 < \cdots < p_k.$$

Example 4.11 Consider the partially ordered set shown in Fig. 4.10a. Since

$$a < c < d, \; a < c < e, \; b < c < d, \; \text{and } b < c < e$$

are the longest chains of inequalities in the partial order, these correspond to the simplices $[a, c, d]$, $[a, c, e]$, $[b, c, d]$, and $[b, c, e]$ that generate the order complex, which is shown in Fig. 4.10b.

Exercise 4.17 Consider the partial order given by the diagram

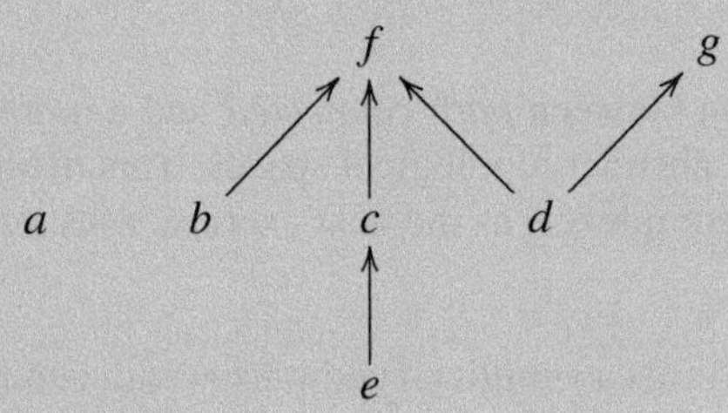

Construct the order complex for this partial order. Is it connected?

Proposition 4.5 *If $(P, \leq)$ is a partially ordered set, each star over a simplex in its order complex $O(P, \leq)$ consists of a chain of inequalities $p_0 < p_1 < \cdots < p_k$ and all of the possible ways this chain can be extended to a longer chain.*

Proof Each simplex $\sigma = [p_0, p_1, \ldots, p_k]$ in $O(P, \leq)$ corresponds to a chain $p_0 < p_1 < \cdots < p_k$ of elements of P. The star over this simplex σ consists of all simplices containing σ, which according to the construction of the order complex is a longer chain of elements of P that has σ as a subset. □

Exercise 4.18 1. Consider the abstract simplicial complex X on $\{a, b, c, d\}$ generated by $[a, b, c]$ and $[a, d]$. Construct a partially ordered set whose order complex is X.
2. Prove or give a counterexample: Every abstract simplicial complex is the order complex for some partially ordered set.

Definition 4.10 The order complex of the face partial order of a simplicial complex X is called the *barycentric subdivision* $Bd(X)$ of X.

$Bd(X)$ has each simplex of X as a vertex. The barycentric subdivision also plays an important role in Sect. 5.4 in Chap. 5.

Exercise 4.19 Describe the structure of the $Bd(X)$ as it relates to X using some small examples.

In addition to the order complex, the *nerve complex* is another, somewhat larger, abstract simplicial complex that can be created from a partial order.

Definition 4.11 Given a preordered set $(P, \leq)$ (see Definition 1.13 in Chap. 1), the *nerve complex* $N(Q, P, \leq)$ is the abstract simplicial complex whose set of vertices is $Q \subseteq P$, and $[q_0, q_1, \ldots, q_k]$ is a simplex of $N(Q, P, \leq)$ if there exists a $p \in P$ such that $q_i \leq p$ for all $i = 0, 1, \ldots, k$. (Notice that p need not be an element of Q.)

Example 4.12 Consider again the partially ordered set from Example 4.11 shown in Fig. 4.10a. The nerve complex using all of the vertices is generated by the two 3-dimensional simplices $[a, b, c, e]$ and $[a, b, c, d]$, and is shown in Fig. 4.10c. Notice that there is no simplex containing both d and e, since any such simplex would require that there be an element x of the partial order with $d \leq x$ and $e \leq x$. Since d and e are maximal elements, no such x exists.

If we only consider vertices in $\{a, b, d, e\}$, then the nerve complex is generated by $[a, b, e]$ and $[a, b, d]$.

Proposition 4.6 *For a partially ordered set $(P, \leq)$, the order complex $O(P, \leq)$ is a subcomplex of the nerve complex $N(P, P, \leq)$.*

Proof Suppose that $[p_0, p_1, \ldots, p_k]$ is a simplex in in $O(P, \leq)$. This means that

$$p_0 < p_1 < \cdots < p_k,$$

of which $p_i \leq p_k$ for $i = 0, \ldots, k$ is a consequence. So $[p_0, p_1, \ldots, p_k]$ is also a simplex of $N(P, P, \leq)$. □

Exercise 4.20 Construct a nerve complex from the partial order given all elements in the diagram

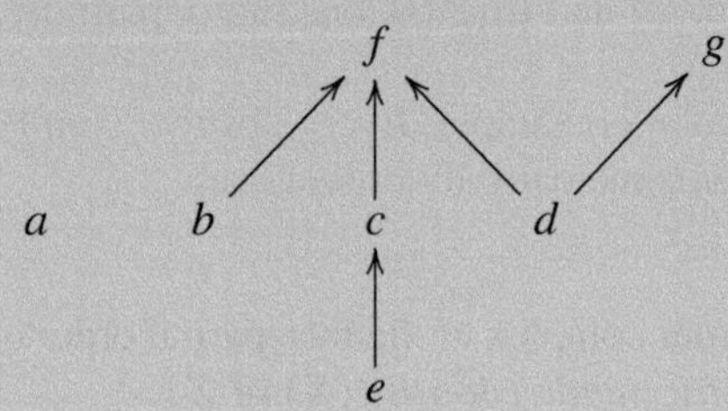

Construct the nerve complex using only $\{a, b, c, d, f\}$.

Exercise 4.21 Consider the simplicial complex generated by $[a, b]$, $[a, c]$, $[b, c]$, and $[a, d]$. Construct a partially ordered set and a subset of its elements that generates this simplicial complex as its nerve complex.

Exercise 4.22 Prove or given a counterexample: Every simplicial complex is the nerve complex of some preordered set. (Hint: compare with Exercise 4.18.)

4.4 Simplicial Maps

Since abstract simplicial complexes are useful mathematical objects to study, it is important to consider transformations that preserve their structure. These transformations consist of functions on vertex sets that preserve the relationship between simplices and are called *simplicial maps*.

Definition 4.12 Suppose X and Y are simplicial complexes with vertex sets V_X and V_Y, respectively. A function $f : V_X \to V_Y$ on vertices is called a *simplicial map* $f : X \to Y$ if it transforms each simplex $[v_0, \cdots, v_k]$ of X into a simplex $[f(v_0), \cdots, f(v_k)]$ of Y, after removing duplicate vertices.

Lemma 4.1 *Let $f : X \to Y$ be a simplicial map. For every pair of simplices σ, τ of X satisfying $\sigma \subseteq \tau$, their images in Y satisfy $f(\sigma) \subseteq f(\tau)$.*

Proof Since f is a simplicial map, $f(\sigma)$ is a simplex of Y and so is $f(\tau)$. If $\sigma \subseteq \tau$, then every vertex v of σ is also a vertex of τ. By the definition of simplicial maps, $f(v)$ is a vertex of both $f(\sigma)$ and $f(\tau)$. Conversely, every vertex of $f(\sigma)$ is the image of some vertex w of σ. □

Example 4.13 Consider the simplicial complex X generated by $[a, b]$, $[a, c]$, $[b, c]$, and $[d]$, and the simplicial complex Y generated by $[e, f, g]$. The function $h : V_X \to V_Y$ given by

$$h(a) := e, \ h(b) := f, \ h(c) := g, \ \text{and } h(d) := g,$$

is a simplicial map since

$$[h(a), h(b)] = [e, f], \ [h(a), h(c)] = [e, g], \ [h(b), h(c)] = [f, g], \ \text{and } [h(d)] = [g]$$

are all simplices of Y. (See Fig. 4.11.) It is not necessary to check all of the simplices that are proper subsets of the generating set for X. We can also interpret the simplicial map as acting on simplices, so that

$$h([a, b]) = [e, f], \ h([a, c]) = [e, g], \ h([b, c]) = [f, g], \ \text{and } h([d]) = [g].$$

Example 4.14 Consider the simplicial complex X generated by $[a, b, c]$ and $[d]$, and the simplicial complex Y generated by $[e, f]$, $[e, g]$, and $[g, f]$. Figure 4.12 shows two vertex functions h and $k : V_X \to V_Y$. The function h given by

$$h(a) := e, \ h(b) := e, \ h(c) := g, \ \text{and } h(d) := g,$$

is a simplicial map since

$$[h(a), h(b)] = [e, e] = [e], \ [h(a), h(c)] = [e, g], \ [h(b), h(c)] = [e, g], \ \text{and } [h(d)] = [g]$$

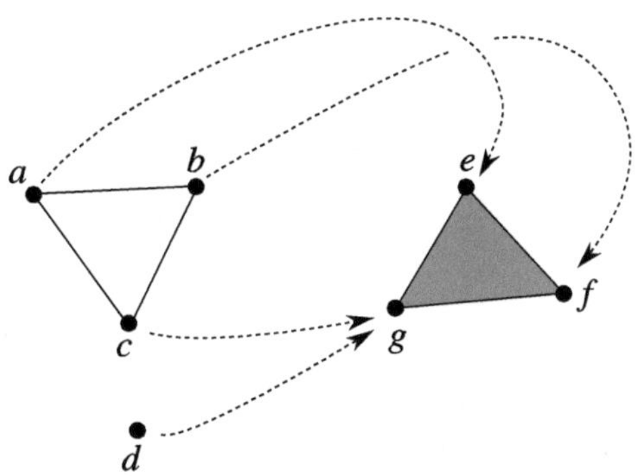

Fig. 4.11 The simplicial map $h : X \to Y$ defined in Example 4.13

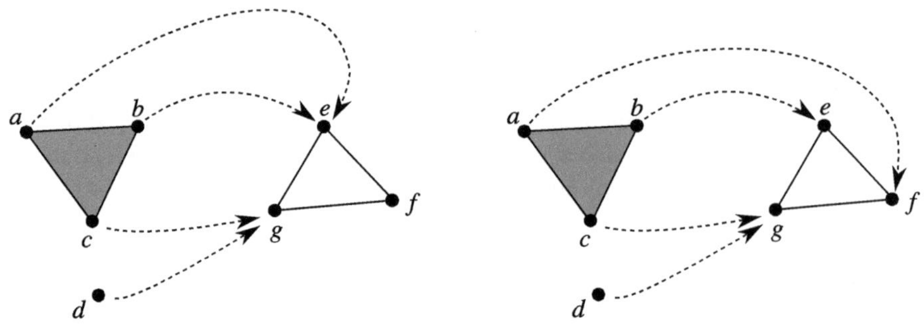

Fig. 4.12 Two vertex functions $h, k : X \to Y$ defined in Example 4.14. The function h is a simplicial map, while k is not

are all simplices of Y. Notice that the image of the 1-dimensional simplex $[a, b]$ through h is the 0-dimensional simplex $[e]$ because of vertex duplication.

To define k, make a slight change to h,

$$k(a) := f,\ k(b) := e,\ k(c) := g,\ \text{and } k(d) := g.$$

In this case, k is not a simplicial map, since the image of $[a, b, c]$ through k is

$$k([a, b, c]) = [k(a), k(b), k(c)] = [f, e, g],$$

which is not a simplex of Y.

Example 4.15 It is possible that a function $h : X \to Y$ that takes simplices from a simplicial complex X to simplices of another simplicial complex Y might not be a simplicial map, because there is no vertex function $V_X \to V_Y$ that induces this function h on simplices. For instance, let X be the simplicial complex generated by $[a, b]$, $[a, c]$, and $[b, c]$ and let Y be the simplicial complex generated by $[e, f]$, $[e, g]$, and $[f, g]$. Consider the function $h : X \to Y$ on *all* simplices by

$$\begin{aligned} h(\emptyset) &:= \emptyset, \\ h([a]) &:= [f, g], \\ h([b]) &:= [e, g], \\ h([c]) &:= [e, f], \\ h([a, b]) &:= [g], \\ h([a, c]) &:= [f], \text{ and} \\ h([b, c]) &:= [e]. \end{aligned}$$

There is no function $V_X \to V_Y$ on vertices that induces h on simplices, because the dimensions of some of the simplices are strictly less than that of their images.

Exercise 4.23 Consider the simplicial complex X generated by $[a, b, e]$, $[a, d, e]$, $[b, c, e]$, $[c, d, e]$, $[a, b, f]$, $[a, d, f]$, $[b, c, f]$, $[c, d, f]$. Verify that the function $h : X \to X$ given by the following

$$\begin{aligned} h(a) &:= a, \\ h(b) &:= b, \\ h(c) &:= c, \\ h(d) &:= d, \\ h(e) &:= e, \\ h(f) &:= e \end{aligned}$$

vertex map is a simplicial map.

Exercise 4.24 Suppose $f : X \to Y$ and $g : Z \to Y$ are simplicial maps. If g is surjective on vertices, is it always the case that there is a simplicial map $h : X \to Z$ such that $g \circ h = f$? Such a map h is called a *lift* and is usually summarized by a diagram

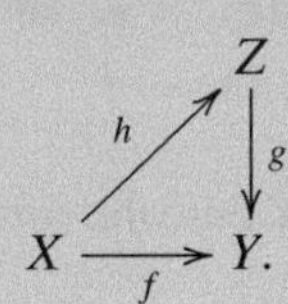

Exercise 4.25 Show that if X is a path connected simplicial complex, then its image through any simplicial map is also path connected.

Exercise 4.26 Suppose that X and Y are simplicial complexes and $f : X \to Y$ be a function taking simplices of X to simplices of Y that

1. Preserves dimension: if $\sigma \in X$ is a simplex, then $\dim \sigma = \dim f(\sigma)$, and
2. Preserves face relations: if $\sigma \subseteq \tau \in X$ then $f(\sigma) \subseteq \tau$.

Show that f is a simplicial map. Does every simplicial map satisfy these two criteria? If not, provide a counterexample.

Exercise 4.27 Consider a simplicial map $f : X \to Y$. Which of the following operations are preserved by simplicial maps for an arbitrary vertex $v \in X$?

1. The star, namely is it true that $f(\text{star } v) = \text{star } f(v)$?
2. The closure, namely is it true that $f(\text{cl } v) = \text{cl } f(v)$?
3. The link, namely is it true that $f(\text{lk } v) = \text{lk } f(v)$?

Definition 4.13 A simplicial map $f : X \to Y$ is called a *retraction* if Y is a subcomplex of X, and f restricted to Y is the identity simplicial map.

Example 4.16 Consider the simplicial complex X generated by $[a, b, c, d]$, $[c, e]$, $[c, d]$, $[d, f]$, and $[e, f]$. The simplicial map given by

$$\begin{aligned} r(a) &:= c, \\ r(b) &:= d, \\ r(c) &:= c, \\ r(d) &:= d, \\ r(e) &:= e, \text{ and} \\ r(f) &:= f, \end{aligned}$$

shown in Fig. 4.13 is a retraction of X onto the subcomplex generated by $[c, e]$, $[c, d]$, $[d, f]$, and $[e, f]$.

On the other hand, the simplicial map given by

$$\begin{aligned} s(a) &:= a, \\ s(b) &:= b, \\ s(c) &:= c, \\ s(d) &:= d, \end{aligned}$$

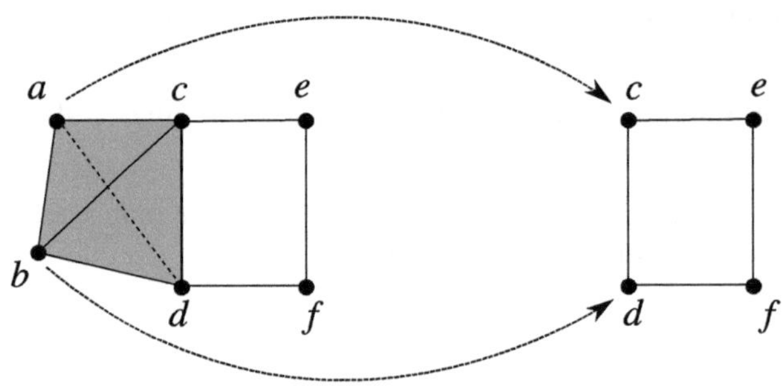

Fig. 4.13 The retraction $r : X \to X$ defined in Example 4.16

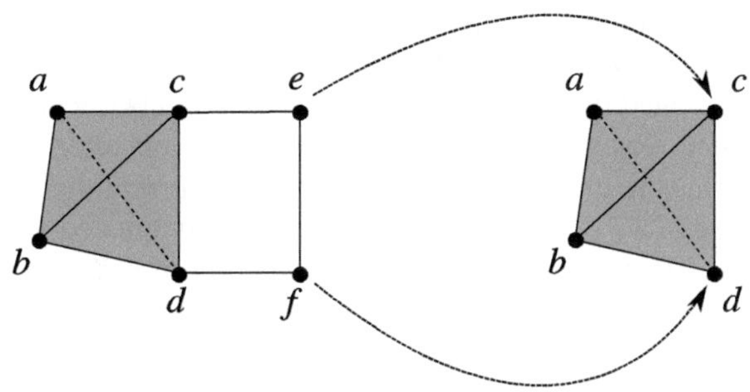

Fig. 4.14 The retraction $s : X \to X$ defined in Example 4.16

$$s(e) := c, \text{ and}$$
$$s(f) := d,$$

shown in Fig. 4.14 is a retraction of X onto the subcomplex generated by $[a, b, c, d]$.

Exercise 4.28 Is there a retraction from the simplicial complex generated by $[a, b, c]$ to the simplicial complex generated by $[a, b]$, $[a, c]$, and $[b, c]$? Prove your answer.

Exercise 4.29 In Sheehy (2012), it is shown that for an abstract simplicial complex X, there is a sequence of simplicial (inclusion) maps of nested subcomplexes of $Bd(X)$, based upon the length of chains in the order complex. Specifically, if we define $S_k(X)$ to be the simplicial subcomplex of $Bd(X)$ consisting of chains of length at least $k - 1$, construct a simplicial map $S_k(X) \to S_{k-1}(X)$.

Table 4.1 shows several operations on simplicial complexes that are used in this chapter and the next. Barycentric subdivision $Bd(X)$ has already been defined in Definition 4.10. We define *coning* and *suspending* below, deferring the definition of the *prism* until Definition 5.8 in Sect. 5.3 of Chap. 5.

Table 4.1 Operations on a simplicial complex X

Name	Symbol	Definition	Fact	Reference (all in Chap. 5)
Cone	CX	Definition 4.14	$\widetilde{H_k}(CX) = 0$	Proposition 5.4
Suspension	SX	Definition 4.14	$\widetilde{H_{k+1}}(SX) \cong \widetilde{H_k}(X)$	Proposition 5.10
Prism	PX	Definition 5.8	$H_k(PX) \cong H_k(X)$	Proposition 5.13
Subdivision	$Bd(X)$	Definition 4.10	$H_k(X) \cong H_k\,(Bd(X))$	Proposition 5.16

Definition 4.14 Suppose that X is an abstract simplicial complex with vertex set V_X. The *cone* on X is the simplicial complex CX on the vertex set $V_{CX} := V_X \sqcup \{*\}$ generated by all simplices of the form $\sigma \sqcup \{*\}$, where σ is a simplex of X and $*$ is some vertex not appearing in V_X. The $\sqcup$ operator is called the *disjoint union*, and forms the union of two sets with the *assertion* that the two sets are disjoint. In the case of V_{CX}, this means that $*$ is a newly created vertex which does not appear in V_X. The vertex $*$ is called the *cone point* of CX.

The *suspension* on X is the simplicial complex SX on the vertex set $V_{SX} := V_X \sqcup \{*, *'\}$ generated by the union of all simplices of the form $\sigma \sqcup \{*\}$ and all simplices of the form $\sigma \sqcup \{*'\}$, where σ is a simplex of X. The two *suspension points* $*$ and $*'$ are chosen so that they do not appear in V_X.

Coning transforms each dimension k simplex of the original abstract simplicial complex into a simplex of dimension $k + 1$. Suspending transforms each dimension k simplex into two simplices, both of dimension $k + 1$.

Example 4.17 Fig. 4.15 shows a simplicial complex X generated by two vertices $[a]$ and $[b]$. The cone CX is generated by the two simplices $[a, *]$ and $[b, *]$. The suspension SX is generated by the four simplices $[a, *]$, $[a, *']$, $[b, *]$, and $[b, *']$.

Iterated suspensions can generate interesting simplicial complexes, as the next example shows.

Example 4.18 Start with the same simplicial complex X as in Example 4.17, generated by two vertices $[a]$ and $[b]$. Repeatedly suspending X generates a sequence of simplicial complexes called *spheres*. Specifically, if we define $S^0 := X$, and then recursively define the k-dimensional sphere by

$$S^{k+1} := SS^k,$$

we obtain a sequence of simplicial complexes, as shown in Fig. 4.16. The spheres have interesting *simplicial homological* structure, as will be discussed in later sections, but some intuition can be gained without any calculation. S^0 consists of two path components, S^1 is a graph with a nontrivial loop (effectively a missing 2-dimensional region), S^2 has a hollow void (a missing 3-dimensional region), S^3 has a hollow 4-dimensional region, and so on.

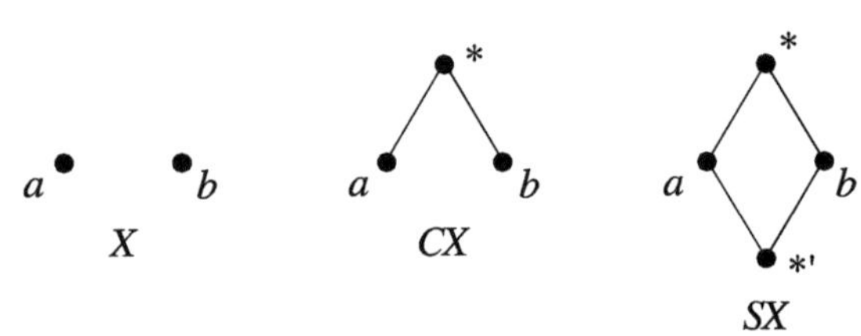

Fig. 4.15 A simplicial complex X, its cone CX, and its suspension SX, in Example 4.17

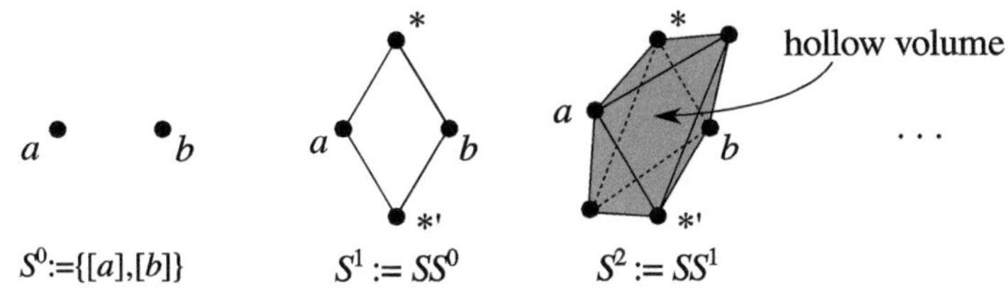

Fig. 4.16 Creating spheres of all dimensions by iterated suspensions in Example 4.18

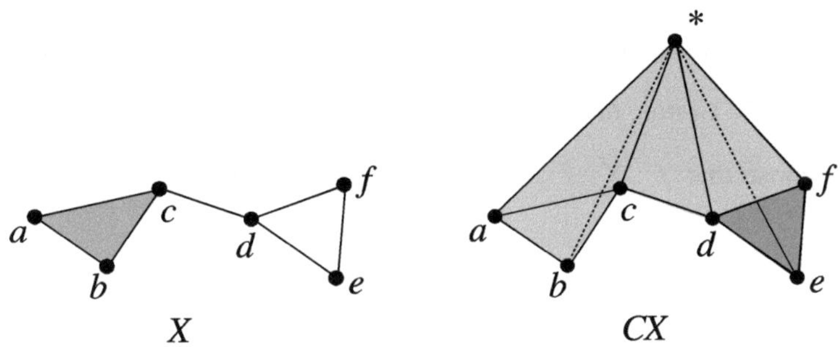

Fig. 4.17 A simplicial complex X and its cone CX in Example 4.19

Example 4.19 Figure 4.17 shows the simplicial complex X generated by $[a, b, c]$, $[c, d]$, $[d, e]$, $[d, f]$, and $[e, f]$. While this complex contains a nontrivial loop—a gap that could be filled by a simplex $[d, e, f]$ that is not present—the cone CX fills this gap with three triangles, $[d, e, *]$, $[d, f, *]$, and $[e, f, *]$. On the other hand, the simplex $[a, b, c]$ is filled into a simplex $[a, b, c, *]$ in the cone.

Exercise 4.30 Construct the cone and suspension of

1. A solid simplex, $[a_0, a_1, \ldots, a_n]$. Find a minimal generating set of simplices for both cone and suspension.
2. A "figure 8," generated by two loops that intersect at a point p: $[p, a_1], \ldots, [a_n, p], [p, b_1], \ldots, [b_m, p]$.

Exercise 4.31 Is there any simplicial complex X for which the operation of coning and suspending commute? Either find a simplicial complex X such that $C(SX) = S(CX)$ or show that such a complex cannot exist.

Lemma 4.2 *The composition of simplicial maps $f : X \to Y$ and $g : Y \to Z$ defined by composing their respective functions on vertices is a simplicial map $h := (g \circ f) : X \to Z$.*

Proof Suppose that $[v_0, \ldots, v_k]$ is a simplex of X. We know that $[f(v_0), \cdots, f(v_k)]$ is a simplex of Y since f is a simplicial map. Since g is also a simplicial map,

$$[g(f(v_0)), \cdots, g(f(v_k))] = [(g \circ f)(v_0), \cdots, (g \circ f)(v_k)]$$

is a simplex of Z. □

Definition 4.15 As a result of the lemma, we can define the *category of abstract simplicial complexes* **Asc**, in which the objects are abstract simplicial complexes and the morphisms are simplicial maps. The identity morphism id_X for an abstract simplicial complex is given by the identity functions on the vertices of X.

Proposition 4.7 *Coning and suspending are covariant functors from* **Asc** *to itself.*

Proof Suppose that $f : X \to Y$ is a simplicial map. To obtain a simplicial map $Cf : CX \to CY$, define

$$(Cf)(v) := \begin{cases} f(v) & \text{if } v \neq * \\ * & \text{otherwise.} \end{cases}$$

This preserves simplices as an immediate consequence of the definition of the cone of X. Since composition of simplicial maps is defined by compositions of vertex sets, this is preserved for cones.

The argument for suspensions is identical, with the added condition that $Sf(*') := *'$. □

Proposition 4.8 *Two abstract simplicial complexes X and Y are isomorphic in* **Asc** *if and only if there is a simplicial map $f : X \to Y$ that is a bijection $f : V_X \to V_Y$ on their vertex sets and is surjective on simplices.*

Abstract simplicial complexes are isomorphic precisely when they consist of the same set of simplices, assembled in the same way, allowing for relabeling of vertices.

Proof Suppose $f : V_X \to V_Y$ is a simplicial map that is also a bijection on vertices. This simplicial map is injective on simplices. Suppose that $\sigma = [s_0, \ldots, s_k]$ and $\tau = [t_0, \ldots, t_m]$ are simplices of X with $f(\sigma) = f(\tau)$. This means that

$$[f(s_0), \ldots, f(s_k)] = [f(t_0), \ldots, f(t_m)].$$

Since f is bijective on vertices, this means that in the above equation there are no duplicate vertices. Therefore, $m = k$ and for each i there is exactly one j such that $f(s_i) = f(t_j)$. Again since f is bijective, we must conclude that for each i, there is exactly one j such that $s_i = t_j$. Thus $\sigma = \tau$. If we additionally assume f induces a surjection on simplices, then this implies that f is bijective on simplices. We then merely need to establish that the map $g = f^{-1} : V_Y \to V_X$ defined on vertices is a simplicial map, since this is the only option for the inverse to f as a simplicial map.

Suppose that $\sigma = [z_0, \ldots, z_n]$ is a simplex of Y. Since f is bijective on vertices, this means that we can find vertices $v_i = f^{-1}(z_i)$ of X for each i. The only question is whether

$$[f^{-1}(z_0), \ldots, f^{-1}(z_n)]$$

is a simplex of X, but this is precisely what the assumption that f is surjective on simplices means.

Conversely, suppose that X and Y are isomorphic. This means that there is a pair of simplicial maps $f : X \to Y$ and $g : Y \to X$ with $f \circ g = \mathrm{id}_Y$ and $g \circ f = \mathrm{id}_X$. These two equations imply that f and g are bijective on simplices. Because of the coverage axiom, every vertex $v \in V_X$ is also a simplex $[v] \in X$. Thus f and g are bijective on the vertices of the simplicial complexes as well. □

The three constructions from Sect. 4.3, namely the face partial order, nerve complex, and order complex, are covariant functors, as Propositions 4.9, 4.10, and 4.11 show.

Proposition 4.9 *The face partial order is a covariant functor* **Asc** $\to$ **Pos**.

Proof The construction of the face partial order from a simplicial complex establishes how the functor transforms objects. Let us denote the face partial order for X by $F(X)$. Lemma 4.1 establishes that every simplicial map $f : X \to Y$ induces an order preserving function (recall Definition 1.23 in Chap. 1) $F(f) : F(X) \to F(Y)$ on the face partial orders for X and Y. To verify that the functor is covariant, suppose that we have another simplicial map $g : Y \to Z$. The composition of these is a simplicial map $g \circ f : X \to Z$ that induces an order preserving map $F(g \circ f) : F(X) \to F(Z)$ on the face partial orders for X and Z. On the other hand, $F(g) \circ F(f) : F(X) \to F(Z)$ is also an order preserving map. Any simplex σ in X can be reinterpreted as an element of $F(X)$, which means that the simplex $(g \circ f)(\sigma)$ of Z corresponds to the same element of $F(Z)$ as does $g(f(\sigma))$, thought of as the image of an element of $F(Y)$. □

Example 4.20 Consider again the simplicial map h defined in Example 4.14 and shown in the left frame of Fig. 4.12. Figure 4.18 shows the face posets of both the domain and codomain of this simplicial map. The solid arrows in Fig. 4.18 show the order relations within the two face partial orders, while the dashed arrows show the order preserving map induced by the simplicial map.

Example 4.21 Consider the simplicial complex X generated by $[a, b]$ and $[c]$, shown in Fig. 4.19a. This has a face partial order shown in Fig. 4.19b. The cone CX adds a vertex $[*]$, three edges $[a, *]$, $[b, *]$, $[c, *]$, and one 2-dimensional simplex $[a, b, *]$, as shown in Fig. 4.19c. The face partial order of CX is shown in Fig. 4.19d. Notice that the effect of coning on the face partial order is that it essentially duplicates the original face partial order, and places the duplicate copy above the original.

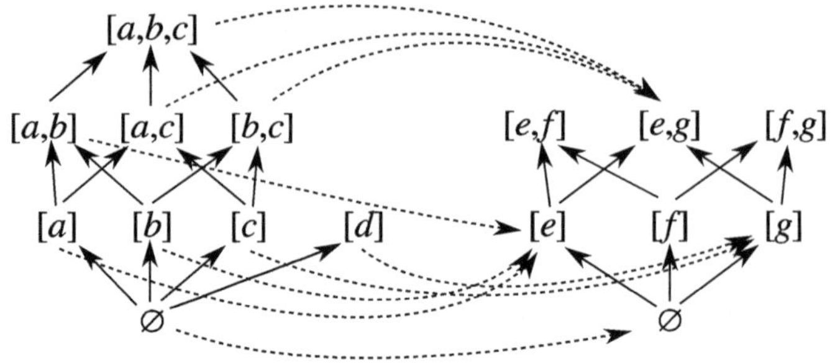

Fig. 4.18 The order preserving map (dashed arrows) induced by the simplicial map $h : X \to Y$ defined on the face partial orders in Example 4.14

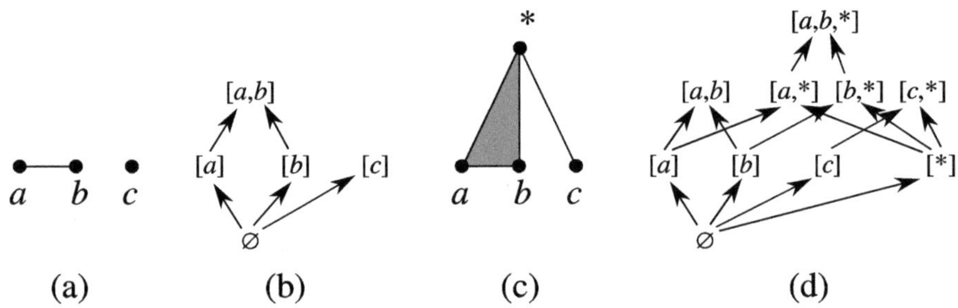

Fig. 4.19 **a** A simplicial complex X, **b** the face partial order of X, **c** the cone CX, and **d** the face partial order of CX. See Example 4.21

Exercise 4.32 Describe what suspending does to face partial orders. Specifically, how does suspending change the order preserving maps induced by a simplicial map?

Exercise 4.33 (*Difficult!*) Which, if any, of the operations on simplicial complexes shown in Table 4.1 are examples of *monads* in the category **Asc**? (You will need to look up the definition of *monad* in a reference on category theory.)

Proposition 4.10 *The nerve complex is a covariant functor* **Preorders** $\to$ **Asc**.

Proof The construction of the nerve complex $N(P, P, \leq)$ from a preordered set $(P, \leq)$ given in Definition 4.11 defines the functor's action on objects. Suppose that $f : (P, \leq_P) \to (Q, \leq_Q)$ is an order preserving function. We need to show that f induces a simplicial map $N(P, P, \leq_P) \to N(Q, Q, \leq_Q)$. By definition, the vertex set for $N(P, P, \leq_P)$ is P and the vertex set for $N(Q, Q, \leq_Q)$ is Q. Suppose $[v_0, \ldots, v_k]$ is a simplex of $N(P, P, \leq_P)$, which means that there is $w \in P$

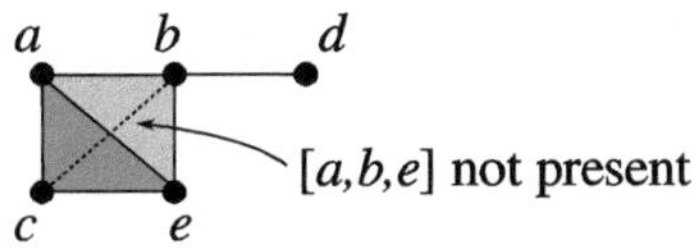

Fig. 4.20 The nerve complex for the partially ordered set in Example 4.22

such that $v_i \leq w$ for all $i = 0, \ldots, k$. Because f is order preserving, this means that $f(v_i) \leq f(w)$ for all $i = 0, \ldots, k$. Thus $[f(v_0), \ldots, f(v_k)]$ is a simplex of $N(Q, Q, \leq_Q)$, so f is a simplicial map.

Now suppose that $g : (Q, \leq_Q) \to (R, \leq_R)$ is another order preserving map. We need to establish that $g \circ f$ induces the composition of simplicial maps induced on the nerve complexes. This is immediate from the fact that f and g as simplicial maps are defined by their action on vertices, and P, Q, and R are the vertex sets in question. □

Example 4.22 Consider the set $P = \{a, b, c, d, e\}$ with the partial order given by the diagram

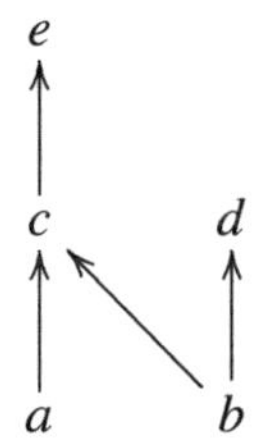

The nerve complex $N(P, P, \leq)$ for this partially ordered set is given in Fig. 4.20. The nerve complex is generated by

$$[a, b, c], [a, b, e], [a, c, e], [b, c, e], [b, d].$$

The order preserving function f given by

$$\begin{aligned} f(a) &:= c, \\ f(b) &:= b, \\ f(c) &:= c, \\ f(d) &:= d, \\ f(e) &:= c, \end{aligned}$$

induces the simplicial map $f : N(P, P, \leq) \to N(P, P, \leq)$ given by the same description. Notice that this induces the following transformations of simplices

$$\begin{aligned} f([a, b, c, e]) &= [c, b, c, c] = [b, c], \\ f([b, d]) &= [b, d]. \end{aligned}$$

Proposition 4.11 *The order complex is a covariant functor* **Pos** $\to$ **Asc**.

Proof This follows from Proposition 4.6, that the order complex is a subcomplex of the nerve complex. □

Exercise 4.34 Construct the simplicial map induced on the order complex by the order preserving function given in Example 4.22.

Exercise 4.35 Show that the Dowker complex defined in Definition 4.4 is actually a covariant functor **Rel** $\to$ **Asc**. Recall from Example 1.28 in Chap. 1 that **Rel** has objects that are relations $R \subseteq X \times Y$ and each morphism $R \to R'$ in **Rel** consist of pairs of functions $f : X \to X'$, $g : Y \to Y'$ such that $(f(x), g(y)) \in R'$ whenever $(x, y) \in R$.

Given the construction of the Dowker complex $D(X, Y, R)$ from $R \subseteq X \times Y$, you must show that

1. Each morphism in **Rel** translates into a simplicial map, and
2. Composition of morphisms in **Rel** translates into composition of simplicial maps.

Exercise 4.36 Show that the composition of the Dowker functor defined in Exercise 4.35 with the face partial order (as a functor **Asc** $\to$ **Pos**) yields the $PosRep$ functor from Exercise 1.37 in Chap. 1. In brief, show that the diagram

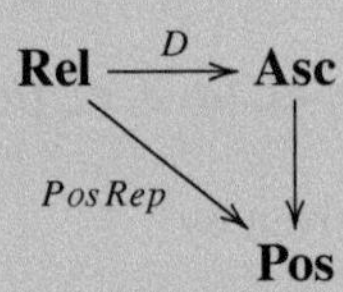

of functors commutes, where the vertical arrow computes the face partial order.

4.5 Simplicial Chain Complexes

This chapter began with the observation that there is an Euler characteristic associated to each abstract simplicial complex. In the previous chapter, the Euler characteristic was defined for sequences of vector spaces. Let us construct a sequence $(C_\bullet(X), \partial_\bullet)$ from a simplicial complex X so that the two definitions of Euler characteristic coincide. The easiest way to ensure that the Euler characteristic of the simplicial complex

X agrees with that of the sequence $(C_\bullet(X), \partial_\bullet)$ is to make the dimension of each space $C_k(X)$ be the number of simplices of X in dimension k. Specifically, let

$$C_k(X) := \begin{cases} \mathbb{R}^{\text{number of } k\text{-dimensional simplices}} & \text{for } k \geq 0, \\ 0 & \text{for } k < 0 \end{cases}$$
$$= \operatorname{span} X^k,$$

where X^k is the subset of all k-dimensional simplices of X, called the *k-skeleton of X*. Each $C_k(X)$ is called the *k-dimensional simplicial chain space*. Each k-dimensional simplex in X corresponds to a basis vector in $C_k(X)$. The Euler characteristic of this sequence now agrees with the one given at the beginning of this chapter.

We are ultimately interested in using $(C_\bullet(X), \partial_\bullet)$ to understand the structure of the simplicial complex X, with the idea that working with $(C_\bullet(X), \partial_\bullet)$ is easier than working with X. While there is some freedom in how we define the $\partial_\bullet$ maps, they should reflect the structure of the simplicial complex. In this section, we will show that each simplicial map $f : X \to Y$ induces a sequence map $C_\bullet(X) \to C_\bullet(Y)$. But if $(C_\bullet(X), \partial_\bullet)$ is only a sequence and not a chain complex, classifying the structure of the induced sequence map $C_\bullet(X) \to C_\bullet(Y)$ will not be any easier than studying f itself. If instead $C_\bullet(X) \to C_\bullet(Y)$ were to be a chain map, we could classify it completely using Theorem 3.1 from Chap. 3, and this is generally much easier than studying the simplicial map f. Furthermore, as we will see in the next chapter, the *simplicial homology* $H_\bullet(C_\bullet(X), \partial_\bullet)$ is determined purely by the topological structure of X.

Therefore, let us set about constructing *simplicial boundary maps* $\partial_\bullet$ to ensure that $(C_\bullet(X), \partial_\bullet)$ is a chain complex, called the *simplicial chain complex*. The indices of this simplicial chain complex most naturally *decrease* because we can enumerate all of the simplices contained in a given simplex by systematically deleting vertices. For instance, if

$$\sigma = [v_0, v_1, \ldots, v_k],$$

is a simplex of dimension k, then we can simply delete each vertex in turn

$$[v_1, \ldots, v_k],\ [v_0, v_2, \ldots, v_k], \ldots [v_0, v_1, \ldots, v_{k-1}]$$

to obtain all the simplices of dimension $k-1$ that are contained within σ. We can then recursively delete vertices from those simplices to obtain all simplices of dimension $k-2$, and so on.

Something interesting happens at dimension $k-2$. There are *two* distinct ways to arrive at each simplex by deleting vertices from σ. This is clearest from a specific example, for instance the 2-dimensional simplex

$$\tau = [v_0, v_1, v_2].$$

If we delete vertices in order, we obtain all of the 1-dimensional simplices contained in τ, namely

$$[v_1, v_2],\ [v_0, v_2],\ [v_0, v_1].$$

These simplices are said to form the *boundary* of τ. Deleting vertices in order from each of these simplices yields the following sequence of 0-dimensional simplices

$$[v_2],\ [v_1],\ [v_2],\ [v_0],\ [v_1],\ [v_0].$$

There are *precisely two* duplicates of each vertex. The duplication arises because there are two possible orders in which the deletions can occur. Deleting v_0 first and then v_1 second from the simplex τ yields the same 0-dimensional simplex $[v_2]$ as if we were to delete v_1 first and v_0 second.

If we use this vertex deletion process on its own to define $\partial_2(\tau)$, then $\partial_1 \circ \partial_2$ will not be zero, and as a result $(C_\bullet(X), \partial_\bullet)$ will not be a chain complex. We can resolve our problem by using *signs* to remember which position the deleted vertices came from. Although the "sign of a simplex σ" does not make sense, we *can* put a sign on a vector in $C_k(X)$ that corresponds to the k-dimensional simplex σ. Let us assign signs based on the indices from which the vertices were deleted, for instance

$$\partial_2\left([v_0, v_1, v_2]\right) = +[v_1, v_2] - [v_0, v_2] + [v_0, v_1],$$

since $(-1)^0 = +1$, $(-1)^1 = -1$, and $(-1)^2 = +1$, respectively. (The equation abuses notation slightly by implicitly treating each simplex *as* a basis vector in $C_\bullet(X)$.) If we do the same process a second time, we obtain the duplicates as before but they come with opposite signs,

$$\begin{aligned}(\partial_1 \circ \partial_2)\left([v_0, v_1, v_2]\right) &= +(+[v_2] - [v_1]) - (+[v_2] - [v_0]) + (+[v_1] - [v_0]) \\ &= [v_0] - [v_0] + [v_1] - [v_1] + [v_2] - [v_2] = 0.\end{aligned}$$

To ensure that this process of assigning signs to indices is well defined for every simplicial complex, it is useful to apply a relatively arbitrary ordering to the vertices and then later (Proposition 4.15) show that the isomorphism class of the chain complex does not depend on the ordering.

Definition 4.16 A *vertex partial order* for a simplicial complex X with vertex set V_X is a partial order $\leq$ on vertices such that every simplex is a totally ordered set. Explicitly, if σ is a simplex and v_i, v_j are vertices in σ, then exactly one of the following holds:

1. $v_i = v_j$, or
2. $v_i \neq v_j$ and $v_i \leq v_j$, or
3. $v_i \neq v_j$ and $v_i \geq v_j$.

By convention, we always write the vertices of a simplex σ in a simplicial complex in ascending order if there is a vertex partial order, so that $\sigma = [v_0, v_1, \ldots, v_k]$ implies

$$v_0 \le v_1 \le \cdots \le v_k$$

and

$$v_i \neq v_j \text{ if } i \neq j.$$

Example 4.23 Suppose $(P, \le)$ is a partial order. Then $\le$ is a vertex partial order for the order complex $O(P, \le)$, because every simplex is totally ordered by $\le$ by definition.

Proposition 4.12 *Every finite abstract simplicial complex has at least one vertex partial order.*

Most simplicial complexes have multiple vertex partial orders, arising from different total orders on their vertices.

Proof Take the union of all simplices; this is a finite set V'. (There may be vertices not appearing in V', though this is a degenerate case.) Since V' is finite, we can impose a completely arbitrary total order on it. Evidently, this ordering results in every subset of V' being totally ordered, so every simplex is thereby totally ordered. □

Proposition 4.13 *Suppose that $f : X \to Y$ is a simplicial map between two simplicial complexes, and that $\le_Y$ is a vertex partial order on Y. There exists a vertex partial order $\le_X$ on X such that f is an order preserving function on vertices.*

Proof Recall that f being order preserving on vertices means that if $x \le_X y$ are two vertices of X, then $f(x) \le_Y f(y)$. With that in mind, suppose that x and y are two arbitrary vertices of X. Exactly one of the following cases occurs:

1. $f(x) \neq f(y)$ and $f(x) \le_Y f(y)$, in which case we define $x \le_X y$,
2. $f(x) \neq f(y)$ and $f(y) \le_Y f(x)$, in which case we define $y \le_X x$,
3. $x \neq y$ but $f(x) = f(y)$, in which case we assert that x and y are not related by $\le_X$, or
4. $f(x)$ and $f(y)$ are not related by $\le_Y$, (which also means that $f(x) \neq f(y)$) in which case no action needs to be taken.

Notice that the above data will result in a well defined relation $\le_X$. This relation is a partial order, because it satisfies

Reflexivity By assumption, and

Antisymmetry The only way that $x \le_X y$ and $y \le_X x$ can happen simultaneously given the conditions above is for $x = y$. Specifically, if $x \le_X y$, either $x = y$ or $f(x) \neq f(y)$ and $f(x) \le_Y f(y)$. Conversely, if $y \le_X x$, either $y = x$ or $f(y) \neq f(x)$ and $f(y) \le_Y f(x)$. Now notice that since $\le_Y$ is a partial order, if we have $f(x) \le_Y f(y)$ and $f(y) \le_Y f(x)$, this implies that $f(x) = f(y)$; hence $x = y$ is the only way to avoid a contradiction.

Transitivity If $x \le_X y$ and $y \le z$, this must have arisen because $f(x) \le_Y f(y) \le_Y f(z)$. Since $\le_Y$ is a partial order, $f(x) \le_Y f(z)$ follows, so $x \le_X z$. □

Exercise 4.37 Prove or give a counterexample to the following statement: If $f : X \to Y$ is a simplicial map between two simplicial complexes, and that $\leq_X$ is a vertex partial order on X, then there exists a vertex partial order $\leq_Y$ on Y such that f is an order preserving function.

With vertex partial orders in hand, we can generalize our boundary map ∂_2 to the *simplicial boundary map* for arbitrary simplices,

$$\begin{aligned}\partial_k([v_0, v_1, \ldots, v_k]) &:= \sum_{i=0}^{k}(-1)^i [v_0, \ldots, v_{i-1}, v_{i+1}, \ldots, v_k] \\ &:= \sum_{i=0}^{k}(-1)^i [v_0, \ldots, \widehat{v_i} \ldots, v_k]\end{aligned}$$

where the hat indicates an omitted index, by extending to a linear map. This expression is well defined because the ordering of the vertices (according to the vertex partial order) is preserved by deletion.

Example 4.24 Consider the abstract simplicial complex X generated by $[a, b]$, $[b, c]$, $[a, c]$. Since there are no simplices whose dimension is higher than 1, the simplicial chain spaces for X will be

$$C_k(X) = 0$$

for $k > 1$. There are three 1-simplices, so $\dim C_1(X) = 3$ with

$$C_1(X) = \operatorname{span}\{[a, b], [b, c], [a, c]\}.$$

There are also a total of three 0-simplices, so

$$C_0(X) = \operatorname{span}\{[a], [b], [c]\}.$$

There is only one nontrivial boundary map, ∂_1. The definition gives that

$$\begin{aligned}\partial_1([a, b]) &= [b] - [a], \\ \partial_1([b, c]) &= [c] - [b], \text{ and} \\ \partial_1([a, c]) &= [c] - [a],\end{aligned}$$

or in matrix form

$$\partial_1 = \begin{pmatrix} -1 & 0 & -1 \\ 1 & -1 & 0 \\ 0 & 1 & 1 \end{pmatrix}$$

in which the rows and columns are listed in the same order as their respective bases are listed.

Proposition 4.14 *The sequence $(C_\bullet(X), \partial_\bullet)$ is a chain complex.*

We will call $(C_\bullet(X), \partial_\bullet)$ the *simplicial chain complex for the abstract simplicial complex X.*

Proof We establish that $\partial_{k-1} \circ \partial_k$ is the zero map by direct calculation,

$$
\begin{aligned}
(\partial_{k-1} \circ \partial_k)([v_0, v_1, \ldots, v_k]) &= \partial_{k-1}\left(\sum_{i=0}^{k}(-1)^i[v_0, \ldots, \widehat{v_i}, \ldots, v_k]\right) \\
&= \sum_{i=0}^{k}(-1)^i \partial_{k-1}([v_0, \ldots, \widehat{v_i} \ldots, v_k]) \\
&= \sum_{i=0}^{k}(-1)^i \left(\sum_{j=0}^{i-1}(-1)^j[v_0, \ldots, \widehat{v_j}, \ldots \widehat{v_i} \ldots, v_k] + \right. \\
&\quad \left. \sum_{j=i+1}^{k}(-1)^{j-1}[v_0, \ldots, \widehat{v_i}, \ldots \widehat{v_j} \ldots, v_k]\right) \\
&= \sum_{i=0}^{k} \sum_{j=0; j\neq i}^{k} (-1)^{i+j}\big([v_0, \ldots, \widehat{v_j}, \ldots \widehat{v_i} \ldots, v_k] + \\
&\quad (-1)[v_0, \ldots, \widehat{v_j}, \ldots \widehat{v_i} \ldots, v_k]\big) \\
&= 0.
\end{aligned}
$$

□

Example 4.25 Consider the abstract simplicial complex Y generated by $[a, b, c]$, $[a, b, d]$, $[a, c, d]$, $[b, c, d]$, $[a, e]$, $[b, e]$. This simplicial complex has no simplices of dimension 3 or higher, so

$$C_k(Y) = 0$$

for $k > 2$. There are four 2-dimensional simplices, eight 1-dimensional simplices and five 0-dimensional simplices. Bases for each of the chain spaces are given by

$$C_2(Y) = \text{span}\{[a, b, c], [a, b, d], [a, c, d], [b, c, d]\},$$

$$C_1(Y) = \text{span}\{[a, b], [a, c], [a, d], [a, e], [b, c], [b, d], [b, e], [c, d]\},$$

and

$$C_0(Y) = \text{span}\{[a], [b], [c], [d], [e]\}.$$

Using these bases (in the order listed) and the vertex partial order $a < b < c < d < e$, we can construct the boundary map $\partial_2 : C_2(Y) \to C_1(Y)$

$$\partial_2 = \begin{pmatrix} 1 & 1 & 0 & 0 \\ -1 & 0 & 1 & 0 \\ 0 & -1 & -1 & 0 \\ 0 & 0 & 0 & 0 \\ 1 & 0 & 0 & 1 \\ 0 & 1 & 0 & -1 \\ 0 & 0 & 0 & 0 \\ 0 & 0 & 1 & 1 \end{pmatrix}$$

and the boundary map $\partial_1 : C_1(Y) \to C_0(Y)$

$$\partial_1 = \begin{pmatrix} -1 & -1 & -1 & -1 & 0 & 0 & 0 & 0 \\ 1 & 0 & 0 & 0 & -1 & -1 & -1 & 0 \\ 0 & 1 & 0 & 0 & 1 & 0 & 0 & -1 \\ 0 & 0 & 1 & 0 & 0 & 1 & 0 & 1 \\ 0 & 0 & 0 & 1 & 0 & 0 & 1 & 0 \end{pmatrix}.$$

Notice in particular that

$$\partial_1 \circ \partial_2 = \begin{pmatrix} 0 & 0 & 0 & 0 \\ 0 & 0 & 0 & 0 \\ 0 & 0 & 0 & 0 \\ 0 & 0 & 0 & 0 \\ 0 & 0 & 0 & 0 \end{pmatrix},$$

which is all that is needed for this sequence to establish that $(C_\bullet, \partial_\bullet)$ is a chain complex.

Exercise 4.38 1. Construct the simplicial chain complex for the simplicial complex generated by $[a, b, c]$ and $[a, c, e]$.
2. Construct the simplicial chain complex for the simplicial complex generated by $[a, b, c, f]$, $[a, c, e, f]$, $[a, b, c, g]$, and $[a, c, e, g]$.
3. How are your answers to (1) and (2) related? Can you obtain one complex as the image of another through a chain map?

Exercise 4.39 Construct the simplicial chain complex for the simplicial complex generated by $[a, b, c]$, $[a, c, e]$, $[c, e, f]$, $[e, f, g]$, and $[a, b, g]$.

Proposition 4.15 *Changing the vertex partial order on a finite abstract simplicial complex induces an isomorphism of simplicial chain complexes.*

Vertex partial orders are therefore only a computational aid and do not impact the simplicial chain complex or its homology.

Proof Suppose that X is an abstract simplicial complex with vertex set V and that $\leq$ and $\leq'$ are two vertex partial orders. To keep the notation simple, we will denote by X' the simplicial complex X in which we use the vertex partial order $\leq'$.

Notice that X and X' contain exactly the same simplices, since simplices are subsets of the vertex set, and subsets have no implied ordering among their elements. However, the order in which the simplices are listed as basis elements for the chain spaces may differ. To account for this difference, we need to consider a set of isomorphisms $p_k : C_k(X) \to C_k(X')$ that permute bases. Permuting the order of the k-dimensional simplices in the basis of C_k and then computing ∂_k or computing ∂_k and then permuting the order of $k-1$-dimensional simplices results in exactly the same set of simplices. Except for the signs, then, ∂_k and p_k would commute.

Let us consider a simplex $\sigma = [v_0, \ldots, v_n]$ in X, where n is the dimension of the simplicial complex X. That same simplex also appears in X', though the vertices will be listed in a different order, $[v_{r(0)}, \ldots, v_{r(n)}]$ for some permutation r of the numbers $0, 1, \ldots, n$. When we compute $\partial_n(\sigma)$, the sign on $[v_1, \ldots, v_n]$ is $+1$ in X. That same simplex will appear in X' with sign $(-1)^{r^{-1}(0)}$, which is to say that it depends on the parity of the slot where 0 was taken by r. This means that we should take

$$p'_{n-1}([v_0, \ldots, \widehat{v_k}, \ldots, v_n]) = (-1)^{r^{-1}(k)}[v_{r(0)}, \ldots, \widehat{v_{r^{-1}(k)}}, \ldots, v_{r(n)}]$$

so that $\partial_n \circ p'_n = p'_{n-1} \circ \partial_n$ with $p'_n = p_n$. We can iterate this process to set the signs of the entries for p'_{n-2} from p_{n-2}, p'_{n-3} from $p_{n-3}, \ldots,$ and p'_0 from p_0. The choice of signs does not alter the rank of these p'_k maps. The resulting chain map $p'_\bullet : C_\bullet(X) \to C_\bullet(X')$ is a chain complex isomorphism. □

Example 4.26 Consider the simplicial complex X generated by $[a, b, c]$ with the vertex partial order $a < b < c$ and the vertex partial order $a <' c <' b$. In the first case, using $<$, we have

$$C_2(X) = \text{span}\,\{[a, b, c]\}\,,$$

$$C_1(X) = \text{span}\,\{[a, b], [a, c], [b, c]\}\,,$$

and

$$C_0(X) = \text{span}\,\{[a], [b], [c]\}\,.$$

The simplicial chain complex $(C_\bullet(X), \partial_\bullet)$ is given by

$$0 \longrightarrow \mathbb{R} \xrightarrow{\begin{pmatrix} 1 \\ -1 \\ 1 \end{pmatrix}} \mathbb{R}^3 \xrightarrow{\begin{pmatrix} -1 & -1 & 0 \\ 1 & 0 & -1 \\ 0 & 1 & 1 \end{pmatrix}} \mathbb{R}^3 \longrightarrow 0$$

In the second case, using $<'$, we have

$$C_2(X') = \text{span}\{[a, c, b]\},$$

$$C_1(X') = \text{span}\{[a, c], [a, b], [c, b]\},$$

and

$$C_0(X') = \text{span}\{[a], [c], [b]\}.$$

The simplicial chain complex $(C_\bullet(X'), \partial'_\bullet)$ is given by

$$0 \longrightarrow \mathbb{R} \xrightarrow{\begin{pmatrix} 1 \\ -1 \\ 1 \end{pmatrix}} \mathbb{R}^3 \xrightarrow{\begin{pmatrix} -1 & -1 & 0 \\ 1 & 0 & -1 \\ 0 & 1 & 1 \end{pmatrix}} \mathbb{R}^3 \longrightarrow 0$$

Clearly, these two simplicial chain complexes are identical, so they are isomorphic. However, the isomorphism induced by the difference in vertex partial orders is not the identity chain map.

The permutation matrices described in the proof of Proposition 4.15 are given as follows,

$$p_2 = (1),$$

since there is exactly one simplex of dimension 2,

$$p_1 = \begin{pmatrix} 0 & 1 & 0 \\ 1 & 0 & 0 \\ 0 & 0 & 1 \end{pmatrix},$$

and

$$p_0 = \begin{pmatrix} 1 & 0 & 0 \\ 0 & 0 & 1 \\ 0 & 1 & 0 \end{pmatrix}.$$

Now to correctly set the signs for the permutation matrix entries, compute

$$\begin{aligned} p'_1([\widehat{a}, b, c]) &= [\widehat{a}, c, b] \\ p'_1([a, \widehat{b}, c]) &= -[a, c, \widehat{b}] \\ p'_1([a, b, \widehat{c}]) &= -[a, \widehat{c}, b] \end{aligned}$$

according to the formula listed in the proof of Proposition 4.15. Therefore, we construct

$$p'_1 = \begin{pmatrix} 0 & -1 & 0 \\ -1 & 0 & 0 \\ 0 & 0 & 1 \end{pmatrix}.$$

Observe that if we set $p_2' = p_2$, then

$$\partial_2' \circ p_2' = \begin{pmatrix} 1 \\ -1 \\ 1 \end{pmatrix} (1) = \begin{pmatrix} 1 \\ -1 \\ 1 \end{pmatrix}$$

and

$$p_1' \circ \partial_2 = \begin{pmatrix} 0 & -1 & 0 \\ -1 & 0 & 0 \\ 0 & 0 & 1 \end{pmatrix} \begin{pmatrix} 1 \\ -1 \\ 1 \end{pmatrix} = \begin{pmatrix} 1 \\ -1 \\ 1 \end{pmatrix}$$

as desired.

For p_0', we again use the formula from the proof but now must accumulate signs from p_1' as well. There are several possible ways to obtain each 0-simplex by deleting vertices,

$$\begin{aligned}
p_0'([b]) &= p_0'([\widehat{a}, b]) = (-1)[\widehat{a}, b] = -[b] \\
p_0'([a]) &= p_0'([a, \widehat{b}]) = (-1)[a, \widehat{b}] = -[a] \\
p_0'([c]) &= p_0'([\widehat{a}, c]) = (-1)[\widehat{a}, c] = -[c] \\
p_0'([a]) &= p_0'([a, \widehat{c}]) = (-1)[a, \widehat{c}] = -[a] \\
p_0'([c]) &= p_0'([\widehat{b}, c]) = -(+1)[c, \widehat{b}] = -[c] \\
p_0'([b]) &= p_0'([b, \widehat{c}]) = -(+1)[\widehat{c}, b] = -[b]
\end{aligned}$$

in which we have marked the signs coming from p_1' with parentheses. Since the duplicate calculations all agree, this means that p_0' is well defined, and is given by the matrix

$$p_0' = \begin{pmatrix} -1 & 0 & 0 \\ 0 & 0 & -1 \\ 0 & -1 & 0 \end{pmatrix}.$$

Under these definitions for p_1' and p_0', we have that

$$\partial_1' \circ p_1' = \begin{pmatrix} -1 & -1 & 0 \\ 1 & 0 & -1 \\ 0 & 1 & 1 \end{pmatrix} \begin{pmatrix} 0 & -1 & 0 \\ -1 & 0 & 0 \\ 0 & 0 & 1 \end{pmatrix} = \begin{pmatrix} 1 & 1 & 0 \\ 0 & -1 & -1 \\ -1 & 0 & 1 \end{pmatrix}$$

which is equal to

$$p_0' \circ \partial_1 = \begin{pmatrix} -1 & 0 & 0 \\ 0 & 0 & -1 \\ 0 & -1 & 0 \end{pmatrix} \begin{pmatrix} -1 & -1 & 0 \\ 1 & 0 & -1 \\ 0 & 1 & 1 \end{pmatrix} = \begin{pmatrix} 1 & 1 & 0 \\ 0 & -1 & -1 \\ -1 & 0 & 1 \end{pmatrix}$$

so that $p_\bullet'$ is the chain isomorphism induced by the change in the vertex partial order. As a diagram, it is

$$
\begin{array}{ccccccccc}
0 & \longrightarrow & \mathbb{R} & \xrightarrow{\begin{pmatrix} 1 \\ -1 \\ 1 \end{pmatrix}} & \mathbb{R}^3 & \xrightarrow{\begin{pmatrix} -1 & -1 & 0 \\ 1 & 0 & -1 \\ 0 & 1 & 1 \end{pmatrix}} & \mathbb{R}^3 & \longrightarrow & 0 \\
 & & \Big\downarrow {\scriptstyle (1)} & & \Big\downarrow {\scriptstyle \begin{pmatrix} 0 & -1 & 0 \\ -1 & 0 & 0 \\ 0 & 0 & 1 \end{pmatrix}} & & \Big\downarrow {\scriptstyle \begin{pmatrix} -1 & 0 & 0 \\ 0 & 0 & -1 \\ 0 & -1 & 0 \end{pmatrix}} & & \\
0 & \longrightarrow & \mathbb{R} & \xrightarrow[\begin{pmatrix} 1 \\ -1 \\ 1 \end{pmatrix}]{} & \mathbb{R}^3 & \xrightarrow[\begin{pmatrix} -1 & -1 & 0 \\ 1 & 0 & -1 \\ 0 & 1 & 1 \end{pmatrix}]{} & \mathbb{R}^3 & \longrightarrow & 0.
\end{array}
$$

Exercise 4.40 1. Construct the simplicial chain complex for the simplicial complex generated by $[a, b, c]$, $[a, d, e]$, and $[b, d, f]$ using the vertex partial order $a < b < c < d < e < f$.
2. Construct the isomorphism induced by changing to the vertex partial order $a < c < b < d < f < e$.

Exercise 4.41 Suppose that $f : X \to Y$ is a simplicial map. Prove or give a counterexample: if $\leq_X$ is a vertex partial order on X, then there exists a vertex partial order $\leq_Y$ on Y making f order preserving on vertices.

Example 4.27 Consider the simplicial complex model X of the *torus* shown in Fig. 4.21. Its simplicial chain complex is given by

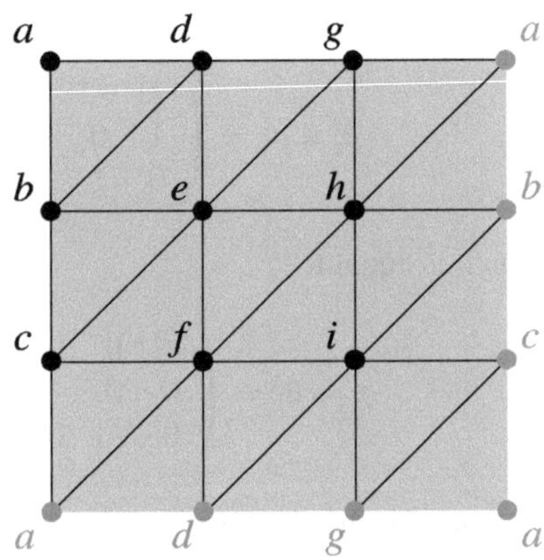

Fig. 4.21 The simplicial complex model of the torus in Example 4.27. Vertices shown in gray are duplicates, to indicate that simplices "reach across" from left to right, and top to bottom

$$0 \longrightarrow C_2(X) = \mathbb{R}^{18} \xrightarrow{\partial_2} C_1(X) = \mathbb{R}^{27} \xrightarrow{\partial_1} C_0(X) = \mathbb{R}^9 \longrightarrow 0.$$

The homology of this chain complex is given by

$$H_0(X) \cong \mathbb{R}, \;\; H_1(X) \cong \mathbb{R}^2, \;\; H_2(X) \cong \mathbb{R}.$$

While this can be established by direct calculation (Exercise 4.42, below), it is insightful to consider a set of basis elements with an easy interpretation. Specifically, let us establish that

$$[a]$$

is an element of coker ∂_1,

$$[a, b] + [b, c] - [a, c] \text{ and } [a, d] + [d, g] - [a, g]$$

are elements of ker ∂_1 and coker ∂_2, and that

$$\begin{aligned}
&[a, b, e] + [b, c, f] - [a, c, d] \\
&+[d, e, h] + [e, f, i] - [d, f, g] \\
&+[b, g, h] + [c, h, i] - [a, g, i] \\
&-[a, d, e] - [b, e, f] + [c, d, f] \\
&-[d, g, h] - [e, h, i] + [f, g, i] \\
&-[a, b, g] - [b, c, h] + [a, c, i]
\end{aligned}$$

is an element of ker ∂_2. This will ensure that the Betti numbers are at least 1, 2, and 1, respectively.

By construction, if there is an element of $p \in C_1(X)$ for which $\partial_1 p = [a]$, this means that vertex a appeared exactly once, in one edge in p. Since edges have exactly two, distinct endpoint vertices, this means that there must be another vertex that appears an odd number of times in p. This vertex will necessarily have a nonzero image in $\partial_1 p$, which contradicts the fact that we want $\partial_1 p = [a]$. Hence $[a]$ cannot be in the image of ∂_1.

As for the kernel of ∂_1, this is verified by direct calculation, namely

$$\partial_1([a, b] + [b, c] - [a, c]) = ([b] - [a]) + ([c] - [b]) - ([c] - [a]) = 0,$$

and

$$\partial_1([a, d] + [d, g] - [a, g]) = ([d] - [a]) + ([g] - [d]) - ([g] - [a]) = 0.$$

Considering $([a, b] + [b, c] - [a, c])$, this arises as the image of $[a, b, c]$ under ∂_2,

$$\partial_2([a, b, c]) = [b, c] - [a, c] + [a, b],$$

but $[a, b, c]$ is not a simplex of X, via inspecting Fig. 4.21. The same logic holds for $([a, d] + [d, g] - [a, g])$; it would have to be the image of $[a, d, g]$, a simplex which is not present in X.

Finally, establishing the kernel of ∂_2 merely requires a lengthy calculation,

$$
\begin{aligned}
\partial_2 \; &([a,b,e]+[b,c,f]-[a,c,d] \\
&+[d,e,h]+[e,f,i]-[d,f,g] \\
&+[b,g,h]+[c,h,i]-[a,g,i] \\
&-[a,d,e]-[b,e,f]+[c,d,f] \\
&-[d,g,h]-[e,h,i]+[f,g,i] \\
&-[a,b,g]-[b,c,h]+[a,c,i]) &&= [b,e]+[c,f]-[c,d] \\
& && \quad +[e,h]+[f,i]-[f,g] \\
\text{(deleting first} & && \quad +[g,h]+[h,i]-[g,i] \\
\text{vertices from each} & && \quad -[d,e]-[e,f]+[d,f] \\
\text{2-simplex above)} & && \quad -[g,h]-[h,i]+[g,i] \\
& && \quad -[b,g]-[c,h]+[c,i] \\
& && \quad -[a,e]-[b,f]+[a,d] \\
& && \quad -[d,h]-[e,i]+[d,g] \\
\text{(deleting second} & && \quad -[b,h]-[c,i]+[a,i] \\
\text{vertices from each} & && \quad +[a,e]+[b,f]-[c,f] \\
\text{2-simplex above)} & && \quad +[d,h]+[e,i]-[f,i] \\
& && \quad +[a,g]+[b,h]-[a,i] \\
& && \quad +[a,b]+[b,c]-[a,c] \\
& && \quad +[d,e]+[e,f]-[d,f] \\
\text{(deleting third} & && \quad +[b,g]+[c,h]-[a,g] \\
\text{vertices from each} & && \quad -[a,d]-[b,e]+[c,d] \\
\text{2-simplex above)} & && \quad -[d,g]-[e,h]+[f,g] \\
& && \quad -[a,b]-[b,c]+[a,c] \\
& &&= 0.
\end{aligned}
$$

A careful check is required to ensure that each of the 1-simplex terms above come in pairs with opposite signs!

Exercise 4.42 Show that the Betti numbers of the torus in Example 4.27 are exactly 1, 2, and 1, as claimed. (Note: in the next chapter, we will give clear topological interpretations of these quantities. This exercise is merely concerned with their calculation.)

We now obtain the most important result of the section. The simplicial chain complex preserves the structure of a simplicial complex, in that each simplicial map induces a chain map. Fortunately, the proof is constructive, which means that computing the induced map can be done algorithmically.

Theorem 4.1 (the simplicial chain complex is a functor) *The simplicial chain complex is a covariant functor* $C_\bullet$: **Asc** $\to$ **Kom**. *Explicitly, each simplicial complex* X *is transformed to a chain complex* $(C_\bullet, \partial_\bullet)$*, and each simplicial map* $f : X \to Y$ *induces a chain map* $(C_\bullet(X), \partial_{X,\bullet}) \to (C_\bullet(Y), \partial_{Y,\bullet})$.

Proof Given an abstract simplicial complex X, Proposition 4.14 shows that $(C_\bullet(X), \partial_{X,\bullet})$ is a chain complex. We now need to construct a chain map $f_\bullet$: $(C_\bullet(X), \partial_{X,\bullet}) \to (C_\bullet(Y), \partial_{Y,\bullet})$ from a simplicial map $f : X \to Y$. Without loss of generality, let us suppose that we have a vertex partial ordering for each simplicial complex. Simplicial maps are defined by their values on vertices but also define a map on simplices. Moreover, if σ is a k-dimensional simplex of X, then it corresponds to an element of the standard basis for $C_k(X)$.

Since f is a simplicial map, if

$$\sigma = [v_0, v_1, \ldots, v_k]$$

is a simplex of X with

$$v_0 < v_1 < \cdots < v_k$$

according to the vertex partial order for X, it might not be the case that f preserves the order. If f does not preserve vertex order, then subsequent applications of ∂_Y will have incorrect signs, meaning that

$$\partial_{Y,k} \circ f_k \neq f_{k-1} \circ \partial_{X,k}.$$

To account for this possible disordering, we may have to perform a sequence of pairwise swaps of $f(v_i)$ with $f(v_j)$ to obtain the correct order for vertices in $f(\sigma)$. Each one of these swaps will result in a sign reversal of $(\partial_{Y,k} \circ f_k)(\sigma)$, because any swap of vertices of

$$[f(v_0), \ldots, f(v_k)]$$

implies a similar swap of vertices *or a sign change* in

$$(-1)^k[f(v_0), \ldots, \widehat{f(v_j)}, \ldots, f(v_k)].$$

While there may be several different ways to perform these swaps, the *parity* depends only on the starting and ending lists of vertices, at least if there are no duplicated vertices. Therefore, the sign required to compensate for the disordering effect is

$$[f : \sigma] := (-1)^{\text{the number of swaps required to sort } [f(v_0), f(v_1), \cdots, f(v_k)]}$$

which does not depend on the particular choice of swaps.

We therefore define

$$f_k(\sigma) := \begin{cases} [f:\sigma]f(\sigma) & \text{if } f(\sigma) \text{ is of dimension } k \\ 0 & \text{otherwise,} \end{cases}$$

where we are interpreting both σ and $f(\sigma)$ as basis elements of $C_k(X)$ and $C_k(Y)$, respectively. This construction satisfies the equation

$$(g \circ f)_\bullet = g_\bullet \circ f_\bullet,$$

if $f : X \to Y$ and $g : Y \to Z$ are simplicial maps, since composition of simplicial maps acts on simplices by composition of functions on their respective vertices.

We must show that $f_\bullet$ is a well defined chain map, namely that

$$\partial_{Y,k} \circ f_k = f_{k-1} \circ \partial_{X,k}$$

for all k. If $f_k(\sigma) \neq 0$, this can be shown by direct calculation,

$$\begin{aligned}
&\left(\partial_{Y,k} \circ f_k\right)(\sigma) \\
&\quad = \left(\partial_{Y,k} \circ f_k\right)([v_0, \ldots, v_k]) \\
&\quad = \partial_{Y,k}\left(f_k([v_0, \ldots, v_k])\right) \\
&\quad = \partial_{Y,k}\left([f:\sigma][f(v_0), \ldots, f(v_k)]\right) \\
&\quad = \sum_{j=0}^{k} (-1)^j [f:\sigma][f(v_0), \ldots, \widehat{f(v_j)}, \ldots, f(v_k)] \\
&\quad = \sum_{j=0}^{k} (-1)^j f_{k-1}\left([v_0, \ldots, \widehat{v_j}, \ldots, v_k]\right) \\
&\quad = f_{k-1}\left(\sum_{j=0}^{k} (-1)^j [v_0, \ldots, \widehat{v_j}, \ldots, v_k]\right) \\
&\quad = \left(f_{k-1} \circ \partial_{X,k}\right)(\sigma).
\end{aligned}$$

If $f_k(\sigma) = 0$, meaning that $f(\sigma)$ is a simplex of dimension less than k, then $f(v_i) = f(v_j)$ for some $i \neq j$. We only need to consider the situation where there is exactly one pair of vertices that is duplicated, since more duplicate vertices will be present in $\partial_{X,k}$, and hence $f_{k-1} \circ \partial_{X,k}$ is zero on such a simplex. Since we may assume that a vertex partial order is available, any pair of duplicate vertices will occur in consecutive indices. Thus, if $f(v_i) = f(v_{i+1})$ for some i,

$$\begin{aligned}
&\sum_{j=0}^{k}(-1)^j[f(v_0),\ldots,\widehat{f(v_j)},\ldots,f(v_k)]\\
&\qquad=\sum_{j=0}^{i-1}(-1)^j[f(v_0),\ldots,\widehat{f(v_j)},\ldots,f(v_k)]\\
&\qquad\quad+\sum_{j=i+2}^{k}(-1)^j[f(v_0),\ldots,\widehat{f(v_j)},\ldots,f(v_k)]\\
&\qquad\quad+(-1)^i[f(v_0),\ldots,\widehat{f(v_i)},\ldots,f(v_k)]\\
&\qquad\quad+(-1)^{i+1}[f(v_0),\ldots,\widehat{f(v_{i+1})},\ldots,f(v_k)]\\
&\qquad=\sum_{j=0}^{i-1}(-1)^j[f(v_0),\ldots,\widehat{f(v_j)},\ldots,f(v_k)]\\
&\qquad\quad+\sum_{j=i+2}^{k}(-1)^j[f(v_0),\ldots,\widehat{f(v_j)},\ldots,f(v_k)]\\
&\qquad=\sum_{j=0}^{i-1}(-1)^j f_{k-1}([v_0,\ldots,\widehat{v_j},\ldots,v_k])\\
&\qquad\quad+\sum_{j=i+2}^{k}(-1)^j f_{k-1}([v_0,\ldots,\widehat{v_j},\ldots,v_k])
\end{aligned}$$

but all of the remaining terms contain duplicate $f(v_i)$ vertices, and so f_{k-1} applied to each of them is zero. □

Example 4.28 Consider the abstract simplicial complex X generated by $[a, b, c]$, $[b, d]$, and $[c, d]$ and the simplicial complex Y generated by $[e, f]$, $[e, g]$, and $[f, g]$. The vertex function $m : V_X \to V_Y$ given by

$$\begin{aligned}
m(a) &:= f,\\
m(b) &:= f,\\
m(c) &:= e, \text{ and}\\
m(d) &:= g
\end{aligned}$$

is a simplicial map $m : X \to Y$ as shown in Fig. 4.22.

Let us choose vertex ordering $a < b < c < d$ and $e < f < g$ for X and Y, respectively. For the chain complex $C_\bullet(X)$, notice that

$$C_2(X) \cong \mathbb{R},\;\; C_1(X) \cong \mathbb{R}^5, \text{ and } C_0(X) \cong \mathbb{R}^4$$

since X has one 2-dimensional simplex, five edges, and 4 vertices. Likewise,

$$C_2(Y) \cong 0,\;\; C_1(Y) \cong \mathbb{R}^3, \text{ and } C_0(Y) \cong \mathbb{R}^3$$

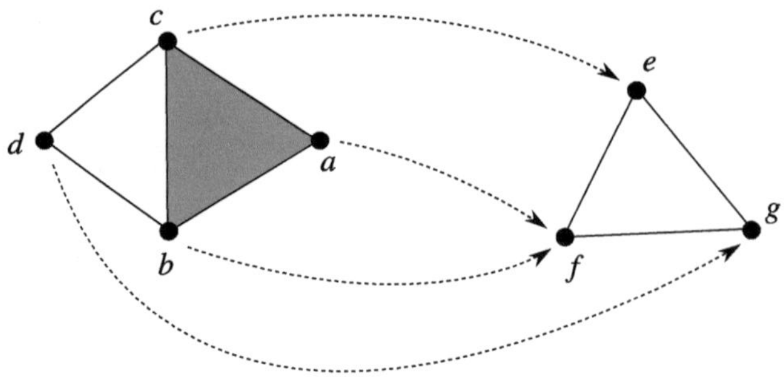

Fig. 4.22 The simplicial map $m : X \to Y$ in Example 4.28

since Y has only three edges and vertices. Using lexicographic ordering on the simplices in each of the simplicial chain spaces makes bookkeeping sense and follows the prescriptions made during the proofs in this section. For instance, the component map $m_0 : C_0(X) \to C_0(Y)$ is simply a restatement of the map m on vertices as a matrix,

$$m_0 = \begin{pmatrix} 0 & 0 & 1 & 0 \\ 1 & 1 & 0 & 0 \\ 0 & 0 & 0 & 1 \end{pmatrix},$$

since both a and b (the first and second columns) are both taken to f (the second row). The second component map $m_1 : C_1(X) \to C_1(Y)$ is given by

$$m_1 = \begin{pmatrix} 0 & -1 & -1 & 0 & 0 \\ 0 & 0 & 0 & 0 & 1 \\ 0 & 0 & 0 & 1 & 0 \end{pmatrix},$$

and is a bit more subtle because m collapses dimensions and does not preserve the vertex partial orders. Notice that

$$m([a, b]) = [f, f] = [f],$$

which means that

$$m_1([a, b]) = 0.$$

This accounts for the first column of m_1. The second column has a minus sign because

$$m([a, c]) = [f, e],$$

which takes one swap to put it in its proper order since $e < f$. Thus $[m : [a, c]] = -1$. This happens again in the third column since

$$m([b, c]) = [f, e].$$

In summary, the chain map is given by the diagram

$$
\begin{array}{ccccccccc}
0 & \longrightarrow & \mathbb{R} & \xrightarrow{\begin{pmatrix} 1 \\ -1 \\ 1 \\ 0 \\ 0 \end{pmatrix}} & \mathbb{R}^5 & \xrightarrow{\begin{pmatrix} -1 & -1 & 0 & 0 & 0 \\ 1 & 0 & -1 & -1 & 0 \\ 0 & 1 & 1 & 0 & -1 \\ 0 & 0 & 0 & 1 & 1 \end{pmatrix}} & \mathbb{R}^4 & \longrightarrow & 0 \\
 & & \downarrow & & \downarrow \begin{pmatrix} 0 & -1 & -1 & 0 & 0 \\ 0 & 0 & 0 & 0 & 1 \\ 0 & 0 & 0 & 1 & 0 \end{pmatrix} & & \downarrow \begin{pmatrix} 0 & 0 & 1 & 0 \\ 1 & 1 & 0 & 0 \\ 0 & 0 & 0 & 1 \end{pmatrix} & & \\
0 & \longrightarrow & 0 & \longrightarrow & \mathbb{R}^3 & \xrightarrow[\begin{pmatrix} -1 & -1 & 0 \\ 1 & 0 & -1 \\ 0 & 1 & 1 \end{pmatrix}]{} & \mathbb{R}^3 & \longrightarrow & 0
\end{array}
$$

Exercise 4.43 Construct the induced simplicial chain map for the simplicial map $m : X \to Y$ given by

$$
\begin{aligned}
m(a) &:= A, \\
m(b) &:= B, \\
m(c) &:= C, \\
m(d) &:= A, \\
m(e) &:= B, \\
m(f) &:= C,
\end{aligned}
$$

where X is the simplicial complex generated by $[a, b]$, $[b, c]$, $[c, d]$, $[d, e]$, $[e, f]$, and $[a, f]$, and Y is the simplicial complex generated by $[A, B]$, $[B, C]$, and $[A, C]$.

Exercise 4.44 Construct the induced simplicial chain map for the simplicial map $f : X \to Y$ given by

$$
\begin{aligned}
f(a) &:= A, \\
f(b) &:= B, \\
f(c) &:= C,
\end{aligned}
$$

where X is the simplicial complex generated by $[a, b]$, $[b, c]$, and $[a, c]$ and $Y = SX$, the suspension of X.

Exercise 4.45 Construct the induced simplicial chain map for the simplicial map $m : X \to X$ given by

$$\begin{aligned} m(a) &:= a, \\ m(b) &:= b, \\ m(c) &:= c, \\ m(d) &:= a, \\ m(e) &:= b, \\ m(f) &:= c, \end{aligned}$$

where X is the simplicial complex generated by $[a, b, d]$, $[b, c, e]$, $[a, c, f]$, $[b, d, e]$, $[c, e, f]$, and $[a, d, f]$.

4.6 Further Reading

Simplicial complexes arise frequently in applications, for instance in neuroscience (Curto and Sanderson 2025; Giusti et al. 2016; Curto 2017), genomics (Roman et al. 2015; Nanda and Sazdanović 2013), signal processing (Krim et al. 2016; Krim and Hamza 2015), and cyber security (Myers et al., 2023; Robinson et al., 2022), among many others that the reader might find interesting. When handling noisy real-world data, one popular strategy is to combine the sequence and chain complex methods of the previous chapters with simplicial complexes. The result is usually called *topological data analysis*. There are many good introductions to the subject, for instance (Dey and Wang 2022; Carlsson and Vejdemo-Johansson 2021; Ziegelmeier 2024; Munch 2017; Lesnick 2013; Cipra 2009; Ghrist 2008; Giunti et al. 2022).

Simplicial complexes are useful from an algorithmic perspective because they are fundamentally combinatorial objects. As we will see in the next chapter, the homology of the simplicial chain complex detects topological features of an abstract simplicial complex. Simplicial chain complexes also have useful geometric information (Chachólski et al. 2023). The concepts we treat here, such as homology, suspension, and many others generalize substantially (Chachólski et al. 2021). Topological properties can be treated independently of simplicial complexes (Barmak 2011; Munkres 1984; Hatcher 2002). Combinatorial properties of topological spaces are interesting, for instance (Bobrowski and Kahle 2018; Kahle 2009). A rather different functor that summarizes properties of a topological space is the *Reeb graph* (Munch and Wang 2016).

The functor that converts an abstract simplicial complex into its simplicial chain complex is an example of *vectorization*, which turns combinatorial data into numerical data. In topological data analysis, there are many other strategies for vectorization, for instance (Luchinsky and Islambekov 2025; Ali et al. 2023; Li et al. 2021; Adams et al. 2017; Topaz et al. 2015).

References

Adams H, Emerson T, Kirby M, Neville R, Peterson C, Shipman P, Chepushtanova S, Hanson E, Motta F, Ziegelmeier L (2017) Persistence images: a stable vector representation of persistent homology. J Mach Learn Res 18(8):1–18
Ali D, Asaad A, Jimenez MJ, Nanda V, Paluzo-Hidalgo E, Soriano-Trigueros M (2023) A survey of vectorization methods in topological data analysis. IEEE Trans Pattern Anal Mach Intell 45(12):14069–14080. https://doi.org/10.1109/tpami.2023.3308391
Barmak JA (2011) Algebraic topology of finite topological spaces and applications. Springer
Bobrowski O, Kahle M (2018) Topology of random geometric complexes: a survey. J Appl Comput Topol 1:331–364
Carlsson G, Vejdemo-Johansson M (2021) Topological data analysis with applications. Cambridge University Press
Chachólski W, Giunti B, Jin A, Landi C, (2023) Decomposing filtered chain complexes: geometry behind barcoding algorithms. Comput Geom 109:101938. https://www.sciencedirect.com/science/article/pii/S0925772122000815
Chachólski W, Giunti B, Landi C (2021) Invariants for tame parametrised chain complexes. Homol Homotopy Appl 23(2):183–213
Cipra BA (2009) Homology: an idea whose time has come. SIAM News 42(10)
Curto C (2017) What can topology tell us about the neural code? Bull Am Math Soc 54(1):63–78
Curto C, Sanderson N (2025) Topological neuroscience: linking circuits to function. Ann Rev Neurosci 48
Dey TK, Wang Y (2022) Computational topology for data analysis. Cambridge University Press
Dowker CH (1952) Homology groups of relations. Ann Math 84–95
Ghrist R (2008) Barcodes: the persistent topology of data. Bull-Am Math Soc 45(1):61
Giunti B, Lazovskis J, Rieck B (2022) DONUT: database of original and non-theoretical uses of topology. https://donut.topology.rocks
Giusti C, Ghrist R, Bassett DS (2016) Two's company, three (or more) is a simplex: algebraic-topological tools for understanding higher-order structure in neural data. J Comput Neurosci 41:1–14
Hatcher A (2002) Algebraic topology. Cambridge University Press
Kahle M (2009) Topology of random clique complexes. Disc Math 309(6):1658–1671
Krim H, Gentimis T, Chintakunta H (2016) Discovering the whole by the coarse: a topological paradigm for data analysis. IEEE Signal Proc Mag 33(2):95–104
Krim H, Hamza AB (2015) Geometric methods in signal and image analysis. Cambridge University Press
Lesnick M (2013) Studying the shape of data using topology. Inst Lett (2013). https://www.ias.edu/ideas/2013/lesnick-topological-data-analysis

Li L, Thompson C, Henselman-Petrusek G, Giusti C, Ziegelmeier L (2021) Minimal cycle representatives in persistent homology using linear programming: an empirical study with user's guide. Front Artif Intell 4. https://doi.org/10.3389/frai.2021.681117

Luchinsky A, Islambekov U (2025) TDAvec: computing vector summaries of persistence diagrams for topological data analysis in R and Python. https://arxiv.org/abs/2411.17340

Munch E (2017) A user's guide to topological data analysis. J Learn Anal 42(2):47–61

Munch E, Wang B (2016) Convergence between categorical representations of Reeb Space and Mapper. In: Fekete S, Lubiw A (eds) 32nd international symposium on computational geometry (SoCG 2016), Leibniz international proceedings in informatics (LIPIcs), vol 51, pp 53:1–53:16. Schloss Dagstuhl–Leibniz-Zentrum für Informatik, Dagstuhl, Germany. https://doi.org/10.4230/LIPIcs.SoCG.2016.53. https://drops.dagstuhl.de/entities/document/10.4230/LIPIcs.SoCG.2016.53

Munkres J (1984) Elements of algebraic topology. Westview Press

Myers A, Bittner A, Aksoy S, Best D, Henselman-Petrusek G, Jenne H, Joslyn C, Kay B, Seppala G, Young SJ et al (2023) Malicious cyber activity detection using zigzag persistence. In: 2023 IEEE conference on dependable and secure computing (DSC). IEEE, pp 1–8

Nanda V, Sazdanović R (2013) Simplicial models and topological inference in biological systems. In: Discrete and topological models in molecular biology, pp 109–141. Springer

Robinson M, Anderson C, Li LW, Huntsman S (2022) Statistical detection of format dialects using the weighted Dowker complex. In: 2022 IEEE security and privacy workshops (SPW), pp 98–112. IEEE

Roman T, Nayyeri A, Fasy BT, Schwartz R (2015) A simplicial complex-based approach to unmixing tumor progression data. BMC Bioinform 16:1–17

Sheehy D (2012) A multicover nerve for geometric inference. In: 24th Canadian conference on computational geometry (CCCG 2012), pp 309–314

Topaz CM, Ziegelmeier L, Halverson T (2015) Topological data analysis of biological aggregation models. PloS One 10(5):e0126383

Ziegelmeier L (2024) A topology scavenger hunt to introduce topological data analysis. Teach Math Cross-Curric Proj 72:285

Chapter 5
Simplicial Homology and Homotopy

Given that every simplicial complex can be transformed into its simplicial chain complex in a structure-preserving way, it is natural to explore the meaning of the homology of the simplicial chain complex in terms of the original simplicial complex. It happens that the *simplicial homology* captures certain *topological* features, which are those properties that are invariant under simplicial maps. Moreover, there is a notion of *simplicial homotopy* that is related to the chain homotopies discussed in Sect. 3.4 of Chap. 3. While chain homotopies were a bit mysterious, *simplicial homotopies* are rather easier to construct and provide a concrete interpretation of the meaning of *simplicial homology*. With these technical tools in hand, the chapter ends by using duality to explore interactions between the algebraic structure of chain complexes and the topology of simplicial complexes.

In this chapter,

Section 5.1 Defines *simplicial homology* as a natural consequence of the previous chapters, and explores a few of its properties.
Section 5.2 Defines a generalization called *relative simplicial homology*.
Section 5.3 Defines elementary simplicial homotopies, which provide concrete examples of chain homotopies induced by simplicial maps.
Section 5.4 Shows that *subdividing* a simplicial complex does not change its simplicial homology.
Section 5.5 Shows how to compose subdivision with simplicial maps.
Section 5.6 Dualizes simplicial homology to obtain *simplicial cohomology*, which leads to additional algebraic structure.
Section 5.7 Demonstrates that simplicial cohomology has a natural product structure, called the *cup product*.

Throughout, we will make use of the theory of chain complexes developed in the previous chapters.

M. Robinson, *Computational Homological Algebra*, Mathematical Engineering,
https://doi.org/10.1007/978-3-032-08634-1_5

5.1 Simplicial Homology

We have already shown that the simplicial chain complex is a covariant functor $C_\bullet$: **Asc** $\to$ **Kom** (Theorem 4.1 in Chap. 4) and that homology is a covariant functor H_k : **Kom** $\to$ **Vec** (Proposition 3.2 in Chap. 3). This suggests the following definition of *simplicial homology.*

Definition 5.1 For a simplicial complex X, its *index k simplicial homology* is the vector space given by

$$H_k(X) := H_k(C_\bullet(X), \partial_{X,\bullet}).$$

The phrase *dimension k simplicial homology* is also frequently used to refer to $H_k(X)$ in the literature.

Corollary 5.1 *Simplicial homology is a covariant functor* H_k : **Asc** $\to$ **Vec** *for each index k, which is to say that each simplicial complex is transformed into a vector space and each simplicial map* $f : X \to Y$ *induces a linear map* $f_k : H_k(X) \to H_k(Y)$.

The simplicial homology $H_\bullet(X)$ measures the presence of certain kinds of structures in a simplicial complex X. Succinctly, according to the barcode decomposition of $C_\bullet(X)$, $H_k(X)$ has a basis consisting of boundaries of $(k+1)$-dimensional simplices. Recall that the path components of a simplicial complex were defined as a result of Proposition 4.3 in Chap. 4. With this in mind, it will suffice to think intuitively for the moment:

1. $H_0(X)$ has a basis consisting of equivalence classes of the path components—boundaries of 1-dimensional simplices (edges)—of X,
2. $H_1(X)$ has a basis consisting of equivalence classes of the loops—boundaries of 2-dimensional simplices (triangles)—of X,
3. $H_2(X)$ has a basis consisting of equivalence classes of the bubbles—boundaries of 3-dimensional simplices (tetrahedra)—of X,
4. and so on.

Example 5.1 Suppose that X is a simplicial complex given by $\{\emptyset, [v_0], [v_1], \ldots, [v_n]\}$. In this case, $C_k(X)$ is the trivial vector space for $k > 1$, and $C_0(X)$ has dimension $(n+1)$. All of the bars in the barcode decomposition for $C_\bullet(X)$ have length 1, and so $H_k(X) \cong C_k(X)$ for all k. In particular, $H_0(X)$ has a basis consisting of vectors of the form

$$0[v_0] + \cdots + 1[v_k] + \cdots + 0[v_n], \tag{5.1}$$

each supported on a single vertex. The nonzero component of each basis element selects a vertex, which is a path component in this special case.

Now let us suppose that Y is given by

$$Y := X \cup \{[v_0, v_1]\},$$

using the same definition of X. Briefly, Y is the same simplicial complex as X with one additional edge between the first two listed vertices. This leaves $C_0(Y) = C_0(X)$ and $C_k(Y) = C_k(X) = 0$ for $k > 1$. But now $C_1(Y) = \mathbb{R}$, spanned by the basis element corresponding to the single edge $[v_0, v_1]$. The simplicial chain complex for Y is given by

$$\cdots \longrightarrow 0 \longrightarrow \mathbb{R} \xrightarrow{\begin{pmatrix} -1 \\ 1 \\ 0 \\ \vdots \\ 0 \end{pmatrix}} \mathbb{R}^{n+1} \longrightarrow 0 \longrightarrow \cdots$$

The barcode decompositions of $C_\bullet(X)$ and $C_\bullet(Y)$ are identical with one exception: one of the bars of $C_\bullet(X)$ (of length 1) has been extended to a bar of length 2 in $C_\bullet(Y)$. That bar does not appear in $H_\bullet(Y)$, and corresponds to some linear combination of $[v_0]$ and $[v_1]$. One convenient basis for $H_\bullet(Y)$ is given by

$$\begin{gathered}
1[v_0] + 1[v_1] + 0[v_2] + \cdots + 0[v_k] + \cdots + 0[v_n], \\
0[v_0] + 0[v_1] + 1[v_2] + \cdots + 0[v_k] + \cdots + 0[v_n], \\
\vdots \\
0[v_0] + 0[v_1] + 0[v_2] + \cdots + 1[v_k] + \cdots + 0[v_n], \\
\vdots \\
0[v_0] + 0[v_1] + 0[v_2] + \cdots + 0[v_k] + \cdots + 1[v_n].
\end{gathered}$$

Notice that the first basis element is supported on the path component containing v_0 and v_1, while all of the other basis elements are supported on path components as well (each of which is single vertex).

Lemma 5.1 *Suppose that X and Y are simplicial complexes with disjoint vertex sets. Their union $X \cup Y$ is a simplicial complex whose homology is given by*

$$H_k(X \cup Y) \cong H_k(X) \oplus H_k(Y)$$

for all k.

Proof Since the vertex sets are disjoint, the set of simplices is disjoint as well. Since each simplex is a basis element in the simplicial chain complex,

$$C_k(X \cup Y) \cong C_k(X) \oplus C_k(Y)$$

for all k, and the boundary matrices are block diagonal

$$\partial_{X \cup Y} = \partial_X \oplus \partial_Y.$$

Since homology preserves direct sums, the result follows. □

Proposition 5.1 *The index* 0 *Betti number* dim $H_0(X)$ *is the number of path components of a simplicial complex* X.

Proof Suppose that two vertices v_0 and v_n of X are in the same path component. There is a path starting at v_0 and ending at v_n. Without loss of generality, let us write this path as the ordered list

$$\{v_0, v_1, v_2, \ldots, v_n\}.$$

Since this is a path, each pair of consecutive vertices v_i, v_{i+1} is contained in the same simplex. Since X contains every subset of every simplex, the existence of the path means that X must contain the following edges,

$$[v_0, v_1], [v_1, v_2], \ldots, [v_{n-1}, v_n].$$

However, some of these edges may not respect whatever vertex partial order we have chosen to use when constructing $C_\bullet(X)$. For instance, if $v_{i+1} < v_i$, then we would have $[v_{i+1}, v_i]$ instead of $[v_i, v_{i+1}]$. Let us translate this path into an element p of $C_1(X)$ by taking a linear combination of each edge in the path, with a coefficient -1 whenever the vertices are taken in an order contrary to the vertex partial order. The effect of the minus signs is that we can effectively compute

$$\begin{aligned}\partial_1 p &= \partial_1 \left([v_0, v_1] + [v_1, v_2] + \cdots + [v_{n-1}, v_n]\right) \\ &= -[v_0] + [v_1] - [v_1] + [v_2] + \cdots - [v_{n-1}] + [v_n] \\ &= [v_n] - [v_0].\end{aligned}$$

Using the interpretation of homology as a quotient

$$H_0(X) \cong \ker \partial_0 / \operatorname{image} \partial_1 \cong C_0(X)/\operatorname{image} \partial_1,$$

which means that $[v_n]$ and $[v_0]$ are in the same ∂_1-coset and so correspond to the same equivalence class of $H_0(X)$.

Before addressing the converse, notice that if v_0 and v_n are the same vertex—in which case p is a path looping back on itself—then the above calculation results in $\partial_1 p = 0$.

Now assume that $[v_0]$ and $[v_n]$ are in the same equivalence class under the quotient. This means that there is a linear combination q of edges in $C_1(X)$ whose image under ∂_1 is $[v_n] - [v_0]$. To complete the argument, it suffices[1] to show that some of the edges listed in q must form a path from v_0 to v_n. Based on our previous observation, we may exclude loops from discussion. This means that v_0 and v_n appear only once when listing the vertices in the edges of q. All of the other vertices must cancel with opposite signs under ∂_1 since we assumed that $\partial_1 q = [v_n] - [v_0]$. The result is that each such vertex appears in exactly two edges, resulting in a path from v_0 to v_n. □

[1] It might be the case that the edges listed in q correspond to a path from v_0 to v_n, but there might be extra edges that form loops unrelated to v_0 or v_n.

Example 5.2 Suppose that X is the simplicial complex generated by $[v_0, v_1]$, $[v_1, v_2]$, and $[v_0, v_2]$. The simplicial chain complex for X is given by the diagram

$$0 \longrightarrow \mathbb{R}^3 \xrightarrow{\begin{pmatrix} -1 & 0 & -1 \\ 1 & -1 & 0 \\ 0 & 1 & 1 \end{pmatrix}} \mathbb{R}^3 \longrightarrow 0.$$

In this chain complex, the columns of the only nontrivial boundary matrix correspond to $[v_0, v_1]$, $[v_1, v_2]$, and $[v_0, v_2]$, respectively. The rows correspond to $[v_0]$, $[v_1]$, and $[v_2]$, respectively.

Row reducing the boundary matrix produces the matrix

$$\begin{pmatrix} -1 & 0 & -1 \\ 0 & 1 & 1 \\ 0 & 0 & 0 \end{pmatrix}$$

which is of rank 2. This means that there are two bars of length 2 starting at index 1, one bar of length 1 starting at index 1, and one bar of length 1 starting at index 0. In summary,

$$H_1(X) \cong \mathbb{R} \text{ and } H_0(X) \cong \mathbb{R}.$$

This indicates that there is one path component of X and one loop. Let us delve into the basis elements to verify this intuition. (See Fig. 5.1 as well.)

- The loop corresponds to the length 1 bar starting at index 1, which has to be an element of the kernel of the boundary matrix. From the basis elements of the chain complex, this corresponds to

$$[v_0, v_1] + [v_1, v_2] - [v_0, v_2] = [v_0, v_1] + [v_1, v_2] + [v_2, v_0].$$

- The single path component corresponds to the other length 1 bar, which starts at index 0. The basis element of $H_0(X)$ corresponding to this bar is a vector outside

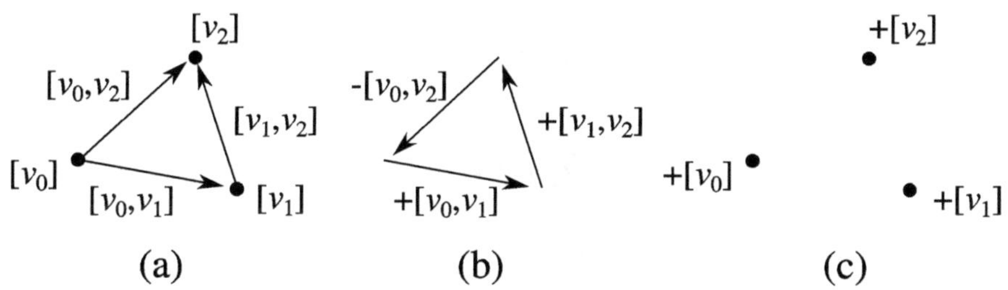

Fig. 5.1 **a** The simplicial complex X, **b** the nontrivial basis element for $H_1(X)$, and **c** the nontrivial basis element for $H_0(X)$ in Example 5.2

the image of the boundary matrix, of which

$$[v_0] + [v_1] + [v_2]$$

is the natural choice.

Exercise 5.1 Extend Example 5.2 to compute the simplicial homology of any polygon, making use of the calculation of $\partial_1 p$ in the proof of Proposition 5.1.

Example 5.3 If each path component is simple enough, the Euler characteristic simply counts path components. Specifically, if we can write a simplicial complex X as a disjoint union of path components

$$X = X_1 \sqcup X_2 \sqcup \cdots \sqcup X_n$$

in which $H_k(X_j) = 0$ for all $k > 0$, then because of Lemma 2.4 and Proposition 2.10 from Chap. 2,

$$\begin{aligned}
\chi(X) &= \chi(C_\bullet(X)) \\
&= \chi\left(C_\bullet\left(\bigsqcup_{j=1}^{n} X_j\right)\right) \\
&= \chi\left(\bigoplus_{j=1}^{n} C_\bullet(X_j)\right) \\
&= \sum_{j=1}^{n} \chi\left(C_\bullet(X_j)\right) \\
&= \sum_{j=1}^{n} \chi\left(H_\bullet(X_j)\right) \\
&= \sum_{j=1}^{n} \sum_{k} (-1)^k \beta_k(X_j) \\
&= \sum_{j=1}^{n} \beta_0(X_j) \\
&= n.
\end{aligned}$$

Example 5.4 Consider the simplicial complex X generated by the simplices in the boundary of $[a, b, c, d]$. Explicitly, all of the simplices are given by

$$\begin{gathered}[b,c,d],[a,c,d],[a,b,d],[a,b,c],\\ [a,b],[a,c],[a,d],[b,c],[b,d],[c,d],\\ [a],[b],[c],[d],\\ \emptyset.\end{gathered}$$

Using this ordering and the vertex partial order $a < b < c < d$, the simplicial chain complex is given by

$$0 \longrightarrow \mathbb{R}^4 \xrightarrow{\partial_2 = \begin{pmatrix} 0 & 0 & 1 & 1 \\ 0 & 1 & 0 & -1 \\ 0 & -1 & -1 & 0 \\ 1 & 0 & 0 & 1 \\ -1 & 0 & 1 & 0 \\ 1 & 1 & 0 & 0 \end{pmatrix}} \mathbb{R}^6 \xrightarrow{\partial_1 = \begin{pmatrix} -1 & -1 & -1 & 0 & 0 & 0 \\ 1 & 0 & 0 & -1 & -1 & 0 \\ 0 & 1 & 0 & 1 & 0 & -1 \\ 0 & 0 & 1 & 0 & 1 & 1 \end{pmatrix}} \mathbb{R}^4 \longrightarrow 0.$$

By row reducing both boundary matrices, we find that

$$\text{rank}\partial_2 = 3, \text{ and rank}\partial_1 = 3.$$

This means that

$$H_2(X) \cong \mathbb{R}, \ H_1(X) \cong 0, \ \text{and } H_0(X) \cong \mathbb{R}.$$

We can also compute the barcode decomposition of $C_\bullet(X)$, which is

$$\left(\mathbb{R}_{(2,1)}\right)^3 \oplus \mathbb{R}_{(2,2)} \oplus \left(\mathbb{R}_{(1,0)}\right)^3 \oplus \mathbb{R}_{(0,0)}.$$

The length 1 bars are what show up in the homology. The first one lies in the kernel of ∂_2, which is spanned by

$$[b,c,d] - [a,c,d] + [a,b,d] + [a,b,c].$$

The second length 1 bar lies outside the image of ∂_1, and corresponds to the single path component, namely

$$[a] + [b] + [c] + [d].$$

Exercise 5.2 Compute the simplicial homology of the simplicial complex generated by $[a,b,c]$,$[a,b,d]$,$[a,d]$, and $[c,d]$. Be sure to list a basis for each homology space.

Exercise 5.3 Compute the simplicial homology of the simplicial complex generated by $[a, b, c]$, $[a, d]$, $[b, d]$, and $[c, d]$. Be sure to list a basis for each homology space.

Example 5.5 The representation of simplicial homology as a quotient

$$H_k(X) = \ker \partial_k / \text{ image } \partial_{k+1}$$

has a simplicial interpretation. In particular, each element of $\ker \partial_k$ is called a *closed chain* and consists of a linear combination of k-simplices whose $(k-1)$-faces in their boundaries "cancel out", leaving no overall boundary. For instance, in Fig. 5.2, there is no collection of 2-dimensional simplices with this property: all 2-dimensional simplices contain edges that cannot be canceled by any other 2-dimensional simplex. There are quite a few collections of 1-dimensional closed chains in Fig. 5.2. For instance,

$$[a, b] + [b, c] + [c, d] + [d, e] - [a, e],$$

$$[a, b] + [b, d] - [a, d],$$

and

$$[a, c] + [c, e] + [b, e] + [b, d] - [a, d]$$

are all closed.

Each element of image ∂_{k+1} lies in the boundary of some linear combination of $(k+1)$-simplices. Naturally enough, these elements are called *boundary chains*. An example of a boundary chain is

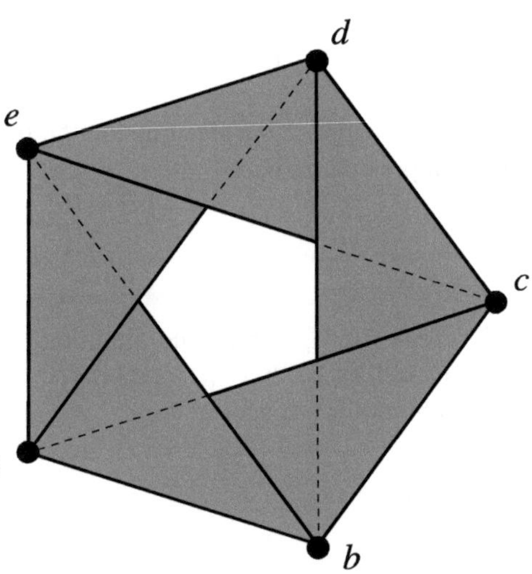

Fig. 5.2 The simplicial complex in Exercise 5.6 and Example 5.5

$$[b, c] + [c, d] - [b, d] + [d, e] - [a, e] + [a, d]$$

which is the boundary of

$$[b, c, d] + [a, d, e],$$

because

$$\partial_2 ([b, c, d] + [a, d, e]) = [b, c] + [c, d] - [b, d] + [d, e] - [a, e] + [a, d].$$

This particular calculation shows that $c_1 = [a, b] + [b, c] + [c, d] + [d, e] - [a, e]$ and $c_2 = [a, b] + [b, d] - [a, d]$ are in the same equivalence class in $H_1(X)$ because

$$\begin{aligned} c_1 - c_2 &= [a, b] + [b, c] + [c, d] + [d, e] - [a, e] - ([a, b] + [b, d] - [a, d]) \\ &= [a, b] + [b, c] + [c, d] + [d, e] - [a, e] - [a, b] - [b, d] + [a, d] \\ &= [b, c] + [c, d] - [b, d] + [d, e] - [a, e] + [a, d] \\ &= \partial_2([b, c, d] + [a, d, e]), \end{aligned}$$

namely the difference between these two closed chains is a boundary chain. Figure 5.3 shows the relationship between c_1 and c_2. The correct intuition is that if c_1 is "deformed" into c_2, the family of curves that are swept out fill the two triangles $[b, c, d]$ and $[a, d, e]$.

An increasingly important consequence of Corollary 5.1 is *persistent homology*. While we will only briefly mention the concept of persistent homology here, the interested reader is directed to Carlsson and Vejdemo-Johansson (2021) for a complete treatment of the subject.

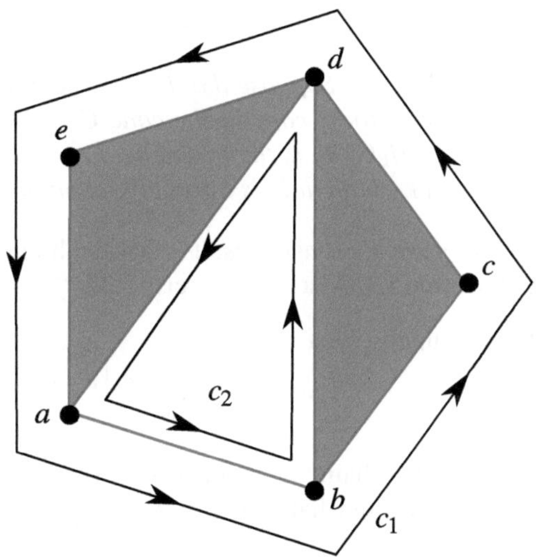

Fig. 5.3 The difference between closed chains c_1 and c_2 is the boundary of two triangles $[a, d, e]$ and $[b, c, d]$, in Example 5.5

Definition 5.2 Consider a sequence of abstract simplicial complexes and simplicial maps (not of vector spaces!)

$$\cdots \to X^0 \to X^1 \to \cdots$$

Because of simplicial homology is a covariant functor (Corollary 5.1), this sequence induces a sequence of vector spaces

$$\cdots \to H_k(X^0) \to H_k(X^1) \to \cdots$$

for each index k, called the *k-persistent homology of the sequence $X^\bullet$*.

The barcode decomposition of persistent homology detects not only features of each abstract simplicial complex X^ℓ, but also how those features are related to $X^{\ell-1}$ and $X^{\ell+1}$.

Exercise 5.4 (The Lefschetz fixed point theorem) Suppose that $f : X \to X$ is a simplicial map and X is a finite simplicial complex. This means that f induces a map $f_k : H_k(X) \to H_k(X)$ on each simplicial homology space. Show that whenever f has no fixed simplices, namely $f(\sigma) \neq \sigma$ for every simplex $\sigma \in X$, then

$$\operatorname{tr} f_k = 0 \text{ for all } k = 0, 1, \ldots$$

where $\operatorname{tr} m$ is the *trace* of a matrix m—the sum of its diagonal entries.

The operations of coning and suspending a simplicial complex have a very definite impact on the resulting simplicial homology. From Proposition 5.1, it should be clear that coning should produce a simplicial complex with 0-Betti number equal to 1.

Proposition 5.2 *Suppose that $i : X \to CX$ is the simplicial map which takes each simplex of X to its copy in the cone CX. The map induced by i on homology $i_k : H_k(X) \to H_k(CX)$ is the trivial map for each $k > 0$. Explicitly, the image of each nontrivial k-chain in X is a trivial k-chain in CX.*

We prove a stronger result—that the homology of CX is completely trivial—in Proposition 5.4. Also see Exercise 5.27.

Proof Suppose that

$$c = \sum_m a_m[v_{m,0}, v_{m,1}, \ldots, v_{m,k}]$$

is a closed k-chain, an element of $C_k(X)$ for which $\partial_k c = 0$. It is also the case that $c \in C_k(CX)$, and that $i(c) = c$. Without loss of generality, assume that the new vertex that is added to CX is $*$ and that this vertex is the minimum in the vertex partial order. Given this, define a related element of $C_{k+1}(CX)$ by

$$d = \sum_m a_m[*, v_{m,0}, v_{m,1}, \ldots, v_{m,k}].$$

A direct calculation shows that $i_k(c)$ is trivial in $H_k(CX)$, because $\partial_{k+1}d = c$, so

$$\begin{aligned}
\partial_{k+1}d &= \partial_{k+1}\sum_m a_m[*, v_{m,0}, v_{m,1}, \ldots, v_{m,k}] \\
&= \sum_m a_m \partial_{k+1}[*, v_{m,0}, v_{m,1}, \ldots, v_{m,k}] \\
&= \sum_m a_m \left([v_{m,0}, v_{m,1}, \ldots, v_{m,k}] - \sum_{n=0}^{k}(-1)^n[*, v_{m,0}, v_{m,1}, \ldots, \widehat{v_{m,n}}, \ldots v_{m,k}] \right) \\
&= c - \sum_m a_m \sum_{n=0}^{k}(-1)^n[*, v_{m,0}, v_{m,1}, \ldots, \widehat{v_{m,n}}, \ldots v_{m,k}] \\
&= c - \sum_{n=0}^{k}(-1)^n \sum_m a_m[*, v_{m,0}, v_{m,1}, \ldots, \widehat{v_{m,n}}, \ldots v_{m,k}]
\end{aligned}$$

The second term on the right is the cone on $\partial_k(c)$, which is to say that it consists of the same coefficients on the same simplices, with the exception that each simplex has $*$ appended to the front of its list of vertices. Therefore, this entire second term vanishes because $\partial_k(c) = 0$, which completes the proof. □

Although we cannot prove it yet (we will in Sect. 5.4), simplicial homology is purely a topological property, not just one that is combinatorial. An inkling of this idea can be found in the fact that some bars of length 2 in the simplicial chain complex correspond to specific simplices called *free faces*. Since bars of length 2 do not impact the homology of a chain complex, these simplices are ignored by simplicial homology.

Definition 5.3 A *free face* of a simplicial complex X is a simplex σ which is contained $\sigma \subset \tau$ in exactly one simplex τ of higher dimension.

Exercise 5.5 Show that the simplicial complex generated by $[a, b]$, $[a, d]$, $[b, c]$, and $[c, d]$ has no free faces.

Exercise 5.6 Identify all of the free faces in the simplicial complex shown in Fig. 5.2.

Proposition 5.3 *Suppose X and Y are simplicial complexes and that $X = Y \cup \{\sigma, \tau\}$. If σ is a free face and $\sigma \subset \tau$, then*

$$H_k(X) = H_k(Y \cup \{\sigma, \tau\}) \cong H_k(Y)$$

for all indices k.

Proof Without loss of generality, suppose that the dimension of σ is n, and that the unique $n+1$-dimensional simplex containing σ is τ. Observe that since X and Y are both simplicial complexes, this means that every subset of τ is contained in Y except σ. Therefore, the only indices at which $C_k(X)$ and $C_k(Y)$ differ are $k = n$ or $k = n+1$. Specifically

$$C_k(X) \cong \begin{cases} C_k(Y) \oplus \mathbb{R} & \text{if } k = n \text{ or } k = n+1 \\ C_k(Y) & \text{otherwise.} \end{cases}$$

This means that we may rewrite $C_{n+1}(Y) \cong C_{n+1}(X)/\mathrm{span}\{\tau\}$ up to isomorphism, which causes the upper triangle in the diagram below to commute,

$$\begin{array}{ccccccc}
\cdots & \longrightarrow & C_{n+1}(X) & \xrightarrow{\partial_{X,n+1}} & C_n(X) & \longrightarrow & \cdots \\
& & \downarrow & \nearrow_{m} & \downarrow & & \\
\cdots & \longrightarrow & C_{n+1}(X)/\mathrm{span}\{\tau\} & \xrightarrow[\partial_{Y,n+1}]{} & C_n(X)/\mathrm{span}\{\sigma\} & \longrightarrow & \cdots
\end{array}$$

Because the only simplex with σ in its boundary is τ, the induced map m cannot contain σ in its image. This means that the lower triangle commutes as well. This diagram therefore describes a chain map that removes the length 2 bar corresponding to σ and τ from $C_\bullet(X)$, completing the proof. □

Removing a free face from a simplicial complex will not change its homology, and may reveal additional free faces.

Example 5.6 Consider the simplicial complex generated by the simplex $[a, b, c, d]$. Each of its 2-dimensional faces is free. Let us remove one of them, say $[b, c, d]$. According to Proposition 5.3, if we delete both the free face and the face containing it, we obtain a new simplicial complex with the same homology. This results in the simplicial complex generated by $[a, b, c]$, $[a, b, d]$, and $[a, c, d]$. Each of the "exposed" edges, $[b, c]$, $[b, d]$, and $[c, d]$ is now free, even though it was not originally free. If we remove one of them, this reveals another free face, and so on. Eventually,

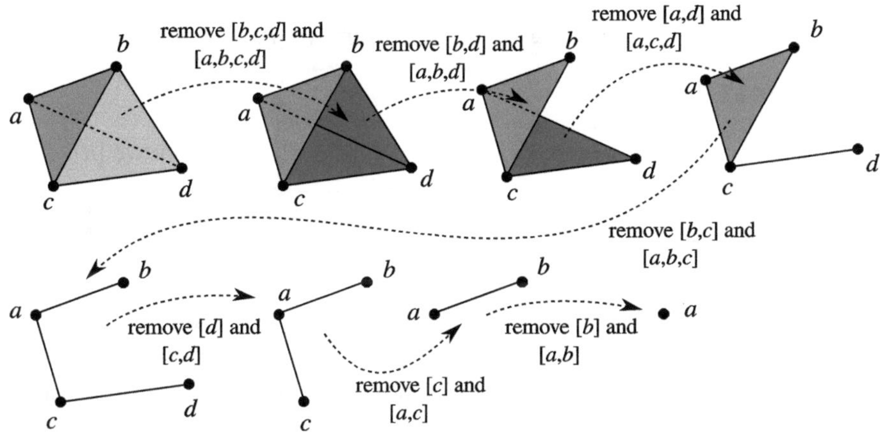

Fig. 5.4 The simplicial complex generated by $[a, b, c, d]$ and a sequence of removals of free faces. See Example 5.6

since the simplicial complex is finite, we will have removed all of the free faces available. One such sequence of free face removals is shown in Fig. 5.4. In the case of this particular simplicial complex, the process ends with a single vertex. This implies that the simplicial homology of the simplicial complex $[a, b, c, d]$ is isomorphic to that of a single vertex.

Sequentially removing free faces (as shown in Example 5.6) can result in a dramatically smaller simplicial complex and as a consequence a dramatically smaller simplicial chain complex. Identifying and removing free faces is nearly always faster than performing the linear algebraic operations necessary to compute homology, and so it is usually a good strategy to remove free faces before computing simplicial homology.

If there are several free faces, the removal of any of them is equivalent from the standpoint of simplicial homology by Proposition 5.3. However, the choice of free face to remove is important from a performance perspective, since one wants to remove as many free faces as possible. Choosing one free face over another may impact those simplices that subsequently become free faces, as Fig. 5.5 shows. This is in general a difficult problem to optimize fully, though there are good strategies discussed in the literature. See Sect. 5.8 for some suggestions.

Exercise 5.7 Consider the simplicial complex X generated by $[a, b, c]$, $[a, d]$, $[a, f]$, $[b, d]$, and $[d, e]$.

1. Reduce the simplicial complex X by repeatedly removing free faces.
2. Construct the simplicial chain complex for both X and the reduction you computed. How much did the dimension of each space decrease?

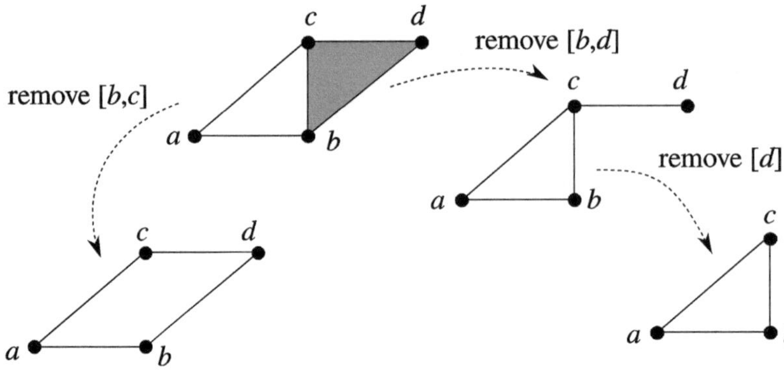

Fig. 5.5 Which free face to remove first impacts the total number of simplices that can be removed

3. Construct the map induced on the simplicial chain complexes by these free face reductions.
4. Compute the homology of the chain complex after these reductions.

Exercise 5.8 Find the maximum number of free faces that can be repeatedly removed from the simplicial complex shown in Fig. 5.6.

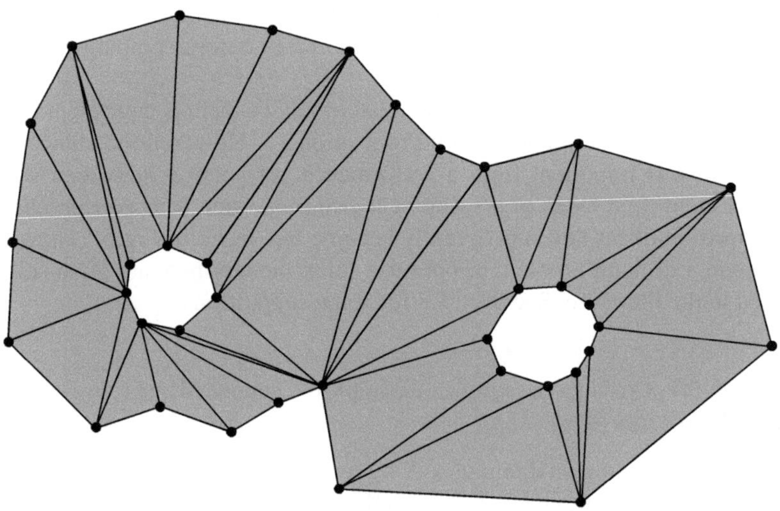

Fig. 5.6 The simplicial complex in Exercise 5.8

Exercise 5.9 Show that $\sigma = [v_0, \ldots, v_k]$ is a free face of a simplicial complex X if and only if $[v_0, \ldots, v_k, *]$ and $[v_0, \ldots, v_k, *']$ are free faces of the suspension SX.

Exercise 5.10 Prove or give a counterexample to the following statement: if X is a simplicial complex and SX is its suspension, all free faces of SX contain either $*$ or $*'$.

Exercise 5.11 Recall the definition of a *retraction* in Definition 4.13 in Chap. 4. Generalizing Proposition 5.3 slightly, show that a retraction $r : X \to Y$ that removes a free face is a quasi-isomorphism on simplicial homology. Specifically, show that the chain map constructed in the proof is the one induced by r.

Exercise 5.12 Even though the simplicial complex X generated by $[a, b]$, $[a, d]$, $[b, c]$, and $[c, d]$ has no free faces, we can still reduce it to a smaller complex, namely the simplicial Y generated by $[a, b]$, $[a, d]$, and $[b, d]$. Show that the map induced on homology by the simplicial map f given by

$$f(a) = a, \; f(b) = b, \; f(c) = b, \; f(d) = d$$

is a quasi-isomorphism. Can you prove a theorem that generalizes this to higher dimensions?

The simplicial chain complex ignores the existence of the empty simplex, since the empty simplex has dimension -1. Sometimes it is useful to include the empty simplex in calculations.

Definition 5.4 The *reduced simplicial chain complex* $(\widetilde{C}_\bullet(X), \widetilde{\partial}_\bullet)$ is defined by

$$\widetilde{C}_k(X) := \begin{cases} \mathbb{R}^{\text{number of } k\text{-dimensional simplices}} & \text{for } k \geq 0, \\ \mathbb{R} & \text{for } k = -1, \\ 0 & \text{for } k < -1. \end{cases}$$

(This can be interpreted as assigning the empty simplex dimension -1.)

The boundary maps $\widetilde{\partial}_k : \widetilde{C_k}(X) \to \widetilde{C_{k-1}}(X)$ are the same as those in the simplicial chain complex for $k > 0$. The map $\widetilde{\partial}_0 := \epsilon : \widetilde{C_0}(X) \to \widetilde{C_{-1}}(X)$ is surjective, and is given by

$$\epsilon([v]) := 1$$

for every vertex v. The map ϵ is called an *augmentation map* and expresses the fact that the empty simplex is a subset of every other simplex.

Diagrammatically, if the simplicial chain complex is

$$\cdots \longrightarrow C_2(X) \xrightarrow{\partial_2} C_1(X) \xrightarrow{\partial_1} C_0(X) \longrightarrow 0,$$

then the reduced simplicial chain complex is

$$\cdots \longrightarrow C_2(X) \xrightarrow{\partial_2} C_1(X) \xrightarrow{\partial_1} C_0(X) \xrightarrow{\epsilon} \mathbb{R} \longrightarrow 0.$$

As a result, the *reduced simplicial homology* is essentially the same as the simplicial homology, except in index 0 where its dimension is reduced by one.

Lemma 5.2 *Suppose that X is a simplicial complex. Its reduced simplicial homology can be computed from its simplicial homology using the formula*

$$\widetilde{H}_k(X) \cong \begin{cases} H_k(X) & \text{for } k \neq 0, \\ H_0(X)/\mathbb{R} & \text{for } k = 0. \end{cases}$$

Proof Construct the following chain map

$$\begin{array}{ccccccccc} \cdots \longrightarrow & C_2(X) & \xrightarrow{\partial_2} & C_1(X) & \xrightarrow{\partial_1} & C_0(X) & \xrightarrow{\epsilon} & \mathbb{R} & \longrightarrow 0 \\ & \downarrow & & \downarrow & & \downarrow & & \downarrow & \\ \cdots \longrightarrow & C_2(X) & \xrightarrow[\partial_2]{} & C_1(X) & \xrightarrow[\partial_1]{} & C_0(X) & \longrightarrow & 0 & \longrightarrow 0 \end{array}$$

in which the vertical maps are all identity maps, except for the rightmost one. Notice that the reduced homology is computed from the top row of this diagram. Since the augmentation map ϵ is of full rank, the last nontrivial bar chain map is of Type 8. This removes one nontrivial basis element from the reduced homology in index 0 as compared to the simplicial homology. □

Proposition 5.4 *If X is a finite simplicial complex and CX is its cone, then $\widetilde{H}_k(CX) = 0$ for all k.*

This is a somewhat stronger statement than Proposition 5.2.

Proof This follows by induction on the simplices of X, partially ordered by subset. As a base case, if X is empty, then $CX = \{[*], \emptyset\}$, which satisfies the statement.

For the induction step, suppose that there is a simplex σ that is not contained in any other simplex of X. Every simplex of X that is not contained in any other simplex is a free face of CX, because CX is generated by simplices of the form $[v_0, \ldots, v_k, *]$. For instance, if $\sigma = [v_0, \ldots, v_k]$ is not contained in any other simplex of X, then it is contained in exactly one simplex of CX, namely $[v_0, \ldots, v_k, *]$. Removing this free face results in a subcomplex of CX that has the same reduced homology as CX, and this satisfies the induction hypothesis that the reduced homology is trivial.

Unless X is empty, there is always at least one simplex of X that is not contained in any other because it is finite. This means that the induction process just defined will eventually remove *all* of the simplices of X as free faces, which establishes the result. □

Whenever suspensions are involved, the reduced simplicial homology is often more convenient to compute.

Proposition 5.5 *Suppose that X is a simplicial complex, then there is a chain map* $m_\bullet : \widetilde{C_\bullet}(X) \to \widetilde{C_{\bullet+1}}(SX)$ *that induces injective maps*

$$\widetilde{H_k}(X) \to \widetilde{H_{k+1}}(SX)$$

on the reduced simplicial homology.

A stronger statement is given in Proposition 5.10 in Sect. 5.2.

Proof Assuming that $*$ and $*'$ are maximal elements in the vertex partial order, define a chain map $m_\bullet : \widetilde{C_\bullet}(X) \to \widetilde{C_{\bullet+1}}(SX)$ by

$$m_k([v_0, v_1, \ldots, v_k]) := [v_0, v_1, \ldots, v_k, *] - [v_0, v_1, \ldots, v_k, *'] \tag{5.2}$$

for each k, resulting in the diagram

$$\begin{array}{ccccccccccc}
\cdots \to & C_{k+1}(X) & \xrightarrow{\partial_{X,k+1}} & C_k(X) & \xrightarrow{\partial_{X,k}} & \cdots \longrightarrow & C_0(X) & \xrightarrow{\epsilon_X} & \mathbb{R} & \longrightarrow & 0 \\
& \downarrow{\scriptstyle m_{k+1}} & & \downarrow{\scriptstyle m_k} & & & \downarrow{\scriptstyle m_0} & & \downarrow{\scriptstyle \alpha} & & \downarrow \\
\cdots \to & C_{k+2}(SX) & \xrightarrow[\partial_{SX,k+2}]{} & C_{k+1}(SX) & \xrightarrow[\partial_{SX,k+1}]{} & \cdots \longrightarrow & C_1(SX) & \xrightarrow[\partial_{SX,1}]{} & C_0(SX) & \xrightarrow[\epsilon_{SX}]{} & \mathbb{R}.
\end{array}$$

The α must be chosen so that the diagram commutes. To that end,

$$\begin{aligned}
\left(\partial_{SX,1} \circ m_0\right)[v] &= \partial_{SX,1}([v, *] - [v, *']) \\
&= [\,*\,] - [v] - [*'] + [v] \\
&= [\,*\,] - [*']
\end{aligned}$$

for any vertex v, so we must have that $\alpha(1) = [*] - [*']$. This definition for α, along with the definition that $\epsilon_{SX}([v]) = 1$ for every vertex $v \in SX$ (including $*$ and $*'$), means that

$$(\epsilon_{SX} \circ \alpha)(1) = \epsilon_{SX}([*] - [*']) = 1 - 1 = 0.$$

In the typical situation when $k > 0$, the diagram commutes because

$$\begin{aligned}
\left(\partial_{SX,k+1} \circ m_k\right)([v_0, v_1, \ldots, v_k]) &= \partial_{SX,k+1}\left([v_0, v_1, \ldots, v_k, *] - [v_0, v_1, \ldots, v_k, *']\right) \\
&= \partial_{SX,k+1}([v_0, v_1, \ldots, v_k, *]) \\
&\quad - \partial_{SX,k+1}\left([v_0, v_1, \ldots, v_k, *']\right) \\
&= \sum_{n=0}^{k} (-1)^n [v_0, v_1, \ldots \widehat{v_n} \ldots, v_k, *] \\
&\quad + \sum_{n=0}^{k} (-1)^{k+1} [v_0, v_1, \ldots, v_k] \\
&\quad - \sum_{m=0}^{k} [v_0, v_1, \ldots \widehat{v_m} \ldots, v_k, *'] \\
&\quad - \sum_{n=0}^{k} (-1)^{k+1} [v_0, v_1, \ldots, v_k] \\
&= \sum_{n=0}^{k} (-1)^n [v_0, v_1, \ldots \widehat{v_n} \ldots, v_k, *] \\
&\quad - \sum_{m=0}^{k} [v_0, v_1, \ldots \widehat{v_m} \ldots, v_k, *'] \\
&= \left(m_{k-1} \circ \partial_{X,k}\right)([v_0, v_1, \ldots, v_k]).
\end{aligned}$$

Observe that m_k (for all k) and α are all injective. The snake lemma for chain complexes (Theorem 3.2 in Chap. 3), implies that the induced map is injective. □

Even though $m_\bullet$ in the above proof is a chain map, it is not induced by a simplicial map, since it maps k-simplices to $k+1$-simplices. An example of the action of $m_\bullet$ is shown in Fig. 5.7. In the Figure, notice that every edge of X remains an edge in SX, but also corresponds to *two triangles* in SX. This combinatorial observation means that the algebraic structure of $m_\bullet$ is rather special as well, in particular the diagram is composed of sub-diagrams of the form (trivial spaces and arrows removed for clarity)

$$\begin{array}{ccccc}
\cdots & \mathbb{R}^p & \longrightarrow & \mathbb{R}^q & \cdots \\
 & & & \downarrow & \\
\cdots & & & \mathbb{R}^{p+2q} & \cdots
\end{array}$$

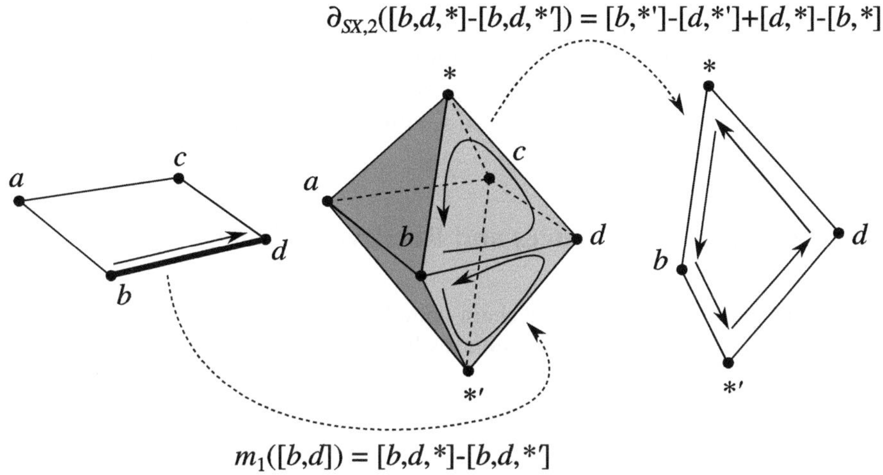

Fig. 5.7 The chain map $m_\bullet$ in the proof of Proposition 5.5 acting on a simplex $[b, d]$ produces a linear combination of simplices in the suspension of $[b, d]$. The arrows follow the vertex partial order, accounting for signs

Example 5.7 Starting with the empty set $\emptyset$, repeated suspensions yield a sequence of spheres

$$S^0 := S\emptyset,\ S^{k+1} := S\,S^k.$$

The chain maps between their reduced simplicial chain complexes are the vertical maps in the diagram

$$\begin{array}{ccccccccccccccc}
\widetilde{C}_\bullet(\emptyset) & & \cdots \longrightarrow & 0 & \longrightarrow & \mathbb{R} & \longrightarrow & 0 & & & & & & \\
\downarrow & & & \downarrow & & \downarrow & & \downarrow & & & & & & \\
\widetilde{C}_\bullet(S^0) & & \cdots \longrightarrow & 0 & \longrightarrow & \mathbb{R}^2 & \longrightarrow & \mathbb{R} & \longrightarrow & 0 & & & & \\
\downarrow & & & \downarrow & & \downarrow & & \downarrow & & \downarrow & & & & \\
\widetilde{C}_\bullet(S^1) & & \cdots \longrightarrow & 0 & \longrightarrow & \mathbb{R}^4 & \longrightarrow & \mathbb{R}^4 & \longrightarrow & \mathbb{R} & \longrightarrow & 0 & & \\
\downarrow & & & \downarrow & & \downarrow & & \downarrow & & \downarrow & & \downarrow & & \\
\widetilde{C}_\bullet(S^2) & & \cdots \longrightarrow & 0 & \longrightarrow & \mathbb{R}^8 & \longrightarrow & \mathbb{R}^{12} & \longrightarrow & \mathbb{R}^6 & \longrightarrow & \mathbb{R} & \longrightarrow & 0 \\
\downarrow & & & \downarrow & & \downarrow & & \downarrow & & \downarrow & & \downarrow & & \downarrow \\
\vdots & & & \vdots & & \vdots & & \vdots & & \vdots & & \vdots & & \vdots
\end{array}$$

from which the combinatorial relationship between terms is quite visible. Each row in the above diagram is the reduced simplicial chain complex of a sphere, taken in order of dimension. The first row is the reduced simplicial chain complex of the empty set, the second row is the reduced simplicial chain complex of S^0, and so on.

5.2 Long Exact Sequences and Relative Homology

Recall that the Snake Lemma for chain complexes, Theorem 3.2 in Chap. 3, starts with an exact sequence of chain complexes like

$$0 \longrightarrow C_\bullet(X) \xrightarrow{f_\bullet} C_\bullet(Y) \xrightarrow{g_\bullet} C_\bullet(Z) \longrightarrow 0.$$

Accounting for the difference in index direction, Theorem 3.2 in Chap. 3 implies that the following sequence

$$\cdots \longrightarrow H_k(X) \longrightarrow H_k(Y) \longrightarrow H_k(Z) \xrightarrow{(\varepsilon_*)_k} H_{k-1}(X) \longrightarrow \cdots$$

of simplicial homology sequences is an exact sequence of vector spaces, where the connecting map $(\varepsilon_*)_\bullet$ is constructed from the chain maps in a specific way.

To use Theorem 3.2 in Chap. 3 in the context of simplicial homology, we need to obtain an exact sequence of chain complexes. One way to start the process is with an appropriate simplicial map $f : X \to Y$, since we can use the induced map $C_\bullet(X) \to C_\bullet(Y)$. A single induced map is not necessarily an exact sequence, but injective and surjective maps can be represented by particular exact sequences. The strategy we will follow constrains f using Lemma 5.3 below to obtain an exact sequence.

Lemma 5.3 *Injective simplicial maps induce simplicial chain maps with injective components. On the other hand, surjective simplicial maps induce simplicial chain maps with surjective components.*

Proof Suppose that $f : X \to Y$ is a simplicial map.

If f is injective, then the only way that $f(\sigma) = f(\tau)$ for two simplices $\sigma, \tau \in X$ is if $\sigma = \tau$. Now consider $f_k : C_k(X) \to C_k(Y)$, the index k component of the simplicial chain map induced by f. Suppose that $x \in C_k(X)$ such that $f_k(x) = 0$. Since the k-dimensional simplices form a basis of the simplicial chain space $C_k(X)$, this means that we can decompose x as a linear combination

$$x = \sum_{\sigma \in X^k} a_\sigma \sigma,$$

where the σ range over the k-skeleton X^k, consisting of all k-dimensional simplices of X. This means that

$$\begin{aligned} 0 &= f_k(x) \\ &= f_k\left(\sum_{\sigma \in X^k} a_\sigma \sigma\right) \\ &= \sum_{\sigma \in X^k} a_\sigma f_k(\sigma) \\ &= \sum_{\sigma \in X^k} a_\sigma f(\sigma). \end{aligned}$$

Since each of the σ is distinct, so are the $f(\sigma)$. This means that each term in the above sum must vanish independently, and $a_\sigma = 0$ for all σ. Consequently $x = 0$, and we conclude f_k is injective.

If f is surjective, every simplex σ of Y has a nonempty preimage $f^{-1}(\sigma)$ in X. To show that this implies $f_k : C_k(X) \to C_k(Y)$ is surjective, suppose that $y \in C_k(Y)$ is arbitrary. Since the k-dimensional simplices form a basis of the simplicial chain spaces $C_k(Y)$, we can decompose y as a linear combination

$$y = \sum_{\sigma \in Y^k} a_\sigma \sigma.$$

But since each simplex σ has a nonempty preimage $f^{-1}(\sigma)$, we simply need to select one simplex $\tau_\sigma \in f^{-1}(\sigma)$ from each preimage to form

$$x = \sum_{\sigma \in Y^k} a_\sigma \tau_\sigma,$$

for which $f_k(x) = y$. □

Consider an injective simplicial map $f : X \to Y$. Lemma 5.3 implies that f induces an exact sequence of simplicial chain complexes

$$0 \longrightarrow C_\bullet(X) \xrightarrow{f_\bullet} C_\bullet(Y).$$

Although this does not seem to be enough to apply Theorem 3.2 in Chap. 1, remember that each $f_k(C_k(X))$ is a vector subspace of $C_k(Y)$. This means that we can form the quotient $C_k(Y)/f_k(C_k(X))$ and the canonical projection $g_k : C_k(Y) \to C_k(Y)/f_k(C_k(X))$. Let us define

$$C_k(Y, X) := C_k(Y)/f_k(C_k(X)).$$

By descending to each quotient (Theorem 1.3 in Chap. 1), we have the larger diagram

$$\begin{array}{ccccccccc}
\cdots \longrightarrow & C_{k+1}(X) & \xrightarrow{\partial_{X,k+1}} & C_k(X) & \xrightarrow{\partial_{X,k}} & C_{k-1}(X) & \xrightarrow{\partial_{X,k-1}} \cdots \\
 & \downarrow f_{k+1} & & \downarrow f_k & & \downarrow f_{k-1} & \\
\cdots \longrightarrow & C_{k+1}(Y) & \xrightarrow{\partial_{Y,k+1}} & C_k(Y) & \xrightarrow{\partial_{Y,k}} & C_{k-1}(Y) & \xrightarrow{\partial_{Y,k-1}} \cdots \\
 & \downarrow g_{k+1} & & \downarrow g_k & & \downarrow g_{k-1} & \\
\cdots \longrightarrow & C_{k+1}(Y, X) & \longrightarrow & C_k(Y, X) & \longrightarrow & C_{k-1}(Y, X) & \longrightarrow \cdots
\end{array}$$

In short, the canonical projections can be interpreted as the components of a chain map $g_\bullet : C_\bullet(Y) \to C_\bullet(Y, X)$. Noting that the canonical projection is always surjective, this yields an exact sequence

$$0 \longrightarrow C_\bullet(X) \xrightarrow{f_\bullet} C_\bullet(Y) \xrightarrow{g_\bullet} C_\bullet(Y, X) \longrightarrow 0$$

of chain complexes. If we define

$$H_\bullet(Y, X) := H_\bullet(C_\bullet(Y, X)),$$

then Theorem 3.2 in Chap. 3 shows that

$$\cdots \longrightarrow H_k(X) \longrightarrow H_k(Y) \longrightarrow H_k(Y, X) \xrightarrow{(\varepsilon_*)_k} H_{k-1}(X) \longrightarrow \cdots$$

is an exact sequence of vector spaces.

Example 5.8 Consider the simplicial complex X generated by $\{[a, b, c], [a, d], [b, d]\}$ and the subset $A = \{a, b, d\}$ shown in Fig. 5.8. Observe that the subset encompasses the loop in X. Using the vertex partial order $a < b < c < d$, the simplicial chain spaces are given by

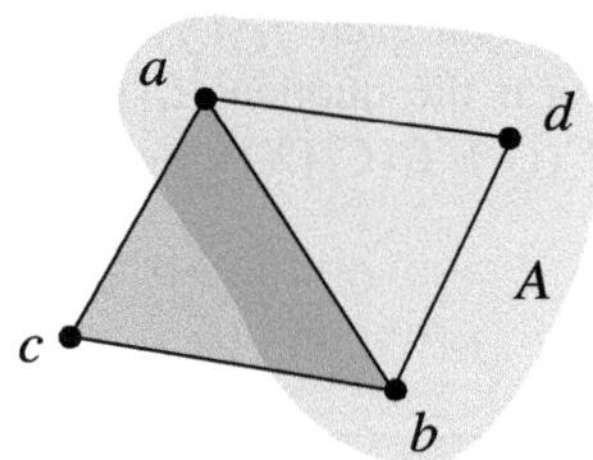

Fig. 5.8 The simplicial complex generated by $\{[a, b, c], [a, d], [b, d]\}$ and the subset $A = \{a, b, d\}$ in Example 5.8

$$C_2(X) = \text{span}\{[a,b,c]\} \cong \mathbb{R},$$

$$C_1(X) = \text{span}\{[a,b],[a,c],[a,d],[b,c],[b,d]\} \cong \mathbb{R}^5,$$

and

$$C_0(X) = \text{span}\{[a],[b],[c],[d]\} \cong \mathbb{R}^4.$$

Using lexicographical ordering, the chain complex $C_\bullet(X)$ is given by

$$0 \longrightarrow \mathbb{R} \xrightarrow{\partial_2 = \begin{pmatrix} 1 \\ -1 \\ 0 \\ 1 \\ 0 \end{pmatrix}} \mathbb{R}^5 \xrightarrow{\partial_1 = \begin{pmatrix} -1 & -1 & -1 & 0 & 0 \\ 1 & 0 & 0 & -1 & -1 \\ 0 & 1 & 0 & 1 & 0 \\ 0 & 0 & 1 & 0 & 1 \end{pmatrix}} \mathbb{R}^4 \longrightarrow 0.$$

A little calculation leads to

$$\begin{aligned} H_1(X) &\cong \text{span}\left\{ \begin{pmatrix} 1 \\ 0 \\ -1 \\ 0 \\ 1 \end{pmatrix} \right\} \\ &\cong \text{span}\{[a,b]-[a,d]+[b,d]\} \\ &\cong \mathbb{R}, \end{aligned}$$

$H_0(X) \cong \mathbb{R}$, and that all other $H_k(X)$ with $k > 1$ are trivial.

Considering just the subset A simply removes the vertex c from consideration. Since the single nontrivial basis vector of $H_1(X)$ does not contain c, we conclude that $H_\bullet(A) \cong H_\bullet(X)$.

If we let $i : A \to X$ be the *inclusion map*, which simply reinterprets elements of A as elements of X, Theorem 3.2 in Chap. 3 asserts that the following sequence is exact

$$\cdots \to H_1(A) \cong \mathbb{R} \to H_1(X) \cong \mathbb{R} \to H_1(X,A) \xrightarrow{(\varepsilon_*)_1} H_0(A) \cong \mathbb{R} \to H_0(X) \cong \mathbb{R}.$$

Given this information, we can conclude that $H_1(X,A) = 0$. This can be verified directly as well. Specifically, $C_\bullet(X,A)$ is obtained from $C_\bullet(X)$ by removing all simplices whose vertices all are contained in A. Evidently this removes the single nontrivial basis vector of $H_1(X)$. More explicitly, $C_2(X,A)$ is unchanged since the single 2-simplex contains the vertex c, which is not in A. The other two spaces are changed, however,

$$C_1(X,A) = \text{span}\{[a,c],[b,c]\} \cong \mathbb{R}^2,$$

and

$$C_0(X, A) = \operatorname{span}\{[c]\} \cong \mathbb{R}.$$

The resulting chain complex is

$$0 \longrightarrow \mathbb{R} \xrightarrow{\partial_2 = \begin{pmatrix} -1 \\ 1 \end{pmatrix}} \mathbb{R}^2 \xrightarrow{\partial_1 = \begin{pmatrix} 1 & 1 \end{pmatrix}} \mathbb{R} \longrightarrow 0.$$

Notice that the kernel of ∂_1 is precisely the image of ∂_2, so $H_1(X, A)$ is trivial. Additionally, $H_2(X, A)$ is trivial because ∂_2 has trivial kernel. Finally, $H_0(X, A)$ is also trivial because ∂_1 is surjective.

Definition 5.5 Given an injective simplicial map $X \to Y$, we call $C_\bullet(Y, X)$ the *simplicial chain complex of* Y *relative to* X and $H_\bullet(Y, X)$ the *simplicial homology of* Y *relative to* X.

It is easy to see that $C_\bullet(X, \emptyset) \cong C_\bullet(X)$. It follows that $H_\bullet(X, \emptyset) \cong H_\bullet(X)$. But relative simplicial homology can change the appearance of a simplicial complex so that it looks like a simplex has been added.

Example 5.9 For instance, if we consider the simplicial complex X generated by $[a, b]$ and $[a, c]$, then $H_1(X)$ is trivial. Specifically, using the vertex partial order $a < b < c$, the simplicial chain complex is

$$0 \longrightarrow \mathbb{R}^2 \xrightarrow{\partial_1} \mathbb{R}^3 \longrightarrow 0$$

where

$$\partial_1 = \begin{pmatrix} -1 & -1 \\ 1 & 0 \\ 0 & 1 \end{pmatrix}.$$

This matrix has trivial kernel, which verifies that $H_1(X)$ is indeed trivial.

On the other hand, let $A = \{b, c\}$ and consider the simplicial chain complex $C_\bullet(X, A)$ of X relative to A. We follow the recipe that all basis vectors corresponding to simplices contained entirely within A are removed from the chain complex. $C_1(X, A)$ is identical to $C_1(X)$ because A contains no edges of X. However, $C_0(X, A)$ is simply spanned by $[a]$. The resulting chain complex is

$$0 \longrightarrow \mathbb{R}^2 \xrightarrow{\partial_1} \mathbb{R} \longrightarrow 0$$

where

$$\partial_1 = \begin{pmatrix} -1 & -1 \end{pmatrix}.$$

Therefore, $H_1(X, A) \cong \mathbb{R}$ just as if we had added an edge $[b, c]$. However, $H_\bullet(X, A)$ is not *quite* the same as what we would have obtained by adding an edge because it is easily seen that $H_0(X, A)$ is trivial.

Reduced homology is a special case of relative homology. This is important to keep in mind when using relative homology to "add" extra simplices like was done in Example 5.9; the result differs from what might be expected in index 0.

Example 5.10 To relate simplicial homology, relative simplicial homology, and reduced simplicial homology, consider the simplicial complex X consisting of a single edge $[a, b]$. The simplicial chain complex is given by

$$0 \longrightarrow \mathbb{R} \xrightarrow{\begin{pmatrix} 1 \\ -1 \end{pmatrix}} \mathbb{R}^2 \longrightarrow 0 \longrightarrow 0.$$

This has one bar of length 2, and one bar of length 1 in index 0. Hence, the only nontrivial homology space is $H_0(X) \cong \mathbb{R}$.

The reduced simplicial chain complex adds a nontrivial vector space in index -1 along with the augmentation map,

$$0 \longrightarrow \mathbb{R} \xrightarrow{\begin{pmatrix} 1 \\ -1 \end{pmatrix}} \mathbb{R}^2 \xrightarrow{(1\ 1)} \mathbb{R} \longrightarrow 0.$$

This has two bars of length 2—an exact sequence—so its homology is completely trivial.

Finally, the relative chain complex $C_\bullet(X, \{a\})$ is given by

$$0 \longrightarrow \mathbb{R} \xrightarrow{(-1)} \mathbb{R} \longrightarrow 0 \longrightarrow 0,$$

which has one bar of length 2. Like the reduced simplicial chain complex, this relative simplicial chain complex is an exact sequence, but it is not *identical* to the reduced simplicial chain complex.

The previous example captures the main idea of the general situation that is characterized by the next proposition.

Proposition 5.6 *If v is a vertex of X, then $H_\bullet(X, \{v\}) \cong \widetilde{H}_\bullet(X)$.*

Proof Because v is a vertex of X, then the only simplex contained entirely within $\{v\}$ is $[v]$. Therefore, $C_k(X, \{v\}) \cong C_k(X)$ for all $k > 0$. For index 0, the situation is slightly altered, in that $C_0(X)$ has one more basis vector—corresponding to $[v]$—than $C_0(X, \{v\})$. This also changes ∂_1: to obtain the matrix for this map in $C_\bullet(X, \{v\})$, the row corresponding to $[v]$ must be deleted from the matrix for $C_\bullet(X)$.

Looking at dimensions of the spaces involved, one should notice that $C_\bullet(X, \{v\})$ is not equal to the reduced simplicial chain complex $\widetilde{C}_\bullet(X)$, since the latter has an additional simplex instead (of dimension -1). However, because $C_0(X, \{v\})$ simply has one less vertex that $\widetilde{C_0}(X)$, there is a chain map $C_\bullet(X, \{v\}) \to \widetilde{C}_\bullet(X)$ whose diagram is

$$\begin{array}{ccccccccc}
\cdots & \longrightarrow & C_1(X, \{v\}) & \longrightarrow & C_0(X, \{v\}) & \longrightarrow & 0 & \longrightarrow & 0 \\
 & & \downarrow{\scriptstyle\cong} & & \downarrow & & \downarrow & & \\
\cdots & \longrightarrow & \widetilde{C_1}(X) & \longrightarrow & \widetilde{C_0}(X) & \xrightarrow{\epsilon} & \mathbb{R} & \longrightarrow & 0
\end{array}$$

The components of this chain map are isomorphisms in all indices greater than 0, and the components are injective in the remaining two indices (0 and -1). Consulting the Escolar-Hiraoka decomposition in Table 3.1 in Chap. 3, we notice that there is no bar chain map of the form

$$\begin{array}{ccc}
\mathbb{R} & \longrightarrow & 0 \\
\downarrow & & \downarrow \\
\mathbb{R} & \longrightarrow & \mathbb{R}
\end{array}$$

Put another way, the augmentation map implies the existence of a Type 3 bar chain map in index 0. This accounts for the difference in dimension between $C_0(X, \{v\})$ and $\widetilde{C_0}(X)$, leaving no room for any other disagreements between the two chain complexes. Since this leaves the bars of length 1 unchanged, the two chain complexes must have the same homology. □

Sometimes it is convenient to interpret relative homology as being a functor from the category of *simplicial pairs*.

Definition 5.6 If A and B are simplicial complexes with $A \subseteq B$, we call (B, A) a *simplicial pair*. If (D, C) is also a simplicial pair, a simplicial map $f : B \to D$ that restricts to a simplicial map $f : A \to C$ is called a *simplicial pair map*. We usually write $f : (B, A) \to (D, C)$ in this case. The *category of simplicial pairs* has the class of simplicial pairs for objects and simplicial pair maps for morphisms, using the usual composition of functions for composition.

Proposition 5.7 *The relative chain complex (and hence also relative homology) is a covariant functor from the category of simplicial pairs. Specifically, each simplicial pair map* $f : (B, A) \to (D, C)$ *induces a chain map* $C_\bullet(B, A) \to C_\bullet(D, A)$.

Proof Suppose that $f : (B, A) \to (D, C)$ is a simplicial pair map. We already know that f induces a chain map $f_\bullet : C_\bullet(B) \to C_\bullet(D)$, and that this chain map restricts to a chain map $f_\bullet : C_\bullet(A) \to C_\bullet(C)$. Therefore, by descending to the quotients, $f_\bullet$ induces a chain map $C_\bullet(B, A) \to C_\bullet(D, A)$. □

Corollary 5.2 **Asc** *is equivalent to a subcategory of the category of simplicial pairs, in which each simplicial complex* X *corresponds to the pair* $(X, \emptyset)$.

Example 5.11 Consider the simplicial complex X generated by $[a, b, c]$, $[b, d]$, and $[c, d]$, and the simplicial complex Y generated by $[a', b', c']$. The function $f : X \to Y$ given by

$$f(a) := a', \ f(b) := b', \ f(c) := c', \ f(d) := c',$$

is a simplicial map. Let A be the subcomplex of X generated by $[b, c]$, $[b, d]$, and $[c, d]$, and B be the subcomplex of Y generated by $[b', c']$. Because the vertices

$$f(A) = f(\{b, c, d\}) = \{b', c'\}$$

are contained in Y, this means that f is a pair map $(X, A) \to (Y, B)$. On the other hand, if C is the subcomplex of Y generated by $[a', b']$, then f is not a pair map $(X, A) \to (Y, C)$.

Exercise 5.13 Construct the relative simplicial chain map induced by the pair map $f : (X, A) \to (Y, B)$ in Example 5.11. Then compute the induced map on homology. Are the two chain complexes isomorphic? quasi-isomorphic?

Exercise 5.14 Consider the same simplicial map $f : X \to Y$ introduced in Example 5.11, but consider different subcomplexes. Let E be the subcomplex of X generated by $[a, b, c]$ and F be the subcomplex of Y generated by $[a', b', c']$. Observe that f is pair map $(X, E) \to (Y, F)$, and compute the relative simplicial chain map induced by this pair map. Compute the induced map on homology. (It is interesting to compare this with the previous exercise, noticing that the underlying simplicial map f is the same.)

Exercise 5.15 If $A \subseteq B \subseteq C$, then we can define $C_\bullet(B, A)$, $C_\bullet(C, B)$, and $C_\bullet(C, A)$.

1. Show that these fit together into an exact sequence of chain complexes

$$0 \longrightarrow C_\bullet(B, A) \longrightarrow C_\bullet(C, A) \longrightarrow C_\bullet(C, B) \longrightarrow 0.$$

2. Using Theorem 3.2 in Chap. 3, derive the long exact sequence

$$\cdots \to H_k(B, A) \to H_k(C, A) \to H_k(C, B) \to H_{k-1}(B, A) \to \cdots$$

Relative simplicial homology is frequently useful to focus one's attention on a particular portion of a simplicial complex. This leads to the concept of *local homology*. Recall (Definition 4.6 of Chap. 4) that an open subset $U \subseteq X$ of an abstract simplicial complex X is a set of simplices which is its own interior,

$$U = \text{int } U = \bigcup_{\sigma \in X \,:\, \text{star}\{\sigma\} \subseteq U} \text{star}\{\sigma\}.$$

Although open sets are not necessarily abstract simplicial complexes, their complements are.

Lemma 5.4 *If U is an open subset of an abstract simplicial complex X, then its complement $X \backslash U$ is an abstract simplicial complex.*

Proof Suppose that σ is a simplex of $X \backslash U$ and that $\tau \subseteq \sigma$. Since U is open, it is equal to the union of the stars over each of its simplices. Therefore, σ is not an element of any star over any of the simplices contained in U. Unpacking this slightly, this means that σ does not contain any simplex of U. Since τ is a subset of σ, τ also cannot contain any simplex of U. Therefore, τ cannot be in the interior of U and so cannot be contained in U. Since X is an abstract simplicial complex, τ is certainly a simplex of X. We have shown that τ is a simplex of $X \backslash U$. □

The Lemma implies that we can form a particular kind of relative homology associated to an open set.

Definition 5.7 (*compare* (Hatcher 2002, *end of Sect. 2.1*), Milnor and Stasheff 1974) For an open subset $U \subseteq X$ of an abstract simplicial complex, the *local homology* at U is $H_k(X, X \backslash U)$. For brevity, we usually write

$$\beta_k(U) = \dim H_k(X, X \backslash U)$$

for the *local k-Betti number at U*.

Proposition 5.8 (Excision for abstract simplicial complexes, compare (Munkres 1984, Lemma 35.1)) *If U is an open set of an abstract simplicial complex X, then $H_k(X, X \backslash U) \cong H_k(\text{cl}\, U, \text{fr}\, U)$.*

Proof It suffices to show that the chain complexes associated to $H_k(X, X \backslash U)$ and $H_k(\text{cl}\, U, \text{fr}\, U)$ are exactly the same. In both cases, the chain spaces $C_k(X, X \backslash U)$ and $C_k(\text{cl}\, U, \text{fr}\, U)$ consist of a vector space whose basis is the set of simplices in U. Also observe that since $\text{cl}\, U$ is a closed subcomplex of X, the boundary map $\partial_k : C_k(X) \to C_{k-1}(X)$ restricts[2] to a map $\partial_k : C_k(\text{cl}\, U) \to C_{k-1}(\text{cl}\, U)$. Likewise, since $\text{fr}\, U$ is a closed subcomplex of $X \backslash U$, the boundary map $\partial_k : C_k(X \backslash U) \to C_{k-1}(X \backslash U)$ restricts to a map $\partial_k : C_k(\text{fr}\, U) \to C_{k+1}(\text{fr}\, U)$. Collecting these facts, we

---

[2] For the reader who knows about singular homology on a general topological space: this particular tactic does not work for the singular chain complex!

conclude that $\partial_k : C_k(X, X\backslash U) \to C_{k-1}(X, X\backslash U)$ restricts to $\partial_k : C_k(\mathrm{cl}\, U, \mathrm{fr}\, U) \to C_{k-1}(\mathrm{cl}\, U, \mathrm{fr}\, U)$. But, we previously established that the domains and ranges of these maps are identical. The maps must in fact be identical. Having shown that the chain complexes are identical, we conclude that their homologies must also be identical. □

Example 5.12 Consider the simplicial complex X generated by $[a, b, c]$, $[a, d]$, and $[b, d]$. The set $U := \{[a, b, c]\}$ is an open set in X, since it is its own star. The complement of U is the closed set

$$X\backslash U = \{[a, b], [a, c], [b, c], [a, d], [b, d], [a], [b], [c], [d], \emptyset\}.$$

Additionally, we have

$$\mathrm{cl}\, U = \{[a, b, c], [a, b], [a, c], [b, c], [a], [b], [c], \emptyset\},$$

and

$$\mathrm{fr}\, U = \{[a, b], [a, c], [b, c], [a], [b], [c], \emptyset\}.$$

Based upon these sets, we can construct the relative simplicial chain complex $C_\bullet(X, X\backslash U)$

$$\cdots \to \mathbb{R} \to 0 \to 0,$$

because

$$\begin{aligned}
C_2(X, X\backslash U) &= C_2(X)/C_2(X\backslash U) \\
&= C_2(X)/0 \cong \mathrm{span}\{[a, b, c]\} \cong \mathbb{R}, \\
C_1(X, X\backslash U) &= C_1(X)/C_1(X\backslash U) \\
&= \mathrm{span}\{[a, b], [a, c], [b, c], [a, d], [b, d]\}/ \\
&\quad \mathrm{span}\{[a, b], [a, c], [b, c], [a, d], [b, d]\} \\
&= 0, \\
C_0(X, X\backslash U) &= C_0(X)/C_0(X\backslash U) \\
&= \mathrm{span}\{[a], [b], [c], [d]\}/\mathrm{span}\{[a], [b], [c], [d]\} \\
&= 0.
\end{aligned}$$

In a similar way, the relative simplicial chain complex $C_\bullet(\mathrm{cl}\, U, \mathrm{fr}\, U)$ is

$$\cdots \to C_2(\mathrm{cl}\, U, \mathrm{fr}\, U) \cong \mathbb{R} \to C_1(\mathrm{cl}\, U, \mathrm{fr}\, U) = 0 \to C_0(\mathrm{cl}\, U, \mathrm{fr}\, U) = 0.$$

The two relative chain complexes are evidently isomorphic. They have isomorphic homology.

Exercise 5.16 Using the simplicial complex X from Example 5.12, verify Proposition 5.8 for the open set $\{[a, d]\}$

Exercise 5.17 Using the simplicial complex X from Example 5.12, the set $\{[a, b]\}$ is not open. Does Proposition 5.8 hold if we take U to be this set?

Local homology generalizes the degree of a graph (Definition 4.1 of Chap. 4) to higher dimensional simplices.

Proposition 5.9 *If X is the 1-dimensional abstract simplicial complex corresponding to a graph $G(V, E)$ (see Example 4.2 in Chap. 4 for a construction) then*

$$\beta_1(\operatorname{star}[v]) = \dim H_1(X, X \setminus \operatorname{star}[v]) = \deg v - 1$$

for each vertex v, where $\deg v$ is the degree of v, the number of edges that contain v.

Proof By excision (Proposition 5.8), we have that

$$\begin{aligned}\beta_1(\operatorname{star}[v]) &= \dim H_1(X, X \setminus \operatorname{star}[v]) \\ &= \dim H_1(\operatorname{cl} \operatorname{star}[v], \operatorname{fr} \operatorname{star}[v]).\end{aligned}$$

Because X is a 1-dimensional abstract simplicial complex, $\operatorname{fr} \operatorname{star}[v] = \operatorname{cl}(\operatorname{star}[v]) \cap \operatorname{cl}(X \setminus \operatorname{star}[v])$ contains no edges, and therefore contains only vertices. (Suppose ϵ were an edge in $\operatorname{fr} \operatorname{star}[v]$. Since the closure of any subset $A \subseteq X$ differs from A only in its vertices, $\epsilon \in \operatorname{star}[v] \cap (X \setminus \operatorname{star}[v]) = \emptyset$, which is a contradiction.) The number of vertices in $\operatorname{fr} \operatorname{star}[v]$ is precisely the degree of v.

Therefore, the long exact sequence for the pair $(\operatorname{cl} \operatorname{star}[v], \operatorname{fr} \operatorname{star}[v])$ is

$$\begin{array}{l} 0 \rightarrow H_1(\operatorname{cl} \operatorname{star}[v]) \rightarrow H_1(\operatorname{cl} \operatorname{star}[v], \operatorname{fr} \operatorname{star}[v]) \longrightarrow \mathbb{R}^{\deg v} \\ \xrightarrow{(i_*)_1} \mathbb{R} \longrightarrow H_0(\operatorname{cl} \operatorname{star}[v], \operatorname{fr} \operatorname{star}[v]) \longrightarrow 0, \end{array}$$

in which the map labeled $(i_*)_1$ above is the connecting map. The connecting map is induced by the inclusion $i : \operatorname{fr} \operatorname{star}[v] \rightarrow \operatorname{cl} \operatorname{star}[v]$, so $(i_*)_1$ is surjective.

We claim that $H_1(\operatorname{cl} \operatorname{star}[v]) = 0$, because this means that $H_1(\operatorname{cl} \operatorname{star}[v], \operatorname{fr} \operatorname{star}[v])$ injects into $\mathbb{R}^{\deg v}$. Because of the surjectivity of $(i_*)_1$, this means that the dimension of $H_1(\operatorname{cl} \operatorname{star}[v], \operatorname{fr} \operatorname{star}[v])$ must be $(\deg v) - 1$ as the proposition claims.

To prove the claim on $H_1(\operatorname{cl} \operatorname{star}[v])$, consider the chain complex,

$$0 \longrightarrow \mathbb{R}^{\deg v} \xrightarrow{\partial_1} \mathbb{R}^{1+\deg v} \longrightarrow 0$$

for which the boundary map is given by

$$\partial_1 = \begin{pmatrix} 1 & 1 & \cdots & 1 \\ -1 & 0 & \cdots & 0 \\ 0 & -1 & \cdots & 0 \\ & & \vdots & \\ 0 & 0 & \cdots & -1 \end{pmatrix}$$

in which the first row corresponds to $[v]$ and each column corresponds to an edge incident to $[v]$. Clearly ∂_1 is injective, so $H_1(\text{cl}\,\text{star}[v]) = 0$. □

One can take $1 + \beta_1(\text{star}\,\sigma)$ to be the *generalized degree* of a simplex σ in an abstract simplicial complex.

Example 5.13 Consider the simplicial complex X generated by $[a, b, c]$, $[c, d]$, $[d, e]$, $[d, f]$, and $[c, g]$ shown in Fig. 5.9a. Notice that if we treat the 1-skeleton of X as a graph, then the graph degree of vertex d is 3 and the graph degree of c is 4.

Although the definition of local homology starts with the entire space X relative to a complement of an open set U, namely $H_\bullet(X, X\backslash U)$, it is usually easier to think without complements via excision. Intuitively the local homology can be though as the simplicial homology of the simplicial complex consisting of U and a "virtual simplex" covering the frontier of U. This means that the generalized degree of $[d]$ can be thought of as considering the star of $[d]$, which is the set $\{[d], [c, d], [d, e], [d, f]\}$ along with a "virtual simplex" $[c, e, f]$ highlighted in Fig. 5.9b. Since the generalized degree is defined to be $1 + \beta_1(\text{star}[d])$, and β_1 counts the number of loops in the space, it is easy to see (by inspection of Fig. 5.9b or by calculation) that $\beta_1(\text{star}[d]) = 2$. Therefore, the generalized degree of $[d]$ agrees with its graph degree, it is also 3.

The generalized degree of a vertex may differ from its graph degree if it is adjacent to non-edge simplices. A consideration of Fig. 5.9c reveals that the simplex $[a, b, c]$ impacts $\beta_1(\text{star}[c])$ just the same as an edge. Therefore, $\beta_1(\text{star}[c]) = 2$, so $[c]$ has generalized degree 3 even though c has graph degree 4.

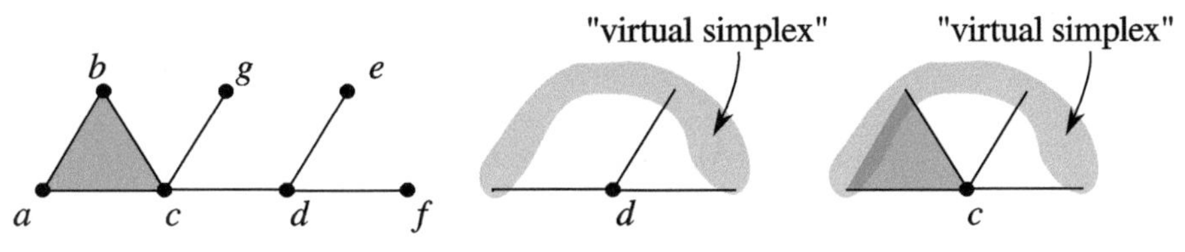

Fig. 5.9 **a** The simplicial complex X, **b** the neighborhood of vertex d, and **c** the neighborhood of vertex c in Example 5.13

Exercise 5.18 Consider the simplicial complex X from Example 5.13, shown in Fig. 5.9a. Compute the generalized degrees of

1. the edge $[c, d]$,
2. the edge $[b, c]$,
3. the simplex $[a, b, c]$, and
4. the vertex $[g]$.

There are a variety of other ways to construct an exact sequence

$$0 \longrightarrow C_\bullet(X) \xrightarrow{f_\bullet} C_\bullet(Y) \xrightarrow{g_\bullet} C_\bullet(Z) \longrightarrow 0$$

of simplicial chain complexes. Of these, the *Mayer-Vietoris exact sequence* is particularly useful. To construct this sequence, let A and B be two subcomplexes of Z with $Z = (A \cup B)$. Take $X = (A \cap B)$ and $Y = (A \sqcup B)$. That is, Z is the union of A and B, X is the intersection of A and B, and Y is the *disjoint* union of A and B. Each of these is an abstract simplicial complex in its own right. $Y = (A \sqcup B)$ is clearly an abstract simplicial complex whenever A and B both are. To show that $X = (A \cap B)$ is an abstract simplicial complex, suppose that σ is a simplex of X. This implies that σ is a simplex of both A and B. Every subset $\tau \subseteq \sigma$ of σ is therefore a simplex of both A and B, so τ must be a simplex of $(A \cap B)$.

The definition of $Y = (A \sqcup B)$ implies that

$$C_\bullet(Y) = C_\bullet(A \sqcup B) = C_\bullet(A) \oplus C_\bullet(B),$$

since each simplex of $A \sqcup B$ is either a simplex of A or a simplex of B but not both.

Noting that $(A \cap B)$ is a subset of A and B, we have two inclusion maps $i_A : (A \cap B) \to A$ and $i_B : (A \cap B) \to B$. We also have one simplicial map $s : (A \sqcup B) \to (A \cup B)$ given by taking each simplex of A or B to its corresponding copy in $Z = (A \cup B)$. The three simplicial maps i_A, i_B, and s induce chain maps

$$(i_A)_\bullet : C_\bullet(A \cap B) \to C_\bullet(A),$$

$$(i_B)_\bullet : C_\bullet(A \cap B) \to C_\bullet(B),$$

and

$$s_\bullet : C_\bullet(A) \oplus C_\bullet(B) \to C_\bullet(A \cup B)$$

on the simplicial chain complexes.

Notice that if σ is a simplex of $(A \cap B)$, then it is in the image of i_A and of i_B. Moreover, σ is also a simplex of $(A \cup B)$. Therefore,

$$(s \circ i_A)(\sigma) = (s \circ i_B)(\sigma).$$

If we reinterpret each k-dimensional simplex σ as a basis element of $C_k(X) = C_k(A \cap B)$, then

$$\begin{aligned} 0 &= (s_k \circ (i_A)_k)(\sigma) - (s_k \circ (i_B)_k)(\sigma) \\ &= s_k\left((i_A)_k(\sigma) - (i_B)_k(\sigma)\right) \\ &= s_k\left(((i_A)_k \oplus (-(i_B))_k)(\sigma)\right) \end{aligned}$$

for each k. Using this fact, the following sequence becomes exact

$$0 \longrightarrow C_\bullet(A \cap B) \xrightarrow{(i_A)_\bullet \oplus (-(i_B))_\bullet} C_\bullet(A) \oplus C_\bullet(B) \xrightarrow{s_\bullet} C_\bullet(A \cup B) \longrightarrow 0.$$

The resulting long exact sequence

$$\cdots \to H_k(A \cap B) \to H_k(A) \oplus H_k(B) \to H_k(A \cup B) \xrightarrow{(\varepsilon_*)_k} H_{k-1}(A \cap B) \to \cdots$$

is called the *Mayer-Vietoris sequence.*

The connecting map $(\varepsilon_*)_k$ in the Mayer-Vietoris sequence is particularly interesting. Recalling from Corollary 3.2 in Chap. 3 that the connecting map is nontrivial precisely when the short exact sequence contains a bar chain map of Type 10, the connecting map is nontrivial when there is a correspondence between a pair of chains in A and B, a chain in $A \cap B$, and a chain in $A \cup B$. Therefore, we can interpret the connecting map as identifying the homology classes in $A \cup B$ that arise from the overlap of A and B.

Exercise 5.19 Consider the simplicial complex X generated by $[a, b]$, $[b, c]$, $[c, d]$, $[a, d]$. Let A be the subcomplex generated by $[a, b]$ and $[b, c]$. Let B be the subcomplex generated by $[c, d]$ and $[a, d]$.

1. Construct the Mayer-Vietoris sequence for $X = A \cup B$, and
2. Interpret the connecting map.

Proposition 5.10 *Suppose that X is a simplicial complex, then*

$$\widetilde{H_{k+1}}(SX) \cong \widetilde{H}_k(X)$$

for all k.

Proof This proposition can be proven directly from the Mayer-Vietoris sequence, if we take both A and B to be the two copies of the cone CX present in SX. This results in an exact sequence[3] of chain maps

$$0 \longrightarrow \widetilde{C_\bullet}(X) \xrightarrow{i_\bullet \oplus (-i_\bullet)} \widetilde{C_\bullet}(CX) \oplus \widetilde{C_\bullet}(CX) \xrightarrow{(j_*)_\bullet + (j_{*'})_\bullet} \widetilde{C_\bullet}(SX) \longrightarrow 0$$

where $i_\bullet$ is the chain map induced by the inclusion map $i : X \to CX$, $(j_*)_\bullet$ is induced by the simplicial map $j_* : CX \to SX$ taking $*$ to $*$, and $(j_{*'})_\bullet$ is induced by the simplicial map $j_* : CX \to SX$ taking $*$ to $*'$. This sequence is exact by construction; every simplex of SX is the image of at least one simplex of one of the two copies of CX. Furthermore, in the sequence above, every simplex of X gets mapped to two copies in the middle chain complex, each with opposite signs. Hence the kernel of $((j_*)_\bullet + (j_{*'})_\bullet)$ is precisely the image of $(i_\bullet \oplus (-i_\bullet))$.

Given this exact sequence of chain complexes, we have the Mayer-Vietoris exact sequence of reduced homologies

$$\cdots \longrightarrow \widetilde{H_{k+1}}(CX) \oplus \widetilde{H_{k+1}}(CX) \xrightarrow{(j_*)_{k+1} + (j_{*'})_{k+1}} \widetilde{H_{k+1}}(SX) \longrightarrow \widetilde{H_k}(X) \xrightarrow{i_k \oplus (-i_k)} \widetilde{H_k}(CX) \oplus \widetilde{H_k}(CX) \longrightarrow \cdots$$

We have seen in Proposition 5.4 that $\widetilde{H_k}(CX) = 0$ for all k. Actually the sequence is

$$\cdots \longrightarrow 0 \longrightarrow \widetilde{H_{k+1}}(SX) \longrightarrow \widetilde{H_k}(X) \longrightarrow 0 \longrightarrow \cdots$$

which asserts that the remaining reduced homologies are isomorphic, precisely as claimed. □

Example 5.14 Consider the simplicial complex X generated by $[a, b]$, $[a, c]$, $[b, c]$. The simplicial chain complex for X has

$$C_1(X) = \text{span}\{[a, b], [a, c], [b, c]\} \cong \mathbb{R}^3,$$
$$C_0(X) = \text{span}\{[a], [b], [c]\} \cong \mathbb{R}^3,$$

with

$$\partial_1 = \begin{pmatrix} -1 & -1 & 0 \\ 1 & 0 & -1 \\ 0 & 1 & 1 \end{pmatrix}.$$

[3] The reader should notice that the chain map taking $\widetilde{C_\bullet}(X) \to \widetilde{C_\bullet}(SX)$ by the composition of the two middle maps in the sequence is intuitively related to the chain map $m_\bullet$ used in Proposition 5.5.

One step of row reduction reveals that this map has a kernel of dimension 1, so that

$$H_1(X) \cong \mathbb{R}, \;\; H_0 \cong \mathbb{R},$$

so the only nontrivial term in the reduced homology is $\widetilde{H}_1(X) \cong \mathbb{R}$.

The suspension SX is the simplicial complex generated by $[a, b, d]$, $[a, c, d]$, $[b, c, d]$, $[a, b, e]$, $[a, c, e]$, $[b, c, e]$, where we have added two vertices d and e, resulting in a somewhat larger simplicial chain complex,

$$\mathbb{R}^6 \xrightarrow{\partial_2} \mathbb{R}^9 \xrightarrow{\partial_1} \mathbb{R}^5 \longrightarrow 0.$$

If we lexicographically sort the simplices, so that $C_2(SX)$ is spanned by $[a, b, d]$, $[a, b, e]$, $[a, c, d]$, $[a, c, e]$, $[b, c, d]$, $[b, c, e]$, and $C_1(SX)$ is spanned by $[a, b]$, $[a, c]$, $[a, d]$, $[a, e]$, $[b, c]$, $[b, d]$, $[b, e]$, $[c, d]$, $[c, e]$, then

$$\partial_2 = \begin{pmatrix} 1 & 1 & 0 & 0 & 0 & 0 \\ 0 & 0 & 1 & 1 & 0 & 0 \\ -1 & 0 & -1 & 0 & 0 & 0 \\ 0 & -1 & 0 & -1 & 0 & 0 \\ 0 & 0 & 0 & 0 & 1 & 1 \\ 1 & 0 & 0 & 0 & -1 & 0 \\ 0 & 1 & 0 & 0 & 0 & -1 \\ 0 & 0 & 1 & 0 & 1 & 0 \\ 0 & 0 & 0 & 1 & 0 & 1 \end{pmatrix}$$

and

$$\partial_1 = \begin{pmatrix} -1 & -1 & -1 & -1 & 0 & 0 & 0 & 0 & 0 \\ 1 & 0 & 0 & 0 & -1 & -1 & -1 & 0 & 0 \\ 0 & 1 & 0 & 0 & 1 & 0 & 0 & -1 & -1 \\ 0 & 0 & 1 & 0 & 0 & 1 & 0 & 1 & 0 \\ 0 & 0 & 0 & 1 & 0 & 0 & 1 & 0 & 1 \end{pmatrix}.$$

A few row operations reveals that the rank of ∂_2 is 5, so that

$$H_2(SX) \cong \mathbb{R}.$$

On the other hand, the rank of ∂_1 is 4. This means that the kernel of ∂_1 is of dimension 5. Since $\partial_1 \circ \partial_2 = 0$, we must conclude that $\ker \partial_1 = \text{image } \partial_2$, so

$$H_1(SX) = 0.$$

Finally, we have that

$$H_0(SX) \cong \mathbb{R}.$$

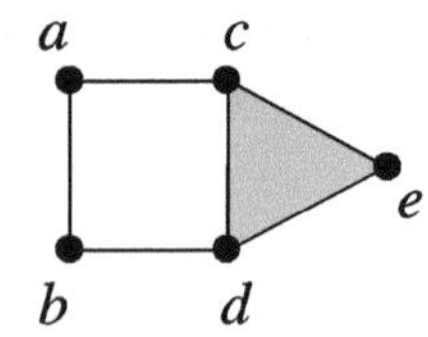

Fig. 5.10 The simplicial complex in Exercise 5.21

This is in accordance with Proposition 5.10, because we have shown that

$$H_2(SX) = \widetilde{H_{1+1}}(SX) \cong \mathbb{R} \cong \widetilde{H_1}(X) = H_1(X),$$

and

$$H_1(SX) = \widetilde{H_{1+0}}(SX) \cong 0 \cong \widetilde{H_0}(X) \cong \mathbb{R}/\mathbb{R}.$$

Exercise 5.20 If X is a simplicial complex with p vertices, SX is a simplicial complex with $p + 2$ vertices. If n is the maximum dimension of a simplex in X, then $n + 1$ is the maximum dimension of SX.

1. Relate the number of simplices of dimension n in X to the number of simplices of dimension $n + 1$ in SX.
2. Relate the number of simplices of dimension $n - 1$ in X to the number of simplices of dimension n in SX.

Exercise 5.21 Compute the simplicial homology of the simplicial complex shown in Fig. 5.10 without doing any linear algebra. This simplicial complex is generated by

$$[a, b], [a, c], [b, d], [c, d, e].$$

(Hint: first apply a free face reduction.)

5.3 Simplicial Homotopies

Because the simplicial chain complex is a functor (Theorem 4.1 in Chap. 4), simplicial maps induce chain maps. Since simplicial maps are rather difficult to classify, we can study them through the lens of their induced chain maps. As discussed in Proposition 3.7 in Chap. 3, chain homotopies are an equivalence between chain maps. In this section we define *simplicial homotopies* between simplicial maps that induce homotopies between their induced simplicial chain maps. *Simplicial homotopies* can be constructed geometrically and describe ways to "deform" one simplicial map into another. We begin by defining a *prism* on a simplicial complex X, which "thickens" X so that simplicial maps with X as a domain can be more easily deformed.

Definition 5.8 Suppose X is an abstract simplicial complex on the vertices V_X. The *prism* PX is the simplicial complex constructed according to the following recipe:

- The vertex set is $V_{PX} := V_X \sqcup V_X$, which contains two copies of the vertices of X. Each vertex $v \in V_X$ of the original complex corresponds to two vertices $v, v' \in V_{PX}$ in the prism.
- The simplices of PX are generated by simplices of the form

$$[v_0, \ldots, v_i, v_i', \ldots, v_k']$$

where $[v_0, \ldots, v_k]$ is a simplex of X and $i = 0, \ldots, k$. Notice that each dimension k simplex of X corresponds to $(k+1)$ different simplices of dimension $(k+1)$ in PX.

Example 5.15 The prism of a single vertex $[v_0]$ is $[v_0, v_0']$. The prism of the simplicial complex generated by $[v_0, v_1]$ is generated by two 2-dimensional simplices, namely

$$[v_0, v_0', v_1'] \text{ and } [v_0, v_1, v_1'].$$

The prism of the simplicial complex generated by $[v_0, v_1, v_2]$ is generated by three 3-dimensional simplices,

$$[v_0, v_0', v_1', v_2'], \ [v_0, v_1, v_1', v_2'], \text{ and } [v_0, v_1, v_2, v_2'],$$

as shown in Fig. 5.11.

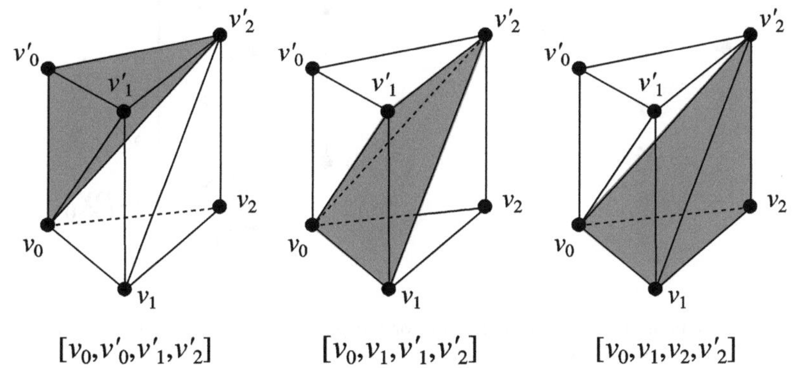

Fig. 5.11 Simplices of the prism over $[v_0, v_1, v_2]$ in Example 5.15

Exercise 5.22 Construct the prism of the simplicial complex generated by $X = \{[a, b, c], [a, d], [b, d]\}$.

Exercise 5.23 Suppose that X is a simplicial complex generated by a single k-dimensional simplex. How many simplices of each dimension does PX have?

Lemma 5.5 *There are two disjoint subcomplexes, PX_0 and PX_1, of PX that are isomorphic to X. The isomorphisms $i_0 : X \to PX_0$ and $i_1 : X \to PX_1$ take the vertices of X to one of the two copies of the vertices in PX.*

Figure 5.12 shows the relationship between PX and the subcomplexes PX_0 and PX_1 for the simplicial complex X generated by $[v_0, v_1, v_2]$ described in Example 5.15.

Proof Suppose that $\sigma = [v_0, \dots, v_k]$ is a simplex of X. By the construction of PX, the following are simplices of PX,

$$[v_0, \dots, v_k, v_k']$$

and

$$[v_0, v_0', \dots, v_k'].$$

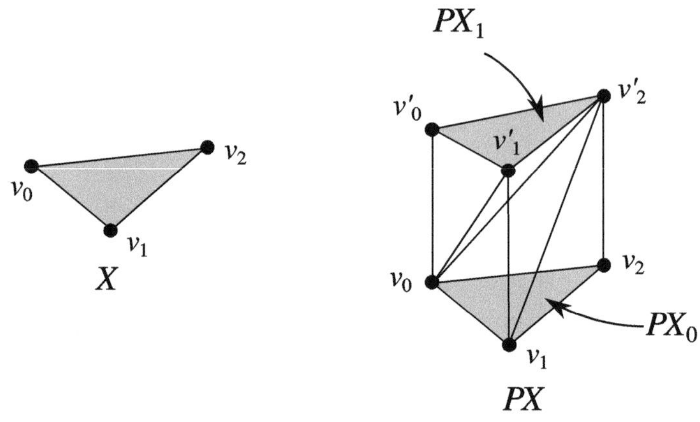

Fig. 5.12 The subcomplexes PX_0 and PX_1 for the prism PX over the simplicial complex X generated by $[v_0, v_1, v_2]$. See Example 5.15

The first of these has σ as a face. The second has $[v_0', \ldots, v_k']$ as a face, which consists of the same vertices as X, just in the other copy of the vertex set. □

Constructing prisms does not disrupt simplicial structure, as the next proposition shows.

Proposition 5.11 *The construction of PX from X defines a covariant functor* **Asc** $\rightarrow$ **Asc**, *which is to say that every simplicial map $X \rightarrow Y$ induces a simplicial map $PX \rightarrow PY$.*

Proof Using a simplicial map $f : X \rightarrow Y$, define a function $Pf : V_{PX} \rightarrow V_{PY}$

$$(Pf)(v) := f(v), \; (Pf)(v') := f(v)'$$

Pf is a simplicial map because

$$[(Pf)(v_0), \ldots, (Pf)(v_i), (Pf)(v_i')', \ldots, (Pf)(v_k')']$$

is a simplex of PY whenever $[v_0, \ldots, v_i, v_i', \ldots, v_k']$ is a simplex of PX.

To show covariance, suppose that $g : Y \rightarrow Z$ is another simplicial map. Transform g into a simplicial map $Pg : PY \rightarrow PZ$ in the same way as was done for Pf. Observe that

$$(Pg \circ Pf)(v) = Pg(Pf(v)) = Pg(f(v)) = (g \circ f)(v) = (P(g \circ f))\,(v)$$

and

$$(Pg \circ Pf)(v') = Pg(Pf(v')) = Pg(f(v)') = g(f(v))' = ((g \circ f)(v))' = (P(g \circ f))\,(v'),$$

which completes the argument. □

The principal use of prisms is to define *simplicial homotopies*, which induce chain homotopies on simplicial chain complexes.

Definition 5.9 An *elementary simplicial homotopy* between two simplicial maps $f, g : X \rightarrow Y$ is a simplicial map $h : PX \rightarrow Y$ such that h restricted to PX_0 is f and PX_1 is g.

Example 5.16 Suppose that X is the simplicial complex generated by $[1, 2]$ and Y is the simplicial complex generated by $[a, b, c]$, and that $f, g : X \rightarrow Y$ are two simplicial maps given by

$$f([1]) := a, \; f([2]) := b,$$

and

$$g([1]) := a, \; g([2]) := c.$$

The vertex function $h : V_{PX} \to V_Y$ given by

$$h(1) := a,\ h(2) := b,\ h(1') := a,\ h(2') := c$$

is a simplicial map (and therefore an elementary simplicial homotopy between f and g) because each of

$$h([1, 1', 2']) = [a, a, c] = [a, c]$$

and

$$h([1, 2, 2']) = [a, b, c]$$

is a simplex of Y.

If we instead consider the simplicial complex Z generated by $[a, b]$, $[a, c]$, $[b, c]$ (the boundary of Y), f and g are still simplicial maps $X \to Z$. However, h is no longer a simplicial map because the image of $[1, 2, 2']$ is not a simplex of Z.

Lemma 5.6 *Suppose that X is a simplicial complex and $i_0 : X \to PX_0$ and $i_1 : X \to PX_1$ are the simplicial isomorphisms constructed in Lemma 5.5. The family of functions (not a chain map) $m_k : C_k(X) \to C_{k+1}(PX)$*

$$m_k([v_0, \ldots, v_k]) := \sum_{i=0}^{k}(-1)^i[v_0, \ldots, v_i, v_i', \ldots, v_k']$$

for each k, is a chain homotopy between the maps $C_\bullet(X) \to C_\bullet(PX)$ induced by i_0 and i_1 on the simplicial chain complexes.

Proof To establish the Lemma, all that is required is to show that

$$\partial_{PX,k+1} \circ m_k + m_{k-1} \circ \partial_{X,k} = (i_1)_k - (i_0)_k$$

for all k. This is simply a long calculation with many cancellations,

$$\begin{aligned}(\partial_{PX,k+1} \circ m_k + m_{k-1} \circ \partial_{X,k})([v_0, \ldots, v_k]) &= \partial_{PX,k+1}\left(\sum_{i=0}^{k}(-1)^i[v_0, \ldots, v_i, v_i', \ldots, v_k']\right)\\ &\quad + m_{k-1}\left(\sum_{j=0}^{k}(-1)^j[v_0, \ldots, \widehat{v_j}, \ldots, v_k]\right)\\ &= \sum_{j=0}^{i}\sum_{i=0}^{k}(-1)^j(-1)^i[v_0, \ldots, \widehat{v_j}, \ldots, v_i, v_i', \ldots, v_k']\\ &\quad + \sum_{j=i}^{k}\sum_{i=0}^{k}(-1)^{j+1}(-1)^i[v_0, \ldots, v_i, v_i', \ldots, \widehat{v_j'} \ldots, v_k']\\ &\quad + \sum_{i=0}^{j-1}\sum_{j=0}^{k}(-1)^i(-1)^j[v_0, \ldots, v_i, v_i' \ldots, \widehat{v_j'}, \ldots, v_k']\end{aligned}$$

$$+\sum_{i=j+1}^{k}\sum_{j=0}^{k}(-1)^{i-1}(-1)^{j}[v_0,\dots\widehat{v_j},\dots v_i,v_i',\dots,v_k']$$
$$=\sum_{i=0}^{k}[v_0,\dots,\widehat{v_i},v_i',\dots,v_k']-\sum_{i=0}^{k}[v_0,\dots,v_i,\widehat{v_i'}\dots,v_k']$$
$$=[v_i',\dots,v_k']-[v_0,\dots,v_i]$$
$$=(i_1)_k-(i_0)_k.$$

□

Proposition 5.12 *Every elementary simplicial homotopy between simplicial maps $f, g : X \to Y$ induces a chain homotopy between induced maps $f_\bullet, g_\bullet : C_\bullet(X) \to C_\bullet(Y)$*

Proof Suppose $h : PX \to Y$ is an elementary simplicial homotopy between f and g. As a consequence of the definition, $f = h \circ i_0$ and $g = h \circ i_1$ where $i_0 : X \to PX_0$ and $i_1 : X \to PX_1$ are the simplicial isomorphisms constructed in Lemma 5.5. This can be expressed as a diagram

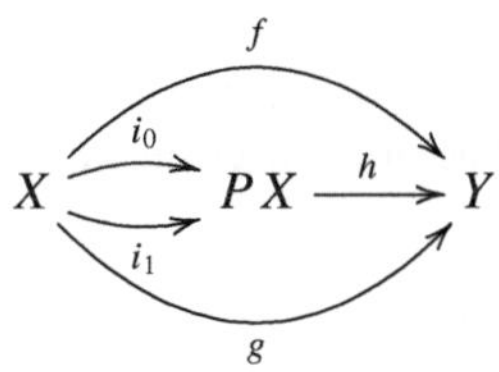

in which the top triangle and the bottom triangle commute but the diagram as a whole does not commute.

According to Lemma 5.6, there is a chain homotopy between the induced maps of i_0 and i_1 given by the functions

$$m_k([v_0,\dots,v_k]) := \sum_{i=0}^{k}(-1)^i[v_0,\dots,v_i,v_i',\dots,v_k'].$$

Inspired by this, define $y_k := h_{k+1} \circ m_k$, which is a family of functions $C_k(X) \to C_{k+1}(Y)$ given by

$$y_k([v_0,\dots,v_k]) := \sum_{i=0}^{k}(-1)^i[f(v_0),\dots,f(v_i),g(v_i),\dots,g(v_k)].$$

Again, y_k does not define a chain map, but it is a chain homotopy. Specifically, following the proof of Lemma 5.6 *mutatis mutandis*,[4] it follows that

$$\partial_{Y,k+1} \circ y_k + y_{k-1} \circ \partial_{X,k} = g_k - f_k,$$

as claimed. □

Given two simplicial maps $f, g : X \to Y$, there is essentially no freedom in how an elementary simplicial homotopy can be defined. Specifically, any elementary simplicial homotopy must have the vertex function $h : V_{PX} \to V_Y$

$$h(v) := f(v), \; h(v') := g(v)$$

where v and v' are the two copies in $V_{PX} = V_X \sqcup V_X$ of the vertex $v \in V_X$. The only question is whether this defines a simplicial map or not.

Proposition 5.13 *Suppose that X is a simplicial complex, and PX is its prism. Then*

$$H_k(PX) \cong H_k(X).$$

In essence, PX works like $X \times [0, 1]$, where PX_0 corresponds to $X \times \{0\}$ and PX_1 corresponds to $X \times \{1\}$. The proof uses an explicit calculation using the retraction $PX \to PX_0$.

Proof Consider the simplicial map $c : PX \to PX$ that collapses V'_X onto V_X, given by its vertex function

$$c(v) = v, \; c(v') = v$$

for each $v \in PX_0$ and each $v' \in PX_1$. This vertex function is evidently simplicial because it takes every simplex to one in PX_0. We claim that there is an elementary simplicial homotopy $h : PPX \to PX$ between c and $\mathrm{id}_{PX} : PX \to PX$. This would suffice since it means that there is a chain homotopy between the identity map on $C_\bullet(PX)$ and the projection of $C_\bullet(PX)$ onto $C_\bullet(X)$, and therefore these two chain complexes are quasi-isomorphic.

The vertex set of PPX is $(V \sqcup V') \sqcup (V'' \sqcup V''')$. Since there is no real freedom in defining h as an elementary simplicial homotopy, the vertex function for h must be c on $(V \sqcup V')$ with image in V only. On $(V'' \sqcup V''')$, h simply "forgets" two of the primes, so that

$$h(v'') = v \text{ and } h(v''') = v'.$$

It only remains to show that h is a simplicial map, namely that every simplex of PPX gets mapped to one in PX. PPX is generated by two kinds of simplices, namely

$$[v_0, \ldots, v_i, v'_i, \ldots, v'_j, v'''_j, \ldots .v'''_k]$$

[4] That is, by making all the substitutions that are clearly required.

and

$$[v_0, \ldots, v_i, v_i'', \ldots, v_k''].$$

The image under h of the first of these is a simplex of PX since

$$\begin{aligned} h([v_0, \ldots, v_i, v_i', \ldots, v_j', v_j''', \ldots .v_k''']) &= [v_0, \ldots, v_i, v_i', \ldots, v_j', v_j', \ldots .v_k'] \\ &= [v_0, \ldots, v_i, v_i', \ldots .v_k'], \end{aligned}$$

duplicate vertices being ignored, and the second has an image in X since

$$h([v_0, \ldots, v_i, v_i'', \ldots, v_k'']) = [v_0, \ldots, v_i, v_i, \ldots .v_k] = [v_0, \ldots, v_k].$$

□

The structure of the proof of Proposition 5.13 suggests that we can form *simplicial homotopy equivalences* in much the same way as homotopy equivalences for chain complexes. However, we must be careful because elementary simplicial homotopies are not transitive.

Example 5.17 Let X be the simplicial complex generated by one vertex $*$ and Y be the simplicial complex generated by $[a, b]$ and $[b, c]$. Consider three simplicial maps $f, g, h : X \to Y$ such that there is an elementary simplicial homotopy h_1 between f and g, and an elementary simplicial homotopy h_2 between g and h, but no elementary simplicial homotopy between f and h. Specifically, let

$$f(*) := a, \;\; g(*) := b, \;\; \text{and } h(*) := c,$$

as shown in Fig. 5.13. Then h_1 is completely determined by $h_1([*, *']) = [a, b]$ and h_2 is completely determined by $h_2([*, *']) = [b, c]$, both of which are simplices of Y. However, any elementary simplicial homotopy h_3 between f and h would have $h_3([*, *']) = [a, c]$, which is not a simplex of Y.

The failure of elementary simplicial homotopies to be transitive is unfortunate, because it breaks the analogy between elementary simplicial homotopies and chain homotopies, since chain homotopies *are* transitive according to Proposition 3.7 in

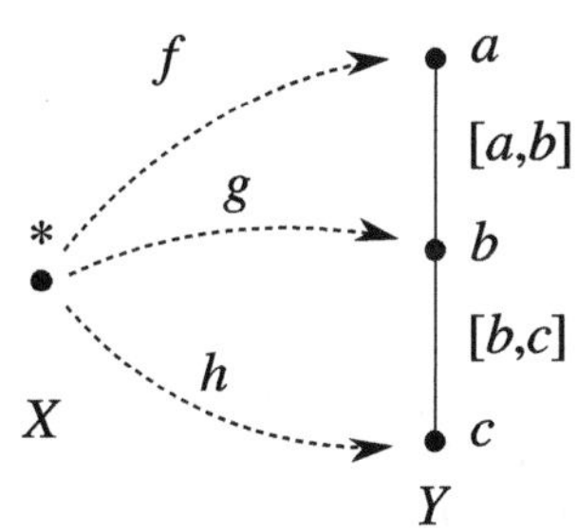

Fig. 5.13 The simplicial maps $f, g, h : X \to Y$ in Example 5.17

Chap. 3. Since elementary simplicial homotopies induce chain homotopies, it is a reasonable strategy to enforce transitivity by enlarging our notion of *simplicial homotopy* to permit transitivity.

Definition 5.10 A *simplicial homotopy* between two maps $f, g : X \to Y$ consists of a sequence of simplicial maps $f_1, f_2, \ldots, f_k : X \to Y$ and a sequence of elementary simplicial homotopies $h_0, h_1, \ldots, h_k$ where h_0 is an elementary simplicial homotopy between f and f_1, h_1 is an elementary simplicial homotopy between f_1 and $f_2, \ldots,$ and h_k is an elementary simplicial homotopy between f_k and g. We say that f and g are *simplicial homotopic* in this case, and will write $f \simeq g$.

Because the construction of simplicial homotopies ensures that they are transitive, they form an equivalence relation on simplicial maps, since elementary simplicial homotopies are symmetric and reflexive.

Corollary 5.3 *If two simplicial maps $f, g : X \to Y$ are simplicial homotopic, then they induce chain homotopic chain maps $f_\bullet, g_\bullet : C_\bullet(X) \to C_\bullet(Y)$.*

Example 5.18 Continuing Example 5.17, the simplicial map $f : X \to Y$ induces the following chain map $f_\bullet : C_\bullet(X) \to C_\bullet(Y)$

$$
\begin{array}{ccccccc}
0 & \longrightarrow & 0 & \longrightarrow & \mathbb{R} & \longrightarrow & 0 \\
 & & \Big\downarrow & & \Big\downarrow \begin{pmatrix} 1 \\ 0 \\ 0 \end{pmatrix} & & \\
0 & \longrightarrow & \mathbb{R}^2 & \xrightarrow[\begin{pmatrix} -1 & 0 \\ 1 & -1 \\ 0 & 1 \end{pmatrix}]{} & \mathbb{R}^3 & \longrightarrow & 0
\end{array}
$$

while the simplicial map $h : X \to Y$ induces the following chain map $h_\bullet : C_\bullet(X) \to C_\bullet(Y)$

$$
\begin{array}{ccccccc}
0 & \longrightarrow & 0 & \longrightarrow & \mathbb{R} & \longrightarrow & 0 \\
 & & \Big\downarrow & & \Big\downarrow \begin{pmatrix} 0 \\ 0 \\ 1 \end{pmatrix} & & \\
0 & \longrightarrow & \mathbb{R}^2 & \xrightarrow[\begin{pmatrix} -1 & 0 \\ 1 & -1 \\ 0 & 1 \end{pmatrix}]{} & \mathbb{R}^3 & \longrightarrow & 0
\end{array}
$$

Although the maps f and h are not elementary simplicial homotopic, they are simplicial homotopic. By Corollary 5.3, their induced chain maps are therefore chain homotopic. This is indeed the case since

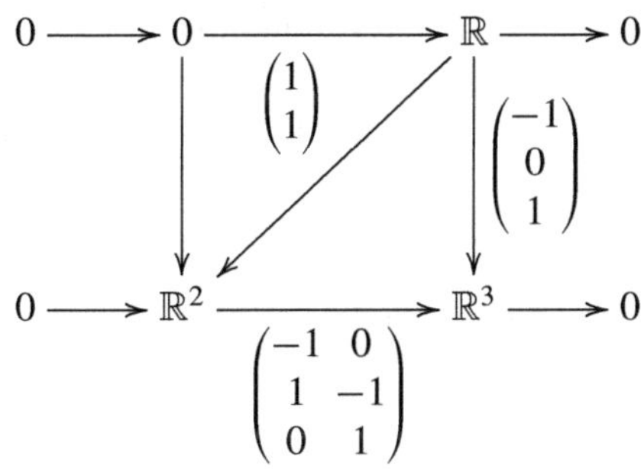

shows that $h_\bullet - f_\bullet$ is null homotopic.

Homotopies and composition work well together, as the next proposition shows.

Proposition 5.14 *Suppose that $f_1, g_1 : X \to Y$ and $f_2, g_2 : Y \to Z$ are simplicial maps satisfying $f_1 \simeq g_1$ and $f_2 \simeq g_2$, then $(f_2 \circ f_1) \simeq (g_2 \circ g_1)$.*

Proof It suffices to prove the result for elementary simplicial homotopies. That is to say if $f_1, g_1 : X \to Y$ and $f_2, g_2 : Y \to Z$ are simplicial maps, h_1 is an elementary simplicial homotopy from f_1 to g_1, and h_2 is an elementary simplicial homotopy from f_2 to g_2 then there is an elementary simplicial homotopy $h_3 : PX \to Z$ from $(f_2 \circ f_1)$ to $(g_2 \circ g_1)$.

Diagrammatically, we have

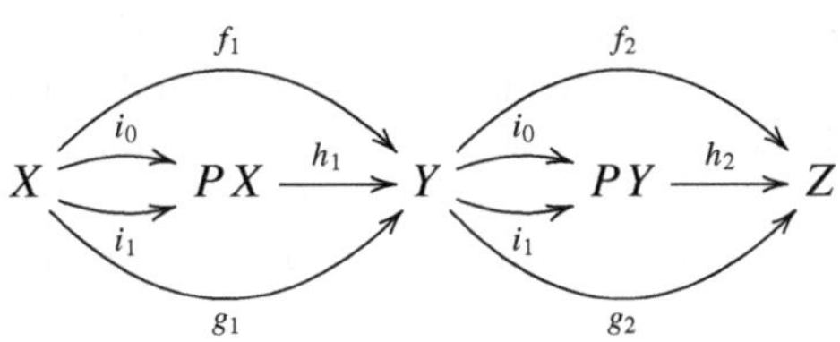

As has been noted earlier, there is no freedom in defining such an elementary simplicial homotopy h_3, since it must be given by

$$h_3(v) := f_2\,(f_1(v))$$

for each vertex v in PX_0 and by

$$h_3(v') := g_2\,(g_1(v))$$

for each vertex v' in PX_1. Let us verify that h_3 is a simplicial map. Each simplex of PX is of the form $\sigma := [v_0, \ldots v_i, v_i', \ldots, v_k']$. Ignoring the second maps for the moment, observe that since h_1 is a simplicial map,

$$[f_1(v_0), \ldots f_1(v_i), g_1(v_i), \ldots, g_1(v_k)]$$

is a simplex of Y, so by construction, it follows that

$$\left[f_1(v_0), \ldots f_1(v_i), (g_1(v_i))', \ldots, (g_1(v_k))'\right]$$

is a simplex of PY. In this way, we can lift $h_1 : PX \to Y$ to a simplicial map $\widetilde{h_1} : PX \to PY$ given by

$$\widetilde{h_1}(v) := f(v)$$

for each vertex v in PX_0 and by

$$\widetilde{h_1}(v') := (g(v))'$$

for each vertex v' in PX_1.

The image of the simplex $\sigma := [v_0, \ldots v_i, v_i', \ldots, v_k']$ through h_3 is

$$\begin{aligned} h_3(\sigma) &= h_3\left([v_0, \ldots v_i, v_i', \ldots, v_k']\right) \\ &= \left[h_3(v_0), \ldots h_3(v_i), h_3(v_i'), \ldots, h_3(v_k')\right] \\ &= [f_2(f_1(v_0)), \ldots f_2(f_1(v_i)), g_2(g_1(v_i)), \ldots, g_2(g_1(v_k))] \\ &= h_2\left(\left[f_1(v_0), \ldots f_1(v_i), (g_1(v_i))', \ldots, (g_1(v_k))'\right]\right) \\ &= (h_2 \circ \widetilde{h_1})(\sigma), \end{aligned}$$

so it follows that h_3 is an elementary simplicial homotopy. □

Definition 5.11 A *loop* in a simplicial complex X is a simplicial map of the form $p : S_k^1 \to X$, where S_k^1 is the simplicial complex generated by the edges $[1, 2]$, $[2, 3]$, $\ldots$, $[k-1, k]$, and $[k, 1]$. (The notation S_k^1 is to be interpreted as "the 1-dimensional sphere with k vertices," consistent with the notation of Example 5.7.)

A simplicial complex X is said to be *simply connected* if it is connected and every pair of loops is simplicial homotopic.

One of the cleanest interpretations of the simplicial homology $H_1(X)$ is that it detects the failure of X to be simply connected.

Proposition 5.15 *If X is simply connected, then $H_1(X)$ is trivial.*

Proof Suppose that c is an element of $C_1(X)$ that is representative of an arbitrary element of $H_1(X)$. (Since $H_1(X)$ is a quotient of a subspace of $C_1(X)$ by the image of ∂_2, it consists of cosets of $C_1(X)$.) Since the elements of $C_1(X)$ are linear combinations of edges, we may think of c as being specified by a set real-valued *weights* on the edges of X. Since c is an element of $H_1(X)$, it lies in the kernel of ∂_1, which means that its boundary—a linear combination of vertices—must be zero. Let us suppose that v_0 is a vertex that is one of the endpoints of one of the edges of c. Without loss of generality, we may assume that the edge containing v_0 has weight 1, since otherwise we simply divide all weights by the weight of that edge to obtain a new element of $H_1(X)$. Notice that v_0 appears as a vertex in c an even number of times, as a consequence of c being in the kernel of ∂_1. Therefore, the presence of c as an element of $H_1(X)$ implies that there is a path p in X of the form

$$v_0, v_1, v_2, \ldots, v_n, v_0,$$

starting and ending at v_0. Evidently, this means that p also corresponds to a loop $\ell : S^1_n \to X$.

Let us now assume that X is simply connected. This means that every loop must be simplicial homotopic to ℓ. For instance, this means that ℓ must be simplicial homotopic to the constant path

$$v_0, v_0, v_0, \ldots, v_0, v_0,$$

which also starts and ends at v_0. That means that there is a sequence of elementary simplicial homotopies $h_1, h_2, \ldots, h_k : PS^1_n \to X$ from ℓ to the constant path. These are elementary simplicial homotopies between $\ell, \ell_1, \ldots, \ell_{k-1}$, and the constant path, respectively. Observe that since each of the h_i are defined as simplicial maps, each of the simplicial maps ℓ_i "between" ℓ and the constant loop must also be loops.

Furthermore, since the image of the constant loop under discussion contains only the vertex v_0, it must induce the zero map $H_1(S^1_n) \to H_1(X)$. Therefore, ℓ_{k-1} must induce the same map on the simplicial homologies. By induction, ℓ must also induce the zero map $H_1(S^1_n) \to H_1(X)$. But the image of ℓ in X is the path p, which means that p must be a trivial element of $H_1(X)$.

After repeating this argument over all path components of c, we must conclude that c is in the image of ∂_2, meaning that c is trivial in $H_1(X)$. □

Showing that two simplicial maps are not simplicial homotopic is much harder than showing they are not elementary simplicial homotopic, since there is considerably more latitude in the definition of simplicial homotopy than in the definition of elementary simplicial homotopy. One strategy is to lift a simplicial map $f : X \to Y$ to a new simplicial map $\widetilde{f} : X \to \widetilde{Y}$ whose homotopies are easier to study. The starting point is a special kind of simplicial map called a *covering map*.

Definition 5.12 A simplicial map $p : X \to Y$ is a *covering map* if for each simplex $\sigma \in Y$, the preimage $p^{-1}(\operatorname{star} \sigma)$ of the star over σ is simplicial isomorphic to a disjoint union of copies of $\operatorname{star} \sigma$. We will allow the number of copies of $\operatorname{star} \sigma$ to be infinite.

Example 5.19 Suppose X is the simplicial complex generated by $[v_0, v_1]$, $[v_1, v_2]$, and $[v_0, v_2]$. First of all, notice that the identity map $\mathrm{id}_X : X \to X$ is a covering map, since the preimage of each star is exactly that same star.

If Y is the simplicial complex generated by $[v_0, v_1]$, $[v_1, v_2]$, $[v_2, v_3]$, $[v_3, v_4]$, $[v_4, v_5]$, $[v_5, v_0]$, and $f : Y \to X$ given by

$$f(v_0) := v_0,\ f(v_1) := v_1,\ f(v_2) := v_2,$$

$$f(v_3) := v_0,\ \ f(v_4) := v_1,\ \ f(v_5) := v_2$$

is a covering map, then there are two typical kinds of stars in X, namely star$[v_0]$ and star$[v_0, v_1]$.

$$\begin{aligned} f^{-1}(\text{star}[v_0]) &= f^{-1}(\{[v_0], [v_0, v_1], [v_0, v_2]\}) \\ &= \{[v_0], [v_0, v_1], [v_0, v_5], [v_3], [v_2, v_3], [v_3, v_4]\} \\ &= \{[v_0], [v_0, v_1], [v_0, v_5]\} \sqcup \{[v_3], [v_2, v_3], [v_3, v_4]\}, \end{aligned}$$

which is simplicial isomorphic (just a relabeling of vertices) of two copies of star$[v_0]$, and

$$\begin{aligned} f^{-1}(\text{star}[v_0, v_1]) &= f^{-1}(\{[v_0, v_1]\}) \\ &= \{[v_0, v_1], [v_3, v_4]\} \\ &= \{[v_0, v_1]\} \sqcup \{[v_3, v_4]\}, \end{aligned}$$

which is simplicial isomorphic to two copies of star$[v_0, v_1]$.

Finally, if Z is the infinite simplicial complex generated by $[n, n+1]$ for all integers n, the map $g : Z \to X$ given by

$$g(3n) := v_0,\ \ g(3n+1) := v_1,\ \ g(3n+2) := v_2,$$

for all integers n is a covering map. The preimage of a star over $[v_0]$ is the disjoint union of an infinite collection of copies of sets of the form $\{[n, n+1], [n+1, n+2], [n]\}$, each of which is isomorphic to star$[v_0]$. Similarly, the preimage over the star of $[v_0, v_1]$ is the disjoint union of a collection of edges of Z.

Exercise 5.24 Consider the simplicial complex X generated by $[a, b]$, $[b, c]$, $[a, c]$, $[c, d]$, $[d, e]$, and $[c, e]$.

1. Following Example 5.19, construct a covering map $p : W \to X$ where W is the simplicial complex generated by all simplices of the form $[2n, 2n+2]$, $[2n, 2n+1]$, $[2n+1, 2n+2]$ for integers n. See Fig. 5.14a.
2. Is your construction in (1) unique? Prove your answer.
3. Suppose that $\mathbb{Z} \times \mathbb{Z}$ (the set of ordered pairs of integers) is the vertex set for a simplicial complex U, in which each is of the form of one of $[(m, 2n), (m+1, 2n)]$, $[(m, 2n), (m, 2n+1)]$, or $[(m, 2n+1), (m, 2n+2)]$ for some integers m and n. Construct a covering map $q : U \to X$. See Fig. 5.14b.

Covering maps are useful because simplicial maps "lift through" them uniquely, as Lemma 5.7 shows.

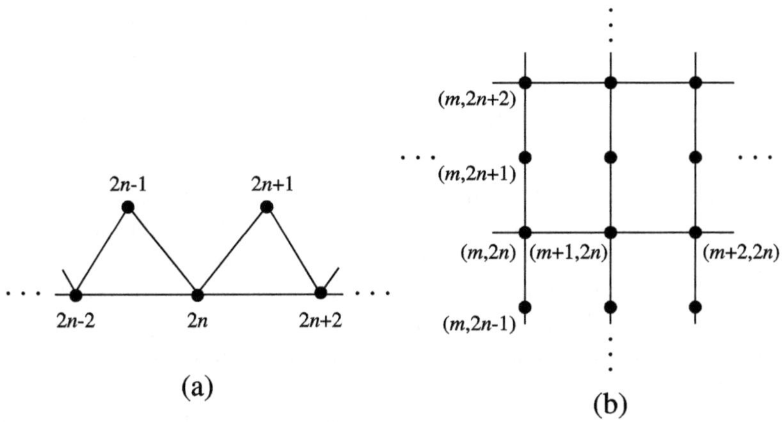

Fig. 5.14 The simplicial complexes in Exercise 5.24: **a** the complex W and **b** the complex U

Lemma 5.7 *Suppose that X is a simply connected simplicial complex. If $f : X \to Y$ is a simplicial map and $p : \widetilde{Y} \to Y$ is a covering map, then for every vertex $x \in X$ and every $\widetilde{y} \in p^{-1}(f(x))$, there is a unique simplicial map $\widetilde{f} : X \to \widetilde{Y}$ such that $\widetilde{f}(x) = \widetilde{y}$ and the diagram*

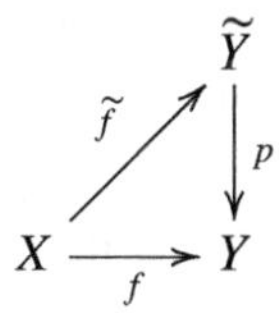

commutes.

Proof Let $x \in X$ be given, and let $\widetilde{f}(x)$ be chosen from $p^{-1}(f(x))$. For any other $x' \in X$, we define $\widetilde{f}(x')$ inductively. Since X is connected, there is a path from x to x' consisting of finitely many vertices $x, x_2, x_3, \ldots, x_{n-1}, x'$. Using this path, we can uniquely determine $\widetilde{f}(x')$.

Base case In the situation that $x' = x_2$, there is a simplex σ containing both x and x' as vertices. This means that $\sigma \in \operatorname{star} x$, which since f is simplicial means that $f(\sigma)$ is a simplex that contains $f(x)$ and $f(x')$. In other words, $f(\sigma)$ is contained in star $f(x)$. Since p is a covering map, $p^{-1}(\operatorname{star} f(x))$ is simplicial isomorphic to a disjoint union of copies of star $f(x)$, and therefore there is a simplex isomorphic to $f(\sigma)$ in each of these copies. Additionally, $\widetilde{f}(x)$ is contained in precisely one of the copies of star $f(x)$ in $p^{-1}(\operatorname{star} f(x))$. Therefore, there is precisely one simplex $\tau \in p^{-1}(\operatorname{star} f(x))$ that is isomorphic to $f(\sigma)$ that also contains $\widetilde{f}(x)$. Since the copies are disjoint, there is exactly one possible choice for $\widetilde{f}(x')$ that results in a simplicial map, namely the element of $p^{-1}(f(x'))$ contained in τ.

Inductive case Assume that the result has been established for all vertices x' for which there is a path consisting of not more than n vertices, starting at x and ending at x'. (The base case established this for paths consisting of 2 vertices.) Let us therefore suppose that there is a path $x, x_2, x_3, \ldots, x_n, x'$ that consists of $n+1$ vertices. The induction hypothesis implies that $\widetilde{f}(x_i)$ is uniquely defined for each $i = 2, 3, \ldots, n$. Therefore, applying the reasoning in the base case using x_n in place of x uniquely defines $\widetilde{f}(x')$.

The proof is not yet complete, since there might be two paths resulting in different choices for $\widetilde{f}(x')$, in which case neither choice works. This cannot happen if X is a simply connected complex. Any two paths in X starting at x and ending x' will be simplicial homotopic. It therefore suffices to show that there is really only one possibility for $\widetilde{f}(x')$ if there are two paths p_1 and p_2 both starting at x and ending at x' that are *elementary* simplicial homotopic, since an arbitrary simplicial homotopy would iterate this construction.

If p_1 and p_2 were to disagree about the value of $\widetilde{f}(x')$, then $\widetilde{f}(x_n)$ and $\widetilde{f}(y_n)$ would lie in disjoint copies of the star of $f(x')$. Suppose that p_1 ends with $\ldots, x_n, x'$, while p_2 ends with $\ldots, y_n, x'$. Since X is simply connected, there is an elementary simplicial homotopy $h : PX \to Y$ that satisfies

$$h([x_n, y_n, x']) = [f(x_n), f(y_n), f(x')].$$

Since this simplex is in the star of $f(x')$, its preimage through p will be a disjoint union of simplicial isomorphic copies. In particular, $[\widetilde{f}(x_n), \widetilde{f}(y_n), \widetilde{f}(x')]$ must be a simplex of Y, meaning that $\widetilde{f}(x_n)$ and $\widetilde{f}(y_n)$ lie in the same copy of star $f(x')$. □

Example 5.20 To illustrate Lemma 5.7, consider the simplicial complex X generated by $[v_0, v_1]$, $[v_1, v_2]$, and $[v_0, v_2]$. If Z is the infinite simplicial complex generated by $[n, n+1]$ for all integers n, the map $g : Z \to X$ given by

$$g(3n) := v_0, \; g(3n+1) := v_1, \; g(3n+2) := v_2,$$

for all integers n was shown to be a covering map in Example 5.19.

Also in Example 5.19, we considered the simplicial complex Y generated by $[v_0, v_1]$, $[v_1, v_2]$, $[v_2, v_3]$, $[v_3, v_4]$, $[v_4, v_5]$, $[v_5, v_0]$. Let $f : Y \to X$ be given by

$$f(v_0) := v_0, \; f(v_1) := v_1, \; f(v_2) := v_2,$$

$$f(v_3) := v_0, \; f(v_4) := v_1, \; f(v_5) := v_2$$

(as in Example 5.19, but that is not relevant here). Notice that Y does not satisfy the hypotheses of Lemma 5.7, because $H_1(Y)$ is not trivial. For instance, if we define

$$\widetilde{f}(v_0) := 0, \; \widetilde{f}(v_1) := 1, \; \widetilde{f}(v_2) := 2,$$

$$\widetilde{f}(v_3) := 3, \; \widetilde{f}(v_4) := 4, \; \widetilde{f}(v_5) := 5,$$

but $\widetilde{f}([v_0, v_5]) = [0, 5]$ is not a simplex of Z.

A small modification to Y, namely the removal of $[v_0, v_5]$ to make a new simplicial complex Y' implies that Y' is simply connected. The above definition of $\widetilde{f} : Y' \to Z$ is then simplicial, and satisfies the equation $f = g \circ \widetilde{f}$.

Two maps that induce chain homotopic maps on simplicial homology need not be simplicial homotopic. The converse of Corollary 5.3 is false. The counterexamples tend to be rather elaborate, but Example 5.21 is about as simple as one can expect.

Example 5.21 Consider the simplicial complexes X and Y shown in Fig. 5.15a–b. The simplicial map f defined by

$$f(v_0) := b,\ f(v_1) := a,\ f(v_2) := c,\ f(v_3) := d,\ f(v_4) := b,$$

$$f(v_5) := c,\ f(v_6) := a,\ f(v_7) := b,\ f(v_8) := d,\ f(v_9) := c,$$

induces the zero map $C_\bullet(X) \to C_\bullet(Y)$ on the simplicial chain complexes because while

$$\begin{aligned} f([v_0, v_1]) &= [b, a], \\ f([v_1, v_2]) &= [a, c], \\ f([v_2, v_3]) &= [c, d], \\ f([v_3, v_4]) &= [d, b], \\ f([v_4, v_5]) &= [b, c], \end{aligned}$$

each of these edges is also traversed in the opposite order (see Fig. 5.15c), because

$$\begin{aligned} f([v_6, v_7]) &= [a, b], \\ f([v_5, v_6]) &= [c, a], \\ f([v_8, v_9]) &= [d, c], \\ f([v_7, v_8]) &= [b, d], \\ f([v_9, v_0]) &= [c, b]. \end{aligned}$$

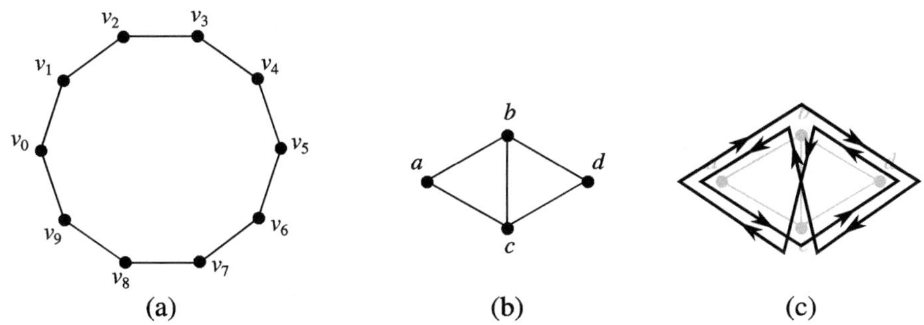

Fig. 5.15 **a** The simplicial complex X, **b** the simplicial complex Y, and **c** the simplicial map $f : X \to Y$ in Example 5.21

On the other hand, the constant map $g : X \to Y$ defined simply by $g(v_i) = b$ for all $i = 0, \ldots, 9$ induces the same (zero) map on the simplicial chain complexes. There is a chain homotopy between h and g because all of the relevant maps are simply zero maps. However, there is no simplicial homotopy between f and g.

Certainly there is no elementary simplicial homotopy h between f and g, since that would require (for instance) that

$$h([v_1, v_2, v_2']) = [f(v_1), f(v_2), g(v_2)] = [a, c, b],$$

which is not a simplex of Y.

Construct the infinite simplicial complex $\widetilde{Y}$ shown in Fig. 5.16. This complex is related to Y through a covering map $p : \widetilde{Y} \to Y$, in which the vertex labels in Fig. 5.16 identify the images of the vertices of $\widetilde{Y}$ under p. Notice that each vertex of Y has infinitely many preimages through p, and moreover that $\widetilde{Y}$ has no closed loops.

Now consider the simplicial complex X' generated by

$$[v_0, v_1], [v_1, v_2], \ldots, [v_9, v_{10}].$$

Notice that if we identify $v_0 = v_{10}$, then we can reinterpret f and g as simplicial maps $X' \to Y$. Since X' is simply connected, we can use Lemma 5.7 to lift the constant map g uniquely to $\widetilde{g} : X' \to \widetilde{Y}$, which is also a constant map. Also using Lemma 5.7, the map f lifts to some non-constant map $\widetilde{f} : X' \to \widetilde{Y}$, as shown in Fig. 5.16.

Let us propose that there is a simplicial homotopy between f and g. This too lifts to a homotopy between $\widetilde{f}$ and $\widetilde{g}$.

Observe that at each step of the simplicial homotopy from $\widetilde{f}$ to $\widetilde{g}$, we will have a simplicial map that is either a loop in $\widetilde{Y}$, or a path whose start and end points are at least 2 edges apart, due to the construction of Y. Starting at $\widetilde{g}$, we cannot get the

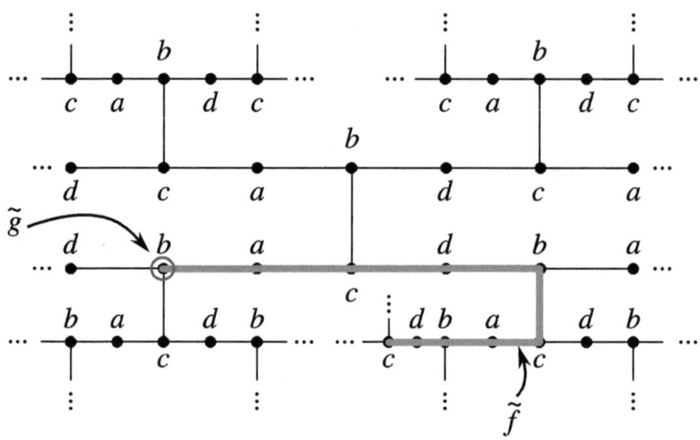

Fig. 5.16 The simplicial complex $\widetilde{Y}$ and the image of the maps $\widetilde{f}$ and $\widetilde{g}$ in Example 5.21

start and end points further than 1 edge apart by an elementary simplicial homotopy. Any simplicial map that is elementary simplicial homotopic to $\widetilde{g}$ must be a loop in $\widetilde{Y}$. But $\widetilde{f}$ is a segment with the endpoints further apart than 3 vertices. There is no way it can be simplicial homotopic to a loop. Therefore, $\widetilde{f}$ and $\widetilde{g}$ cannot be simplicial homotopic. Thus, f cannot be simplicial homotopic to g.

Exercise 5.25 Find another example of two simplicial maps $f, g : X \to Y$ whose induced maps $C_\bullet(X) \to C_\bullet(Y)$ are chain homotopic but f and g are not simplicial homotopic.

Exercise 5.26 Let X be the simplicial complex generated by $[a, b]$, $[b, c]$, $[c, d]$, $[d, e]$, $[e, f]$, and $[a, f]$, and Y be the simplicial complex generated by $[A, B]$, $[B, C]$, and $[A, C]$. Consider the simplicial map $m_1 : X \to Y$ given by

$$\begin{aligned} m_1(a) &:= A, \\ m_1(b) &:= B, \\ m_1(c) &:= C, \\ m_1(d) &:= A, \\ m_1(e) &:= B, \\ m_1(f) &:= C. \end{aligned}$$

Show that this cannot be simplicial homotopic to the simplicial map $m_2 : X \to Y$ given by

$$\begin{aligned} m_2(a) &:= A, \\ m_2(b) &:= A, \\ m_2(c) &:= B, \\ m_2(d) &:= B, \\ m_2(e) &:= C, \\ m_2(f) &:= C. \end{aligned}$$

Hint: show that the induced maps on homology are different, then use Corollary 5.3.

Exercise 5.27 Suppose that X is a simplicial complex, and CX is its cone. Use simplicial homotopies to show that

$$H_k(CX) \cong \begin{cases} \mathbb{R} & \text{if } k = 0, \\ 0 & \text{otherwise.} \end{cases}$$

Notice that Proposition 5.2 is an immediate consequence of this fact, though it was proved a bit differently. Construct $f : CX \to \{*\}$ by the only option, and let $g : \{*\} \to CX$ map the vertex $*$ to the cone point of CX.

1. Show that f and g are both simplicial maps
2. Show that $(f \circ g) : \{*\} \to \{*\}$ is homotopic to the identity map.
3. Show that $(g \circ f) : CX \to CX$ (taking all simplices to the cone point) is homotopic to the identity map. (Hint: this requires constructing PCX, which is a bit messy, but not really troublesome.)

Since simplicial homotopies are transitive, we can define *simplicial homotopy equivalence* in a manner analogous to homotopy equivalence for chain complexes.

Definition 5.13 Two simplicial complexes X and Y are *simplicial homotopy equivalent* if there exist simplicial maps $f : X \to Y$ and $g : Y \to X$ such that $(g \circ f) \simeq \text{id}_X$ and $(f \circ g) \simeq \text{id}_Y$.

According to the definition of homotopy equivalence, every simplicial complex X is simplicial homotopy equivalent to its prism PX by Proposition 5.13.

Lemma 5.8 *Simplicial homotopy equivalent is an equivalence relation on simplicial complexes.*

Proof Reflexivity For any simplicial complex X, choosing $f = \text{id}_X$ and $g = \text{id}_X$ results in a simplicial homotopy equivalence of X with itself.

Symmetry This follows right from the definition, since the roles of f and g can be reversed without any change.

Transitivity Suppose that X, Y, and Z are simplicial complexes and that $f : X \to Y$, $g : Y \to X$, $h : Y \to Z$, and $k : Z \to Y$ are simplicial maps such that

$$(g \circ f) \simeq \text{id}_X, \ (f \circ g) \simeq \text{id}_Y,$$

and

$$(k \circ h) \simeq \text{id}_Y, \ (h \circ k) \simeq \text{id}_Z .$$

Then $(h \circ f)$ and $(g \circ k)$ are simplicial maps such that

$$(h \circ f) \circ (g \circ k) = h \circ (f \circ g) \circ k$$

$$
\begin{aligned}
&\simeq h \circ \mathrm{id}_Y \circ k \\
&\simeq h \circ k \\
&\simeq \mathrm{id}_Z
\end{aligned}
$$

and

$$
\begin{aligned}
(g \circ k) \circ (h \circ f) = &\; g \circ (k \circ h) \circ f \\
\simeq &\; g \circ \mathrm{id}_Y \circ f \\
\simeq &\; g \circ f \\
\simeq &\; \mathrm{id}_X
\end{aligned}
$$

so X and Z are simplicial homotopy equivalent. □

Corollary 5.4 *If two simplicial complexes are simplicial homotopy equivalent, then their simplicial chain complexes are homotopy equivalent as well. As a result, they have isomorphic simplicial homology sequences.*

5.4 Subdivision

Although theoretically pleasing, simplicial homotopy equivalence is rather too fine for many purposes. Worse, finding homotopies is extremely computationally difficult in general Attali et al. (2016). In short, the requirement that for X and Y to be simplicial homotopy equivalent there must be simplicial maps $X \to Y$ and $Y \to X$ is substantially more stringent than the algebraic requirement of chain homotopy equivalence on the induced chain maps $C_\bullet(X) \to C_\bullet(Y)$ and $C_\bullet(Y) \to C_\bullet(X)$.

For instance, consider the case of a triangle and a square. Although their simplicial chain complexes are not isomorphic, they are chain homotopic. Intuitively, this encodes the statement that the topology of both complexes is the same, namely one path component and one loop. However, they are not simplicial homotopy equivalent as the next example shows.

Example 5.22 Consider the simplicial complexes X and Y shown in Fig. 5.17. It is easy to define a simplicial map $g : Y \to X$. However, there is no simplicial map $f : X \to Y$ that would result in a simplicial homotopy equivalence.

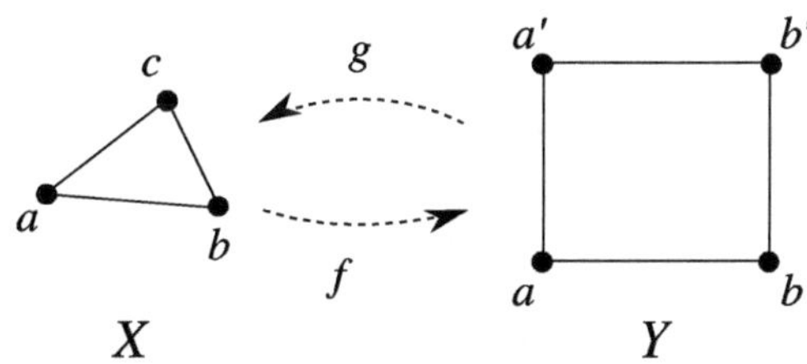

Fig. 5.17 The two simplicial complexes X and Y, and the two simplicial maps f and g in Example 5.22

In particular, no simplicial map $f : X \to Y$ can be injective. If f were to be injective, it would have an injective vertex function. Since X has three vertices and Y has four, there would be one vertex of Y left outside the image of f. Of the remaining vertices of Y, one is between the others. Without loss of generality, suppose that the image of f contains the vertices a, b, a'. Since f is assumed to be injective, this means that the preimages of each of these vertices is distinct. Since X contains all possible edges among three vertices, this means that the image of f also must contain all such edges if f is to be a simplicial map. Evidently $[b, a']$ is not an edge of Y, which is a contradiction.

The consequence of f being non-injective means that regardless of what g is, $(f \circ g) : Y \to Y$ cannot be injective either. Since there are only four vertices, $(f \circ g)$ also cannot be surjective. This spoils any chance of $(f \circ g)$ being simplicial homotopic to the identity map. Although somewhat intuitive-sounding, this statement is a bit subtle because of the way simplicial homotopies are defined; we would need to rule out sequences of elementary simplicial homotopies. It is much easier to approach the problem indirectly using simplicial homology!

The simplicial chain complex for Y is

$$0 \longrightarrow \mathbb{R}^4 \xrightarrow{\begin{pmatrix} -1 & -1 & 0 & 0 \\ 1 & 0 & -1 & 0 \\ 0 & 1 & 0 & -1 \\ 0 & 0 & 1 & 1 \end{pmatrix}} \mathbb{R}^4 \longrightarrow 0$$

under the vertex partial order $a < b < a' < b'$. The only nontrivial basis element of $H_1(Y)$ is (up to rescaling by a constant)

$$[a, b] - [a, a'] + [b, b'] - [a', b'].$$

On the other hand, a non-surjective vertex function cannot have this as an image, since all four distinct vertices are represented. This means that $(f \circ g)$ induces the zero map on $H_1(Y) \to H_1(Y)$. Since this differs from what is induced by the identity map, Corollary 5.3 allows us to conclude that $(f \circ g)$ cannot be homotopic to the identity map. Therefore, X and Y cannot be simplicial homotopy equivalent.

Example 5.23 Continuing Example 5.22, let us examine the simplicial chain map induced by $g : Y \to X$ when

$$g(a) := a, \; g(b) := b, \; g(a') := a, \; g(b') := c.$$

Applying g to all of the simplices of Y,

$$g([a, b]) = [a, b], \; g([a, a']) = [a], \; g([b, b']) = [b, c], \; g([a', b']) = [a, c].$$

Therefore, the diagram of the induced chain map is

$$\begin{array}{ccccccc}
0 & \longrightarrow & \mathbb{R}^4 & \xrightarrow{\begin{pmatrix}-1 & -1 & 0 & 0\\ 1 & 0 & -1 & 0\\ 0 & 1 & 0 & -1\\ 0 & 0 & 1 & 1\end{pmatrix}} & \mathbb{R}^4 & \longrightarrow & 0\\
 & & \Big\downarrow \begin{pmatrix}1&0&0&0\\0&0&0&1\\0&0&1&0\end{pmatrix} & & \Big\downarrow \begin{pmatrix}1&0&1&0\\0&1&0&0\\0&0&0&1\end{pmatrix} & & \\
0 & \longrightarrow & \mathbb{R}^3 & \xrightarrow[\begin{pmatrix}-1 & -1 & 0\\ 1 & 0 & -1\\ 0 & 1 & 1\end{pmatrix}]{} & \mathbb{R}^3 & \longrightarrow & 0,
\end{array}$$

using the same vertex partial order $a < b < a' < b'$ as before. This chain map decomposes into

- a bar chain map of Type 5 at index 1,
- a bar chain map of Type 7 at index 1,
- two bar chain maps of Type 9 at index 1, and
- a bar chain map of Type 5 at index 0,

which means that this is a quasi-isomorphism.

Exercise 5.28 Consider the simplicial complexes X and Y shown in Fig. 5.18. The complex X is the suspension of a square, namely the complex generated by

$$[a, b, f],\ [a, c, f],\ [b, d, f],\ [c, d, f],$$
$$[a, b, e],\ [a, c, e],\ [b, d, e],\ [c, d, e].$$

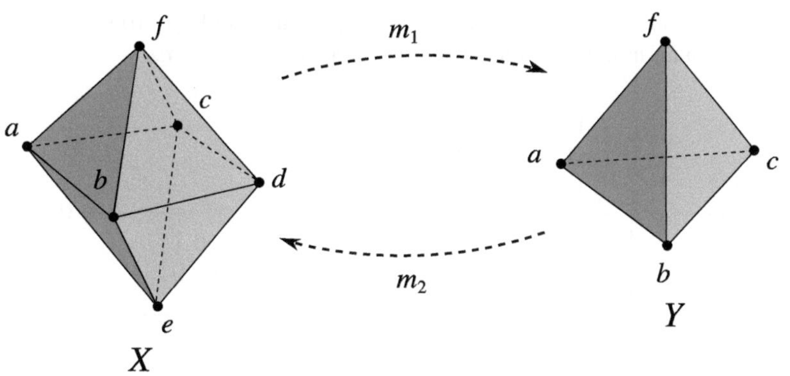

Fig. 5.18 The two simplicial complexes in Exercise 5.28

The complex Y is the boundary of the 3-dimensional simplex, and is generated by

$$[a, b, f],\ [a, c, f],\ [b, c, f],\ [a, b, c].$$

Show that these two complexes are not simplicial homotopy equivalent.
Hint: although you can easily define m_1 by

$$m_1(a) := a,\ m_1(b) := b,\ m_1(c) := c,$$

$$m_1(d) := c,\ m_1(e) := c,\ m_1(f) := f,$$

you need to show that any simplicial map m_2 will not have $m_2 \circ m_1 \simeq \mathrm{id}_X$.

Even though the square and triangle are not simplicial homotopy equivalent, they do have the same simplicial homology. This is a consequence of the fact that there is a simplicial map from the square to the triangle that *collapses* one of the edges into a vertex.

Definition 5.14 An *elementary collapse map* is a simplicial map $f : X \to Y$ such that for each $\sigma \in Y$, the subset

$$\bigcup_{y \in \sigma} f^{-1}(y)$$

of X is a simplex of X. A *collapse map* is a simplicial map $f : X \to Y$ that can be written as the composition $f = f_n \circ \cdots \circ f_1$ of finitely many elementary collapse maps f_i.

Example 5.24 Consider the simplicial complex X generated by $[a, b, c]$ and the simplicial complex Y generated by $[a, b]$. The simplicial map $f : X \to Y$ given by

$$f(a) := a,\ f(b) := b,\ f(c) := a,$$

is an elementary collapse map, because every subset of $\{a, b\}$ is already a simplex of X.

The map f induces an isomorphism on simplicial homology. To see this fact, let us write the chain map under the vertex partial order $a < b < c$,

$$\begin{array}{ccccccccc}
0 & \longrightarrow & \mathbb{R} & \xrightarrow{\begin{pmatrix} 1 \\ -1 \\ 1 \end{pmatrix}} & \mathbb{R}^3 & \xrightarrow{\begin{pmatrix} -1 & -1 & 0 \\ 1 & 0 & -1 \\ 0 & 1 & 1 \end{pmatrix}} & \mathbb{R}^3 & \longrightarrow & 0 \\
 & & \downarrow & & \downarrow{\scriptstyle (1\ 0\ -1)} & & \downarrow{\scriptstyle \begin{pmatrix} 1 & 0 & 1 \\ 0 & 1 & 0 \end{pmatrix}} & & \\
0 & \longrightarrow & 0 & \longrightarrow & \mathbb{R} & \xrightarrow[\begin{pmatrix} -1 \\ 1 \end{pmatrix}]{} & \mathbb{R}^2 & \longrightarrow & 0.
\end{array}$$

It is easy to verify that this is a chain map because

$$(1\ 0\ -1)\begin{pmatrix}1\\-1\\1\end{pmatrix} = 0,$$

and

$$\begin{pmatrix}-1\\1\end{pmatrix}(1\ 0\ -1) = \begin{pmatrix}1\ 0\ 1\\0\ 1\ 0\end{pmatrix}\begin{pmatrix}-1 & -1 & 0\\1 & 0 & -1\\0 & 1 & 1\end{pmatrix},$$

so the diagram commutes.

It is straightforward to use Algorithm 1 in Chap. 3 to see that its chain map decomposition contains

- One bar chain map of Type 5 starting in index 0,
- One bar chain map of Type 9 starting in index 1, and
- Two bar chain maps of Type 7, one that starts in index 2 and one that starts in index 1,

which implies that the chain map is a quasi-isomorphism.

The interpretation of this decomposition is that the bar chain map of Type 5 represents the single path component present in both X and Y. The bar chain map of Type 9 represents the edge $[a, b]$. The bar chain map of Type 7 starting in index 2 represents the simplices $[a, b, c]$, and the other bar chain map of Type 7 represents the simplex $[a, c]$ that was collapsed by the simplicial map f.

Exercise 5.29 Show that every elementary simplicial homotopy $h : PX \to Y$ can be written as the composition $(c \circ f)$ of a simplicial map $f : PX \to PY$ followed by a collapse map $c : PY \to Y$ of the copied vertices $V_Y \sqcup V'_Y \to V_Y$.

Example 5.25 Similar to the previous examples, consider the simplicial complex X generated by $[a, b, c]$ and the simplicial complex Z generated by $[a]$, in which $g : X \to Z$ is the only possible simplicial map. Evidently g is an elementary collapse map. This example actually seems to be a bit more subtle than it might appear, because both the 2-dimensional *and* the 1-dimensional simplices of X are collapsed by g. The chain map has the following diagram

$$\begin{array}{ccccccccc}
0 & \longrightarrow & \mathbb{R} & \xrightarrow{\begin{pmatrix}1\\-1\\1\end{pmatrix}} & \mathbb{R}^3 & \xrightarrow{\begin{pmatrix}-1 & -1 & 0\\ 1 & 0 & -1\\ 0 & 1 & 1\end{pmatrix}} & \mathbb{R}^3 & \longrightarrow & 0\\
 & & \downarrow & & \downarrow & & \downarrow{\scriptstyle (1\ 1\ 1)} & & \\
0 & \longrightarrow & 0 & \longrightarrow & 0 & \longrightarrow & \mathbb{R} & \longrightarrow & 0
\end{array}$$

that is easily seen to commute.

The chain map decomposition can be computed by inspection in this case, and it contains

- One bar chain map of Type 5 starting in index 0, and
- Three bar chain maps of Type 7, one that starts in index 2 and two that start in index 1.

Therefore, the chain map is a quasi-isomorphism.

As in the previous example, the bar chain map of Type 5 represents the single path component of both X and Z. The others represent simplices that where collapsed by the simplicial map g.

Exercise 5.30 Consider the simplicial complex X generated by the 3-simplex $[a, b, c, d]$ and the simplicial complex Y generated by $[a, b]$. Consider the simplicial map $h : X \to Y$ given by

$$h(a) := a, \ h(b) := b, \ h(c) := a, \ h(d) := b.$$

Show that

1. h is a collapse map, and
2. h induces isomorphisms $H_k(X) \to H_k(Y)$ on simplicial homology for all k.

As should be inferred from the previous few examples, collapse maps preserve homology. This is actually a very useful (and rather famous) theorem.

Theorem 5.1 (the Vietoris-Begle mapping theorem) *Every collapse map $f : X \to Y$ induces an isomorphism $H_k(X) \to H_k(Y)$ on simplicial homology.*

Proof It suffices to work with elementary collapse maps, since compositions of quasi-isomorphisms are still quasi-isomorphisms. Elementary collapse maps necessarily induce a surjection on simplices. Therefore, as far as the induced simplicial chain map goes, each component $f_k : C_k(X) \to C_k(Y)$ is a surjective linear map. Considering the chain map decomposition of $f_\bullet : C_\bullet(X) \to C_\bullet(Y)$, the theorem claims that the decomposition can only contain bar chain maps of Types 1, 3, 5, 7, 9, and 10. Surjectivity of the components rules out chain maps of Types 2, 3, 6, or 10. To complete the proof, we just have to rule out Types 4 and 8 to get a quasi-isomorphism.

Let us begin by ruling out a Type 4 bar chain map by supposing that we have the end of a bar in the codomain that maps to zero in the domain. (This happens for Types 4, 7, 8, and 10.) Without loss of generality, such a bar in X is an element $C_k(X)$ that can be written as a linear combination $\sum a_i \sigma_i$ such that $f(\sigma_i)$ is the same simplex of Y regardless of i and $\sum a_i = 0$. The only way the $f(\sigma_i)$ can all be the same simplex of Y regardless of i is if f collapsed some of the vertices together. Therefore, $f^{-1}(f(\sigma_i))$ is the same for all i, too. This is by assumption a simplex τ of X, whose dimension is larger than that of σ_i. Every σ_i is a face of τ. The subcomplex of X generated by τ contains σ_i for every i. That means that the chain map $C_k(\tau) \to C_k(X)$ induced by including each vertex of τ in X is both injective and contains $\sum a_i \sigma_i$ in its image. On the other hand, $\widetilde{H}_k(\tau) = 0$ by either direct calculation or an application of Proposition 5.4 noting that τ is a cone on its faces. Unless $k = 0$, we conclude that $\sum a_i \sigma_i$ corresponds to a trivial element of homology in X. On the other hand, an elementary collapse map cannot change the number of path components. If $k = 0$, then $\sum a_i \sigma_i$ must be trivial since we additionally assumed that the component map f_k took it to zero in $C_0(Y)$.

The argument thus far rules out a Type 4 bar chain map. But since $f(\tau)$ is a lower-dimensional simplex than τ (some of its vertices are necessarily collapsed by f), the component map will take τ to zero. It cannot be Type 8 either. Hence bar chain map that looks like it ends like a Type 4 actually extends to a Type 7 or 10. □

The reader is cautioned that not every simplicial map that appears to collapse vertices is a collapse map. There may be other quasi-isomorphisms involved, such as those arising from free faces! Indeed, the collapse of the square to a triangle, introduced earlier in the section, is one such example.

Example 5.26 The preimages of vertices under the simplicial map g defined in Example 5.23 are all simplices of X, as is exhibited by

$$g^{-1}([a]) = [a, a'],\ g^{-1}([b]) = [b],\ g^{-1}([c]) = [b'].$$

However, $g^{-1}([a, b]) = \{a, b, a'\}$, which is not a simplex of Y. (Y only has simplices of dimensions 0 and 1 in any case.) This means that g is not an elementary collapse. However, the simplicial complex Y' generated by $[a, b, a']$, $[a, b, b']$, $[b, a', b']$, and $[a, a', b']$ is quasi-isomorphic to Y because each of the 2-dimensional faces is a free face. The map $g' : Y' \to X$ defined by the same vertex function as g is an elementary collapse map.

The result is a commutative diagram of the form

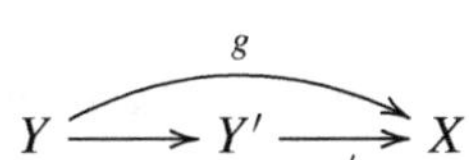

where the arrow $Y \to Y'$ represents the inclusion, which induces a quasi-isomorphism by Proposition 5.3. The elementary collapse map g' induces a quasi-isomorphism by Theorem 5.1. Therefore, g induces a quasi-isomorphism as well.

A convenient way to interpret the example of collapsing the square to the triangle after adding additional free faces is to interpret it in reverse: as subdividing one of the edges of the triangle. Recall that the barycentric subdivision $Bd(X)$ of an abstract simplicial complex X is the order complex of the face partial order of X and has each simplex of X as a vertex (Definition 4.10). Associated to each barycentric subdivision is a simplicial map called the *elementary subdivision map* that relates $Bd(X) \to X$. It stands to reason that while barycentric subdivision does not usually result in simplicial homotopy equivalence, it can induce isomorphisms on simplicial homology (as will be shown in Proposition 5.16).

Definition 5.15 If X is a simplicial complex with a vertex partial order $\leq$, then the vertex function $c : V_{Bd(X)} \to V_X$ that takes each simplex $[w_0, \ldots, w_k]$ (a vertex of $Bd(X)$) to the smallest vertex w_0 is called its *elementary subdivision map*. The composition of a sequence of elementary subdivision maps is called *subdivision map*.

Example 5.27 Consider the simplicial complex X generated by $[a, b, c]$. Each of the 7 nonempty simplices of X correspond to vertices in $Bd(X)$, namely

$$[a], [b], [c], [a, b], [a, c], [b, c], [a, b, c]$$

Each of the face inclusions in X correspond to 1-simplices, of which there are 12, namely

$$\begin{aligned}
[a] \subset [a, b], [a] \subset\ & [a, c], [a] \subset [a, b, c],\\
[b] \subset [a, b], [b] \subset\ & [b, c], [b] \subset [a, b, c],\\
[c] \subset [a, c], [c] \subset\ & [b, c], [c] \subset [a, b, c],\\
[a, b] \subset\ & [a, b, c],\\
[a, c] \subset\ & [a, b, c],\\
[b, c] \subset\ & [a, b, c].
\end{aligned}$$

Finally, sequences of face inclusions in X correspond to 2-simplices, of which there are 6, namely

$$\begin{aligned}
&[a] \subset [a, b] \subset [a, b, c],\\
&[a] \subset [a, c] \subset [a, b, c],\\
&[b] \subset [a, b] \subset [a, b, c],\\
&[b] \subset [b, c] \subset [a, b, c],\\
&[c] \subset [a, c] \subset [a, b, c],\\
&[c] \subset [b, c] \subset [a, b, c].
\end{aligned}$$

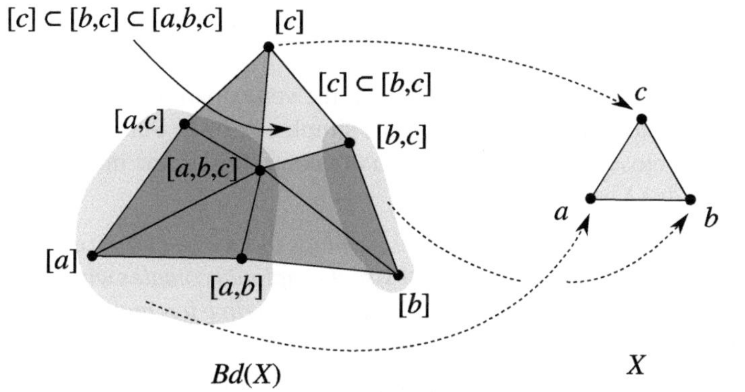

Fig. 5.19 A simplicial complex X generated by a single 2-dimensional simplex, its barycentric subdivision $Bd(X)$, and the elementary subdivision map $Bd(X) \to X$. See Example 5.27

The elementary subdivision map $s : Bd(X) \to X$ is shown in Fig. 5.19, and takes all of the vertices of $Bd(X)$ to the first vertex of X listed, namely

$$\begin{aligned} s([a]) &= a, \\ s([b]) &= b, \\ s([c]) &= c, \\ s([a,b]) &= a, \\ s([a,c]) &= a, \\ s([b,c]) &= b, \\ s([a,b,c]) &= a. \end{aligned}$$

This is a simplicial map because, for instance,

$$s([c] \subset [b,c] \subset [a,b,c]) = [s([c]), s([b,c]), s([a,b,c])] = [c, b, a],$$

which is a simplex of X.

Lemma 5.9 *Elementary subdivision maps are simplicial maps.*

Proof Recalling that the barycentric subdivision is the order complex of the face partial order, each simplex $[\sigma_0, \sigma_1, \ldots, \sigma_k]$ of $Bd(X)$ is a chain of simplices of X

$$\sigma_0 \subset \sigma_1 \subset \cdots \subset \sigma_k$$

where each one is a subset of the next. On the other hand, each of these simplices of X is a vertex of $Bd(X)$. If $c : V_{Bd(X)} \to V_X$, then

$$c(\sigma) := [c(\sigma_0), c(\sigma_1), \ldots, c(\sigma_k)] = [v_0, v_1, \ldots, v_k],$$

where each v_i is the smallest vertex in σ_i in the vertex partial order on X. The result is a set of vertices all drawn from the same simplex σ_k of X. Since abstract simplicial complexes are closed under the operation of taking subsets, the image under c of this simplex of $Bd(X)$ is a simplex of X. □

Proposition 5.16 *Elementary subdivision maps induce isomorphisms on simplicial homology. Specifically, if X is a finite abstract simplicial complex with a vertex partial order* $\leq$, *and* $c : Bd(X) \to X$ *is its elementary subdivision map, then c induces an isomorphism* $H_k(Bd(X)) \cong H_k(X)$ *for each k.*

Proof We show that an elementary subdivision map $c : Bd(X) \to X$ can be factored into the composition $Bd(X) \to (Bd(x))' \to X$ of an elementary collapse map after a free face inclusion, which therefore induces isomorphisms on simplicial homology. (Intuitively, the free faces we need to add are visible as the preimages of each simplex shown in Figure 5.19.)

Construct a new simplicial complex

$$(Bd(X))' := Bd(X) \cup \{\tau \subseteq X : \tau \subseteq c^{-1}(\sigma) \text{ for some } \sigma \in X\}$$

by adding to the barycentric subdivision the preimage of each simplex (and all its faces) of the original simplex.

If we consider a dimension k simplex $\sigma \in X$, its preimage $c^{-1}(\sigma)$ that we just added as a new simplex cannot be contained in the preimage simplex $c^{-1}(\tau)$ of any other k-dimensional simplex τ of X. Because c is surjective, whenever its preimages are nested $c^{-1}(\sigma) \subseteq c^{-1}(\tau)$, its corresponding images must also be nested $\sigma \subseteq \tau$. Since X is a finite abstract simplicial complex, this means that every simplex σ of X not contained in any other simplex will correspond to a free face of $(Bd(X))'$. Proposition 5.3 implies that the removal of this preimage simplex (and all its faces)—and the preimages of all other simplices of the same dimension of σ—will result in quasi-isomorphic simplicial homology. Therefore, we can recursively reduce all of these new simplices starting at the maximum dimension of X and work down in dimension without disrupting the simplicial homology. Therefore, $(Bd(X))'$ and $Bd(X)$ have isomorphic simplicial homology.

On the other hand, the simplicial map $(Bd(X))' \to X$ given by extending the elementary subdivision map c to the additional simplices is a collapse map, which therefore induces an isomorphism on simplicial homology by Theorem 5.1. □

Example 5.28 Consider the simplicial complex X generated by $[a, b]$, $[a, c]$, $[b, c]$, under the vertex partial order $a < b < c$. This has simplicial homology $H_0(X) \cong \mathbb{R}$ and $H_1(X) \cong \mathbb{R}$.

According to Proposition 5.16, $H_k(Bd(X)) \cong H_k(X)$ is induced by the elementary subdivision map $h : Bd(X) \to X$. Let us verify this by direct computation, where $Bd(X)$ is generated by $[a] \subset [a, b]$, $[b] \subset [a, b]$, $[a] \subset [a, c]$, $[c] \subset [a, c]$, $[b] \subset [b, c]$, $[c] \subset [b, c]$. See the map labeled h in Fig. 5.20.

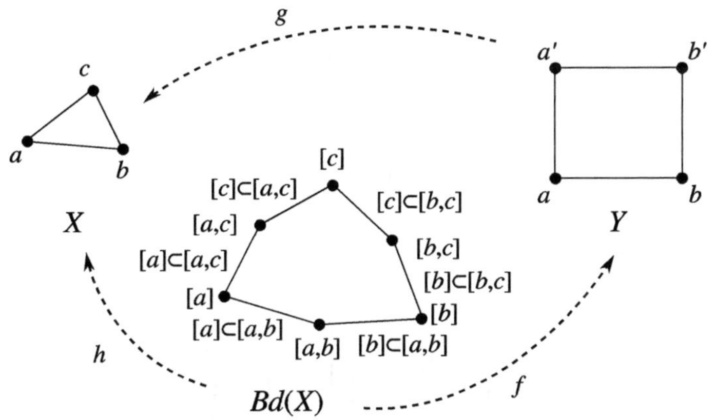

Fig. 5.20 Inserting a vertex into a triangle by way of a preliminary subdivision. See Example 5.29

Specifically, $h_\bullet : C_\bullet(Bd(X)) \to C_\bullet(X)$ is given by the diagram

$$\begin{array}{ccccccc}
0 \longrightarrow & \mathbb{R}^6 & \xrightarrow{\partial_{Bd(X),1}=\begin{pmatrix}-1&-1&0&0&0&0\\0&0&-1&-1&0&0\\0&0&0&0&-1&-1\\1&0&1&0&0&0\\0&1&0&0&1&0\\0&0&0&1&0&1\end{pmatrix}} & \mathbb{R}^6 & \longrightarrow 0 \\
& \Big\downarrow h_1=\begin{pmatrix}0&0&-1&0&0&0\\0&0&0&0&-1&0\\0&0&0&0&0&-1\end{pmatrix} & & \Big\downarrow h_0=\begin{pmatrix}1&0&0&1&1&0\\0&1&0&0&0&1\\0&0&1&0&0&0\end{pmatrix} & \\
0 \longrightarrow & \mathbb{R}^3 & \xrightarrow[\partial_{X,1}=\begin{pmatrix}-1&-1&0\\1&0&-1\\0&1&1\end{pmatrix}]{} & \mathbb{R}^3 & \longrightarrow 0
\end{array}$$

in which lexicographical ordering is used throughout. Explicitly, the components of $C_0(Bd(X))$ are given in the following order: $[a]$, $[b]$, $[c]$, $[a,b]$, $[a,c]$, $[b,c]$. The components of $C_1(Bd(X))$ are given in the following order: $[a] \subset [a,b]$, $[a] \subset [a,c]$, $[b] \subset [a,b]$, $[b] \subset [b,c]$, $[c] \subset [a,c]$, $[c] \subset [b,c]$. Although this is not a chain isomorphism, $H_1(Bd(X))$ is generated by the vector

$$\begin{pmatrix} -1 \\ 1 \\ 1 \\ -1 \\ -1 \\ 1 \end{pmatrix}$$

whose image under h_1 is

$$\begin{pmatrix} 0 & 0 & -1 & 0 & 0 & 0 \\ 0 & 0 & 0 & 0 & -1 & 0 \\ 0 & 0 & 0 & 0 & 0 & -1 \end{pmatrix} \begin{pmatrix} -1 \\ 1 \\ 1 \\ -1 \\ -1 \\ 1 \end{pmatrix} = \begin{pmatrix} 1 \\ -1 \\ 1 \end{pmatrix}$$

which generates the $H_1(X)$. Any vector in $C_0(Bd(X))$ with only one nonzero entry (a vector supported on a single vertex) can be taken as a basis for $H_0(Bd(X))$. This is evidently mapped by h_0 to a similar vector in $C_0(X)$. Therefore, the induced maps on simplicial homology are simply 1×1 identity matrices $(h_*)_k = (1)$, which are invertible—hence h induces an isomorphism on simplicial homology.

Exercise 5.31 One can sometimes use subdivision to simplify a calculation of simplicial homology. Consider the hollow tetrahedron, written as the simplicial complex X generated by $\{[a, b, c], [a, b, d], [a, c, d], [b, c, d]\}$.

1. Show that if one of the 2-dimensional faces of X is subdivided, the resulting simplicial complex is a suspension of a triangle.
2. Show that this process can be iterated for a hollow simplex of arbitrary dimension k.
3. Use your answers to (1) and (2) above (see also Proposition 5.10 and Example 5.14) to compute the simplicial homology of a hollow simplex of arbitrary dimension k.

Definition 5.16 It is convenient to use $Bd^n(X)$ to refer to n iterations of barycentric subdivision of X. Specifically, let

$$Bd^0(X) := X$$

and

$$Bd^{n+1}(X) := Bd(Bd^n(X)).$$

Since barycentric subdivision does not change the simplicial homology of a simplicial complex, it is immediate from Proposition 5.16 that

$$H_k(Bd^n(X)) \cong H_k(X)$$

for all n and k.

Proposition 5.17 *Barycentric subdivision is a covariant functor Bd : **Asc** $\to$ **Asc**, which means that every simplicial map $f : X \to Y$ induces a simplicial map $Bd(f) : Bd(X) \to Bd(Y)$ that preserves composition. Moreover, this map $Bd(f)$ makes the following diagram of simplicial maps commute*

$$\begin{array}{ccc} Bd(X) & \xrightarrow{Bd(f)} & Bd(Y) \\ \downarrow & & \downarrow \\ X & \xrightarrow[f]{} & Y \end{array}$$

in which the vertical maps are elementary subdivision maps.

Proof Although the functoriality follows from Propositions 4.9 and 4.11 in Chap. 4 establishing that the face partial order and order complex are both covariant functors, the construction of $Bd(f)$ is itself rather straightforward. Simply observe that each vertex of $Bd(X)$ is a simplex σ of X. The vertex function for $Bd(X)$ takes σ to $f(\sigma)$, namely by applying f to each vertex of σ. □

5.5 Simplicial Rooves

We have now established that the barycentric subdivision of a simplicial complex adds new simplices to a simplicial complex in a way that preserves simplicial homology. In this section, we construct a new category in which the morphisms are simplicial maps "up to subdivision".

Example 5.29 Returning to Examples 5.22 and 5.23, let us recall the simplicial map $g : Y \to X$ from the square Y to the triangle X given by

$$g(a) := a, \ g(b) := b, \ g(a') := a, \ g(b') := c.$$

This simplicial map induces a simplicial chain map $g_\bullet$ that is a quasi-isomorphism. Constructing a simplicial map that is an inverse to g is certainly impossible, as Example 5.22 explained, and any simplicial map $X \to Y$ cannot induce an isomorphism on simplicial homology. However, we do the next best thing using the barycentric subdivision $Bd(X)$ of X. Supposing that $h : Bd(X) \to X$ is the elementary subdivision under the vertex partial order $a < b < c$, we can define a simplicial map

$f : Bd(X) \to Y$ whose induced map on simplicial homology is an inverse to the map induced by g. As shown in Fig. 5.20 in Sect. 5.4, let

$$f([a]) := a, \;\; f([b]) := b, \;\; f([c]) := b',$$

$$f([a,b]) := a, \;\; f([a,c]) := a', \;\; f([b,c]) := b.$$

While this certainly does not result in an inverse to g—the domain is wrong—the maps do satisfy

$$h = g \circ f,$$

which induces an isomorphism on simplicial homology by Proposition 5.16 because h is a subdivision map.

Although h cannot be inverted, its induced map on simplicial homology $(h_*)_\bullet : H_\bullet(Bd(X)) \to H_\bullet(X)$ does have an inverse $(h_*)_\bullet^{-1}$ according to Example 5.28.

According to the calculation below, the induced maps $(f_*)_\bullet : H_\bullet(Bd(X)) \to H_\bullet(Y)$ and $(g_*)_\bullet : H_\bullet(Y) \to H_\bullet(X)$ on simplicial homology compose to yield an isomorphism

$$(f_*)_\bullet \circ (h_*)_\bullet^{-1} \circ (g_*)_\bullet.$$

Because of these facts, it makes sense to treat f like an inverse to g, "up to subdivision."

Let us now verify the calculation of $(f_*)_\bullet \circ (h_*)_\bullet^{-1} \circ (g_*)_\bullet$. To that end, $f_\bullet : C_\bullet(Bd(X)) \to C_\bullet(Y)$ is given by the diagram

$$\begin{array}{ccccccc} 0 & \longrightarrow & \mathbb{R}^6 & \xrightarrow{\quad \partial_{Bd(X),1} \quad} & \mathbb{R}^6 & \longrightarrow & 0 \\ & & \Big\downarrow f_1 & & \Big\downarrow f_0 & & \\ 0 & \longrightarrow & \mathbb{R}^4 & \xrightarrow{\quad \partial_{Y,1} \quad} & \mathbb{R}^4 & \longrightarrow & 0, \end{array}$$

$$\partial_{Bd(X),1} = \begin{pmatrix} -1 & -1 & 0 & 0 & 0 & 0 \\ 0 & 0 & -1 & -1 & 0 & 0 \\ 0 & 0 & 0 & 0 & -1 & -1 \\ 1 & 0 & 1 & 0 & 0 & 0 \\ 0 & 1 & 0 & 0 & 1 & 0 \\ 0 & 0 & 0 & 1 & 0 & 1 \end{pmatrix}$$

$$f_1 = \begin{pmatrix} 0 & 0 & -1 & 0 & 0 & 0 \\ 0 & 1 & 0 & 0 & 0 & 0 \\ 0 & 0 & 0 & 0 & 0 & -1 \\ 0 & 0 & 0 & 0 & -1 & 0 \end{pmatrix}, \qquad f_0 = \begin{pmatrix} 1 & 0 & 0 & 1 & 0 & 0 \\ 0 & 1 & 0 & 0 & 0 & 1 \\ 0 & 0 & 0 & 0 & 1 & 0 \\ 0 & 0 & 1 & 0 & 0 & 0 \end{pmatrix}$$

$$\partial_{Y,1} = \begin{pmatrix} -1 & -1 & 0 & 0 \\ 1 & 0 & -1 & 0 \\ 0 & 1 & 0 & -1 \\ 0 & 0 & 1 & 1 \end{pmatrix}$$

and $g_\bullet : C_\bullet(Y) \to C_\bullet(X)$ given by the diagram

$$
\begin{array}{ccccccc}
 & & & \partial_{Y,1}=\begin{pmatrix} -1 & -1 & 0 & 0 \\ 1 & 0 & -1 & 0 \\ 0 & 1 & 0 & -1 \\ 0 & 0 & 1 & 1 \end{pmatrix} & & & \\
0 & \longrightarrow & \mathbb{R}^4 & \xrightarrow{\qquad\qquad\qquad} & \mathbb{R}^4 & \longrightarrow & 0 \\
 & & \Big\downarrow {\scriptstyle g_1=\begin{pmatrix} 1&0&0&0 \\ 0&0&0&1 \\ 0&0&1&0 \end{pmatrix}} & & \Big\downarrow {\scriptstyle g_0=\begin{pmatrix} 1&0&1&0 \\ 0&1&0&0 \\ 0&0&0&1 \end{pmatrix}} & & \\
0 & \longrightarrow & \mathbb{R}^3 & \xrightarrow{\qquad\qquad\qquad} & \mathbb{R}^3 & \longrightarrow & 0. \\
 & & & \partial_{X,1}=\begin{pmatrix} -1 & -1 & 0 \\ 1 & 0 & -1 \\ 0 & 1 & 1 \end{pmatrix} & & &
\end{array}
$$

Considering that $H_1(Y)$ is generated by the vector

$$\begin{pmatrix} 1 \\ -1 \\ 1 \\ -1 \end{pmatrix}$$

and the image of this vector through g_1 is

$$\begin{pmatrix} 1 \\ -1 \\ 1 \end{pmatrix},$$

we already know that this has a unique preimage through h_1, which is

$$\begin{pmatrix} -1 \\ 1 \\ 1 \\ -1 \\ -1 \\ 1 \end{pmatrix}.$$

Finally, mapping this vector through f_1 yields the following equation,

$$\left(f_1 \circ h_1^{-1} \circ g_1\right) \begin{pmatrix} 1 \\ -1 \\ 1 \\ -1 \end{pmatrix} = \begin{pmatrix} -1 \\ 1 \\ -1 \\ 1 \end{pmatrix},$$

where we are of course abusing notation slightly because h_1^{-1} is not a function! Since $H_1(Y)$ is 1-dimensional, this means that the map induced by $f_1 \circ h_1^{-1} \circ g_1$ on simplicial homology is an isomorphism. To show that the map induced by $f_0 \circ h_0^{-1} \circ g_0$ on $H_0(Y)$ is an isomorphism, it suffices to select the vector

$$\begin{pmatrix} 0 \\ 0 \\ 0 \\ 1 \end{pmatrix},$$

since this corresponds to the vertex b' which is carried through the calculation unchanged.

Example 5.29 suggests that we should extend our idea of a mapping between simplicial complexes to include a *simplicial roof* of the form

$$X \xleftarrow{f} Bd(X) \xrightarrow{g} Y, \tag{5.3}$$

which we should regard as a "mapping" $X \to Y$ when f is a subdivision map. This is not an unreasonable idea, because many pairs of simplicial complexes with isomorphic simplicial homology, such as those related by the insertion or removal of vertices, would become "isomorphic" if we made this extension, as we will show in Proposition 5.20. The correct way to make this extension is to supplement the category **Asc** with additional morphisms corresponding to *simplicial roof* diagrams.

Before we extend **Asc** to include *simplicial roof diagrams*, though, we need to handle a small technical detail. Specifically, if we can construct a *simplicial roof* like the one shown in (5.3), we can certainly also construct another one whose diagram is

$$X \xleftarrow{f'} Bd^2(X) \xrightarrow{g'} Y, \tag{5.4}$$

by inserting many more vertices. This diagram can be factored into a diagram involving the original *simplicial roof*,

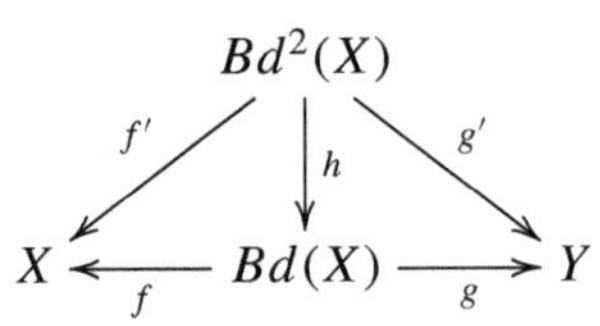

where h is another elementary subdivision. We really should not regard (5.3) and (5.4) as being different "maps", since the only difference between them is how many subdivisions on X we perform as an intermediate step. This motivates the following definition.

Definition 5.17 A diagram of the form

$$X \xleftarrow{f} Bd^n(X) \xrightarrow{g} Y$$

is called a *simplicial roof*, when X and Y are abstract simplicial complexes, g is a simplicial map, and f is a subdivision map.

Two simplicial rooves

$$X \xleftarrow{f_1} Bd^m(X) \xrightarrow{g_1} Y,$$

and

$$X \xleftarrow{f_2} Bd^n(X) \xrightarrow{g_2} Y,$$

are to be regarded as equivalent if there is a simplicial roof

$$X \xleftarrow{f_3} Bd^p(X) \xrightarrow{g_3} Y,$$

such that the diagram

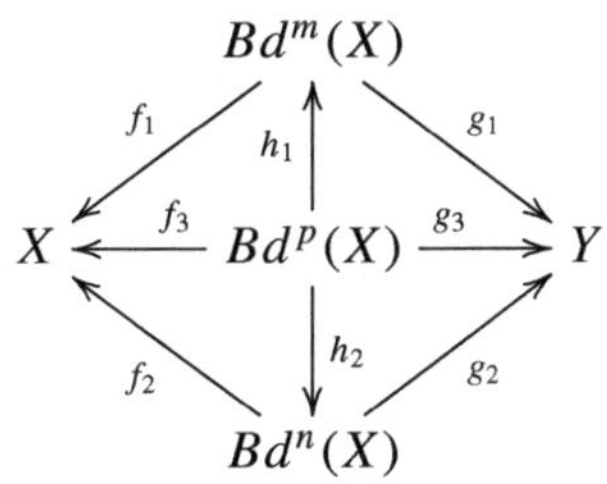

commutes, where h_1 and h_2 are subdivision maps.

We usually write $(f, g) : X \rightsquigarrow Y$ for the equivalence class of simplicial rooves represented by

$$X \xleftarrow{f} Bd^n(X) \xrightarrow{g} Y.$$

Remark 5.1 A diagram of the form

$$X \leftarrow Y \rightarrow Z$$

is usually called a *span* in the category theory literature. If X, Y, and Z are vector spaces and the arrows are projection maps, then Y is indeed the vector space span of X and Z. The term *roof* is reserved for when the arrow on the left is some kind of equivalence, like a subdivision. In Chap. exrefchap:spectral, we will permit the arrow on the left to be a quasi-isomorphism, which is somewhat more general than subdivisions.

Lemma 5.10 *Equivalence of simplicial rooves is a well defined equivalence relation.*

Proof Reflexivity and symmetry of roof equivalence follow immediately from the definition. Transitivity is straightforward, but requires a somewhat larger diagram. Suppose that we have three simplicial rooves

$$X \xleftarrow{f_1} Bd^m(X) \xrightarrow{g_1} Y,$$

$$X \xleftarrow{f_2} Bd^n(X) \xrightarrow{g_2} Y,$$

and

$$X \xleftarrow{f_3} Bd^p(X) \xrightarrow{g_3} Y,$$

such that the pairs (f_1, g_1) and (f_2, g_2), (f_2, g_2) and (f_3, g_3) are equivalent simplicial rooves. We want to show that (f_1, g_1) and (f_3, g_3) are also equivalent. The hypotheses thus far mean that for some integers q and r, the following diagram commutes

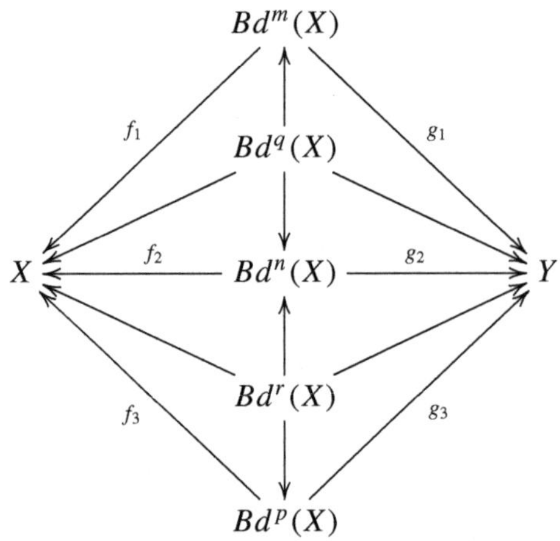

Without loss of generality, assume $q < r$. This means that there is a subdivision map $Bd^r(X) \to Bd^q(X)$. Adding this map to the diagram above and removing the irrelevant arrows yields

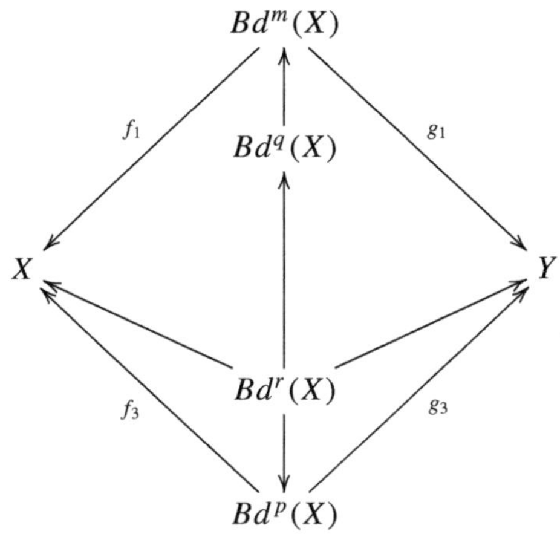

which means that (f_1, g_1) and (f_3, g_3) are also equivalent simplicial rooves. □

With these results in hand, we can extend the category **Asc** by changing the structure of its morphisms so that the isomorphism classes are built by repeated subdivisions. All of the old morphisms—simplicial maps—can be restructured to fit the new model, but there are many new morphisms that are not present in **Asc**. This new category is called the *category of subdivided simplicial complexes*.

Definition 5.18 The *category of subdivided simplicial complexes* **BdAsc** has the class of simplicial complexes as its objects (just like **Asc**), but the morphisms consist of equivalence classes of simplicial rooves $X \rightsquigarrow Y$.

Composition of a roof $X \rightsquigarrow Y$ with a roof $Y \rightsquigarrow Z$ is defined to be the roof $X \rightsquigarrow Z$ given by the diagram

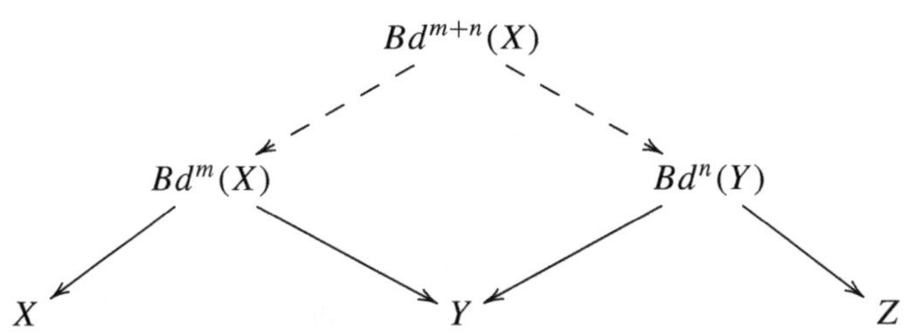

using Proposition 5.17 to define the dashed arrows. The composition of the two leftward arrows is a composition of subdivision maps, which is itself a subdivision map. This yields a new roof $X \rightsquigarrow Z$.

Proposition 5.18 *The category* **BdAsc** *of subdivided simplicial complexes is a well defined category.*

Proof Given a simplicial complex X, the following roof is an identity morphism

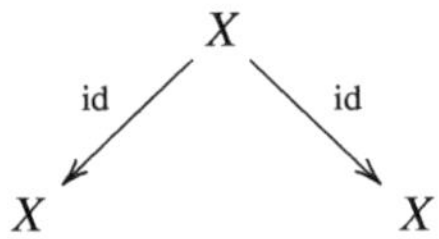

in that composition with any other roof

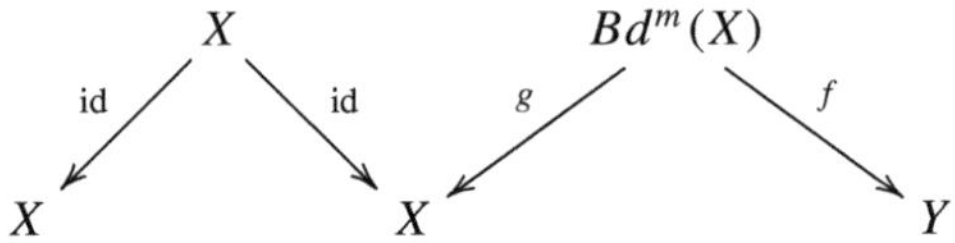

yields the larger diagram

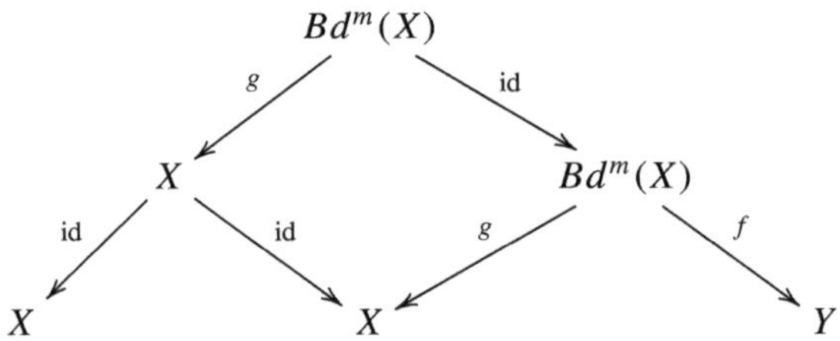

for which the top two paths compose to the original roof.

Composition of the identity morphism on the right yields much the same diagram, namely

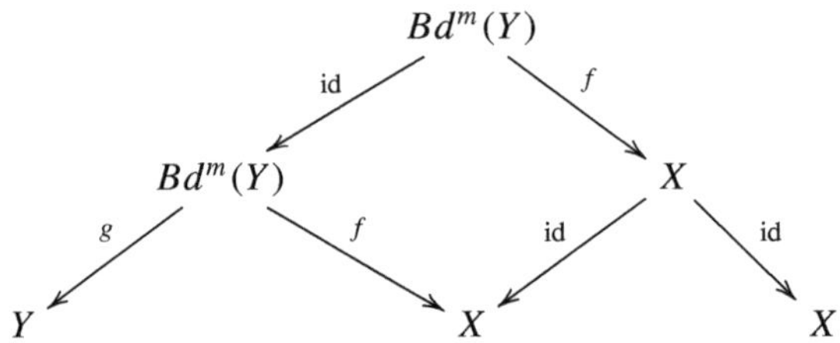

Again, the top two paths compose to the original roof.

Now to show that composition is associative, suppose that we have three rooves $(f_1, g_1) : W \rightsquigarrow X$, $(f_2, g_2) : X \rightsquigarrow Y$, and $(f_3, g_3) : Y \rightsquigarrow Z$. The composition $(f_3, g_3) \circ ((f_2, g_2) \circ (f_1, g_1))$ is constructed from the diagram

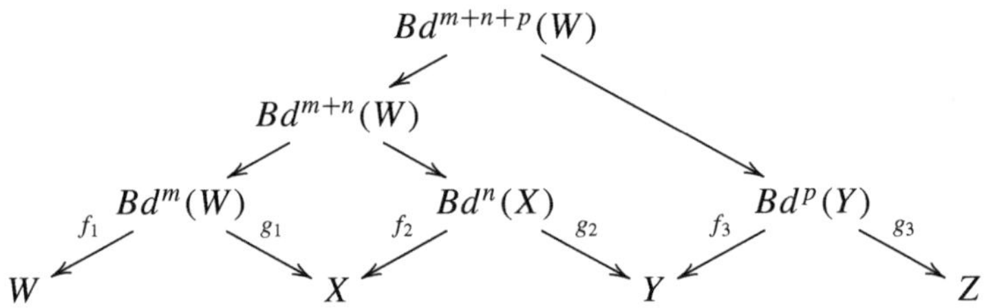

while the $((f_3, g_3) \circ (f_2, g_2)) \circ (f_1, g_1)$ is constructed from the diagram

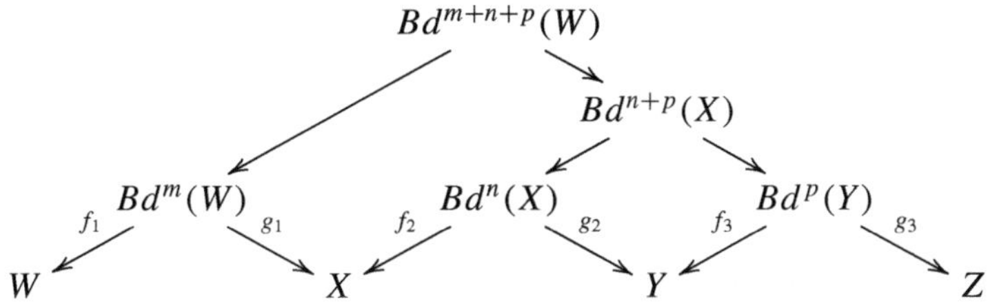

Both of these diagrams are completely determined by the larger diagram

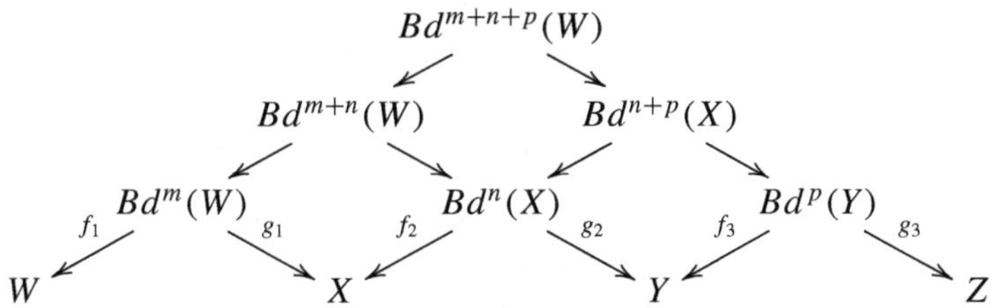

which establishes that the two compositions must therefore be the same. □

Proposition 5.19 *Simplicial rooves induce maps on the simplicial homology sequence. Specifically, if $(f, g) : X \rightsquigarrow Y$ is a simplicial roof, then this induces a map*

$$(g_*)_k \circ (f_*)_k^{-1} : H_k(X) \to H_k(Y)$$

for each k.

Proof Since f is a subdivision map by assumption, it induces isomorphisms on the simplicial homology sequence by Proposition 5.16. Therefore, $(f_*)_k$ can be inverted. Composing this with the map $(g_*)_k$ induced by the simplicial map g completes the construction.

We now need to establish that this is well defined, which is to say that if $(f', g') : X \rightsquigarrow Y$ is an equivalent simplicial roof, then $(g'_*)_k \circ (f'_*)_k^{-1} = (g_*)_k \circ (f_*)_k^{-1}$ for all k. To that end, the equivalence of these two simplicial rooves is expressed by the diagram

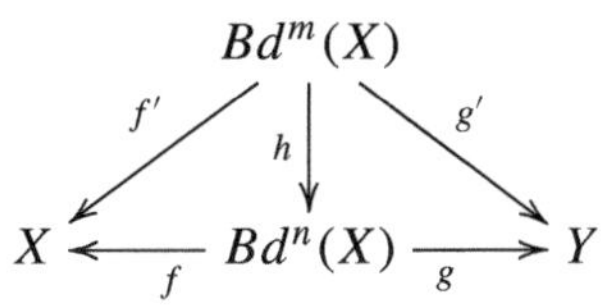

where h is a subdivision map. This means that

$$f' = f \circ h \text{ and } g' = g \circ h.$$

Therefore,

$$\begin{aligned}(g'_*)_k \circ (f'_*)_k^{-1} &= ((g \circ h)_*)_k \circ ((f \circ h)_*)_k^{-1} \\ &= (g_*)_k \circ (h_*)_k \circ ((f_*)_k \circ (h_*)_k))^{-1} \\ &= (g_*)_k \circ (h_*)_k \circ (h_*)_k^{-1} \circ (f_*)_k^{-1} \\ &= (g_*)_k \circ (f_*)_k^{-1},\end{aligned}$$

which completes the argument. □

Every simplicial map $f : X \to Y$ corresponds to a roof $(\mathrm{id}_X, f) : X \rightsquigarrow Y$, simply by not subdividing X. Therefore, we may regard **Asc** as a non-full subcategory of **BdAsc** without confusion, and imagine that **BdAsc** simply has "more" morphisms that **Asc**. What are these new morphisms? For the most part they consist of inverse morphisms for subdivision maps, as the next proposition shows.

Proposition 5.20 *Every subdivision map has an inverse in* **BdAsc**.

Proof Suppose that $c : Bd^n(X) \to X$ is a subdivision map. Regarding this as the roof $(\mathrm{id}_{Bd^n(X)}, c) : Bd^n(X) \rightsquigarrow X$, observe that there is another roof $(c, \mathrm{id}_{Bd^n(X)}) : X \rightsquigarrow Bd^n(X)$. If we form the composition $(c, \mathrm{id}_{Bd^n(X)}) \circ (\mathrm{id}_{Bd^n(X)}, c)$, this is given by the diagram

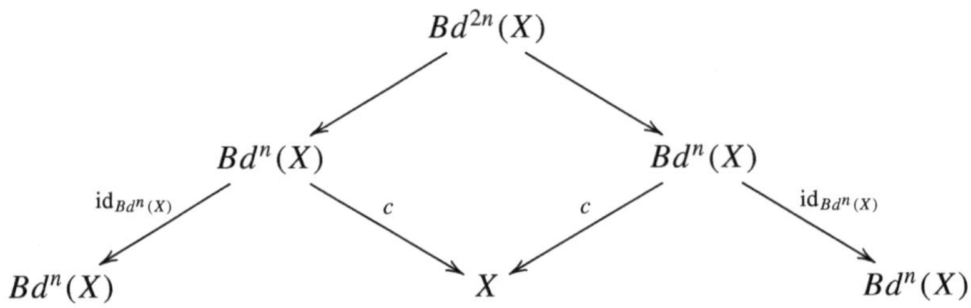

This roof is not equal to the identity roof

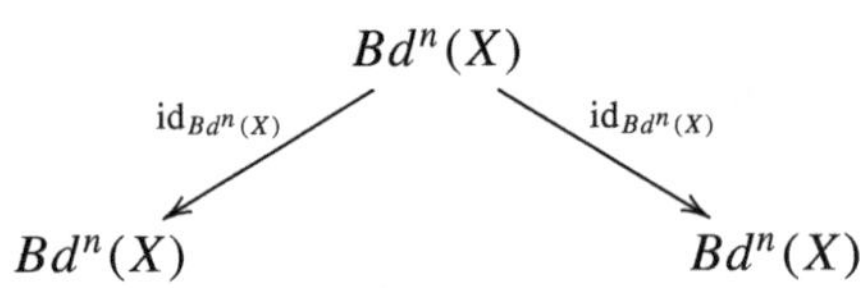

but it is equivalent to that identity roof, since all subdivision maps preserve the original vertices. Diagrammatically, this is shown by

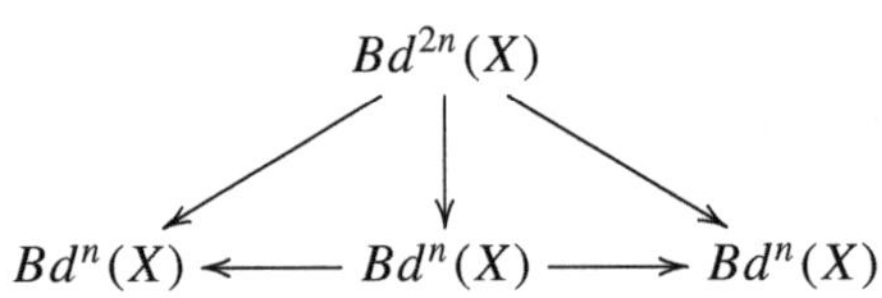

where each arrow is a subdivision map or an identity map.

By making a nearly identical argument, we find that $(\mathrm{id}_{Bd^n(X)}, c) \circ (c, \mathrm{id}_{Bd^n(X)})$ is equivalent to $(\mathrm{id}_X, \mathrm{id}_X)$ as rooves (morphisms) in **BdAsc**. □

Exercise 5.32 Show that when thought of as two objects of **BdAsc**, the two simplicial complexes discussed in Exercise 5.28 and shown in Fig. 5.18 are isomorphic.

Proposition 5.21 *If two abstract simplicial complexes are isomorphic in* **BdAsc**, *then their simplicial chain complexes are chain homotopy equivalent.*

Proof As a start, the identity roof $(\mathrm{id}_X, \mathrm{id}_X) : X \rightsquigarrow X$ induces identity maps on the simplicial homology of X according to Proposition 5.19. For the general case, suppose that X and Y are simplicial complexes for which $(f, g) : X \rightsquigarrow Y$ and $(h, k) : Y \rightsquigarrow X$ are inverses in **BdAsc**. That means that $(f, g) \circ (h, k)$ and $(h, k) \circ (f, g)$ are rooves that are equivalent to the identity roof on Y and X, respectively. According to Proposition 5.19, this means that

$$\left((g_*)_\bullet \circ (f_*)_\bullet^{-1}\right) \circ \left((k_*)_\bullet \circ (h_*)_\bullet^{-1}\right) = \mathrm{id}_{H_\bullet(Y)}$$

and

$$\left((k_*)_\bullet \circ (h_*)_\bullet^{-1}\right) \circ \left((g_*)_\bullet \circ (f_*)_\bullet^{-1}\right) = \mathrm{id}_{H_\bullet(X)},$$

which imply that

$$(g_*)_\bullet \circ (f_*)_\bullet^{-1} \text{ and } (k_*)_\bullet \circ (h_*)_\bullet^{-1}$$

are isomorphisms.

Applying Theorem 3.4 in Chap. 3 completes the argument, since quasi-isomorphic chain complexes of vector spaces are chain homotopy equivalent. □

Exercise 5.33 Does an arbitrary elementary collapse $f : X \to Y$ always become invertible in **BdAsc**? Specifically, does any roof equivalent to $(\mathrm{id}_X, f) : X \rightsquigarrow Y$ have an inverse when f is an elementary collapse? (Hint: Propose that the inverse of an elementary collapse can be written as a factorization $Y \to Bd(Y) \to X$.)

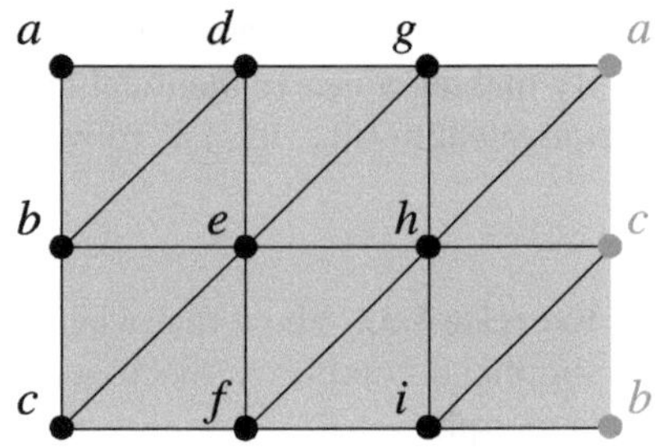

Fig. 5.21 The simplicial complex model of the Möbius strip in Exercise 5.34. Vertices shown in gray are duplicates, to indicate that simplices "reach across" from left to right

Exercise 5.34 Consider the simplicial complex model of the Möbius strip shown in Fig. 5.21.

1. List a generating set of simplices for this simplicial complex.
2. Identify any elementary collapses that are available to reduce the size of the simplicial complex.
3. Use your answers to the previous two parts to compute the simplicial homology of the simplicial complex.

Exercise 5.35 Consider the simplicial complex model of the Klein bottle shown in Fig. 5.22.

1. List a generating set of simplices for this simplicial complex.
2. Identify any elementary collapses that are available to reduce the size of the simplicial complex.
3. Use your answers to the previous two parts to compute the simplicial homology of the simplicial complex.

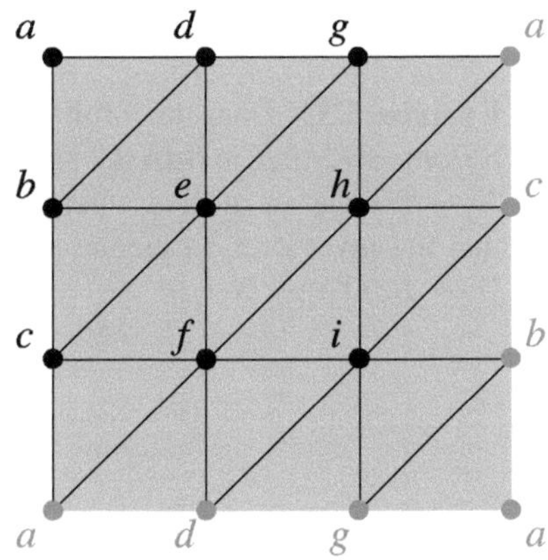

Fig. 5.22 The simplicial complex model of the Klein bottle in Exercise 5.35. Vertices shown in gray are duplicates, to indicate that simplices "reach across" from top to bottom and left to right

Exercise 5.36 Consider a directed graph $G = (V, E)$. We can use the order complex construction (Definition 4.9 in Chap. 4) to build a simplicial complex from G if we permit a vertex to appear in a path multiple times. Construct an example of a graph such that its order complex has nontrivial H_2.

5.6 Simplicial Cohomology

To every chain complex of vector spaces $(C_\bullet, c_\bullet)$ there is an associated dual chain complex $(C^\bullet, c^\bullet)$, defined by taking the vector space dual of each space in the chain complex and the dual (transpose)[5] of each linear map. We already developed the general algebraic properties of the dual chain complex, culminating in Theorem 3.5 (the Hodge theorem) in Chap. 3. The *simplicial* chain complex has more structure than a chain complex of vector spaces, so there is more that we can say here.

Let us define the *simplicial cochain complex* $(C^\bullet(X), d^\bullet) := ((C_\bullet(X))^T, (\partial_{\bullet+1})^T)$ to be the chain complex that is dual to the simplicial chain complex $(C_\bullet(X), \partial_\bullet)$ for a simplicial complex X. It is traditional to call $d^\bullet$ the *coboundary map*. As we will soon see, $d^\bullet$ has a close connection to computing differences between values of functions on X.

Elements of $C^k(X)$ can be thought of as functions that label every k-dimensional simplex of X with a real number. This is also the case for $C_k(X)$, of course! What is the difference between $C^k(X)$ and $C_k(X)$? When considered individually, there is no difference between $C^k(X)$ and $C_k(X)$, since they are isomorphic vector spaces, but together they interact.

At a glance, if τ is a k-dimensional simplex of X, then the distinction between τ and τ^T is merely notational, because

$$(\tau^T)(\sigma) = \begin{cases} 1 & \text{if } \sigma = \tau \\ 0 & \text{otherwise.} \end{cases}$$

Usually, such a function τ^T is called the *indicator function on* τ.

However, since τ^T is an element of $C^k(X) = (C_k(X))^T$, this means that τ^T is a linear function $\tau^T : C_k(X) \to \mathbb{R}$. Being an element of the dual basis means that the values of τ^T are determined by extending by linearity.

To make the calculations in this section explicit, let us assume that the isomorphism between $C^k(X)$ and $C_k(X)$ is given by $g_k : C^k(X) \to C_k(X)$, so that

$$g_k(\tau^T) = \tau$$

[5] It may be helpful to recall Definition 1.10 in Chap. 1 for the dual map, since this will be used frequently in this section.

for every k-simplex $\tau \in X$. We will write elements of $C_k(X)$ as linear combinations of bracketed lists of vertices, like $4[v_0, v_1, v_2] + 3[v_1, v_3, v_5]$, but will write elements of $C^k(X)$ as functions like

$$f([v_0, v_1, v_2]) = 4, \ \ f([v_1, v_3, v_5]) = 3.$$

For instance, suppose that an element a of $C_k(X)$ is written as an abstract linear combination of k-dimensional simplices,

$$a = \sum_{\sigma \in X^k} a_\sigma \sigma.$$

In the expression above, we interpret each k-dimensional simplex $\sigma \in X$ as being a basis vector $\sigma \in C_k(X)$. Similarly any $f \in C^k(X)$ can be written as a linear combination of functions of linear combinations of k-dimensional simplices,

$$f = \sum_{\tau \in X^k} f^\tau \tau^T,$$

in which $f^\tau \in \mathbb{R}$ and τ^T is the indicator function on τ. Therefore, since f is an element of $C^k(X)$, it too is a linear function $C_k(X) \to \mathbb{R}$, and its value on $a \in C_k(X)$ is given by

$$\begin{aligned} f(a) &= f\left(\sum_{\sigma \in X^k} a_\sigma \sigma\right) \\ &= \sum_{\sigma \in X^k} a_\sigma f(\sigma) \\ &= \sum_{\sigma \in X^k} a_\sigma \left(\sum_{\tau \in X^k} f^\tau \tau^T\right)(\sigma) \\ &= \sum_{\sigma \in X^k} a_\sigma \sum_{\tau \in X^k} f^\tau \tau^T(\sigma) \\ &= \sum_{\sigma \in X^k} a_\sigma f^\sigma. \end{aligned}$$

Example 5.30 Consider the simplicial complex X generated by the triangle $[v_0, v_1, v_2]$ shown in Fig. 5.23. Each element of $C^0(X)$ is given by real values on each of the three vertices. For instance, if f is an element of $C^0(X)$, it could assign $f([v_0]) = x$, $f([v_1]) = y$, and $f([v_2]) = z$. Then since $d^0 = (\partial_1)^T$,

$$(d^0 f)([v_0, v_1]) = f(\partial_1([v_0, v_1])) = f([v_0] - [v_1]) = f([v_0]) - f([v_1]) = x - y.$$

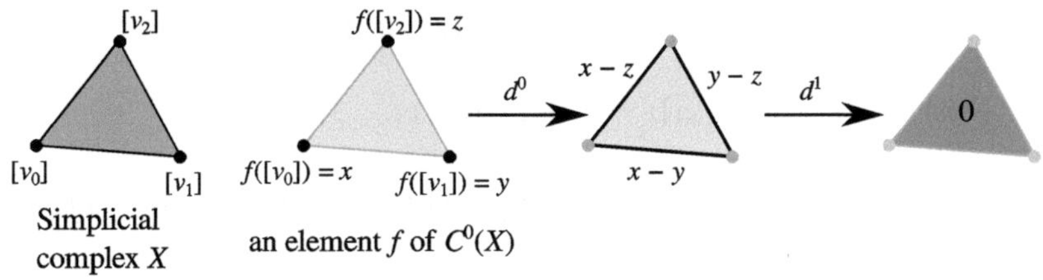

Fig. 5.23 A simplicial complex X, an element f of $C^0(X)$, and its images under d^0 and $(d^1 \circ d^0)$, in Example 5.30

In short, d^0 computes the difference between the values assigned by f to the endpoints of each edge, and places these differences on their corresponding edges, as shown in Fig. 5.23.

The fact that $(C_\bullet(X), \partial_\bullet)$ is a chain complex implies that $(C^\bullet(X), d^\bullet)$ is also a chain complex. In particular, $d^1 \circ d^0 = 0$. This can be verified directly for f, by

$$\begin{aligned}((d^1 \circ d^0)f)([v_0, v_1, v_2]) &= (d^0 f)(\partial_2([v_0, v_1, v_2]))\\ &= (d^0 f)([v_1, v_2] - [v_0, v_2] + [v_0, v_1])\\ &= (d^0 f)([v_1, v_2]) - (d^0 f)([v_0, v_2]) + (d^0 f)([v_0, v_1])\\ &= f(\partial_1([v_1, v_2])) - f(\partial_1([v_0, v_2])) + f(\partial_1([v_0, v_1]))\\ &= f([v_1] - [v_2]) - f([v_0] - [v_2]) + f([v_0] - [v_1])\\ &= f([v_1]) - f([v_2]) - f([v_0]) + f([v_2]) + f([v_0]) - f([v_1])\\ &= y - z - x + z + x - y = 0.\end{aligned}$$

This calculation also makes it clear that if g is an element of $C^1(X)$ that is outside the image of d^0, labeling the edges of X with values that are not differences between the values on their corresponding endpoints, then $d^1 g \neq 0$.

Lemma 5.11 *For $f \in C^k(X)$, $d^k f$ is the element of $C^{k+1}(X)$ given by the formula*

$$(d^k f)([v_0, \ldots, v_{k+1}]) = \sum_{i=0}^{k+1} (-1)^i f([v_0, \ldots, v_{i-1}, v_{i+1}, \ldots, v_{k+1}]),$$

for any $[v_0, \ldots, v_{k+1}] \in X$.

By a similar argument as for $\partial_k \circ \partial_{k+1} = 0$, this formula implies that $d^{k+1} \circ d^k = 0$.

Proof Notice that $d^k f$ is clearly an element of $C^{k+1}(X)$ if this formula holds. Recall that

$$\partial_{k+1}([v_0, v_1, \ldots, v_{k+1}]) = \sum_{i=0}^{k+1} (-1)^i [v_0, \ldots, v_{i-1}, v_{i+1}, \ldots, v_{k+1}].$$

We must show that $d^k = (\partial_{k+1})^T$. This is simply calculation

$$\begin{aligned}\left((\partial_{k+1})^T f\right)([v_0, v_1, \ldots, v_{k+1}]) &= f\left(\partial_{k+1}([v_0, v_1, \ldots, v_{k+1}])\right)\\ &= f\left(\sum_{i=0}^{k+1}(-1)^i[v_0, \ldots, v_{i-1}, v_{i+1}, \ldots, v_{k+1}]\right)\\ &= \sum_{i=0}^{k+1}(-1)^i f([v_0, \ldots, v_{i-1}, v_{i+1}, \ldots, v_{k+1}])\\ &= (d^k f)([v_0, \ldots, v_{k+1}]).\end{aligned}$$

□

Recall that Sect. 3.5 in Chap. 3 defined cohomology via duality. The same situation applies here.

Definition 5.19 The *simplicial cohomology* $H^\bullet(X)$ for a simplicial complex X is the cohomology of the simplicial chain complex, or equivalently, is the homology of $(C^\bullet(X), d^\bullet)$.

Corollary 5.5 *For an abstract simplicial complex X, for every k, $H^k(X) \cong H_k(X)$.*

Remark 5.2 Corollary 5.5 may be false if the simplicial chain complex is not constructed with finite dimensional vector spaces.

Example 5.31 Consider the simplicial complex Y generated by

$$[a, b], [b, c], [a, c], [a, d], [d, g], [a, g], [a, c, e], [a, c, g], [a, e, g], [c, e, g],$$

which is shown in Fig. 5.24. Its simplicial chain complex is given by

$$0 \longrightarrow C_2(Y) = \mathbb{R}^4 \xrightarrow{\partial_2} C_1(Y) = \mathbb{R}^{10} \xrightarrow{\partial_1} C_0(Y) = \mathbb{R}^6 \longrightarrow 0.$$

The homology of this chain complex is given by

$$H_0(Y) \cong \mathbb{R}, \;\; H_1(Y) \cong \mathbb{R}^2, \;\; H_2(Y) \cong \mathbb{R},$$

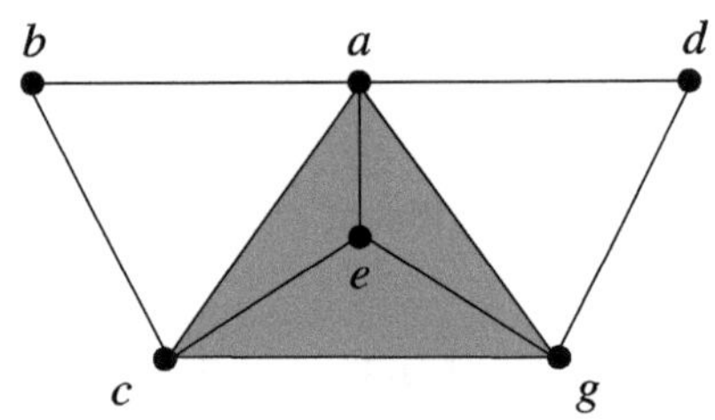

Fig. 5.24 The simplicial complex Y in Example 5.31

in which $H_0(Y)$ is generated by

$$[a],$$

$H_1(Y)$ is generated by

$$[a, b] + [b, c] - [a, c] \text{ and } [a, d] + [d, g] - [a, g],$$

and $H_2(Y)$ is generated by

$$[a, c, e] - [a, c, g] + [a, e, g] - [c, e, g].$$

Now let us consider the cohomology of Y. Applying transposition at each index to the simplicial chain complex, the simplicial cochain complex is

$$0 \longleftarrow C^2(Y) = \mathbb{R}^4 \xleftarrow{d^1} C^1(Y) = \mathbb{R}^{10} \xleftarrow{d^0} C^0(Y) = \mathbb{R}^6 \longleftarrow 0$$

This does not mean that each basis element for $H_\bullet(Y)$ translates directly to a basis element of $H^\bullet(Y)$. For instance, although $[a]$ is in coker ∂_1, $[a]^T$ is not an element of $\ker d^0$, since

$$d^0\left([a]^T\right)([a, e]) = [a]^T([e] - [a]) = -1 \neq 0.$$

To find a nontrivial basis element for $H^0(Y)$, it is useful to understand its topological interpretation. Recall that according to Proposition 5.1, $H_0(Y)$ can be interpreted as identifying path components of Y. Since there is only one path component, $[a]$ is a vertex somewhere on that path component, though any other vertex would work equally well. On the other hand, since $H_0(Y)$ consists of the length 1 bars at $C_0(Y)$, dualization will not change that. Thus $H^0(Y)$ will also correspond to path components of Y, but in a different way than $H_0(Y)$ does. To that end, consider the constant function on vertices, namely

$$\alpha := [a]^T + [b]^T + [c]^T + [d]^T + [e]^T + [f]^T + [g]^T.$$

We can compute for any edge $[x, y]$ in $C_1(Y)$,

$$\begin{aligned} d^0\alpha([x, y]) &= \alpha(\partial_1[x, y]) \\ &= \alpha([y] - [x]) \\ &= \alpha([y]) - \alpha([x]) \\ &= 1 - 1 = 0. \end{aligned}$$

Similarly, notice that the transposes of the basis elements exhibited above for $H_1(Y)$ are not elements of $\ker d^1$. The problem is with the triangles, since

$$
\begin{aligned}
d^1\left([a,b]^T+[b,c]^T\ -[a,c]^T\right)([a,c,e]) &= \left([a,b]^T+[b,c]^T-[a,c]^T\right)(\partial_2[a,c,e])\\
&= \left([a,b]^T+[b,c]^T-[a,c]^T\right)([c,e]-[a,e]+[a,c])\\
&= -1 \neq 0.
\end{aligned}
$$

The solution is to compensate for the triangles by adding a few more terms to the expression for the basis elements. Since each edge in the central portion of Y borders on two triangles, it is necessary to adjust the coefficients by a factor of $1/2$ and choose the signs carefully. This results in two elements of $C^1(Y)$,

$$
\beta := [a,b]^T+[b,c]^T-[a,c]^T-\frac{1}{2}[a,e]^T+\frac{1}{2}[c,e]^T-\frac{1}{2}[a,g]^T+\frac{1}{2}[c,g]^T,
$$

and

$$
\gamma := [a,d]^T+[d,g]^T-[a,g]^T-\frac{1}{2}[a,e]^T-\frac{1}{2}[e,g]^T-\frac{1}{2}[a,c]^T-\frac{1}{2}[c,g]^T.
$$

These are both elements of $\ker d^1$, since

$$
\begin{aligned}
d^1\beta([a,c,e]) &= \beta(\partial_2[a,c,e]) = \beta([c,e]-[a,e]+[a,c]) = \frac{1}{2}-\left(-\frac{1}{2}\right)-1=0,\\
d^1\beta([a,c,g]) &= \beta(\partial_2[a,c,g]) = \beta([c,g]-[a,g]+[a,c]) = \frac{1}{2}-\left(-\frac{1}{2}\right)-1=0,\\
d^1\beta([a,e,g]) &= \beta(\partial_2[a,e,g]) = \beta([e,g]-[a,g]+[a,e]) = 0-\left(-\frac{1}{2}\right)-\frac{1}{2}=0,\\
d^1\beta([c,e,g]) &= \beta(\partial_2[c,e,g]) = \beta([e,g]-[c,g]+[c,e]) = 0-\frac{1}{2}+\frac{1}{2}=0.
\end{aligned}
$$

The situation for γ is analogous,

$$
\begin{aligned}
d^1\gamma([a,c,e]) &= \gamma(\partial_2[a,c,e]) = \gamma([c,e]-[a,e]+[a,c]) = 0-\left(-\frac{1}{2}\right)-\frac{1}{2}=0,\\
d^1\gamma([a,c,g]) &= \gamma(\partial_2[a,c,g]) = \gamma([c,g]-[a,g]+[a,c]) = -\frac{1}{2}-(-1)-\frac{1}{2}=0,\\
d^1\gamma([a,e,g]) &= \gamma(\partial_2[a,e,g]) = \gamma([e,g]-[a,g]+[a,e]) = -\frac{1}{2}-(-1)-\frac{1}{2}=0,\\
d^1\gamma([c,e,g]) &= \gamma(\partial_2[c,e,g]) = \gamma([e,g]-[c,g]+[c,e]) = -\frac{1}{2}-\left(-\frac{1}{2}\right)+0=0.
\end{aligned}
$$

Neither β nor γ are in the image of d^0, since if $\zeta \in C^0(Y)$, then

$$
\begin{aligned}
d^0\zeta([a,b]+[b,c]-[a,c]) &= \zeta(\partial_1([a,b]+[b,c]-[a,c]))\\
&= \zeta([b]-[a]+[c]-[b]+[a]-[c])\\
&= \zeta(0) = 0.
\end{aligned}
$$

But

$$\beta([a,b]+[b,c]-[a,c]) = 3 \neq 0.$$

Therefore, we have established that β and γ are representatives of nontrivial elements of $H^1(Y)$. Moreover, since

$$\beta - \gamma = [a,b]^T + [b,c]^T - \frac{1}{2}[a,c]^T - [a,d]^T + [c,g]^T - [d,g]^T + \frac{1}{2}[a,g]^T + \frac{1}{2}[c,e]^T + \frac{1}{2}[e,g]^T$$

is nonzero when evaluated on $[a,b]+[b,c]-[a,c]$, we must conclude that β and γ are representatives of different equivalence classes in $H^1(Y)$.

Finally, regarding $H^2(Y)$, transposing the sole basis element for $H_2(Y)$ happens to work, because if $\zeta \in C^1(Y)$,

$$\begin{aligned} d^1\zeta([a,c,e]-[a,c,g]+[a,e,g]-[c,e,g]) &= \zeta(\partial_2([a,c,e]-[a,c,g] \\ &\quad + [a,e,g]-[c,e,g])) \\ &= \zeta([c,e]-[a,e]+[a,c] \\ &\quad - [c,g]+[a,g]-[a,c] \\ &\quad + [e,g]-[a,g]+[a,e] \\ &\quad - [e,g]+[c,g]-[c,e])) \\ &= 0, \end{aligned}$$

but

$$\left([a,c,e]^T - [a,c,g]^T + [a,e,g]^T - [c,e,g]^T\right)([a,c,e]-[a,c,g]+[a,e,g]-[c,e,g]) = 4 \neq 0.$$

The simplicial complex Y from Example 5.31 has isomorphic cohomology to a simplicial model X of the torus, as shown by Example 5.32 below. Ultimately these two complexes can be distinguished cohomologically using the *cup product* structure discussed in Sect. 5.7, as will be established in Example 5.40.

Example 5.32 Recall Example 4.27 in Chap. 4, in which we considered the homology of the torus. The simplicial complex model X for the torus is shown in Fig. 5.25, and its simplicial cochain complex is given by

$$0 \longrightarrow C^0(X) = \mathbb{R}^9 \xrightarrow{d^0} C^1(X) = \mathbb{R}^{27} \xrightarrow{d^1} C^2(X) = \mathbb{R}^{18} \longrightarrow 0$$

By duality and the results of Example 4.27 in Chap. 4,

$$H^0(X) \cong \mathbb{R}, \; H^1(X) \cong \mathbb{R}^2, \; H^2(X) \cong \mathbb{R}.$$

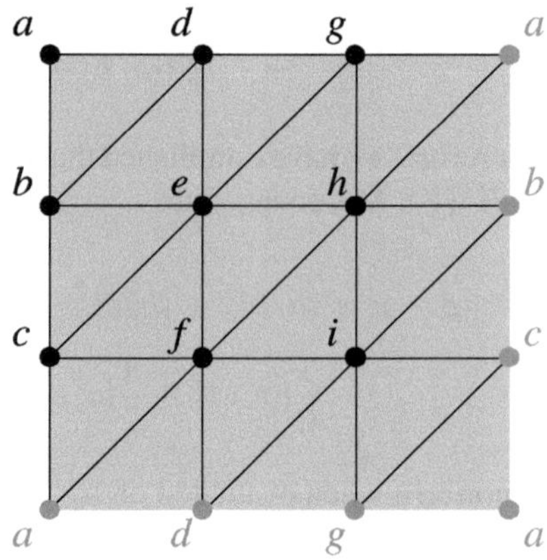

Fig. 5.25 The simplicial complex model of the torus in Example 5.32. Vertices shown in gray are duplicates, to indicate that simplices "reach across" from left to right. and top to bottom

We will express basis elements for these spaces as representatives in $C^\bullet(X)$. As in Example 5.31, directly transposing the basis elements for $H_\bullet(X)$ (which were previously computed in Example 4.27 in Chap. 4) does not work. For instance, we have that $[a, b] + [b, c] - [a, c]$ is a basis element for $H_1(X)$. Therefore, one might expect that the element of $C^1(X)$ given by

$$[a, b]^T + [b, c]^T - [a, c]^T$$

could be a basis element for $H^1(X)$. However, notice that

$$\begin{aligned}\left(d^1([a, b]^T + [b, c]^T \right. & \left. -[a, c]^T)\right)([a, b, c]) \\ &= \left([a, b]^T + [b, c]^T - [a, c]^T\right)(\partial_2[a, b, c]) \\ &= \left([a, b]^T + [b, c]^T - [a, c]^T\right)(\partial_2[b, c] - [a, c] + [a, b]) \\ &= 1 + 1 + 1 = 3 \neq 0.\end{aligned}$$

Therefore, $[a, b]^T + [b, c]^T - [a, c]^T$ is not an element of $\ker d^1$.

Following the intuition that basis elements of $H_0(X)$ select path components of X, it is similarly true that $H^0(X)$ is spanned by linear functions that are constant on each path component of X. Since X is path connected, $H^0(X)$ is spanned by the constant function, namely

$$\alpha := \sum_{v \in \{a,b,c,d,e,f,g,h,i\}} [v]^T.$$

Like the case of Example 5.31, a short calculation shows that $d^0\alpha$ is the zero function on $C^1(X)$. For example, if $[x, y] \in C^1(X)$, then

$$d^0\alpha([x, y]) = \alpha(\partial_1[x, y]) = \alpha([y]) - \alpha([x]) = 0.$$

On the other hand, any element α' of $C^0(X)$ that is not a constant function will have $d^0\alpha' \neq 0$ by the same logic.

The basis elements for $H^1(X)$ and $H^2(X)$ can be found explicitly, just as in Example 5.31. $H^1(X)$ spanned by

$$\begin{aligned}\beta :=&[a,b]^T+[b,c]^T-[a,c]^T-[b,d]^T-[c,e]^T-[a,f]^T\\&+[d,e]^T+[e,f]^T-[d,f]^T-[e,g]^T-[f,h]^T-[d,i]^T\\&+[g,h]^T+[h,i]^T-[g,i]^T+[a,h]^T+[b,i]^T+[c,g]^T,\end{aligned}$$

and

$$\begin{aligned}\gamma :=&[a,d]^T+[d,g]^T-[a,g]^T+[b,d]^T+[e,g]^T-[a,h]^T\\&+[b,e]^T+[e,h]^T-[b,h]^T+[c,e]^T+[f,h]^T-[b,i]^T\\&+[c,f]^T+[f,i]^T-[c,i]^T+[a,f]^T+[d,i]^T-[c,g]^T,\end{aligned}$$

(Notice that the first few terms of β and γ are identical to their counterparts in Example 5.31. This is intentional and plays an important role in distinguishing X and Y using cup products in Example 5.40 in Sect. 5.7.) To verify that these are nontrivial elements of $H^1(X)$, we need to verify that $d^1\beta = 0$ and β is not in the image of d^0, and likewise show the same for γ.

The direct way to show $d^1\beta = 0$ is to verify that $d^1\beta([x,y,z]) = 0$ for every 2-simplex $[x,y,z]$ in X. Although there are 18 such simplices, the construction of X is quite symmetric. Because of that, it suffices to test two simplices,

$$\begin{aligned}d^1\beta([a,b,d]) &= \beta([b,d]-[a,d]+[a,b]) = (-1)-(0)+(1) = 0\\d^1\beta([b,d,e]) &= \beta([d,e]-[b,e]+[b,d]) = (1)-(0)+(-1) = 0,\end{aligned}$$

and

$$\begin{aligned}d^1\gamma([a,b,d]) &= \gamma([b,d]-[a,d]+[a,b]) = (1)-(1)+(0) = 0\\d^1\gamma([b,d,e]) &= \gamma([d,e]-[b,e]+[b,d]) = (0)-(1)+(1) = 0.\end{aligned}$$

We can show that β and γ are outside the image of d^0 using the same idea as in Example 5.31. If $\zeta \in C^0(X)$, this means that

$$\begin{aligned}d^0\zeta([a,b]+[b,c]-[a,c]) &= \zeta\,(\partial_1([a,b]+[b,c]-[a,c]))\\&= \zeta([b]-[a]+[c]-[b]+[a]-[c]) = \zeta(0) = 0,\\d^0\zeta([a,d]+[d,g]-[a,g]) &= \zeta\,(\partial_1([a,d]+[d,g]-[a,g]))\\&= \zeta([d]-[a]+[g]-[d]+[a]-[g]) = \zeta(0) = 0.\end{aligned}$$

However,

$$\begin{aligned}\beta([a,b]+[b,c]-[a,c]) &= 1+1+-(-1) = 3 \neq 0,\\\gamma([a,d]+[d,g]-[a,g]) &= 1+1+-(-1) = 3 \neq 0,\end{aligned}$$

so neither β nor γ are the image of any $\zeta \in C^0(X)$.

Lastly, $H^2(X)$ is spanned by

$$\begin{aligned}\delta :=&[a,b,d]^T + [d,e,g]^T + [a,g,h]^T\\ &+ [b,c,e]^T + [e,f,h]^T + [b,h,i]^T\\ &- [a,c,f]^T - [d,f,i]^T - [c,g,i]^T\\ &- [b,d,e]^T - [e,g,h]^T - [a,b,h]^T\\ &- [c,e,f]^T - [f,h,i]^T - [b,c,i]^T\\ &+ [a,d,f]^T + [d,g,i]^T + [a,c,g]^T.\end{aligned}$$

We merely need to establish that δ is not the image of any $\zeta \in C^1(X)$. Notice that δ is the transpose of the following element of $C_2(X)$

$$\begin{aligned}\Delta :=&\ [a,b,d] + [d,e,g] + [a,g,h]\\ &+ [b,c,e] + [e,f,h] + [b,h,i]\\ &- [a,c,f] - [d,f,i] - [c,g,i]\\ &- [b,d,e] - [e,g,h] - [a,b,h]\\ &- [c,e,f] - [f,h,i] - [b,c,i]\\ &+ [a,d,f] + [d,g,i] + [a,c,g].\end{aligned}$$

This element is special because

$$\begin{aligned}\partial_2(\Delta) =& +[b,d] - [a,d] + [a,b] + [e,g] - [d,g] + [d,e] + [g,h] - [a,h] + [a,g]\\ &+ [c,e] - [b,e] + [b,c] + [f,h] - [e,h] + [e,f] + [h,i] - [b,i] + [b,h]\\ &- [c,f] + [a,f] - [a,c] - [f,i] + [d,i] - [d,f] - [g,i] + [c,i] - [c,g]\\ &- [d,e] + [b,e] - [b,d] - [g,h] + [e,h] - [e,g] - [b,h] + [a,h] - [a,b]\\ &- [e,f] + [c,f] - [c,e] - [h,i] + [f,i] - [f,h] - [c,i] + [b,i] - [b,c]\\ &+ [d,f] - [a,f] + [a,d] + [g,i] - [d,i] + [d,g] + [c,g] - [a,g] + [a,c]\\ =&\ 0.\end{aligned}$$

We must therefore have $d^1\zeta(\Delta) = 0$ for any $\zeta \in C^1(X)$. On the other hand,

$$\delta(\Delta) = 18 \neq 0.$$

As should be expected at this point, the simplicial cochain complex $C^\bullet(X)$ is functorial, which is to say that simplicial maps induce chain maps between their respective cochain complexes. Because the vector space dual is a contravariant functor, this contravariance gets carried through to the simplicial cochain complex as well.

Proposition 5.22 *The simplicial cochain complex is a contravariant functor* $C^\bullet$: **Asc** $\to$ **Kom**, *in which each simplicial map* $f : X \to Y$ *induces a chain map* $f^\bullet : C^\bullet(Y) \to C^\bullet(X)$ *given by*

$$(f^k a)(\sigma) := a(f_k(\sigma))$$

for each $a \in C^k(Y)$ *and each* k*-dimensional simplex* $\sigma \in X$.

Proof The definition of $(f^\bullet a)$ is simply an application of the definition of the dual map. If $\sigma \in X$, then $f_k(\sigma)$ is indeed an element of $C_k(Y)$ because f is a simplicial map. If $a \in C^k(Y)$ as the hypothesis indicates, these facts imply that $a(f_k(\sigma))$ is well defined, making $(f^k a)$ a linear function $C_k(X) \to \mathbb{R}$, which is an element of $C^k(X)$. □

The following Corollary now follows immediately from Proposition 3.10 in Chap. 3.

Corollary 5.6 *Simplicial cohomology* H^k *is a contravariant functor* **Asc** $\to$ **Vec**.

Exercise 5.37 Consider the simplicial complexes X and Y shown in Fig. 5.26, where X is generated by $[a, b, c], [a, d], [c, d]$ and Y is generated by $[a', b', c']$, and the simplicial map $m : X \to Y$ defined by

$$m(a) := a', \; m(b) := b', \; m(c) := c', \; m(d) := c'.$$

Consider the simplicial cochain $f \in C^1(Y)$ given by extending the following by linearity,

$$f([a', b']) := 1, \; f([a', c']) := 3, \; f([b', c']) := 2,$$

as is also shown in Fig. 5.26. Compute the simplicial cochain $m^\bullet(f) \in C^1(X)$.

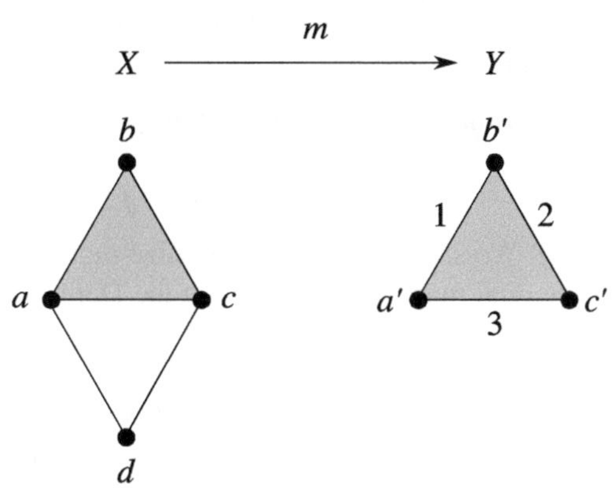

Fig. 5.26 The simplicial map $m : X \to Y$ in Exercise 5.37. Vertices are labeled with letters, the cochain in $f \in C^1(Y)$ is shown by its numerical values on the relevant simplices

Given that an element of $a \in C_k(X)$ is a linear combination of k-simplices, and an element $f \in C^k(X)$ is a function on all k-simplices, it is useful to interpret the evaluation $f(a)$ as evaluating the integral of f along a,

$$f(a) = \int_a f.$$

This notation is suggestive, because the expression is linear in f since $C^k(X)$ is a vector space. Specifically, this leads to the mnemonic

$$(f+g)(a) = \int_a (f+g) = \int_a f + \int_a g = f(a) + g(a).$$

Additionally, reversing the orientation of a path $a \in C_1(X)$ is achieved by switching its sign

$$-\int_a f = \int_{-a} f = \int_a (-f).$$

Exercise 5.38 Verify that the interpretation of evaluating $f(a)$ for $f \in C^k(X)$ and $a, b \in C_k(X)$ satisfies the property that

$$\int_{a+b} f = \int_a f + \int_b f.$$

Explain how the pairing between $C^k(X)$ and $C_k(X)$ can be thought of as integrating this function against some weighted k-dimensional surface.

If we interpret the relationship between $C_k(X)$ and $C^k(X)$ as integration, the boundary and coboundary maps are related through a simplicial analog of Stokes' theorem.

Proposition 5.23 *(simplicial Stokes' theorem) If $a \in C_k(X)$ and $f \in C^{k-1}(X)$, then*

$$(d^{k-1}f)(a) = f(\partial_k a).$$

Proof We simply need to recognize that since $d^{k-1} = (\partial_k)^T$, this means that d^{k-1} is the dual map of ∂_k. Following the Definition 1.10 in Chap. 1 of the dual map, we have

$$\begin{aligned}(d^{k-1}f)(a) &= \left((\partial_k)^T f\right)(a)\\ &= f(\partial_k a).\end{aligned}$$

□

Using more suggestive notation, we might write Proposition 5.23 as

$$\int_a d^{k-1} f = \int_{\partial_k a} f.$$

When exploring the relationship between the simplicial chain complex $C_\bullet(X)$ and its dual cochain complex $C^\bullet(X)$ for an abstract simplicial complex X, it is useful to recall the Hodge Laplacian $L^k : C^k(X) \to C^k(X)$ given in Definition 3.8 of Chap. 3 by

$$L^k = g_k^{-1} \circ \partial_{k+1} \circ g_{k+1} \circ d^k - d^{k-1} \circ g_{k-1}^{-1} \circ \partial_k \circ g_k,$$

where $g_k : C^k(X) \to C_k(X)$ is an isomorphism relating the basis of k-dimensional simplices in $C_k(X)$ to their duals in $C^k(X)$. (Recall that g_k is not unique!)

As a trivial case, consider a simplicial complex that contains only vertices and no simplices of higher dimension. For such a simplicial complex, the Hodge Laplacians are all zero maps. For graphs, which can be thought of as 1-dimensional simplicial complexes, the Hodge 0-Laplacian has more interesting structure. It is called the *graph Laplacian*, because it reduces to

$$L^0 = g_0^{-1} \circ \partial_1 \circ g_1 \circ d^0,$$

which depends only on the boundary and coboundary maps for edges. In particular, L^0 relates values of a function on vertices to each other based on the edges connecting them.

Exercise 5.39 If we write elements of $C^k(X)$ as column vectors, and take g_k to be simply the matrix transpose, show that the graph Laplacian L^0 has the matrix representation

$$L^0 = \partial_1 \circ (\partial_1)^T.$$

Notice that since $L^0 : C^0(X) \to C^0(X)$, we should interpret L^0 as transforming a real-valued function of the vertices of X into another real-valued function of the vertices. It is enlightening to construct an explicit formula for the graph Laplacian. Suppose that $f \in C^0(X)$ is some real valued function on the vertices of X. We can then compute $(L^0 f)(v)$ for any vertex $v \in V_X$, using the definition,

$$(L^0 f)(v) = \left(\left(g_0^{-1} \circ \partial_1 \circ g_1 \circ d^0\right)(f)\right)(v).$$

In what follows, we will assume a vertex partial order in which $v_i < v_j$ and will evaluate the expression for $L^0 f$ in stages. We begin with the value of $d^0 f$ on an edge $[v_i, v_j]$ of X,

$$(d^0 f)([v_i, v_j]) = f([v_j]) - f([v_i]).$$

The function $d^0 f$ is a function on edges, which corresponds to an element of $C_1(X)$ by way of a linear combination,

$$(g_1(d^0 f)) = \sum_{[v_i, v_j] \in X} \big(f([v_j]) - f([v_i])\big)\,[v_i, v_j].$$

With these facts in hand, we can compute the next step in the formula,

$$\begin{aligned}
\partial_1(g_1(d^0 f)) &= \partial_1 \left(\sum_{i=1}^{n} \left(\sum_{j=i+1:\,[v_i,v_j]\in X}^{n} \big(f([v_j]) - f([v_i])\big)\,[v_i, v_j] \right) \right) \\
&= \sum_{i=1}^{n} \left(\sum_{j=i+1:\,[v_i,v_j]\in X}^{n} \big(f([v_j]) - f([v_i])\big)\,\partial_1([v_i, v_j]) \right) \\
&= \sum_{i=1}^{n} \left(\sum_{j=i+1:\,[v_i,v_j]\in X}^{n} \big(f([v_j]) - f([v_i])\big)\,([v_j] - [v_i]) \right) \\
&= \sum_{i=1}^{n} \left(\sum_{\substack{j=i+1: \\ [v_i,v_j]\in X}}^{n} \big(f([v_j])[v_j] - f([v_j])[v_i] - f([v_i])[v_j] + f([v_i])[v_i]\big) \right).
\end{aligned}$$

Recognize that the first and last terms inside the sum are duplicated according to the degree of the vertex v_i, so

$$\partial_1(g_1(d^0 f)) = \sum_{i=1}^{n} \left(\deg v_i\, f([v_i])[v_i] - \sum_{j=i+1:\,[v_i,v_j]\in X}^{n} \big(f([v_j])[v_i] + f([v_i])[v_j]\big) \right).$$

The remaining two terms inside the second sum are basically of the same kind, but account for the fact that an edge $[v_i, v_j]$ has two endpoints. The vertex partial order ensures that this edge is counted once in the sum, but both endpoints count! We can split and rearrange the last sum,

$$\begin{aligned}
\partial_1(g_1(d^0 f)) = &\sum_{i=1}^{n} \left(\deg v_i\, f([v_i])[v_i] - \sum_{j=i+1:\,[v_i,v_j]\in X}^{n} f([v_j])[v_i] \right) \\
&- \sum_{i=1}^{n} \left(\sum_{j=i+1:\,[v_i,v_j]\in X}^{n} f([v_i])[v_j] \right)
\end{aligned}$$

$$= \sum_{i=1}^{n} \left(\deg v_i \, f([v_i])[v_i] - \sum_{j=i+1:\, [v_i,v_j]\in X}^{n} f([v_j])[v_i] \right)$$
$$- \sum_{j=1}^{n} \left(\sum_{i=j+1:\, [v_j,v_i]\in X}^{n} f([v_j])[v_i] \right),$$

and reindex it before finally recombining,

$$\begin{aligned}
\partial_1(g_1(d^0 f)) &= \sum_{i=1}^{n} \left(\deg v_i \, f([v_i])[v_i] - \sum_{j=i+1:\, [v_i,v_j]\in X}^{n} f([v_j])[v_i] \right) \\
&\quad - \sum_{i=1}^{n} \left(\sum_{j=1:\, [v_j,v_i]\in X}^{i-1} f([v_j])[v_i] \right) \\
&= \sum_{i=1}^{n} \left(\deg v_i \, f([v_i])[v_i] - \sum_{\substack{j=i+1: \\ [v_i,v_j]\in X}}^{n} f([v_j])[v_i] - \sum_{\substack{j=1: \\ [v_j,v_i]\in X}}^{i-1} f([v_j])[v_i] \right) \\
&= \sum_{i=1}^{n} \left(\deg v_i \, f([v_i])[v_i] - \sum_{\substack{j=1:\, [v_i,v_j]\in X \\ \text{or } [v_j,v_i]\in X}}^{n} f([v_j])[v_i] \right) \\
&= \sum_{i=1}^{n} [v_i] \left(\deg v_i \, f([v_i]) - \sum_{\substack{j=1:\, [v_i,v_j]\in X \\ \text{or } [v_j,v_i]\in X}}^{n} f([v_j]) \right).
\end{aligned}$$

The result of this calculation is that we have a succinct formula for the graph Laplacian,

$$(L^0 f)(v) = \deg_v f([v]) - \sum_{w\in V_X:[v,w]\in X \text{ or } [w,v]\in X} f([w]). \tag{5.5}$$

Recalling that harmonic elements are those in the kernel of L^k, harmonic elements of $C^0(X)$ consist of functions f on vertices that satisfy

$$f([v]) = \frac{1}{\deg_v} \sum_{w\in V_X:[v,w]\in X \text{ or } [w,v]\in X} f([w]).$$

Rather succinctly, harmonic elements are those functions on the vertices of a graph for which the value on any vertex is the mean of the values on adjacent vertices.

Example 5.33 Consider the graph X generated by $[a, b]$, $[b, c]$, $[d, e]$ as shown in Fig. 5.27a. Interpreted as a simplicial complex, the simplicial chain complex of X

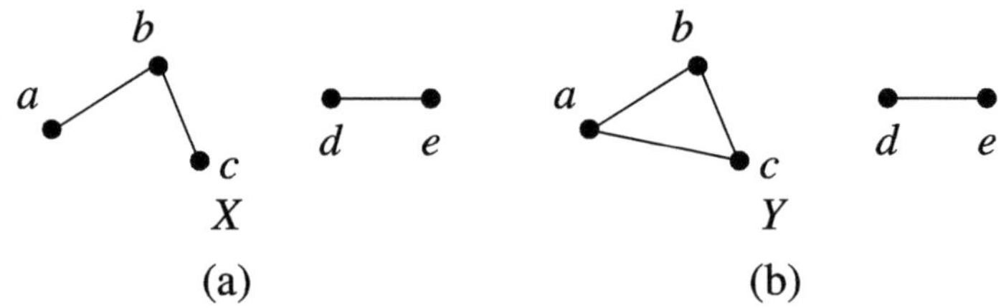

Fig. 5.27 **a** The graph X generated by $[a, b]$, $[b, c]$, $[d, e]$ in Example 5.33 and **b** the graph Y generated by $[a, b]$, $[b, c]$, $[a, c]$, $[d, e]$ in Example 5.34

has only one nontrivial boundary map, namely

$$\partial_1 = \begin{pmatrix} -1 & 0 & 0 \\ 1 & -1 & 0 \\ 0 & 1 & 0 \\ 0 & 0 & -1 \\ 0 & 0 & 1 \end{pmatrix},$$

assuming we use the vertex partial order $a < b < c < d < e$.

If we use the isomorphism $g_0 : (\mathbb{R}^5)^T \to \mathbb{R}^5$ given by matrix transpose, then we can simply construct the Laplacians from the definition.

$$L^0 = g_0^{-1} \circ \partial_1 \circ g_1 \circ d^0 = \begin{pmatrix} -1 & 0 & 0 \\ 1 & -1 & 0 \\ 0 & 1 & 0 \\ 0 & 0 & -1 \\ 0 & 0 & 1 \end{pmatrix} \begin{pmatrix} -1 & 1 & 0 & 0 & 0 \\ 0 & -1 & 1 & 0 & 0 \\ 0 & 0 & 0 & -1 & 1 \end{pmatrix} = \begin{pmatrix} 1 & -1 & 0 & 0 & 0 \\ -1 & 2 & -1 & 0 & 0 \\ 0 & -1 & 1 & 0 & 0 \\ 0 & 0 & 0 & 1 & -1 \\ 0 & 0 & 0 & -1 & 1 \end{pmatrix}$$

and

$$L^1 = -d^0 \circ g_0^{-1} \circ \partial_1 \circ g_1 = -\begin{pmatrix} -1 & 1 & 0 & 0 & 0 \\ 0 & -1 & 1 & 0 & 0 \\ 0 & 0 & 0 & -1 & 1 \end{pmatrix} \begin{pmatrix} -1 & 0 & 0 \\ 1 & -1 & 0 \\ 0 & 1 & 0 \\ 0 & 0 & -1 \\ 0 & 0 & 1 \end{pmatrix} = \begin{pmatrix} -2 & 1 & 0 \\ 1 & -2 & 0 \\ 0 & 0 & -2 \end{pmatrix}.$$

The formula (5.5) for L^0 is consistent with the above matrix representation because vertices a, c, d, and e are all of degree 1, while vertex b is of degree 2. This matches the pattern of values $1, 2, 1, 1, 1$ on the diagonal of L^0. Furthermore, the off-diagonal entries are all 0 or -1 as required by (5.5), and simply record the adjacent vertices.

The kernel of L^0 is given by

$$\ker L^0 = \text{span}\left\{\begin{pmatrix}1\\1\\1\\0\\0\end{pmatrix}, \begin{pmatrix}0\\0\\0\\1\\1\end{pmatrix}\right\}.$$

This implies that the 0-harmonic functions are locally constant functions, spanned by a function supported on $[a]$, $[b]$, $[c]$ and a function supported on $[d]$, $[e]$. In other words, the harmonic functions are locally constant functions: functions that are constant on each path component of X. According to the Hodge theorem, this means that $\beta_0 = 2$, which aligns with our interpretation of being the number of path components of X.

The kernel of L^1 is trivial, which means that there are no loops in the graph.

Example 5.34 Consider the graph Y generated by $[a, b]$, $[b, c]$, $[a, c]$, $[d, e]$, which has path components (one with a loop) as shown in Fig. 5.27b. The simplicial chain complex only consists of

$$\partial_1 = \begin{pmatrix}-1 & 0 & -1 & 0\\ 1 & -1 & 0 & 0\\ 0 & 1 & 1 & 0\\ 0 & 0 & 0 & -1\\ 0 & 0 & 0 & 1\end{pmatrix}.$$

We construct the Laplacians from their definitions,

$$\begin{aligned}
L^0 &= g_0^{-1} \circ \partial_1 \circ g_1 \circ d^0\\
&= \begin{pmatrix}-1 & 0 & -1 & 0\\ 1 & -1 & 0 & 0\\ 0 & 1 & 1 & 0\\ 0 & 0 & 0 & -1\\ 0 & 0 & 0 & 1\end{pmatrix}\begin{pmatrix}-1 & 1 & 0 & 0 & 0\\ 0 & -1 & 1 & 0 & 0\\ -1 & 0 & 1 & 0 & 0\\ 0 & 0 & 0 & -1 & 1\end{pmatrix}\\
&= \begin{pmatrix}2 & -1 & -1 & 0 & 0\\ -1 & 2 & -1 & 0 & 0\\ -1 & -1 & 2 & 0 & 0\\ 0 & 0 & 0 & 1 & -1\\ 0 & 0 & 0 & -1 & 1\end{pmatrix}
\end{aligned}$$

and

$$L^1 = -d^0 \circ g_0^{-1} \circ \partial_1 \circ g_1$$
$$= -\begin{pmatrix} -1 & 1 & 0 & 0 & 0 \\ 0 & -1 & 1 & 0 & 0 \\ -1 & 0 & 1 & 0 & 0 \\ 0 & 0 & 0 & -1 & 1 \end{pmatrix} \begin{pmatrix} -1 & 0 & -1 & 0 \\ 1 & -1 & 0 & 0 \\ 0 & 1 & 1 & 0 \\ 0 & 0 & 0 & -1 \\ 0 & 0 & 0 & 1 \end{pmatrix}$$
$$= \begin{pmatrix} -2 & 1 & -1 & 0 \\ 1 & -2 & -1 & 0 \\ -1 & -1 & -2 & 0 \\ 0 & 0 & 0 & -2 \end{pmatrix}.$$

The formula (5.5) for L^0 is consistent with the above matrix representation because vertices d and e are both of degree 1, while vertices a, b, c are of degree 2. This matches the pattern of values 2, 2, 2, 1, 1 on the diagonal of L^0. The off-diagonal entries are all -1 wherever there is an adjacency, as required by (5.5).

As in Example 5.33, the kernel of L^0 is given by

$$\ker L^0 = \operatorname{span}\left\{ \begin{pmatrix} 1 \\ 1 \\ 1 \\ 0 \\ 0 \end{pmatrix}, \begin{pmatrix} 0 \\ 0 \\ 0 \\ 1 \\ 1 \end{pmatrix} \right\},$$

since the path components consist of exactly the same sets of vertices.

The kernel of L^1 is given by

$$\ker L^1 = \operatorname{span}\left\{ \begin{pmatrix} -1 \\ -1 \\ 1 \\ 0 \end{pmatrix} \right\},$$

which corresponds to the oriented loop

$$-[a,b] - [b,c] + [a,c].$$

Notice the sign change on $[a,c]$, which indicates that the edge to be traversed in the opposite direction from the others.

Exercise 5.40 Construct the graph Laplacians for the graphs W and Z as defined below. Diagonalize the graph Laplacians, and interpret their eigenvectors in terms of the graph.

1. W is generated by $[a, b], [b, c], [c, d], [d, e], [e, f]$,
2. Z is generated by $[a, b], [b, c], [c, d], [d, e], [e, f], [d, f]$.

Let us now revisit the general k-Laplacian L^k, given by

$$L^k = g_k^{-1} \circ \partial_{k+1} \circ g_{k+1} \circ d^k - d^{k-1} \circ g_{k-1}^{-1} \circ \partial_k \circ g_k.$$

In the case of $k = 0$, we have found that the Laplacian computes the difference between a value on a vertex and the mean value of its neighboring vertices. Intuitively, L^0 measures how far a given function on vertices is from being a constant function.

For a general k, the Laplacian is a linear map $C^k(X) \to C^k(X)$, and so by the formula above, must form linear combinations of values on neighboring k-simplices. Because of Theorem 3.5 (the Hodge theorem) of Chap. 3, we already know that the kernel of L^k is isomorphic of $H^k(X)$ (and by Corollary 5.6 this is isomorphic to $H_k(X)$). Therefore, L^k can be interpreted as computing the difference between an arbitrary k-cochain and $H^k(X)$. Briefly, L^k measures how far an arbitrary k-cochain is from being a k-cohomology class.

Example 5.35 Consider the simplicial complex X generated by $[a, b]$,$[a, c]$,$[b, c]$, $[c, d]$, $[b, d]$. In this case, the formula for L^1 simplifies to

$$\begin{aligned} L^1 &= -d^{k-1} \circ g_{k-1}^{-1} \circ \partial_k \circ g_k \\ &= -(\partial_k)^T \circ \partial_k, \end{aligned}$$

where we have chosen to use the standard bases so that $d^{k-1} = (\partial_k)^T$.

It is then easy to see that using the vertex order $a < b < c < d$,

$$\begin{aligned} L^1 &= - \begin{pmatrix} -1 & 1 & 0 & 0 \\ -1 & 0 & 1 & 0 \\ 0 & -1 & 1 & 0 \\ 0 & -1 & 0 & 1 \\ 0 & 0 & -1 & 1 \end{pmatrix} \begin{pmatrix} -1 & -1 & 0 & 0 & 0 \\ 1 & 0 & -1 & -1 & 0 \\ 0 & 1 & 1 & 0 & -1 \\ 0 & 0 & 0 & 1 & 1 \end{pmatrix} \\ &= \begin{pmatrix} -2 & -1 & 1 & 1 & 0 \\ -1 & -2 & -1 & 0 & 1 \\ 1 & -1 & -2 & -1 & 1 \\ 1 & 0 & -1 & -2 & -1 \\ 0 & 1 & 1 & -1 & -2 \end{pmatrix}. \end{aligned}$$

Now consider a 1-cochain given by

$$f([a, b]) = 1,\ \ f([a, c]) = -1,\ \ f([b, c]) = x,\ \ f([b, d]) = 1,\ \ f([c, d]) = -1,$$

for some x. We can then compute

$$L^1 f = \begin{pmatrix} -2 & -1 & 1 & 1 & 0 \\ -1 & -2 & -1 & 0 & 1 \\ 1 & -1 & -2 & -1 & 1 \\ 1 & 0 & -1 & -2 & -1 \\ 0 & 1 & 1 & -1 & -2 \end{pmatrix} \begin{pmatrix} 1 \\ -1 \\ x \\ 1 \\ -1 \end{pmatrix}$$

$$= x \begin{pmatrix} 1 \\ -1 \\ -2 \\ -1 \\ 1 \end{pmatrix},$$

which clearly only vanishes if $x = 0$. On the other hand, notice that if $x = 0$, then

$$\partial_1 \begin{pmatrix} 1 \\ -1 \\ 0 \\ 1 \\ -1 \end{pmatrix} = \begin{pmatrix} -1 & -1 & 0 & 0 & 0 \\ 1 & 0 & -1 & -1 & 0 \\ 0 & 1 & 1 & 0 & -1 \\ 0 & 0 & 0 & 1 & 1 \end{pmatrix} \begin{pmatrix} 1 \\ -1 \\ 0 \\ 1 \\ -1 \end{pmatrix} = \begin{pmatrix} 0 \\ 0 \\ 0 \\ 0 \end{pmatrix},$$

so this means that f is in the cokernel of $d^0 = (\partial_1)^T$, hence it is a nontrivial cohomology class precisely when $x = 0$.

Exercise 5.41 Consider the simplicial complex generated by $[a, b]$, $[a, c]$, $[c, d, e]$, $[b, d, e]$, $[b, c, e]$, $[b, c, d]$. Assuming that $g_\bullet$ maps are all identity maps,

1. Sketch the simplicial complex.
2. Construct the matrix representation for L^0, L^1, and L^2.
3. Find the nontrivial elements of the kernel of L^1.

5.7 Cup Products

The structure exposed by the Hodge k-Laplacian of the simplicial chain complex is useful but is somewhat limited. Although each element of $f \in C^k(X)$ is a linear function that is determined by its values on each k-simplex σ of X, the coboundary maps do not directly relate these values to those on simplices whose dimension is larger than $k + 1$ or smaller than $k - 1$. The *cup product* is an algebraic way to capture some of the relationships between simplices whose dimensions differ by more than 1. While cup products initially feel a bit like getting something for nothing, they are not defined for general chain complexes. An abstract simplicial complex has a bit more structure than what appears in its simplicial cohomology, and this extra structure is what the cup product captures. Much of this structure is hidden when we consider the simplicial cochain complex, as the cleanest properties of the cup product apply to simplicial cohomology only.

If we consider a simplex $\sigma \in C^{p+q}(X)$ given by its vertices (using a vertex partial ordering $v_0 < v_1 < \cdots < v_{p+q}$)

$$\sigma = [v_0, v_1, \ldots, v_{p+q}],$$

then the fact that X is an abstract simplicial complex means that both

$$[v_0, \ldots, v_p] \text{ and } [v_p, \ldots, v_{p+q}]$$

are also simplices of X. We can turn this information around to define the *cup product* $a \cup b$ of $a \in C^p(X)$ and $b \in C^q(X)$ as the element of $C^{p+q}(X)$ given by

$$(a \cup b)([v_0, v_1, \ldots, v_{p+q}]) := a([v_0, \ldots, v_p])b([v_p, \ldots, v_{p+q}]).$$

Proposition 5.24 *Suppose that* $a \in C^p(X)$, $b \in C^q(X)$, $c \in C^q(X)$, *and* $d \in C^r(X)$. *The cup product is*

Associative $(a \cup b) \cup d = a \cup (b \cup d)$,
Distributive $a \cup (b + c) = (a \cup b) + (a \cup c)$ *and* $(b + c) \cup a = (b \cup a) + (c \cup a)$,

and it has a

Multiplicative unit $a = a \cup \left(\sum_{v \in V_X} [v]^T\right) = \left(\sum_{v \in V_X} [v]^T\right) \cup a$.

When combined with the fact that $C^\bullet(X)$ is a vector space, these axioms are those of a *graded ring*.

Proof Associative Observe that $(a \cup b) \cup d$ and $a \cup (b \cup d)$ are both elements of $C^{(p+q)+r}(X) = C^{p+(q+r)}(X)$. Given a simplex $[v_0, \ldots, v_{p+q+r}]$, we have

$$\begin{aligned}
((a \cup b) \cup d)([v_0, \ldots, v_{p+q+r}]) &= (a \cup b)([v_0, \ldots, v_{p+q}])d([v_{p+q}, \ldots, v_{p+q+r}]) \\
&= \big(a([v_0, \ldots, v_p])b([v_p, \ldots, v_{p+q}])\big)\, d([v_{p+q}, \ldots, v_{p+q+r}]) \\
&= a([v_0, \ldots, v_p])\, \big(b([v_p, \ldots, v_{p+q}])d([v_{p+q}, \ldots, v_{p+q+r}])\big) \\
&= a([v_0, \ldots, v_p])(b \cup d)([v_p, \ldots, v_{p+q+r}]) \\
&= (a \cup (b \cup d))([v_0, \ldots, v_{p+q+r}]).
\end{aligned}$$

Distributive This is simply computation. For the first equality,

$$(a \cup (b + c))([v_0, \ldots, v_{p+q}]) = a([v_0, \ldots, v_p])(b + c)([v_p, \ldots, v_{p+q}])$$

$$\begin{aligned}
&= a([v_0, \ldots, v_p])(b([v_p, \ldots, v_{p+q}]) + c([v_p, \ldots, v_{p+q}])) \\
&= a([v_0, \ldots, v_p])b([v_p, \ldots, v_{p+q}] \\
&\quad + a([v_0, \ldots, v_p])c([v_p, \ldots, v_{p+q}]) \\
&= (a \cup b)([v_0, \ldots, v_{p+q}]) + (a \cup c)([v_0, \ldots, v_{p+q}]).
\end{aligned}$$

For the second equality

$$\begin{aligned}
((b+c) \cup a)([v_0, \ldots, v_{p+q}])a &= (b+c)([v_0, \ldots, v_q])a([v_q, \ldots, v_{p+q}]) \\
&= (b([v_0, \ldots, v_q]) + c([v_0, \ldots, v_q]))a([v_q, \ldots, v_{p+q}]) \\
&= b([v_0, \ldots, v_q])a([v_q, \ldots, v_{p+q}]) \\
&\quad + c([v_0, \ldots, v_q])a([v_q, \ldots, v_{p+q}]) \\
&= (b \cup a)([v_0, \ldots, v_{p+q}]) + (c \cup a)([v_0, \ldots, v_{p+q}]).
\end{aligned}$$

Multiplicative unit For each vertex $v \in V_X$, $[v]^T$ is an element of $C^0(X)$. The notation is helpful here, since $[v]$ is the 0-dimensional simplex consisting of just v. For a k-simplex σ, its dual $\sigma^T \in C^k(X)$ is the indicator function on σ, the linear map that takes the value 1 on σ and 0 on all other k-simplices. This means that the cup product $a \cup [v]^T$ is an element of $C^{p+0}(X) = C^p(X)$. We then compute

$$\begin{aligned}
\left(a \cup \left(\sum_{v \in V_X} [v]^T\right)\right)([v_0, \ldots, v_p]) &= a([v_0, \ldots, v_p]) \left(\sum_{v \in V_X} [v]^T\right)([v_p]) \\
&= a([v_0, \ldots, v_p]) \left(\sum_{v \in V_X} [v]^T([v_p])\right) \\
&= a([v_0, \ldots, v_p]) \left([v_p]^T([v_p])\right) \\
&= a([v_0, \ldots, v_p]).
\end{aligned}$$

The other equality follows *mutatis mutandis*.

□

Remark 5.3 The cup product is defined on simplicial cochains, rather than simplicial chains. While it is possible to define a dual construction using just the simplices, which requires studying the prism PX, the resulting object does not behave algebraically as a product. In any event, we can define the cup product as the composition of a map $m^\bullet : C^\bullet(X) \oplus C^\bullet(X) \to C^\bullet(X \times X)$ which multiplies two cochains. If we dualized everything, we would end up with a map $C_\bullet(X) \to C_\bullet(X) \oplus C_\bullet(X)$ which recounts the Mayer-Vietoris exact sequence discussed in Sect. 5.2.

Example 5.36 Consider the simplicial complex model of the torus X shown in Fig. 5.28, and recall Example 5.32. If we denote the cochain on the left side of Fig. 5.28 by

$$\bar{\beta} := [a, d]^T + [d, g]^T - [a, g]^T$$

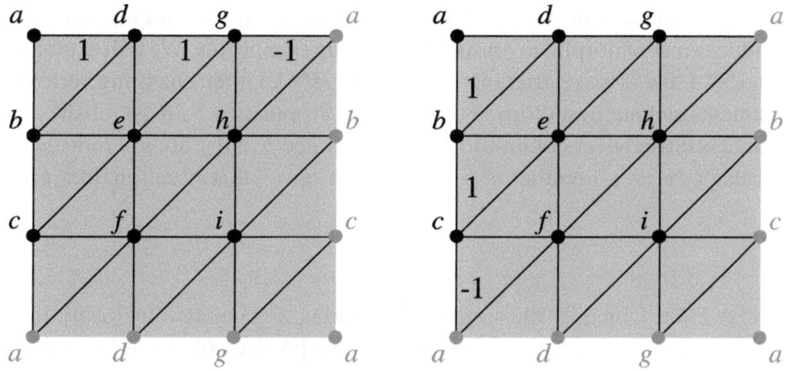

Fig. 5.28 Two simplicial cohains in $C^1(X)$ on the torus X in Example 5.36. Values not shown are zero

and the cochain on the right side of Fig. 5.28 by

$$\bar{\gamma} := [a, b]^T + [b, c]^T - [a, c]^T,$$

we can compute $\bar{\beta} \cup \bar{\gamma}$. Unlike β and γ in Example 5.32 (which we will revisit once more in Example 5.40), notice that neither $\bar{\beta}$ nor $\bar{\gamma}$ are elements of $\ker d^1$, since

$$\begin{aligned} d^1\bar{\beta}([a, b, d]) &= \bar{\beta}(\partial_2[a, b, d]) \\ &= \bar{\beta}([b, d] - [a, d] + [a, b]) \\ &= 0 - 1 + 0 \neq 0, \end{aligned}$$

and

$$\begin{aligned} d^1\bar{\gamma}([a, b, d]) &= \bar{\gamma}(\partial_2[a, b, d]) \\ &= \bar{\gamma}([b, d] - [a, d] + [a, b]) \\ &= 0 - 0 + 1 \neq 0. \end{aligned}$$

There is only one possible 2-simplex where the cup product $\bar{\beta} \cup \bar{\gamma}$ could be nontrivial, namely the simplex $[a, b, d]$, since this is the only place where two edges have nonzero values of $\bar{\beta}$ or $\bar{\gamma}$. Using the usual vertex partial order, $a < b < c < \cdots$,

$$(\bar{\beta} \cup \bar{\gamma})([a, b, d]) = \bar{\beta}([a, b])\bar{\gamma}([b, d]) = 1 \cdot 0 = 0.$$

Therefore, the cup product $\bar{\beta} \cup \bar{\gamma} = 0$.

Interestingly, the cup product is sensitive to the vertex partial order. If we have $d < a < b$, then the only possible nonzero value of $\bar{\beta} \cup \bar{\gamma}$ occurs on $[d, a, b]$,

$$(\bar{\beta} \cup \bar{\gamma})([d, a, b]) = \bar{\beta}([d, a])\bar{\gamma}([a, b]) = -\bar{\beta}([a, d])\bar{\gamma}([a, b]) = -1 \cdot 1 = -1.$$

We already know from Proposition 4.15 in Chap. 4 that changing vertex partial orders induces an isomorphism on simplicial chain complexes. We will establish later in Lemma 5.14 that if we restrict our attention to $H^\bullet(X)$, then changing vertex partial orders induces a chain homotopy. Furthermore, Proposition 5.26 establishes that the cup product is closed over cohomology classes. Since $\bar{\beta}$ and $\bar{\gamma}$ are *not* representatives of elements of $H^1(X)$—neither $d^1\bar{\beta}$ nor $d^1\bar{\gamma}$ are zero—this situation does not apply here.

Exercise 5.42 Consider the simplicial complex X generated by $[a, b]$, $[a, c]$, $[b, c]$, $[d, c]$, $[d, e]$, $[c, e]$. Compute the cup product of the following two elements of $C^1(X)$,

$$\alpha := [a, b]^T + [a, c]^T + [b, c]^T,$$

and

$$\beta := [d, c]^T + [d, e]^T + [c, e]^T.$$

Lemma 5.12 *If $f : X \to Y$ is a simplicial map that preserves the vertex partial order, $a \in C^p(Y)$, and $b \in C^q(Y)$, then*

$$f^{p+q}(a \cup b) = (f^p a) \cup (f^q b)$$

as elements of $C^\bullet(X)$. Briefly, one can say that $f^\bullet : C^\bullet(Y) \to C^\bullet(X)$ is a ring homomorphism.

Proof This statement is proven by calculation. Suppose that $[v_0, \ldots, v_{p+q}]$ is a simplex of X. To start, suppose that $f(v_i) \neq f(v_j)$ whenever $i \neq j$, and compute via Proposition 5.22 in Sect. 5.6

$$\begin{aligned}
(f^{p+q}(a \cup b))([v_0, \ldots, v_{p+q}]) &= (a \cup b)(f_{p+q}([v_0, \ldots, v_{p+q}])) \\
&= (a \cup b)([f(v_0), \ldots, f(v_{p+q})]) \\
&= a([f(v_0), \ldots, f(v_p)])b([f(v_p), \ldots, f(v_{p+q})]) \\
&= a(f_p([v_0, \ldots, v_p]))b(f_q([v_p, \ldots, v_{p+q}])) \\
&= (f^p a)([v_0, \ldots, v_p])(f^q b)([v_p, \ldots, v_{p+q}]) \\
&= ((f^p a) \cup (f^q b)([v_0, \ldots, v_{p+q}]).
\end{aligned}$$

Note well that if f does not preserve the vertex partial order, then the third equality does not necessary follow from the second, because the simplices on which a and b are evaluated will be different!

On the other hand, if $f(v_i) = f(v_j)$ for some $i \neq j$, then

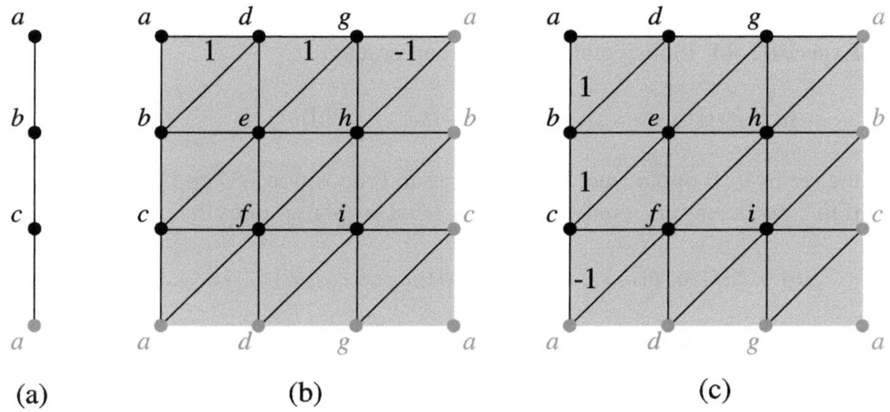

Fig. 5.29 **a** The simplicial complex Z, and **b–c** two simplicial cohains in $C^1(X)$ on the torus X in Example 5.37. Values not shown are zero

$$f_{p+q}([v_0, \ldots, v_{p+q}]) = 0$$

and at least one of

$$f_p([v_0, \ldots, v_p]) = 0, \text{ or } f_q([v_p, \ldots, v_{p+q}]) = 0$$

as well, so the equation still holds. □

Example 5.37 (This example is a continuation of Example 5.36.) Suppose that X is the simplicial complex model of the torus shown in Fig. 5.29b–c, and that Z is the abstract simplicial complex generated by $[a, b]$, $[b, c]$, $[a, c]$ shown in Fig. 5.29a, so that there is a natural inclusion $i : Z \to X$ given simply by matching the names of the vertices.

The map $C^1(X) \to C^1(Z)$ induced by i transforms a function defined on the edges of X to one defined on the edges of Z simply by restriction, and likewise $C^2(X) \to C^2(Z)$ induced by i restricts functions defined on triangles.

As a result, of the two cochains shown in Fig. 5.29, only the one in Fig. 5.29c is nontrivial when passed to $C^1(Z)$ by the map induced by i. Hence their cup product is immediately zero.

Exercise 5.43 Construct a counterexample to Lemma 5.12 by way of a simplicial map that does not preserve the vertex partial orders. (Hint: consider Example 5.36 closely.)

Exercise 5.44 In the definition of the cup product

$$(a \cup b)\,([v_0, v_1, \ldots, v_{p+q}]) = a([v_0, \ldots, v_p])b([v_p, \ldots, v_{p+q}]),$$

the vertex v_p is duplicated. What happens to Proposition 5.24 and Lemma 5.12 if this duplication is removed? That is, what properties does the operation

$$(a \vee b)\,([v_0, v_1, \ldots, v_{p+q}]) = a([v_0, \ldots, v_p])b([v_{p+1}, \ldots, v_{p+q}])$$

have when compared to the cup product?

We next show that the cup product and the coboundary maps are compatible. This means that the cup product can be restricted from $C^\bullet(X)$ to $H^\bullet(X)$. We begin first with an example.

Example 5.38 Suppose X is the simplicial complex generated by the simplex $[v_0, v_1, v_2, v_3, v_4]$, that $a \in C^1(X)$, and that $b \in C^2(X)$. Therefore, $a \cup b \in C^3(X)$, and $d^3(a \cup b) \in C^4(X)$. We can explicitly compute $d^3(a \cup b)([v_0, v_1, v_2, v_3, v_4])$,

$$\begin{aligned}
d^3(a \cup b)([v_0, v_1, v_2, v_3, v_4]) &= (a \cup b)(\partial_5[v_0, v_1, v_2, v_3, v_4]) \\
&= (a \cup b)([v_1, v_2, v_3, v_4] - [v_0, v_2, v_3, v_4] + [v_0, v_1, v_3, v_4] \\
&\quad - [v_0, v_1, v_2, v_4] + [v_0, v_1, v_2, v_3]) \\
&= a([v_1, v_2])b([v_2, v_3, v_4]) - a([v_0, v_2])b([v_2, v_3, v_4]) \\
&\quad + a([v_0, v_1])b([v_1, v_3, v_4]) - a([v_0, v_1])b([v_1, v_2, v_4]) \\
&\quad + a([v_0, v_1])b([v_1, v_2, v_3]) \\
&= (a([v_1, v_2]) - a([v_0, v_2]))\, b([v_2, v_3, v_4]) \\
&\quad + a([v_0, v_1])\, (b([v_1, v_3, v_4]) - b([v_1, v_2, v_4]) \\
&\quad + b([v_1, v_2, v_3]))\,.
\end{aligned}$$

This looks like it ought to simplify with boundary maps, but it is missing a term in each of the parenthesized expressions. Fortunately, the missing terms are the same and of opposite signs, so they may simply be supplied. The calculation proceeds,

$$\begin{aligned}
d^3(a \cup b)([v_0, v_1, v_2, v_3, v_4]) &= (a([v_1, v_2]) - a([v_0, v_2]))\, b([v_2, v_3, v_4]) \\
&\quad + a([v_0, v_1])\, (b([v_1, v_3, v_4]) - b([v_1, v_2, v_4]) + b([v_1, v_2, v_3])) \\
&\quad + a([v_0, v_1])b([v_2, v_3, v_4]) - a([v_0, v_1])b([v_2, v_3, v_4]) \\
&= (a([v_1, v_2]) - a([v_0, v_2]) + a([v_0, v_1]))\, b([v_2, v_3, v_4]) \\
&\quad + a([v_0, v_1])\, (-b([v_2, v_3, v_4]) + b([v_1, v_3, v_4]) - b([v_1, v_2, v_4]) \\
&\qquad + b([v_1, v_2, v_3])) \\
&= a(\partial_2([v_0, v_1, v_2])b([v_2, v_3, v_4]) - a([v_0, v_1])b(\partial_3([v_1, v_2, v_3, v_4]))
\end{aligned}$$

$$= ((d^1a) \cup b)([v_0, v_2, v_3, v_4]) - (a \cup (d^2b))([v_0, v_1, v_2, v_3, v_4]))$$
$$= \big(((d^1a) \cup b) - (a \cup (d^2b))\big)\,([v_0, v_1, v_2, v_3, v_4]).$$

We can generalize the calculation from the previous example to arbitrary cup products.

Proposition 5.25 *Suppose that $a \in C^p(X)$ and that $b \in C^q(X)$. The coboundary of their cup product is the element of $C^{p+q+1}(X)$ given by*

$$d^{p+q}(a \cup b) = (d^p a) \cup b + (-1)^p a \cup (d^q b).$$

Proof This is simply an elaborate calculation,

$$\begin{aligned}
\big(d^{p+q}(a \cup b)\big)([v_0, \ldots, v_{p+q+1}]) &= (a \cup b)\big(\partial_{p+q+1}[v_0, \ldots, v_{p+q+1}]\big)\\
&= (a \cup b)\left(\sum_{k=0}^{p+q+1} (-1)^k [v_0, \ldots, v_{k-1}, v_{k+1}, \ldots, v_{p+q+1}]\right)\\
&= (a \cup b)\left(\sum_{k=0}^{p} (-1)^k [v_0, \ldots, v_{k-1}, v_{k+1}, \ldots, v_{p+q+1}]\right.\\
&\quad \left. + (-1)^{p+1} \sum_{k=p+1}^{p+q+1} (-1)^k [v_0, \ldots, v_{k-1}, v_{k+1}, \ldots, v_{p+q+1}]\right)\\
&= (a \cup b)\left(\sum_{k=0}^{p} (-1)^k [v_0, \ldots, v_{k-1}, v_{k+1}, \ldots, v_p, \ldots, v_{p+q+1}]\right.\\
&\quad \left. + (-1)^{p+1} \sum_{k=p+1}^{p+q+1} (-1)^k [v_0, \ldots, v_p, \ldots, v_{k-1}, v_{k+1}, \ldots, v_{p+q+1}]\right)\\
&= \sum_{k=0}^{p} (-1)^k a([v_0, \ldots, v_{k-1}, v_{k+1}, \ldots, v_{p+1}])b([v_{p+1}, \ldots, v_{p+q+1}])\\
&\quad + \sum_{k=p+1}^{p+q+1} (-1)^{k+p+1} a([v_0, \ldots, v_p])b([v_p, \ldots, v_{k-1}, v_{k+1}, \ldots, v_{p+q+1}]).
\end{aligned}$$

As in the calculation from Example 5.38, we need to supply two missing terms, namely

$$(-1)^{p+1} a([v_0, \ldots, v_p])b([v_{p+1}, \ldots, v_{p+q+1}]) +$$
$$(-1)^{p+2} a([v_0, \ldots, v_p])b([v_{p+1}, \ldots, v_{p+q+1}]),$$

so that the calculation can proceed

$$\begin{aligned}\left(d^{p+q}(a\cup b)\right)([v_0,\ldots,v_{p+q+1}]) &= a(\partial_{p+1}[v_0,\ldots,v_{p+1}])b([v_{p+1},\ldots,v_{p+q+1}])\\ &\quad +(-1)^{p+2}a([v_0,\ldots,v_p])b(\partial_{q+1}[v_p,\ldots,v_{p+q+1}])\\ &= (d^p a\cup b)([v_0,\ldots,v_{p+q+1}])\\ &\quad +(-1)^p(a\cup d^q b)([v_0,\ldots,v_{p+q+1}]).\end{aligned}$$

□

Exercise 5.45 Continuing Example 5.38, suppose that X is the simplicial complex generated by the simplex $[v_0, v_1, v_2, v_3, v_4]$, that $a \in C^1(X)$ is given by

$$a := [v_0, v_1]^T - [v_0, v_2]^T + [v_1, v_2]^T,$$

and that $b \in C^2(X)$ is given by

$$b := [v_1, v_2, v_3]^T + [v_2, v_3, v_4]^T.$$

Verify Proposition 5.25 holds by direct calculation of

$$d^3(a\cup b) \text{ and } \left(((d^1a)\cup b) - (a\cup(d^2b))\right).$$

Proposition 5.26 *The cup product of two length 1 bars is another length 1 bar. Therefore, the cup product is closed over cohomology classes: the cup of two cohomology classes is another cohomology class.*

Therefore, $H^\bullet(X)$ is a sub-ring of $C^\bullet(X)$—a ring unto itself—called the *cohomology ring*.

Proof Suppose that $a \in C^p(X)$ and $b \in C^q(X)$. If both a and b are at the ends of their respective bars, then $a \in \ker d^p$ and $b \in \ker d^q$, and

$$\begin{aligned}d^{p+q}(a\cup b) &= (d^p a)\cup b + (-1)^p a\cup(d^q b)\\ &= 0\cup b + (-1)^p a\cup 0 = 0.\end{aligned}$$

Now suppose additionally that a is at the end of a length 2 bar. There is an $a' \in C^{p-1}(X)$ such that $d^{p-1}(a') = a$. We can form $(a')\cup b \in C^{p+q-1}(X)$, and compute

$$\begin{aligned}d^{p+q-1}((a')\cup b) &= (d^{p-1}a')\cup b + (-1)^{p-1}(a')\cup(d^q b)\\ &= a\cup b + (-1)^{p-1}(a')\cup 0\\ &= a\cup b,\end{aligned}$$

which implies that $a\cup b$ is at the end of a length 2 bar.

Finally, suppose the reverse, that a is at the end of its bar (of whatever length) and that b is at the end of a length 2 bar. There is a $b' \in C^{q-1}(X)$ such that $d^{q-1}(b') = b$. This implies

$$\begin{aligned} d^{p+q-1}(a \cup b') &= (d^p a) \cup b' + (-1)^p a \cup (d^{-1}(b')) \\ &= 0 \cup b' + (-1)^p a \cup b \\ &= (-1)^p a \cup b. \end{aligned}$$

This too implies that $a \cup b$ is at the end of a length 2 bar. Assembling these three pieces, if a and b are both at the ends of length 1 bars, then $a \cup b$ is also at the end of a length 1 bar. □

Exercise 5.46 Show that the cup product is commutative on $C^0(X)$. That is, for a and b in $C^0(X)$, show that $a \cup b = b \cup a$.

The above exercise notwithstanding, the cup product is not commutative on $C^\bullet(X)$. Consider the simplicial complex X generated by a single edge $[v_0, v_1]$. Let a be the element of $C^0(X)$ defined by

$$a([v_0]) = a_0 \text{ and } a([v_1]) = a_1.$$

Let b be the element of $C^1(X)$ defined by $b([v_0, v_1]) = b_0$. The cup product $a \cup b$ is an element of $C^1(X)$ and is defined by its value on $[v_0, v_1]$,

$$(a \cup b)([v_0, v_1]) = a([v_0])b([v_0, v_1]) = a_0 b_0.$$

On the other hand, $b \cup a$ is also an element of $C^1(X)$, but its value is

$$(b \cup a)([v_0, v_1]) = b([v_0, v_1])a([v_1]) = b_0 a_1.$$

Unless $a_0 = a_1$, the values of $a \cup b$ and $b \cup a$ will be different. Rearranging slightly, if $a \in \ker d^0$, then

$$0 = (d^0 a)([v_0, v_1]) = a([v_1]) - a([v_0]),$$

so that $a \cup b = b \cup a$ in this specific case.

At stake with commutativity in the cup product is the fact that in $C^\bullet(X)$, the two cup products

$$(a \cup b)([v_0, \ldots, v_p]) \text{ and } (b \cup a)([v_0, \ldots, v_p])$$

have nothing to do with one another even though they are defined on the same simplex. The second expression can be interpreted as being related to

$$(a \cup b)([v_p, \ldots, v_0])$$

by a possible sign change, since exchanging vertices in a vertex flips its sign according to the convention stated in Sect. 4.5 of Chap. 4. The reader can be forgiven for thinking this matter is a bit out of hand! Strictly speaking, the constraint in the definition of the cup product that the vertex partial order be followed means that this expression is not even defined! This annoying defect is remedied by restricting our attention to the cup product on $H^\bullet(X)$ instead of $C^\bullet(X)$, because the values of cohomology classes on nearby simplices are related to one another. The key insight is captured in the following Lemma.

Lemma 5.13 *Suppose that X is an abstract simplicial complex with two vertex partial orders, $\le$ and $\le'$ such that for every pair of vertices x, y, we have that $x \le y$ if and only if $y \le' x$. Let us write $C_\bullet(X)$ for the simplicial chain complex constructed using $\le$ and $C'_\bullet(X)$ for the simplicial chain complex constructed using $\le'$.*

The identity simplicial map $\mathrm{id}_X : X \to X$ induces the chain map $r_\bullet : C_\bullet(X) \to C'_\bullet(X)$ given by

$$r_k([v_0, \dots, v_k]) = (-1)^{k(k+1)/2}[v_k, \dots, v_0].$$

Proof According to Proposition 4.15 in Chap. 4, changing to a different vertex partial order induces an isomorphism on the simplicial chain complex, whose action is to flip the signs of some simplices. The present statement follows from an inspection of the proof of that proposition, noting that a single transposition induces a single additional sign flip. Since it takes $k(k+1)/2$ transpositions to reverse a sequence of length $k+1$, the proof is complete. □

Although $C'_k(X)$ and $C_k(X)$ are isomorphic vector spaces for each k, one might ask whether $[v_0, \dots, v_k]$ and $r_k([v_0, \dots, v_k])$ are somehow the "same simplices" in $C_k(X)$. Unless they are vertices, they are not, since the vertex partial ordering prohibits one or the other representation.

However, if we restrict our attention to homology classes only, then $[v_0, \dots, v_k]$ and $r_k([v_0, \dots, v_k])$ are different representatives of the same homology class. This is proven by way of a somewhat elaborate argument using chain homotopies.

There is one technical issue, though, and that is that there is not quite enough "room" in X for both vertex partial orders. It happens that the prism PX is sufficiently "roomy": we use the original vertex partial order $\le$ on PX_0 and the other one $\le'$ on PX_1. Furthermore, Proposition 5.13 in Sect. 5.3 establishes that $H_\bullet(X) \cong H_\bullet(PX)$.

Lemma 5.14 *Consider the vertex partial order on PX given by $\le$ on PX_0 and $\le'$ on PX_1, where these two partial orders are defined as in Lemma 5.13. Composing the simplicial map r defined in Lemma 5.13 with the inclusion $i_1 : X \to PX_1 \subseteq PX$ induces a simplicial chain map $r_\bullet : C_\bullet(X) \to C_\bullet(PX)$. The simplicial chain map $r_\bullet$ is chain homotopic to the chain map $i_\bullet : C_\bullet(X) \to C_\bullet(PX)$ induced by the inclusion $i_0 : X \to PX_0 \subset PX$. Namely, there are linear maps $h_k : C_k(X) \to C'_{k+1}(PX)$ such that*

$$\partial'_{k+1} \circ h_k + h_{k-1} \circ \partial_k = r_k - i_k$$

for each k.

This chain homotopy can be summarized by the *non-commutative* diagram

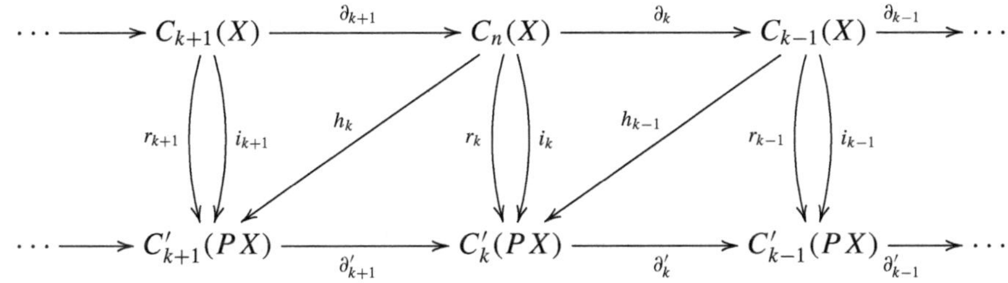

Proof Recall that every vertex v of X has two copies in PX, namely v and v'. Additionally, every simplex of X has several copies in PX as well. Abusing notation slightly (due to several vertex partial orders being in play!), and using Lemma 5.13 to handle signs, the appropriate definition for the chain homotopy maps happens to be

$$h_k([v_0, \ldots, v_k]) := \sum_{i=0}^{k} (-1)^i (-1)^{(k-i)(k-i+1)/2} [v_0, \ldots, v_i, v'_k, \ldots, v'_i].$$

The action of h_k is to collect all possible reversals of the latter portion of each simplex in X. Since the output of h_k is in the simplicial chain complex of the prism PX, h_k is able to increase the index (the length of the input simplex is $k+1$, while the length of the output simplex is $k+2$) by duplicating the v_i vertex.

This definition for h_k succeeds because the very first term of the sum consists of no reversal at all (just duplication of v_k, while the very last term of the sum reverses all of the vertices (with a duplication of v_0). With the definition of h_k in hand, the rest of the proof is calculation.

To this end, let us compute

$$\begin{aligned}
(\partial'_{k+1} \circ h_k)([v_0, \ldots, v_k]) &= \partial'_{k+1} \left(\sum_{i=0}^{k} (-1)^i (-1)^{(k-i)(k-i+1)/2} [v_0, \ldots, v_i, v'_k, \ldots, v'_i] \right) \\
&= \sum_{i=0}^{k} (-1)^i (-1)^{(k-i)(k-i+1)/2} \partial'_{k+1}([v_0, \ldots, v_i, v'_k, \ldots, v'_i]) \\
&= \sum_{i=0}^{k} (-1)^i (-1)^{(k-i)(k-i+1)/2} \sum_{\ell=0}^{i} (-1)^\ell [v_0, \ldots, \widehat{v_\ell}, \ldots, v_i, v'_k, \ldots, v'_i] \\
&+ \sum_{i=0}^{k} (-1)^i (-1)^{(k-i)(k-i+1)/2} \sum_{\ell=i}^{k} (-1)^{\ell+1} [v_0, \ldots, v_i, v'_k, \ldots, \widehat{v'_\ell}, \ldots, v'_i] \\
&= \sum_{i=0}^{k} \sum_{\ell=0}^{i} (-1)^{i+\ell} (-1)^{(k-i)(k-i+1)/2} [v_0, \ldots, \widehat{v_\ell}, \ldots, v_i, v'_k, \ldots, v'_i] \\
&- \sum_{i=0}^{k} \sum_{\ell=i}^{k} (-1)^{i+\ell} (-1)^{(k-i)(k-i+1)/2} [v_0, \ldots, v_i, v'_k, \ldots, \widehat{v'_\ell}, \ldots, v'_i].
\end{aligned}$$

We also must compute

$$
\begin{aligned}
(h_{k-1} \circ \partial_k)([v_0, \dots, v_k]) &= h_{k-1}\left(\sum_{i=0}^{k} (-1)^i [v_0, \dots, \widehat{v_i}, \dots, v_k]\right) \\
&= \sum_{\ell=0}^{k} (-1)^\ell h_{k-1}\left([v_0, \dots, \widehat{v_\ell}, \dots, v_k]\right) \\
&= \sum_{\ell=0}^{k} (-1)^\ell \sum_{i=0}^{\ell-1} (-1)^{i+(k-i)(k-i+1)/2} [v_0, \dots, v_i, v'_k, \dots, \widehat{v'_\ell}, \dots, v'_i] \\
&+ \sum_{\ell=0}^{k} (-1)^\ell \sum_{i=\ell+1}^{k} (-1)^{i+1+(k-i)(k-i+1)/2} [v_0, \dots, \widehat{v_\ell}, \dots, v_i, v'_k, \dots, v'_i] \\
&= \sum_{\ell=0}^{k} \sum_{i=0}^{\ell-1} (-1)^{i+\ell} (-1)^{(k-i)(k-i+1)/2} [v_0, \dots, v_i, v'_k, \dots, \widehat{v'_\ell}, \dots, v'_i] \\
&- \sum_{\ell=0}^{k} \sum_{i=\ell+1}^{k} (-1)^{\ell+i} (-1)^{(k-i)(k-i+1)/2} [v_0, \dots, \widehat{v_\ell}, \dots, v_i, v'_k, \dots, v'_i] \\
&= \sum_{i=0}^{k-1} \sum_{\ell=i}^{k} (-1)^{i+\ell} (-1)^{(k-i)(k-i+1)/2} [v_0, \dots, v_i, v'_k, \dots, \widehat{v'_\ell}, \dots, v'_i] \\
&- \sum_{i=1}^{k} \sum_{\ell=0}^{i} (-1)^{i+\ell} (-1)^{(k-i)(k-i+1)/2} [v_0, \dots, \widehat{v_\ell}, \dots, v_i, v'_k, \dots, v'_i].
\end{aligned}
$$

Carefully comparing $(\partial'_{k+1} \circ h_k)$ with $(h_{k-1} \circ \partial_k)$, it can be seen that the only terms that do not cancel in their sum come from $(\partial'_{k+1} \circ h_k)$ when $i = 0$ and when $i = k$. Collecting this information yields

$$
\begin{aligned}
\partial'_{k+1} \circ h_k + h_{k-1} \circ \partial_k &= (-1)^{k(k+1)/2} [v'_k, \dots, v'_i] - [v_0, \dots, v_k] \\
&= r_k([v_0, \dots, v_k]) - i_k([v_0, \dots, v_k])
\end{aligned}
$$

as claimed. □

The conclusion from Lemma 5.14 is that reversing the order of the vertices induces an isomorphism on homology classes, because $i_\bullet$ is a chain isomorphism and Proposition 3.6 in Chap. 3 asserts that homotopic maps induce identical maps on homology. This is slightly more specific than Proposition 4.15 in Chap. 4, which merely establishes that a simplicial chain complex isomorphism exists.

Theorem 5.2 (the cup product distributes over simplicial maps) *If $f : X \to Y$ is a simplicial map, $a \in H^p(Y)$, and $b \in H^q(Y)$, then*

$$
f^{p+q}(a \cup b) = (f^p a) \cup (f^q b)
$$

as elements of $H^\bullet(X)$. One can say that $f^\bullet : H^\bullet(Y) \to H^\bullet(X)$ is a ring homomorphism.

This statement removes the restriction in Lemma 5.12 that f preserves the vertex partial order.

Proof Given that we have already shown most of the calculation necessary in Lemma 5.12, we merely need to handle the case where the vertex partial order is not respected.

Suppose that X and Y have arbitrary vertex partial orders, $\leq_X$ and $\leq_Y$, respectively. Proposition 4.13 in Chap. 4 constructs a new vertex partial order $\leq'_X$ on X such that $f : (X, \leq'_X) \to (Y, \leq_Y)$ is order preserving. Let us denote by X' the simplicial complex X using the vertex partial order $\leq'_X$. Lemma 5.14 constructs an isomorphism between $H_\bullet(X) \to H_\bullet(X')$ induced by changing only the vertex partial order. This implies that there is also an isomorphism $i : H^\bullet(X') \to H^\bullet(X)$, noting well the contravariance.

Since the vertices of X and X' are identical (along with all the other simplices, too!), the map f can be thought of as a map $X \to Y$ or equally well as $X' \to Y$. In the latter case, $f : X' \to Y$, the vertex partial order is preserved by construction. Hence Lemma 5.12 gives that $C^\bullet(Y) \to C^\bullet(X')$ is a ring homomorphism, from which it follows that $H^\bullet(Y) \to H^\bullet(X')$ is also a ring homomorphism. Composing this with i yields the desired result, namely a ring homomorphism $H^\bullet(Y) \to H^\bullet(X)$.

Explicitly, suppose that $[a] \in H^p(Y)$ and $[b] \in H^q(Y)$, for which $a \in C^p(Y)$ and $b \in C^q(Y)$ are representatives. Consider a homology class $[x] \in H_{p+q}(X')$. By construction of X' and its vertex partial order $\leq'_X$, we have from Lemma 5.12 that

$$f^\bullet\left((a \cup b)\right)(x) = \left(f^\bullet(a) \cup f^\bullet(b)\right)(x).$$

Since $[x]$ was arbitrary, this implies that

$$f^\bullet\left((a \cup b)\right) = f^\bullet(a) \cup f^\bullet(b) \tag{5.6}$$

as elements of $C^{p+q}(X')$ and hence also as cohomology classes in $H^{p+q}(X')$. It is also the case that both $f^\bullet\left((a \cup b)\right)$ and $f^\bullet(a) \cup f^\bullet(b)$ define cohomology classes in $H^{p+q}(X)$. Since we have that the only difference between X and X' is the vertex partial order, and that we have the $H^{p+q}(X)$ and $H^{p+q}(X')$ are isomorphic via the change in this vertex partial order *only*, it must be the case that Eq. (5.6) holds in $H^{p+q}(X)$ as well. □

Example 5.39 Let us consider the cup product of $a \in H^1(X)$ and $b \in H^2(X)$ on the simplicial complex X generated by $[v_0, v_1, v_2, v_3]$, shown in Fig. 5.30. Specifically, we can compute

$$(a \cup b)([v_0, v_1, v_2, v_3]) = a([v_0, v_1])b([v_1, v_2, v_3])$$

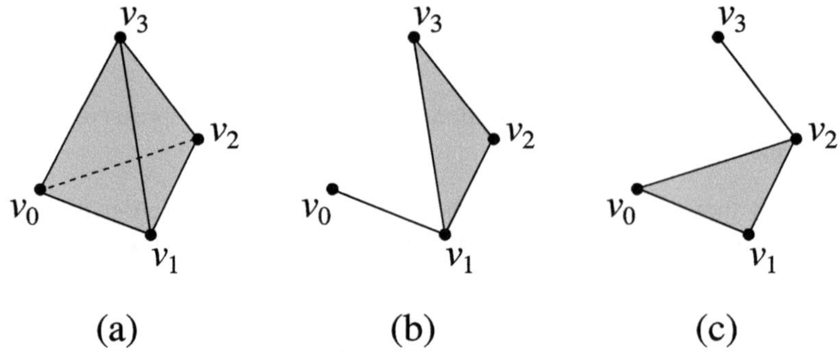

Fig. 5.30 **a** The simplicial complex X, **b** the faces upon which the cup product $(a \cup b)([v_0, v_1, v_2, v_3])$ is evaluated, and **c** the faces upon which $(b \cup a)([v_0, v_1, v_2, v_3])$ is evaluated, in Example 5.39

and compare this with

$$(b \cup a)([v_0, v_1, v_2, v_3]) = b([v_0, v_1, v_2])a([v_2, v_3]).$$

According to Lemma 5.14, when we permute the order of the vertices, we permute the sign of the cup product, which means that in the present situation up to homology,

$$\begin{aligned}
(a \cup b)([v_0, v_1, v_2, v_3]) &= (a \cup b)(r_3([v_0, v_1, v_2, v_3])) \\
&= (a \cup b)([v_3, v_2, v_1, v_0]) \\
&= a([v_3, v_2])b([v_2, v_1, v_0]) \\
&= (-a([v_2, v_3]))\,(-b([v_0, v_1, v_2])) \\
&= b([v_0, v_1, v_2])a([v_2, v_3]) \\
&= (b \cup a)([v_0, v_1, v_2, v_3]).
\end{aligned}$$

The number of sign flips in the previous example is sensitive to the number of vertices being exchanged, but the idea of reversing the order of vertices generalizes neatly.

Proposition 5.27 *The cup product is* anticommutative

$$a \cup b = (-1)^{pq} b \cup a$$

on cohomology classes $a \in H^p(X)$, $b \in H^q(X)$.

Proof Lemma 5.14 asserts that exchanging vertices in a simplex switches the sign of that simplex in $H_\bullet(X)$. This means that $[v_0, \ldots, v_p]$ and

$$r_p([v_0, \dots, v_p]) = (-1)^{\frac{p(p+1)}{2}} [v_p, \dots, v_0]$$

are in the same homology class in $H_p(X)$.

Therefore, if we use equality to represent equality in $H^{p+q}(X)$, we compute

$$\begin{aligned}
(a \cup b)([v_0 &, v_1, \dots, v_p, v_{p+1}, \dots, v_{p+q}]) \\
&= (a \cup b)(r_{p+q}([v_0, v_1, \dots, v_p, v_{p+1}, \dots, v_{p+q}])) \\
&= (a \cup b)((-1)^k [v_{p+q}, \dots, v_{p+1}, v_p, \dots, v_0]) \\
&= (-1)^k (a \cup b)([v_{p+q}, \dots, v_{p+1}, v_p, \dots, v_0]) \\
&= (-1)^k a([v_{p+q}, \dots, v_{p+1}, v_p]) b([v_p, \dots, v_0]) \\
&= (-1)^k a((-1)^\ell [v_p, \dots, v_{p+q}]) b((-1)^m [v_0, \dots, v_p]) \\
&= (-1)^{k+\ell+m} a([v_p, \dots, v_{p+q}]) b([v_0, \dots, v_p]) \\
&= (-1)^{k+\ell+m} b([v_0, \dots, v_p]) a([v_p, \dots, v_{p+q}]) \\
&= (-1)^{k+\ell+m} (b \cup a)([v_0, \dots, v_p, \dots, v_{p+q}]).
\end{aligned}$$

It remains to show that

$$(-1)^{k+\ell+m} = (-1)^{pq},$$

namely that the parity of $k + \ell + m$ is the same as pq. To that end, observe that

$$k = \frac{(p+q)(p+q+1)}{2}, \quad \ell = \frac{q(q+1)}{2}, \quad \text{and } m = \frac{p(p+1)}{2}.$$

Therefore

$$\begin{aligned}
k + \ell + m &= \frac{(p+q)(p+q+1)}{2} + \frac{q(q+1)}{2} + \frac{p(p+1)}{2} \\
&= \frac{p^2 + q^2 + 2pq + p + q}{2} + \frac{q^2 + q}{2} + \frac{p^2 + p}{2} \\
&= \frac{2p^2 + 2q^2 + 2pq + 2p + 2q}{2} \\
&= p^2 + q^2 + p + q + pq \\
&= p(p+1) + q(q+1) + pq.
\end{aligned}$$

Since $p(p+1)$ and $q(q+1)$ are always even, they do not change the parity of a number when added. This implies that

$$(-1)^{k+\ell+m} = (-1)^{pq}.$$

□

Example 5.40 Consider the simplicial complex Y from Example 5.31 and Fig. 5.24 in Sect. 5.6. The cup product between $\beta \cup \gamma$ is an element of $C^2(Y)$ given by

$$(\beta \cup \gamma)([a, c, e]) = \beta([a, c])\gamma([c, e]) = (-1)(0) = 0$$
$$(\beta \cup \gamma)([a, c, g]) = \beta([a, c])\gamma([c, g]) = (-1)(-1/2) = 1/2$$
$$(\beta \cup \gamma)([a, e, g]) = \beta([a, e])\gamma([e, g]) = (-1/2)(-1/2) = 1/4$$
$$(\beta \cup \gamma)([c, e, g]) = \beta([c, e])\gamma([e, g]) = (1/2)(-1/2) = -1/4$$

While this is nonzero in $C^2(Y)$, it is trivial in $H^2(Y)$ because it is the image of

$$\zeta := \left(\frac{1}{4}[a, e] - \frac{1}{4}[c, e] + \frac{1}{2}[a, c]\right),$$

because

$$d^1\zeta([a, c, e]) = \zeta([c, e] - [a, e] + [a, c]) = -1/4 - 1/4 + 1/2 = 0,$$
$$d^1\zeta([a, c, g]) = \zeta([c, g] - [a, g] + [a, c]) = 0 + 0 + 1/2 = 1/2,$$
$$d^1\zeta([a, e, g]) = \zeta([e, g] - [a, g] + [a, e]) = 0 + 0 + 1/4 = 1/4,$$
$$d^1\zeta([c, e, g]) = \zeta([e, g] - [c, g] + [c, e]) = 0 + 0 - 1/4 = -1/4,$$

namely

$$d^1\zeta = \beta \cup \gamma.$$

On the other hand, if we consider the analogous β and γ from Example 5.32 and Fig. 5.25 in Sect. 5.6, we have

$$(\beta \cup \gamma)([a, b, d]) = \beta([a, b])\gamma([b, d]) = (1)(1) = 1$$
$$(\beta \cup \gamma)([d, e, g]) = \beta([d, e])\gamma([e, g]) = (1)(1) = 1$$
$$(\beta \cup \gamma)([a, g, h]) = \beta([a, g])\gamma([g, h]) = (0)(0) = 0$$
$$(\beta \cup \gamma)([b, c, e]) = \beta([b, c])\gamma([c, e]) = (1)(1) = 1$$
$$(\beta \cup \gamma)([e, f, h]) = \beta([e, f])\gamma([f, h]) = (1)(1) = 1$$
$$(\beta \cup \gamma)([b, h, i]) = \beta([b, h])\gamma([h, i]) = (0)(0) = 0$$
$$(\beta \cup \gamma)([a, c, f]) = \beta([a, c])\gamma([c, f]) = (-1)(1) = -1$$
$$(\beta \cup \gamma)([d, f, i]) = \beta([d, f])\gamma([f, i]) = (-1)(1) = -1$$
$$(\beta \cup \gamma)([c, g, i]) = \beta([c, g])\gamma([g, i]) = (1)(0) = 0$$
$$(\beta \cup \gamma)([b, d, e]) = \beta([b, d])\gamma([d, e]) = (-1)(0) = 0$$
$$(\beta \cup \gamma)([e, g, h]) = \beta([e, g])\gamma([g, h]) = (-1)(0) = 0$$
$$(\beta \cup \gamma)([a, b, h]) = \beta([a, b])\gamma([b, h]) = (1)(-1) = -1$$
$$(\beta \cup \gamma)([c, e, f]) = \beta([c, e])\gamma([e, f]) = (-1)(0) = 0$$
$$(\beta \cup \gamma)([f, h, i]) = \beta([f, h])\gamma([h, i]) = (-1)(0) = 0$$
$$(\beta \cup \gamma)([b, c, i]) = \beta([b, c])\gamma([c, i]) = (1)(-1) = -1$$
$$(\beta \cup \gamma)([a, d, f]) = \beta([a, d])\gamma([d, f]) = (0)(0) = 0$$
$$(\beta \cup \gamma)([d, g, i]) = \beta([d, g])\gamma([g, i]) = (-1)(0) = 0$$
$$(\beta \cup \gamma)([a, c, g]) = \beta([a, c])\gamma([c, g]) = (-1)(-1) = 1$$

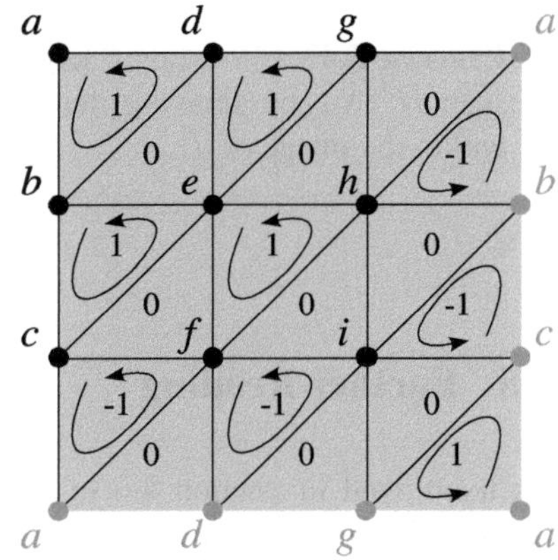

Fig. 5.31 Values of the cup product $\beta \cup \gamma$ computed on the torus X in Example 5.40. Ordering of the vertices within each 2-simplex accounting for signs is shown by arrows

which can be written more succinctly as

$$\begin{aligned}\beta \cup \gamma =&[a, b, d]^T + [d, e, g]^T + [b, c, e]^T \\ &+ [e, f, h]^T - [a, c, f]^T - [d, f, i]^T \\ &- [a, b, h]^T - [b, c, i]^T + [a, c, g]^T,\end{aligned}$$

as is summarized in Fig. 5.31. Notice in particular that once the vertices in each 2-simplex are ordered accounting for the signs, the vertices are always traversed in a counter-clockwise fashion. We will show that $\beta \cup \gamma$ is a nontrivial cohomology class, using much the same procedure as in Example 5.32. Specifically, we will use the element $\Delta \in C_2(X)$ given by

$$\begin{aligned}\Delta := &[a, b, d] + [d, e, g] + [a, g, h] \\ &+ [b, c, e] + [e, f, h] + [b, h, i] \\ &- [a, c, f] - [d, f, i] - [c, g, i] \\ &- [b, d, e] - [e, g, h] - [a, b, h] \\ &- [c, e, f] - [f, h, i] - [b, c, i] \\ &+ [a, d, f] + [d, g, i] + [a, c, g],\end{aligned}$$

with the observation that every element of $\zeta \in C^1(X)$ satisfies $d^1\zeta(\Delta) = 0$. However, $(\beta \cup \gamma)(\Delta) = 9 \neq 0$, so $\beta \cup \gamma$ is a nontrivial element of $H^2(X)$.

To summarize, we have shown that although $H^k(X)$ and $H^k(Y)$ are isomorphic for each k, the cup product of the two basis elements in $H^1(X)$ is nontrivial, while the cup product for the two basis elements in $H^1(Y)$ is trivial. Therefore, these two abstract simplicial complexes can be distinguished by their cohomology rings.

Exercise 5.47 Show that $\beta \cup \gamma$ in Example 5.40 and δ in Example 5.32 both span $H^2(X)$, by showing that a particular linear combination of them lies within the image of d^1.

5.8 Further Reading

As mentioned in Section 4.6 in Chap. 4, topological data analysis makes use of simplicial complexes to identify topological features of datasets. The usual tool in topological data analysis is persistent homology (Definition 5.2). In addition to topological features, persistent homology also detects geometric features (Gómez and Mémoli 2024; Bubenik and Wagner 2020; Bubenik et al. 2020; Jaquette and Schweinhart 2020; Turner et al. 2014).

While local homology has been studied since the early 1920s, it has not been studied systematically (Robinson et al. 2018). The time seems ripe for considering applications, since there are only a few, for instance (Wang et al. 2024; Robinson 2014). Relative homology has recently gotten more attention, with important results being established regarding functoriality and the development of efficient algorithms (Bauer et al. 2023; Bauer and Schmahl 2023; de Silva et al. 2011).

Applications require efficient algorithms. These tend to come from generalizing free face reduction and subdivision strategies. Free face reduction strategies (mentioned in Sect. 5.1) are discussed in the literature. A good overview can be found in Bauer (2021); Edelsbrunner and Harer (2009); Kaczynski et al. (2004). Additionally, discrete Morse theory (Harker et al. 2014; Mischaikow and Nanda 2013) is a popular approach to guide the selection of free faces. Also, pruning algorithms operating on the level of sequences of simplicial chain complexes are available (Bauer et al. 2024; Giunti and Lazovskis 2023).

Much of what is treated in Sect. 5.4 is described in Jardine (2004), where the focus is on the *simplicial approximation* of topological spaces. One can instead start with finite topological spaces, is discussed in Barmak and Minian (2012); Barmak (2011). From the perspective of topology, simplicial complexes are seen as simplified models of topological spaces that become more accurate as more subdivisions are taken. In contrast, our focus has been algebraic instead. Here, homology is taken as fundamental, and simplicial complexes (and simplicial maps) are viewed through that lens. Subdivision for us is not so much an approximation, as it is more of a way to bring the combinatorial model of a simplicial complex a little closer into line with the algebra. Of course, these two viewpoints coincide: what homology "actually" detects are topological features!

If we choose to construct the simplicial chain complex with modules over a ring instead of working with vector spaces, then $H^k(X)$ and $H_k(X)$ may differ. This more general perspective is also closer to the underlying topology of X. If we use integer coefficients instead of real numbers in defining $H^1(X)$, then elements of $H^1(X)$ correspond to homotopy classes of continuous maps $X \to S^1$. This leads to the fascinating study of *Eilenberg-MacLane spaces*.

The process of supplying inverses to some of the morphisms of a category is called *localization*. A succinct way of expressing Proposition 5.20 is that **BdAsc** is the localization $\mathbf{Asc}[W^{-1}]$ of **Asc** on the class W of subdivision maps. We do not need notation for different localizations of **Asc** in this book. We will use the somewhat more suggestive **BdAsc** notation instead.

General localizations can be tricky to handle. Instead of localizing on subdivision maps, we might want to localize on those simplicial maps that induce chain homotopy equivalences on the simplicial chain complexes, but to do this properly requires defining the product $X \times Y$ of two simplicial complexes X and Y. This can be done, resulting in the *simplicial homotopy category*, but it is quite an involved construction. This construction also takes us far afield from the algebraic structure of homology! Furthermore, there are many more kinds of simplicial maps that induce simplicial chain homotopy equivalences beyond the few kinds discussed thus far (free face reductions, simplicial homotopy equivalences, collapses, and subdivisions). Completely describing the simplicial homotopy category is rather difficult (Goerss and Jardine 2009).

At the time of writing, there are a number of software packages that aid in the computation of homology for topological spaces, including simplicial homology and its variants. Without any claim to being exhaustive, the reader might consider:

- https://openappliedtopology.github.io,
- http://ripser.org/,
- https://cran.r-project.org/web/packages/TDA/index.html,
- https://docs.scikit-tda.org/en/latest/,
- https://giotto-ai.github.io/gtda-docs/latest/index.html,
- https://gregoryhenselman.org/eirene/,
- https://appliedtopology.github.io/javaplex/,
- Tierny et al. (2017), Bin Masood et al. (2019), Guillou et al. (2024).

References

Attali D, Devillers O, Glisse M, Lazard S (2016) Recognizing shrinkable complexes is NP-complete. J Comput Geometry 7(1). https://doi.org/10.20382/jocg.v7i1a18
Barmak JA (2011) Algebraic topology of finite topological spaces and applications. Springer
Barmak JA, Minian EG (2012) Strong homotopy types, nerves and collapses. Discret Comput Geom 47(2):301–328. https://doi.org/10.1007/s00454-011-9357-5.pdf
Bauer U (2021) Ripser: efficient computation of Vietoris-Rips persistence barcodes. J Appl Comput Topol 5(3):391–423
Bauer U, Masood TB, Giunti B, Houry G, Kerber M, Rathod A (2024) Keeping it sparse: computing persistent homology revisited. Comput Geom Topol 3(1):1–6

Bauer U, Lenzen F, Lesnick M (2023) Efficient two-parameter persistence computation via cohomology. In: Chambers EW, Gudmundsson J (eds) 39th international symposium on computational geometry (SoCG 2023), Leibniz international proceedings in informatics (LIPIcs), vol 258. Schloss Dagstuhl–Leibniz-Zentrum für Informatik, Dagstuhl, Germany, pp 15:1–15:17. DOI https://doi.org/10.4230/LIPIcs.SoCG.2023.15. https://drops.dagstuhl.de/entities/document/10.4230/LIPIcs.SoCG.2023.15

Bauer U, Schmahl M (2023) Lifespan functors and natural dualities in persistent homology. Homol Homotopy Appl 25(2)

Bin Masood T, Budin J, Falk M, Favelier G, Garth C, Gueunet C, Guillou P, Hofmann L, Hristov P, Kamakshidasan A, Kappe C, Klacansky P, Laurin P, Levine J, Lukasczyk J, Sakurai D, Soler M, Steneteg P, Tierny J, Usher W, Vidal J, Wozniak M (2019) An overview of the topology toolkit. In: TopoInVis

Bubenik P, Wagner A (2020) Embeddings of persistence diagrams into Hilbert spaces. J Appl Comput Topol 4(3):339–351

Bubenik P, Hull M, Patel D, Whittle B (2020) Persistent homology detects curvature. Inverse Probl 36(2):025008

Carlsson G, Vejdemo-Johansson M (2021) Topological data analysis with applications. Cambridge University Press

de Silva V, Morozov D, Vejdemo-Johansson M (2011) Dualities in persistent (co)homology. Inverse Probl 27(12):124003. https://doi.org/10.1088/0266-5611/27/12/124003

Edelsbrunner H, Harer J (2009) Computational topology: an introduction. American Mathematical Society

Giunti B, Lazovskis J (2023) Pruning vineyards: updating barcodes by removing simplices. arXiv:2312.03925

Goerss PG, Jardine JF (2009) Simplicial homotopy theory. Springer Science & Business Media

Gómez M, Mémoli F (2024) Curvature sets over persistence diagrams. Discret Comput Geom 72(1):91–180

Guillou EL, Will M, Guillou P, Lukasczyk J, Fortin P, Garth C, Tierny J (2024) TTK is getting MPI-ready. IEEE Trans Vis Comput Graph

Harker S, Mischaikow K, Mrozek M, Nanda V (2014) Discrete Morse theoretic algorithms for computing homology of complexes and maps. Found Comput Math 14(1):151–184

Hatcher A (2002) Algebraic topology. Cambridge University Press

Jaquette J, Schweinhart B (2020) Fractal dimension estimation with persistent homology: a comparative study. Commun Nonlinear Sci Numer Simul 84:105163

Jardine R (2004) Simplicial approximation. Theory Appl Categ 12(2):34–72

Kaczynski T, Mischaikow K, Mrozek M (2004) Computational homology. Springer

Milnor J, Stasheff JD (1974) Characteristic classes. Princeton University Press

Mischaikow K, Nanda V (2013) Morse theory for filtrations and efficient computation of persistent homology. Discret Comput Geom 50(2):330–353

Munkres J (1984) Elements of algebraic topology. Westview Press

Robinson M (2014) Analyzing wireless communication network vulnerability with homological invariants. In: 2014 IEEE global conference on signal and information processing (GlobalSIP), pp 900–904. https://doi.org/10.1109/GlobalSIP.2014.7032250

Robinson M, Capraro C, Joslyn C, Purvine E, Praggastis B, Ranshous S, Sathanur A (2018) Local homology of abstract simplicial complexes. https://arxiv.org/abs/1805.11547

Tierny J, Favelier G, Levine JA, Gueunet C, Michaux M (2017) The topology toolkit. In: IEEE transactions on visualization and computer graphics (Proceedings of IEEE VIS). https://topology-tool-kit.github.io/

Turner K, Mukherjee S, Boyer DM (2014) Persistent homology transform for modeling shapes and surfaces. Inf Inference J IMA 3(4):310–344

Wang M, Yan H, Huang Z, Wang D, Xu J (2024) Persistent local homology in graph learning. Trans Mach Learn Res

Chapter 6
Sequences and Chain Complexes of Sequences

The business of homological algebra is to ascribe a homology theory to algebraic objects, including those that may not initially be amenable to such a theory. For instance, one may ascribe a homology theory to a *module*, which is a generalization of a vector space. This book does not assume familiarity with module theory, but one can still explore the beginnings of the theory by instead studying sequences *of sequences*, since these are examples of modules. Sequences of sequences are strictly richer than sequences of vector spaces but provide enough concrete structure to reveal the outlines of the general theory. To study a given module homologically, one starts by constructing *resolutions*, which are exact sequences that either start or end at the given module of interest.

As with chain complexes of vector spaces, chain complexes of sequences (and of chain complexes) have a well defined notion of homology. Recalling that a quasi-isomorphism is a chain map that induces isomorphisms on homology, it is fruitful to characterize chain complexes (of sequences or of chain complexes) up to quasi-isomorphism. In the case of chain complexes of vector spaces, we found that quasi-isomorphisms correspond to homotopy equivalences (Theorem 3.4 in Chap. 3), which are in turn related to simplicial homotopies (Corollary 5.3 in Chap. 5). This equivalence is special: homotopy equivalence and quasi-isomorphism are different relations for chain complexes of sequences (as will be shown in Example 6.22 in Sect. 6.6).

In order to tease apart the difference between quasi-isomorphism and homotopy equivalence, we will encounter several new categories as outlined in Table 6.1. Identifying quasi-isomorphic chain complexes of sequences, one can determine exactly what is measured by their homology. This leads directly to a number of useful, but rather technical, constructions, culminating in the *derived category*. These constructions are related to the construction of simplicial rooves in Sect. 5.4 of Chap. 5. Although a full exploration of the derived category is out of scope for this book, this book ends with the a brief treatment of *derived functors*, which generalize the idea of an induced map on homology into this new context. In brief, they transform a single functor into a collection of functors that measure homological properties.

M. Robinson, *Computational Homological Algebra*, Mathematical Engineering,
https://doi.org/10.1007/978-3-032-08634-1_6

Table 6.1 Categories of chain complexes discussed in this chapter

Name	Symbol		Category **C**		
			Vec	**Seq**	**Kom**
Chain complexes	**Kom**(**C**)	Objects (up to iso.)	Chain complexes of vector spaces (= **Kom**)	Chain complexes of sequences	Double complexes
		Morphisms	Chain maps		
		Reference	Chapters 2 and 3	Section 6.2	Section 6.5
Homotopy category	$K(\mathbf{C})$	Objects (up to iso.)	Homotopy equivalence (= quasi-isomorphism)	Homotopy equivalence	
		Morphisms	Homotopy classes of chain maps		
		Reference	Section 3.4 in Chap. 3	Section 6.6	
Derived category	$D(\mathbf{C})$	Objects (up to iso.)	Homotopy equivalence (= quasi-isomorphism)	Quasi-isomorphism	
		Morphisms	Homotopy classes of rooves		
			("chain maps up to homology")		
		Reference	Section 6.7		

In this chapter,

Section 6.1 Explores many variations of categories of sequences of sequences, finding that there are a number of symmetries that reduce the number of possibilities considerably.
Section 6.2 Reveals more symmetries present for chain complexes of sequences.
Section 6.3 Defines *projective* and *injective* sequences that have certain properties making them useful "yardsticks" in our study of sequences of sequences.
Section 6.4 Uses projective and injective sequences to derive exact *resolutions* of a given sequence of interest.
Section 6.5 Proves some special results for *double complexes*, which are simply chain complexes of chain complexes.
Section 6.6 Shows that although resolutions of a given sequence are not uniquely defined, they are unique up to chain homotopy (Proposition 6.19).
Section 6.7 Defines a notion of homology and quasi-isomorphism using resolutions.
Section 6.8 Shows that in the derived category, the analog of induced maps on homology is the notion of *derived functors*.

6.1 Sequences of Sequences

Recall that there are two indexing conventions used in this book: (1) the indices increase when following the arrows, and (2) the indices decrease when following the arrows. These two conventions are not really that different. In Sect. 3.5 of Chap. 3, we noted that there is a duality between them. We now need to take a closer look at this.

Denote by $\mathbf{Seq}^{\bullet}$ the category of sequences in which the indices increase when following the maps, like

$$\cdots \xrightarrow{a^0} A^1 \xrightarrow{a^1} A^2 \xrightarrow{a^2} \cdots$$

Similarly, let us denote by $\mathbf{Seq}_{\bullet}$ the category of sequences in which the indices decrease when following the maps, like

$$\cdots \xrightarrow{a_2} A_1 \xrightarrow{a_1} A_0 \xrightarrow{a_0} \cdots$$

These categories are really the same because the indices merely facilitate the notation.

As the reader should suspect, there is a notion of isomorphism between categories, given by the following definition. Isomorphisms between categories tend to be rare. More commonly one encounters equivalences of categories (Definition 1.18 in Chap. 1).

Definition 6.1 For a category $\mathbf{C}$, the *identity functor on* $\mathbf{C}$ is the covariant functor $\mathrm{id}_{\mathbf{C}} : \mathbf{C} \to \mathbf{C}$ that leaves both objects and morphisms unchanged.

A pair of functors $F : \mathbf{C} \to \mathbf{D}$ and $G : \mathbf{D} \to \mathbf{C}$ between categories $\mathbf{C}$ and $\mathbf{D}$ forms an *isomorphism* if

$$G \circ F = \mathrm{id}_{\mathbf{C}}$$

and

$$F \circ G = \mathrm{id}_{\mathbf{D}} .$$

As is usual, if an isomorphism exists, then we say $\mathbf{C}$ and $\mathbf{D}$ are *isomorphic*.

Proposition 6.1 *The categories* $\mathbf{Seq}^\bullet$ *and* $\mathbf{Seq}_\bullet$ *are isomorphic.*

Proof One merely needs to construct two functors $F : \mathbf{Seq}^\bullet \to \mathbf{Seq}_\bullet$ and $G : \mathbf{Seq}_\bullet \to \mathbf{Seq}^\bullet$ that are each others' inverses on both objects and morphisms. To that end, define

$$F\left((A^\bullet, a^\bullet)\right) := (A_\bullet, a_\bullet),$$

where

$$A_k := A^{-k} \text{ and } a_k := a^{-k}.$$

Observe that $a_k = a^{-k} : A^{-k} \to A^{-k+1}$, which is also a function $A_k \to A_{k-1}$. Thus $(A_\bullet, a_\bullet)$ is an object of $\mathbf{Seq}_\bullet$. Before addressing morphisms, notice that

$$G\left((B_\bullet, b_\bullet)\right) := (B^\bullet, b^\bullet),$$

where

$$B^k := B_{-k} \text{ and } b^k := b_{-k}$$

is evidently the inverse of F on objects.

Now suppose that we have a morphism $m^\bullet : (A^\bullet, a^\bullet) \to (B^\bullet, b^\bullet)$. The correct definition for $F(m^\bullet)$ is

$$F(m^\bullet) := m_\bullet,$$

where

$$m_k := m^{-k}.$$

This definition ensures that $m_k : A_k \to B_k$ because $m^{-k} : A^{-k} \to B^{-k}$. Moreover, the commutativity of the diagram from $m^\bullet$ ensures that the corresponding diagram for $m_\bullet$ commutes. There is nothing additional required to show that F respects composition of chain maps, and so F is a covariant functor.

The action of G on morphisms can defined to be just the inverse of that for F, which completes the proof. □

Exercise 6.1 Proposition 6.1 is a bit of notational gymnastics that is necessary because of indexing. If $(A^\bullet, a^\bullet)$ is the following sequence

$$0 \to \mathbb{R}^2 \to \mathbb{R} \to \mathbb{R} \to 0,$$

where the second arrow is an arbitrary matrix, compute $F((A^\bullet, a^\bullet))$ (using the definition of F from the proof of Proposition 6.1). Justify your computation.

The proof of Proposition 6.1 is not the only isomorphism between these two categories. The dualization functor T defined in Proposition 3.9 in Chap. 3 is another isomorphism between $\mathbf{Seq}^\bullet$ and $\mathbf{Seq}_\bullet$.

Proposition 6.2 *Transposing every map in a sequence, given by*

$$T(A^\bullet, a^\bullet) := \left((A^\bullet)^T, (a^{\bullet-1})^T\right)$$

induces an isomorphism of categories $T : \mathbf{Seq}^\bullet \to \mathbf{Seq}_\bullet$.

Proof We already established that dualization T is a contravariant functor in Proposition 3.9 in Chap. 3. For every finite dimensional vector space V, Theorem 1.5 in Chap. 1 establishes that $T(T(V)) \cong V$. Therefore, applying T twice returns each space and each map in a sequence to itself. □

Remark 6.1 Proposition 6.2 is not necessarily true for sequences of infinite dimensional vector spaces!

Example 6.1 Proposition 6.2 establishes an isomorphism of categories, not of sequences. Consider the sequence given by

$$\cdots \longrightarrow \mathbb{R}^3 \xrightarrow{\begin{pmatrix} 3 & -2 & 0 \\ 1 & 4 & 2 \end{pmatrix}} \mathbb{R}^2 \xrightarrow{\begin{pmatrix} 1 & -1 \\ 2 & 3 \\ -1 & 1 \end{pmatrix}} \mathbb{R}^3 \longrightarrow \cdots$$

Applying T to this sequence yields

$$\cdots \longleftarrow \mathbb{R}^3 \xleftarrow{\begin{pmatrix} 3 & 1 \\ -2 & 4 \\ 0 & 2 \end{pmatrix}} \mathbb{R}^2 \xleftarrow{\begin{pmatrix} 1 & 2 & -1 \\ -1 & 3 & 1 \end{pmatrix}} \mathbb{R}^3 \longleftarrow \cdots$$

Evidently the two sequences are *not* isomorphic. However, if we apply T once more, we obtain the original sequence.

Now let us consider sequences of sequences. A typical sequence of sequences can be expressed by the diagram

$$\cdots \xrightarrow{b^{\bullet,k-1}} (A^{\bullet,k}, a^{\bullet,k}) \xrightarrow{b^{\bullet,k}} (A^{\bullet,k+1}, a^{\bullet,k+1}) \xrightarrow{b^{\bullet,k+1}} \cdots$$

There are two kinds of maps. The $a^{\bullet,\bullet}$ maps are those *within* each term, while the $b^{\bullet,\bullet}$ maps are those *between* terms. Each space and map in a sequence of sequences has *two* indices. We will use the first index to represent the index within each term. We will use the second index to select the terms individually.

A different representation for a sequence of sequences is a *grid diagram*, in which the spaces and maps are arranged in a square grid. For consistency with earlier constructions in this book (most notably Section 3.3 in Chap. 3), we will choose to make the columns of such a grid correspond to the terms in the sequence. Using this convention, the above sequence can be written

$$\begin{array}{ccccccc}
 & & \vdots & & \vdots & & \\
 & & \big\downarrow{\scriptstyle a^{0,1}} & & \big\downarrow{\scriptstyle a^{0,2}} & & \\
\cdots & \xrightarrow{b^{1,0}} & A^{1,1} & \xrightarrow{b^{1,1}} & A^{1,2} & \xrightarrow{b^{1,2}} & \cdots \\
 & & \big\downarrow{\scriptstyle a^{1,1}} & & \big\downarrow{\scriptstyle a^{1,2}} & & \\
\cdots & \xrightarrow{b^{2,0}} & A^{2,1} & \xrightarrow{b^{2,1}} & A^{2,2} & \xrightarrow{b^{2,2}} & \cdots \\
 & & \big\downarrow{\scriptstyle a^{2,1}} & & \big\downarrow{\scriptstyle a^{2,2}} & & \\
 & & \vdots & & \vdots & &
\end{array}$$

Notice that the first index denotes the row and the second index denotes the column of each term in the diagram.

A simple summary of a grid diagram is the dimension of the space at each index. Since the indices are ordered pairs of integers, this is captured by the *Hilbert function* $h : \mathbb{Z}^2 \to \mathbb{N}$

$$h\left((A^{\bullet,\bullet}, a^{\bullet,\bullet}, b^{\bullet,\bullet})\right)(m, n) := \dim A^{m,n}.$$

(See Oudot and Scoccola 2024 for a thorough treatment of the Hilbert function.)

Exercise 6.2 Exhibit two non-isomorphic grid diagrams with the same Hilbert function.

Exercise 6.3 Prove or give a counterexample to the following statement: Two isomorphic grid diagrams necessarily have the same Hilbert function.

Since the indices for both the spaces and the maps can either increase or decrease when the maps are followed, there are four possible ways that the maps and their respective indices can be constructed. Each of these ways corresponds to a different category, which will be defined formally after we examine the structure of their typical objects (Definition 6.2).

We will continue to use superscripts for indices that increase as the maps are followed and subscripts for indices that decrease. The category in which the indices increase *within* each terms, but for which the indices *between* the terms decrease will be donated as $\mathbf{Seq}_\bullet(\mathbf{Seq}^\bullet)$. Similarly, the category in which indices increase both within and between the terms will be denoted as $\mathbf{Seq}^\bullet(\mathbf{Seq}^\bullet)$. Informal definitions for the other two categories are similar.

In the two grid diagrams below, a typical object of $\mathbf{Seq}^\bullet(\mathbf{Seq}^\bullet)$ is shown on the left and an object of $\mathbf{Seq}_\bullet(\mathbf{Seq}_\bullet)$ is shown on the right,

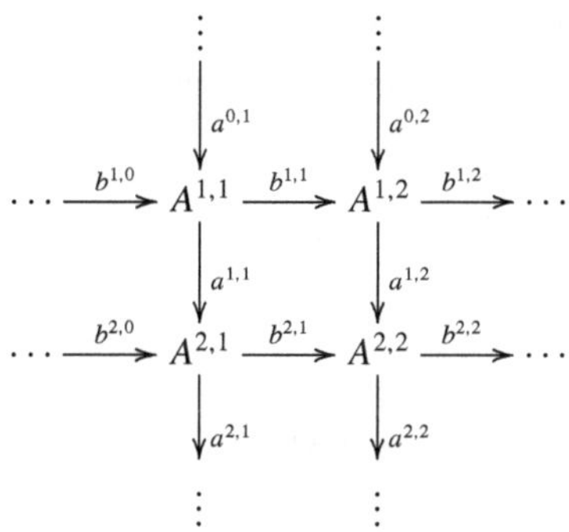

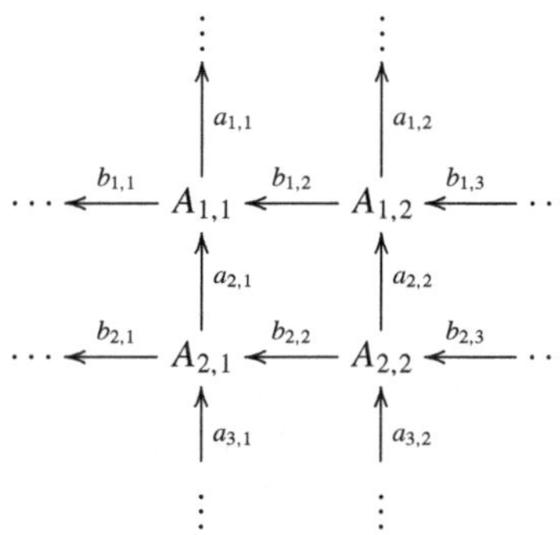

The other two possibilities, where the row and column maps "go" in opposite directions are given by the grid diagrams below. The spaces and maps will be notated with indices that are mixed sub- and superscripts. An element of $\mathbf{Seq}^\bullet(\mathbf{Seq}_\bullet)$ is shown on the left and an object of $\mathbf{Seq}_\bullet(\mathbf{Seq}^\bullet)$ is shown on the right,

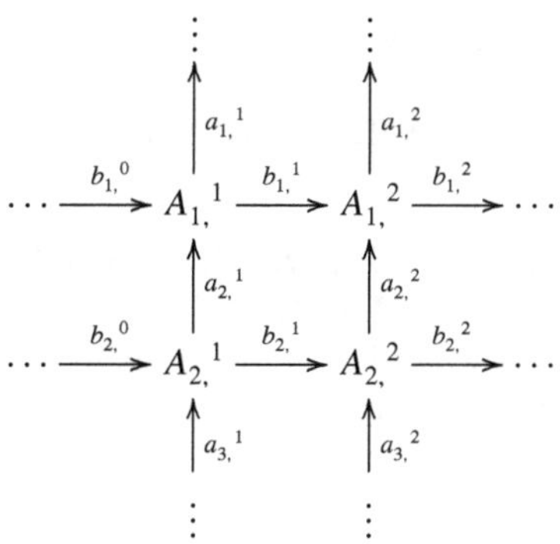

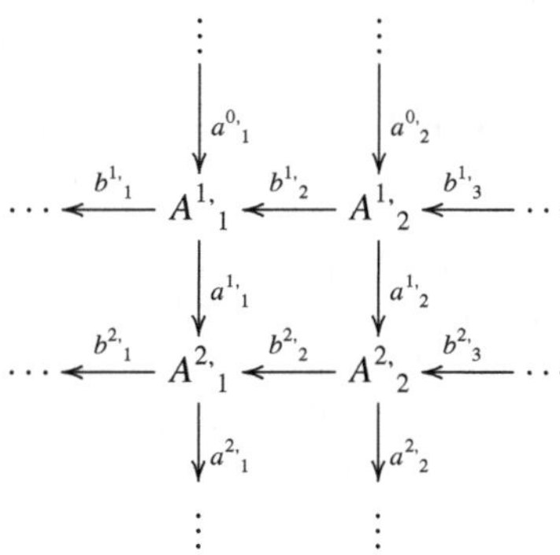

The formal definitions of the corresponding four categories is inspired by the structure of these four grid diagrams.

Definition 6.2 The category $\mathbf{Seq}^\bullet(\mathbf{Seq}^\bullet)$ has for its objects sequences of sequences of vector spaces. Each term is therefore a sequence, which has its own terms. The objects of this category are given by tuples $(A^{\bullet,\bullet}, a^{\bullet,\bullet}, b^{\bullet,\bullet})$ of spaces, maps within terms, and maps between terms, respectively.

More explicitly, for every $i, j \in \mathbb{Z}$, each object of $\mathbf{Seq}^\bullet(\mathbf{Seq}^\bullet)$ has a space $A^{i,j}$, a map $a^{i,j} : A^{i,j} \to A^{i+1,j}$, and a map $b^{i,j} : A^{i,j} \to A^{i,j+1}$.
We require that the resulting diagram of all $a^{\bullet,\bullet}$ and $b^{\bullet,\bullet}$ maps commutes, which means that

$$b^{i+1,j} \circ a^{i,j} = a^{i,j+1} \circ b^{i,j}$$

for all i and j.

Each morphism $m^{\bullet,\bullet} : (A^{\bullet,\bullet}, a^{\bullet,\bullet}, b^{\bullet,\bullet}) \to (C^{\bullet,\bullet}, c^{\bullet,\bullet}, d^{\bullet,\bullet})$ of $\mathbf{Seq}^\bullet(\mathbf{Seq}^\bullet)$ is defined by maps $m^{i,j} : A^{i,j} \to C^{i,j}$ such that the resulting (large!) diagram consisting of all $a^{\bullet,\bullet}, b^{\bullet,\bullet}, c^{\bullet,\bullet}, d^{\bullet,\bullet}$, and $m^{\bullet,\bullet}$ maps commutes. In addition to the equation above, the following two equations must also hold

$$m^{i+1,j} \circ a^{i,j} = a^{i,j} \circ m^{i,j} \text{ and } m^{i,j+1} \circ b^{i,j} = b^{i,j} \circ m^{i,j}$$

for all i and j.

The category $\mathbf{Seq}_\bullet(\mathbf{Seq}_\bullet)$ is defined similarly, but the $+1$ expressions in the indices are replaced with -1. Namely, an object of $\mathbf{Seq}_\bullet(\mathbf{Seq}_\bullet)$ is given by spaces $A_{i,j}$, maps $a_{i,j} : A_{i,j} \to A_{i-1,j}$, and maps $b_{i,j} : A_{i,j} \to A_{i,j-1}$.

An object of $\mathbf{Seq}^\bullet(\mathbf{Seq}_\bullet)$ is given by spaces $A_{i,}{}^{j}$, maps $a_{i,}{}^{j} : A_{i,}{}^{j} \to A_{i-1,}{}^{j}$, and maps $b_{i,}{}^{j} : A_{i,}{}^{j} \to A_{i,}{}^{j+1}$.

Finally, an object of $\mathbf{Seq}_\bullet(\mathbf{Seq}^\bullet)$ is given by spaces $A^{i,}{}_{j}$, maps $a^{i,}{}_{j} : A^{i,}{}_{j} \to A^{i+1,}{}_{j}$, and maps $b^{i,}{}_{j} : A^{i,}{}_{j} \to A^{i,}{}_{j-1}$.

Fortunately, one does not usually have to handle this dizzying collection of possible variations of diagrams and indices! The key insight is that an object of any of these categories can be converted to one of any other by reflecting its grid diagram.

Proposition 6.3 *The categories* $\mathbf{Seq}^\bullet(\mathbf{Seq}^\bullet)$, $\mathbf{Seq}^\bullet(\mathbf{Seq}_\bullet)$, $\mathbf{Seq}_\bullet(\mathbf{Seq}^\bullet)$, *and* $\mathbf{Seq}_\bullet(\mathbf{Seq}_\bullet)$ *are all isomorphic to each other.*

Proof Concretely, let us construct an isomorphism $F : \mathbf{Seq}^\bullet(\mathbf{Seq}^\bullet) \to \mathbf{Seq}^\bullet(\mathbf{Seq}_\bullet)$. Consider the morphism $m^{\bullet,\bullet} : (A^{\bullet,\bullet}, a^{\bullet,\bullet}, b^{\bullet,\bullet}) \to (C^{\bullet,\bullet}, c^{\bullet,\bullet}, d^{\bullet,\bullet})$ of $\mathbf{Seq}^\bullet(\mathbf{Seq}^\bullet)$. First, handle the objects. The domain of m is shown at left below in its usual form and reflected vertically "into position". It is ready for transforming into an object of $\mathbf{Seq}^\bullet(\mathbf{Seq}_\bullet)$ at right,

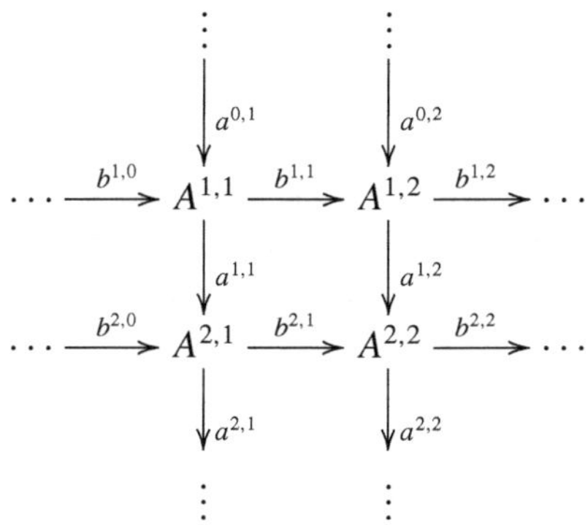

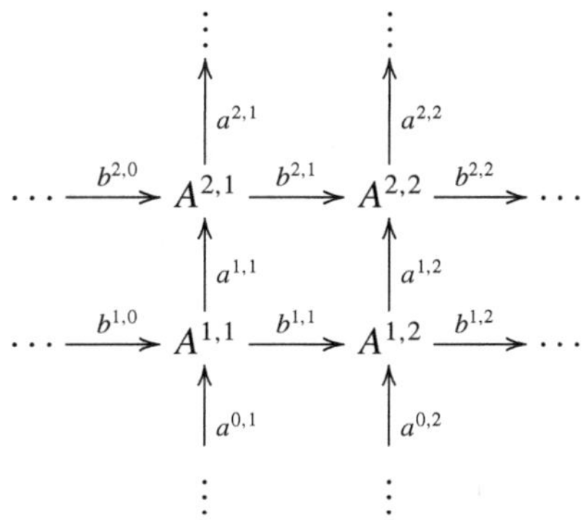

The reader should notice that this is merely a cosmetic change! As a second step, let us define

$$F((A^{\bullet,\bullet}, a^{\bullet,\bullet}, b^{\bullet,\bullet})) = (E_{\bullet,}{}^{\bullet}, e_{\bullet,}{}^{\bullet}, f_{\bullet,}{}^{\bullet})$$

diagrammatically, by simply relabeling indices accordingly. That is, the domain of F is shown at left below, while the codomain is on the right,

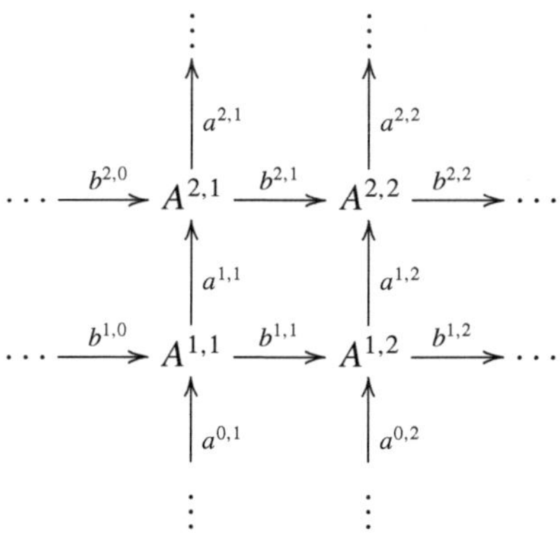

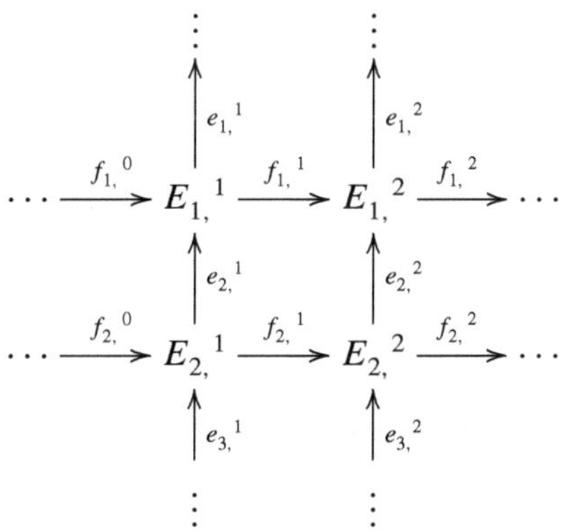

Rather explicitly,

$$E_{i,}{}^{j} := A^{3-i,j}, \; e_{i,}{}^{j} := a^{3-i,j}, \text{ and } f_{i,}{}^{j} := b^{3-i,j}.$$

(Note that the 3s appearing in this definition are rather arbitrary!)

To finish off the definition of the functor F, we need to handle the morphisms. With the previous definitions in hand, the same reindexing can be applied to the component maps of the morphism $m^{\bullet,\bullet}$. That is,

$$(F(m^{\bullet,\bullet}))_{i,}{}^{j} := m^{3-i,j}.$$

Functoriality of F follows without any further issues, since composition of functions does not disrupt the indices. Additionally, it should be clear that F is bijective on both objects and morphisms, and moreover that it is its own inverse. Thus, F is an isomorphism of categories.

Although $F : \mathbf{Seq}^{\bullet}(\mathbf{Seq}^{\bullet}) \to \mathbf{Seq}^{\bullet}(\mathbf{Seq}_{\bullet})$ only applied a vertical reflection, it should be clear to the reader that the other two categories are isomorphic to $\mathbf{Seq}^{\bullet}(\mathbf{Seq}^{\bullet})$ via a very similar argument and hence are isomorphic to each other. □

Exercise 6.4 Supply the remaining constructions to complete the proof of Proposition 6.3.

Proposition 6.3 means that when not performing specific calculations, it is not necessary to distinguish between the four possible categories. We will usually therefore speak of $\mathbf{Seq}(\mathbf{Seq})$ without any ambiguity. Should calculations be necessary, we will sometimes use $\mathbf{Seq}^{\bullet}(\mathbf{Seq}^{\bullet})$ for notational consistency.

Proposition 6.4 *If two sequences of sequences are isomorphic in* $\mathbf{Seq}(\mathbf{Seq})$ *then each corresponding pair of terms is isomorphic in* $\mathbf{Seq}$.

Proof This is merely the definition, because an isomorphism of $\mathbf{Seq}(\mathbf{Seq})$ is of the form

$$\begin{array}{ccccccc} \cdots & \longrightarrow & (A^{\bullet,1}, a^{\bullet,1}) & \longrightarrow & (A^{\bullet,2}, a^{\bullet,2}) & \longrightarrow & \cdots \\ & & \downarrow & & \downarrow & & \\ \cdots & \longrightarrow & (B^{\bullet,1}, b^{\bullet,1}) & \longrightarrow & (B^{\bullet,2}, b^{\bullet,2}) & \longrightarrow & \cdots \end{array}$$

where the vertical arrows are isomorphisms in $\mathbf{Seq}$. □

Note that the converse of Proposition 6.4 is not true, as the next example shows.

Example 6.2 Consider the following two diagrams

$$\begin{array}{ccc} \mathbb{R}^2 & \xrightarrow{(1\ 0)} & \mathbb{R} \\ {\scriptstyle (0\ 1)}\downarrow & & \downarrow \\ \mathbb{R} & \longrightarrow & 0 \end{array} \qquad \begin{array}{ccc} \mathbb{R}^2 & \xrightarrow{(1\ 0)} & \mathbb{R} \\ {\scriptstyle (1\ 0)}\downarrow & & \downarrow \\ \mathbb{R} & \longrightarrow & 0 \end{array}$$

which can be thought of as objects in $\mathbf{Seq}(\mathbf{Seq})$ by extending with zero spaces and maps. Notice that these two diagrams are term-by-term isomorphic. Moreover, if we only consider the rows, they have the same barcode decomposition, namely

$$\begin{pmatrix} \mathbb{R} \longrightarrow \mathbb{R} \\ \\ 0 \longrightarrow 0 \end{pmatrix} \oplus \begin{pmatrix} \mathbb{R} \longrightarrow 0 \\ \\ 0 \longrightarrow 0 \end{pmatrix} \oplus \begin{pmatrix} 0 \longrightarrow 0 \\ \\ \mathbb{R} \longrightarrow 0 \end{pmatrix}$$

Additionally, they have the same column-wise barcode decompositions as well, namely

$$\begin{pmatrix} \mathbb{R} & 0 \\ \downarrow & \downarrow \\ \mathbb{R} & 0 \end{pmatrix} \oplus \begin{pmatrix} \mathbb{R} & 0 \\ \downarrow & \downarrow \\ 0 & 0 \end{pmatrix} \oplus \begin{pmatrix} 0 & \mathbb{R} \\ \downarrow & \downarrow \\ 0 & 0 \end{pmatrix}$$

Nevertheless, these are not isomorphic objects in **Seq**(**Seq**). To see this, "straighten out" the two nontrivial edges in the original diagrams. The resulting diagrams are

$$\mathbb{R} \xleftarrow{(1\ 0)} \mathbb{R}^2 \xrightarrow{(0\ 1)} \mathbb{R} \qquad\qquad \mathbb{R} \xleftarrow{(1\ 0)} \mathbb{R}^2 \xrightarrow{(1\ 0)} \mathbb{R}$$

If the two objects were to be isomorphic, we should be able to find an isomorphism $m : \mathbb{R}^2 \to \mathbb{R}^2$ such that the diagram below commutes

$$\begin{array}{ccccc} \mathbb{R} & \xleftarrow{(1\ 0)} & \mathbb{R}^2 & \xrightarrow{(0\ 1)} & \mathbb{R} \\ \downarrow{\scriptstyle \mathrm{id}} & & \downarrow{\scriptstyle m} & & \downarrow{\scriptstyle \mathrm{id}} \\ \mathbb{R} & \xleftarrow{(1\ 0)} & \mathbb{R}^2 & \xrightarrow{(1\ 0)} & \mathbb{R} \end{array}$$

This condition implies that

$$(1\ 0) \circ m = (1\ 0)$$

and

$$(1\ 0) \circ m = (0\ 1),$$

which is clearly a contradiction.

Exercise 6.5 Construct another pair of grid diagrams like those in Example 6.2. They must have the same Hilbert function, same row-wise and column-wise barcode decompositions, yet they are not isomorphic.

It is known that one can write a sequence of sequences as a direct sum of *indecomposable* sequences of sequences (Crawley-Boevey 2015). These indecomposables are analogous to bars in the barcode decomposition of a sequence of vector spaces.

One might hope that a generalization of bars into rectangles might yield all of these indecomposables, but in fact there are many indecomposables that are not rectangles.

Example 6.3 (Botnan et al. 2024) It is not possible to write the diagram below as a direct sum of two nontrivial grid diagrams,

$$
\begin{array}{ccccc}
0 & \longrightarrow & \mathbb{R} & \xrightarrow{\text{id}} & \mathbb{R} \\
\big\downarrow & & \big\downarrow{\scriptstyle \begin{pmatrix}0\\1\end{pmatrix}} & & \big\downarrow{\scriptstyle \text{id}} \\
\mathbb{R} & \xrightarrow{\begin{pmatrix}1\\0\end{pmatrix}} & \mathbb{R}^2 & \xrightarrow{(1\ 1)} & \mathbb{R} \\
{\scriptstyle \text{id}}\big\downarrow & & \big\downarrow{\scriptstyle (1\ 0)} & & \big\downarrow \\
\mathbb{R} & \xrightarrow[\text{id}]{} & \mathbb{R} & \longrightarrow & 0
\end{array}
$$

Any direct sum decomposition of the above grid diagram will also yield a nontrivial direct sum decomposition of both its rows and columns. These row-wise and column-wise decompositions are the usual barcode decompositions. The columns have barcode decomposition

$$
\begin{pmatrix}
0 & 0 & 0 \\
\downarrow & \downarrow & \downarrow \\
\mathbb{R} & \mathbb{R} & 0 \\
\downarrow & \downarrow & \downarrow \\
\mathbb{R} & \mathbb{R} & 0
\end{pmatrix}
\oplus
\begin{pmatrix}
0 & \mathbb{R} & \mathbb{R} \\
\downarrow & \downarrow & \downarrow \\
0 & \mathbb{R} & \mathbb{R} \\
\downarrow & \downarrow & \downarrow \\
0 & 0 & 0
\end{pmatrix}
$$

while the rows have the barcode decomposition

$$
\begin{pmatrix}
0 \longrightarrow 0 \longrightarrow 0 \\
\\
\mathbb{R} \longrightarrow \mathbb{R} \longrightarrow \mathbb{R} \\
\\
\mathbb{R} \longrightarrow \mathbb{R} \longrightarrow 0
\end{pmatrix}
\oplus
\begin{pmatrix}
0 \longrightarrow \mathbb{R} \longrightarrow \mathbb{R} \\
\\
0 \longrightarrow \mathbb{R} \longrightarrow 0 \\
\\
0 \longrightarrow 0 \longrightarrow 0
\end{pmatrix}
$$

These two decompositions together imply that the original grid diagram should decompose as

$$\left(\begin{array}{ccccc} 0 & \to & 0 & \to & 0 \\ \downarrow & & \downarrow & & \downarrow \\ \mathbb{R} & \to & \mathbb{R} & \to & \mathbb{R} \\ \downarrow & & \downarrow & & \downarrow \\ \mathbb{R} & \to & \mathbb{R} & \to & 0 \end{array}\right) \oplus \left(\begin{array}{ccccc} 0 & \to & \mathbb{R} & \to & \mathbb{R} \\ \downarrow & & \downarrow & & \downarrow \\ 0 & \to & \mathbb{R} & \to & 0 \\ \downarrow & & \downarrow & & \downarrow \\ 0 & \to & 0 & \to & 0 \end{array}\right)$$

or as

$$\left(\begin{array}{ccccc} 0 & \to & \mathbb{R} & \to & \mathbb{R} \\ \downarrow & & \downarrow & & \downarrow \\ \mathbb{R} & \to & \mathbb{R} & \to & \mathbb{R} \\ \downarrow & & \downarrow & & \downarrow \\ 0 & \to & 0 & \to & 0 \end{array}\right) \oplus \left(\begin{array}{ccccc} 0 & \to & 0 & \to & 0 \\ \downarrow & & \downarrow & & \downarrow \\ 0 & \to & \mathbb{R} & \to & 0 \\ \downarrow & & \downarrow & & \downarrow \\ \mathbb{R} & \to & \mathbb{R} & \to & 0 \end{array}\right)$$

(among several other, less plausible options) but neither of these recovers the original grid diagram including its maps.

Exercise 6.6 (An example of *antenna attachment* Bauer and Scoccola 2025, Definition 27) Show that it is not possible to write the following grid diagram

$$\begin{array}{ccccccc} 0 & \to & \mathbb{R} & \to & 0 & \to & 0 \\ \downarrow & & \downarrow & & \downarrow & & \downarrow \\ 0 & \to & \mathbb{R} & \to & 0 & \to & 0 \\ \downarrow & & \downarrow & & \downarrow & & \downarrow \\ 0 & \to & \mathbb{R} & \to & 0 & \to & 0 \\ \downarrow & & \downarrow & & \downarrow & & \downarrow \\ 0 & \to & \mathbb{R} & \to & \mathbb{R} & \to & \mathbb{R} \\ \downarrow & & \downarrow & & \downarrow & & \downarrow \\ 0 & \to & 0 & \to & \mathbb{R} & \to & \mathbb{R} \end{array}$$

(where each map is (as usual) assumed to have maximal rank) as a direct sum of two nontrivial grid diagrams.

Remark 6.2 In some sense, **Seq**(**Seq**) is difficult to study because it is quite expressive. If we generalize away from real vector spaces, every finite dimensional algebra over a field $\mathbb{K}$ can be represented as an object in **Seq**(**Seq**) with terms being $\mathbb{K}$-vector spaces (Simson 1993).

6.2 Chain Complexes of Sequences

The previous section was a preamble to the exploration of **Kom**(**Seq**), the category of chain complexes of sequences. There is another potential ambiguity, since we could also study **Seq**(**Kom**), the category of sequences of chain complexes. How do $\mathbf{Seq}^\bullet(\mathbf{Kom}^\bullet)$ and $\mathbf{Kom}^\bullet(\mathbf{Seq}^\bullet)$ differ? Graphically, the chain complex condition is applied either column-wise (to the $a^{\bullet,\bullet}$ maps) or row-wise (to the $b^{\bullet,\bullet}$ maps), respectively. Once again, these two categories are isomorphic but not identical. The isomorphism amounts to transposing the grid of maps, instead of reflecting vertically or horizontally as was done in Proposition 6.3. Exchanging the positions of **Seq** and **Kom** along with the indexing leads to eight different possible categories. Fortunately all of these possibilities are isomorphic to each other.

Proposition 6.5 *Every category of sequences of chain complexes and every category of chain complexes of sequences is isomorphic to* $\mathbf{Kom}^\bullet(\mathbf{Seq}^\bullet)$.

Proof There are 28 pairs of categories that need to be shown isomorphic. These fall into two groups. Clearly the following are pairs of isomorphic categories via diagram transposition. The argument works just like the proof of Proposition 6.3, *mutatis mutandis*:

- $\mathbf{Seq}^\bullet(\mathbf{Kom}^\bullet)$ and $\mathbf{Kom}^\bullet(\mathbf{Seq}^\bullet)$
- $\mathbf{Seq}_\bullet(\mathbf{Kom}_\bullet)$ and $\mathbf{Kom}_\bullet(\mathbf{Seq}_\bullet)$
- $\mathbf{Seq}_\bullet(\mathbf{Kom}^\bullet)$ and $\mathbf{Kom}^\bullet(\mathbf{Seq}_\bullet)$
- $\mathbf{Seq}^\bullet(\mathbf{Kom}_\bullet)$ and $\mathbf{Kom}_\bullet(\mathbf{Seq}^\bullet)$

Now let us consider the other 24 pairs of categories, where the indexing direction on the chain complex part differs, like

- $\mathbf{Seq}^\bullet(\mathbf{Kom}_\bullet)$ and $\mathbf{Kom}^\bullet(\mathbf{Seq}_\bullet)$
- $\mathbf{Seq}_\bullet(\mathbf{Kom}^\bullet)$ and $\mathbf{Kom}_\bullet(\mathbf{Seq}^\bullet)$
- and so on.

These pairs are *also* isomorphic, though in addition to reflecting the grid diagrams, we must reverse the indexing in either the rows or columns. That each of these reversals yields an isomorphism is a consequence of Proposition 6.1. □

Based on Proposition 6.5, we need only focus on one of these categories. We will prefer to use **Kom**(**Seq**) to align with the notation used in the Snake Lemma (Sect. 3.3 of Chap. 3), namely that short exact sequences of sequences yield long exact sequences of sequences or of vector spaces.

Suppose that we have an object **Kom(Seq)**,

$$\cdots \xrightarrow{b^{\bullet,k-1}} (A^{\bullet,k}, a^{\bullet,k}) \xrightarrow{b^{\bullet,k}} (A^{\bullet,k+1}, a^{\bullet,k+1}) \xrightarrow{b^{\bullet,k+1}} \cdots$$

also given by the grid diagram

$$\begin{array}{ccccccc}
 & & \vdots & & \vdots & & \\
 & & \downarrow{\scriptstyle a^{0,1}} & & \downarrow{\scriptstyle a^{0,2}} & & \\
\cdots & \xrightarrow{b^{1,0}} & A^{1,1} & \xrightarrow{b^{1,1}} & A^{1,2} & \xrightarrow{b^{1,2}} & \cdots \\
 & & \downarrow{\scriptstyle a^{1,1}} & & \downarrow{\scriptstyle a^{1,2}} & & \\
\cdots & \xrightarrow{b^{2,0}} & A^{2,1} & \xrightarrow{b^{2,1}} & A^{2,2} & \xrightarrow{b^{2,2}} & \cdots \\
 & & \downarrow{\scriptstyle a^{2,1}} & & \downarrow{\scriptstyle a^{2,2}} & & \\
 & & \vdots & & \vdots & &
\end{array}$$

which means that $b^{i,j+1} \circ b^{i,j} = 0$ for all i and j. It is possible to take homology row-wise. Due to the functoriality of homology, the $a^{\bullet,\bullet}$ maps induce maps on the resulting homology spaces. This can be expressed diagrammatically as

$$\begin{array}{cccc}
 & \vdots & \vdots & \\
 & \downarrow{\scriptstyle a_*^{0,1}} & \downarrow{\scriptstyle a_*^{0,2}} & \\
\cdots & H^1(A^{1,\bullet}, b^{1,\bullet}) & H^2(A^{1,\bullet}, b^{1,\bullet}) & \cdots \\
 & \downarrow{\scriptstyle a_*^{1,1}} & \downarrow{\scriptstyle a_*^{1,2}} & \\
\cdots & H^1(A^{2,\bullet}, b^{2,\bullet}) & H^2(A^{2,\bullet}, b^{2,\bullet}) & \cdots \\
 & \downarrow{\scriptstyle a_*^{2,1}} & \downarrow{\scriptstyle a_*^{2,2}} & \\
 & \vdots & \vdots &
\end{array}$$

where we have omitted the horizontal zero maps. This diagram can be summarized as

$$\cdots \qquad (H^1(A^{\diamond,\bullet}, b^{\diamond,\bullet}), a_*^{\diamond,1}) \qquad (H^2(A^{\diamond,\bullet}, b^{\diamond,\bullet}), a_*^{\diamond,2}) \qquad \cdots$$

using two bullet styles to disambiguate the indices. The homology acts "in the direction of" the round bullet $\bullet$.

Exercise 6.7 Consider the half-infinite bar sequence $\mathbb{R}^{(-\infty,k)}$, which has the diagram

$$\cdots \to \mathbb{R} \to \mathbb{R} \cdots \to \mathbb{R} \to 0 \to \cdots$$

in which the last nontrivial term is given index k.

1. Show that there is exactly one way to construct a chain complex of sequences the form

$$\cdots \to \mathbb{R}^{(-\infty,k)} \to \mathbb{R}^{(-\infty,k+1)} \to \cdots$$

2. Compute all homology spaces for the chain complex in (1).
3. On the other hand, show that there are many ways to construct a chain complex of sequences of the form

$$\cdots \to \mathbb{R}^{(-\infty,k)} \to \mathbb{R}^{(-\infty,k-1)} \to \cdots$$

4. Does homology completely characterize all of the possibilities in (3)? Prove your answer.

Proposition 6.6 *For each k, homology is a covariant functor* H^k : **Kom**(**Seq**) $\to$ **Seq**, *defined by*

$$H^k(A^{\diamond,\bullet}, a^{\diamond,\bullet}, b^{\diamond,\bullet}) := (H^{\diamond,k}, a_*^{\diamond,k}),$$

where

$$H^{i,k} := H^k(A^{i,\bullet}, b^{i,\bullet}),$$

and $a_*^{i,k}$ *is the map induced on* $H^k(A^{i,\bullet}, b^{i,\bullet}) \to H^k(A^{i+1,\bullet}, b^{i+1,\bullet})$ *by* $a^{i,k} : A^{i,k} \to A^{i+1,k}$.

Based upon our convention of writing a grid diagram in **Kom**(**Seq**) so that the rows are chain complexes, we will usually call $H^k(A^{i,\bullet}, b^{i,\bullet})$ *row-wise homology*, even if sometimes the grid diagram is transposed to conserve space on the page.

Proof We need to check that this construction takes morphisms in **Kom**(**Seq**) to morphisms in **Seq**, namely that given a morphism $m^{\diamond,\bullet} : (A^{\diamond,\bullet}, a^{\diamond,\bullet}, b^{\diamond,\bullet}) \to (C^{\diamond,\bullet}, c^{\diamond,\bullet}, d^{\diamond,\bullet})$ the diagram

$$\begin{array}{ccccccc}
\cdots \xrightarrow{a_*^{i-1,\bullet}} & H^k(A^{i,\bullet}, b^{i,\bullet}) & \xrightarrow{a_*^{i,\bullet}} & H^k(A^{i+1,\bullet}, b^{i+1,\bullet}) & \xrightarrow{a_*^{i+1,\bullet}} \cdots \\
& \downarrow{\scriptstyle m_*^{i,\bullet}} & & \downarrow{\scriptstyle m_*^{i+1,\bullet}} & \\
\cdots \xrightarrow{c_*^{i-1,\bullet}} & H^k(C^{i,\bullet}, d^{i,\bullet}) & \xrightarrow{c_*^{i,\bullet}} & H^k(C^{i+1,\bullet}, d^{i+1,\bullet}) & \xrightarrow{c_*^{i+1,\bullet}} \cdots
\end{array}$$

commutes, where the $m_*^{\diamond,\bullet}$ maps are the ones induced on homology by $m^{\diamond,\bullet}$. This is merely a consequence of the functoriality of homology for chain complexes of vector spaces. □

Following the constructions from earlier chapters, it is sometimes useful to bundle all of the H^k functors together, so that we have a single *homology functor* H : **Kom**(**Seq**) → **Kom**(**Seq**) that retains the zero maps in the earlier diagrams.

Definition 6.3 *Homology* is a functor H : **Kom**(**Seq**) → **Kom**(**Seq**) that takes an object $(A^{\bullet,\bullet}, a^{\bullet,\bullet}, b^{\bullet,\bullet})$ to

$$H\left((A^{\bullet,\bullet}, a^{\bullet,\bullet}, b^{\bullet,\bullet})\right) := (H^{\square}(A^{\diamond,\bullet}, b^{\diamond,\bullet}), a_*^{\diamond,\square}, 0).$$

Although we can use the Escolar-Hiraoka decomposition to identify bar chain maps in each row of a chain complex of sequences, the rows can interact to produce much more complexity than in **Kom**. We have already seen an instance of this in Example 2.8 in Chap. 2. There is even more "room" for complexity in **Kom**(**Seq**) as the next example shows.

Example 6.4 Consider the following two diagrams defining objects of **Kom**(**Seq**) in which every arrow represents a linear map of maximal rank

$$\begin{array}{ccccc}
& \vdots & & \vdots & \\
& \downarrow & & \downarrow & \\
\cdots \longrightarrow & \mathbb{R} & \longrightarrow & \mathbb{R} & \longrightarrow \cdots \\
& \downarrow & & \downarrow & \\
\cdots \longrightarrow & \mathbb{R} & \longrightarrow & 0 & \longrightarrow \cdots \\
& \downarrow & & \downarrow & \\
\cdots \longrightarrow & \mathbb{R} & \longrightarrow & 0 & \longrightarrow \cdots \\
& \downarrow & & \downarrow & \\
\cdots \longrightarrow & \mathbb{R} & \longrightarrow & 0 & \longrightarrow \cdots \\
& \downarrow & & \downarrow & \\
& \vdots & & \vdots &
\end{array}
\qquad\qquad
\begin{array}{ccccc}
& \vdots & & \vdots & \\
& \downarrow & & \downarrow & \\
\cdots \longrightarrow & 0 & \longrightarrow & 0 & \longrightarrow \cdots \\
& \downarrow & & \downarrow & \\
\cdots \longrightarrow & \mathbb{R} & \longrightarrow & 0 & \longrightarrow \cdots \\
& \downarrow & & \downarrow & \\
\cdots \longrightarrow & \mathbb{R} & \longrightarrow & 0 & \longrightarrow \cdots \\
& \downarrow & & \downarrow & \\
\cdots \longrightarrow & 0 & \longrightarrow & 0 & \longrightarrow \cdots \\
& \downarrow & & \downarrow & \\
& \vdots & & \vdots &
\end{array}$$

A morphism between these two objects can be defined by the more elaborate diagram

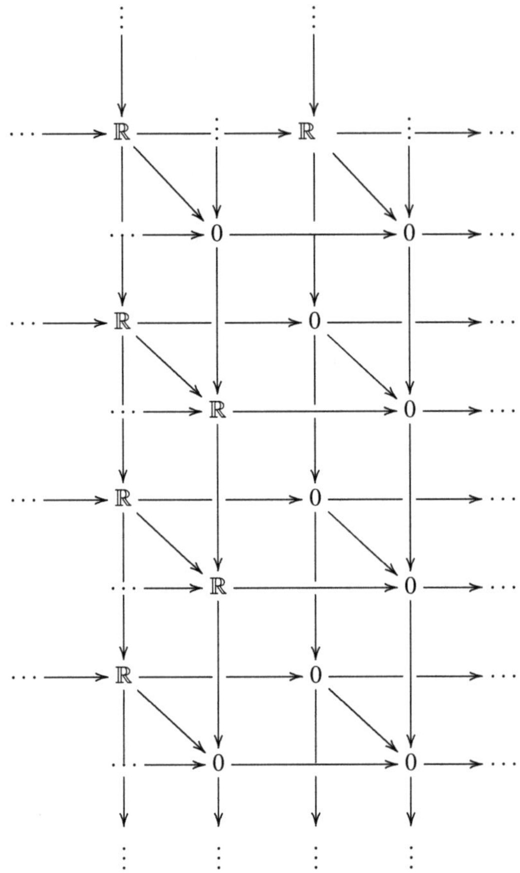

This diagram is effectively a *single* bar chain map for **Kom**(**Seq**). Specifically, it is *indecomposable*, which means that there is no direct sum decomposition of this diagram into smaller such diagrams (those with more trivial terms).

To obtain other examples of chain maps in **Kom**(**Seq**) that cannot be decomposed further, one can extend the diagram in the above example in the vertical direction by adding as many copies of $\mathbb{R}$ as desired. Unlike the case of **Kom**, in which every chain map decomposes into a direct sum of a finite list of possible bar chain maps, such a list for **Kom**(**Seq**) of possible direct summands must be infinite. We express this situation by saying that **Kom** is *representation finite* whereas **Kom**(**Seq**) is *representation infinite*.

Exercise 6.8 Show that **Seq** is representation infinite.

Exercise 6.9 The fact that **Kom**(**Seq**) is representation infinite appears to come from the fact that **Seq** is representation infinite (see previous exercise). This is misleading, however. Show that **Kom**(**Kom**) is also representation infinite.

Exercise 6.10 Inspired by the previous exercise, construct two distinct infinite families of indecomposables in **Kom**(**Kom**).

Exercise 6.11 Generalize the antenna attachment construction in Exercise 6.6 (to three dimensions!) to create another indecomposable morphism of **Kom** (**Seq**).

In order to simplify the characterization of objects in **Kom**(**Seq**), we will loosen the notion of isomorphism a bit, as we did for chain complexes of vector spaces. In Sect. 3.4 of Chap. 3 this led to the notion of a *quasi-isomorphism* in Definition 3.3 in Chap. 3. That notion applies to chain complexes of sequences virtually without change.

Definition 6.4 A chain map between chain complexes of sequences (a morphism of **Kom**(**Seq**)) is a *quasi-isomorphism* if the sequence map induced by the homology functor is an isomorphism in **Kom**(**Seq**). We will say that two chain complexes of sequences are *quasi-isomorphic* if a quasi-isomorphism between them exists.

Lemma 6.1 *The existence of a quasi-isomorphism between two objects of* **Kom**(**Seq**) *is an equivalence relation. Specifically,*

Reflexivity *The identity chain map* $\mathrm{id}_{(A^{\bullet,\bullet},a^{\bullet,\bullet},b^{\bullet,\bullet})} : (A^{\bullet,\bullet}, a^{\bullet,\bullet}, b^{\bullet,\bullet}) \to (A^{\bullet,\bullet}, a^{\bullet,\bullet}, b^{\bullet,\bullet})$ *for any chain complex is a quasi-isomorphism,*

Symmetry *If there is a quasi-isomorphism* $q_1^{\bullet,\bullet} : (A^{\bullet,\bullet}, a^{\bullet,\bullet}, b^{\bullet,\bullet}) \to (C^{\bullet,\bullet}, c^{\bullet,\bullet}, d^{\bullet,\bullet})$*, then there is also a quasi-isomorphism* $q_2^{\bullet,\bullet} : (C^{\bullet,\bullet}, c^{\bullet,\bullet}, d^{\bullet,\bullet}) \to (A^{\bullet,\bullet}, a^{\bullet,\bullet}, b^{\bullet,\bullet})$*, and*

Transitivity *If there is a quasi-isomorphism* $q_1^{\bullet,\bullet} : (A^{\bullet,\bullet}, a^{\bullet,\bullet}, b^{\bullet,\bullet}) \to (C^{\bullet,\bullet}, c^{\bullet,\bullet}, d^{\bullet,\bullet})$ *and a quasi-isomorphism* $q_2^{\bullet,\bullet} : (C^{\bullet,\bullet}, c^{\bullet,\bullet}, d^{\bullet,\bullet}) \to (E^{\bullet,\bullet}, e^{\bullet,\bullet}, f^{\bullet,\bullet})$*, then their composition* $q_2^{\bullet,\bullet} \circ q_1^{\bullet,\bullet} : (A^{\bullet,\bullet}, a^{\bullet,\bullet}, b^{\bullet,\bullet}) \to (E^{\bullet,\bullet}, e^{\bullet,\bullet}, f^{\bullet,\bullet})$ *is also a quasi-isomorphism.*

The statement is reminiscent of Proposition 3.5 in Chap. 3. The proofs for reflexivity and transitivity are basically the same as in the case of chain complexes of vector spaces. The proof for symmetry in the case of chain complexes of vector spaces used bar chain map decompositions directly. That technique does not work for **Kom**(**Seq**) for two reasons: (1) **Kom**(**Seq**) is representation infinite, and (2) **Seq** morphisms need not respect the barcode decomposition (see Example 2.8 in Chap.2). We would need to check an infinite number of possibilities, many of which are not obvious maps between bars!

Proof Reflexivity Functoriality (Proposition 6.6) ensures that the identity chain map is a quasi-isomorphism.

Transitivity Follows from the functoriality of homology (Proposition 6.6), since the induced maps on the homology sequence are isomorphisms, their composition must also be an isomorphism.

Symmetry Apply Theorem 1.4 in Chap. 1 to the inverse of each induced map, namely

$$\left(q_1^{i,k}\right)_*^{-1} : H^k(C^{i,\bullet}, d^{i,\bullet}) \to H^k(A^{i,\bullet}, b^{i,\bullet})$$

so as to define a map $q_2^{i,k} : \ker d^{i,k} \to \ker b^{i,k}$. This definition is not unique, but the construction in Theorem 1.4 in Chap. 1 is well defined. Moreover, it yields a commutative diagram

$$\begin{array}{ccccccc} \cdots & \longrightarrow & \ker d^{i,k} & \xrightarrow{c^{i,k}} & \ker d^{i+1,k} & \longrightarrow & \cdots \\ & & \downarrow{\scriptstyle q_2^{i,k}} & & \downarrow{\scriptstyle q_2^{i+1,k}} & & \\ \cdots & \longrightarrow & \ker b^{i,k} & \xrightarrow{a^{i,k}} & \ker b^{i+1,k} & \longrightarrow & \cdots \end{array}$$

To see this, consider that a typical element of $\ker d^{i,k}$ can be written $x + \operatorname{image} d^{i,k-1}$, where $x \in \operatorname{coker} d^{i,k-1}$. The construction in Theorem 1.4 in Chap. 1 means that

$$\begin{aligned} a^{i,k}\left(q_2^{i,k}(x + \operatorname{image} d^{i,k-1})\right) &= a^{i,k}\left(q_2^{i,k}(x) + 0\right) \\ &= a^{i,k}\left(\left(q_1^{i,k}\right)^{-1}(x)\right). \end{aligned}$$

On the other hand,

$$\begin{aligned} q_2^{i+1,k}\left(c^{i,k}(x + \operatorname{image} d^{i,k-1})\right) &= q_2^{i+1,k}\left(c^{i,k}(x) + c^{i,k}(\operatorname{image} d^{i,k-1})\right) \\ &\subseteq q_2^{i+1,k}\left(c^{i,k}(x) + \operatorname{image} d^{i+1,k-1}\right) \\ &= q_2^{i+1,k}\left(c^{i,k}(x)\right) + 0 \\ &= \left(q_1^{i+1,k}\right)^{-1}\left(c^{i,k}(x)\right). \end{aligned}$$

The fact that q_1 is a morphism establishes that these two values are equal.

To complete the construction, we extend each $q_2^{i,k}$ to the rest of $C^{i,k}$ by direct summing with the zero map. With this definition, $q_2^{\bullet,\bullet}$ induces $(q_1^{\bullet,\bullet})_*^{-1}$ on homology. It is indeed a quasi-isomorphism. □

A somewhat more explicit way to define a quasi-isomorphism is that the sequence map induced by H^k is an isomorphism of **Seq** for each k. Intuitively, this can be stated easily.

Proposition 6.7 *Quasi-isomorphism classes of* **Kom**(**Seq**) *are characterized by column-wise barcodes for their homology sequences. Moreover, if two chain complexes of sequences are isomorphic, then they are also quasi-isomorphic.*

While Example 6.2 shows that isomorphism classes of **Seq**(**Seq**) can be quite subtle to characterize, Proposition 6.7 ensures that the quasi-isomorphism classes of **Kom**(**Seq**) are easier to handle.

Proof The first statement follows as a consequence of Lemma 6.1.

The second statement follows from Proposition 6.4, which implies that the row-wise homologies of both chain complexes of sequences are the same. □

One should not be too blithe about Proposition 6.7. It *does not* say that identical column-wise barcode decompositions are necessary, nor are they sufficient, for quasi-isomorphisms. The next two examples provide counterexamples.

Example 6.5 There are chain complexes of sequences that have identical column-wise barcode decompositions, but are not quasi-isomorphic. (They are not isomorphic, either.) For instance, two such chain complexes of sequences are given by extending the diagrams below by zeros

$$\begin{array}{ccc} \mathbb{R} & \xrightarrow{\text{id}} & \mathbb{R} \\ {\scriptstyle \text{id}}\downarrow & & \downarrow \\ \mathbb{R} & \longrightarrow & 0 \end{array} \qquad \begin{array}{ccc} \mathbb{R} & \xrightarrow{0} & \mathbb{R} \\ {\scriptstyle \text{id}}\downarrow & & \downarrow \\ \mathbb{R} & \longrightarrow & 0 \end{array}$$

Although these two diagrams have the same column-wise barcode decomposition, the homologies of the top rows differ. The homology sequences are given by the diagrams

$$\left(\begin{array}{cc} 0 & 0 \\ \downarrow & \downarrow \\ \mathbb{R} & 0 \end{array}\right) \text{ and } \left(\begin{array}{cc} \mathbb{R} & \mathbb{R} \\ \downarrow & \downarrow \\ \mathbb{R} & 0 \end{array}\right).$$

There are also quasi-isomorphic chain complexes of sequences that do not have identical column-wise barcodes. These are perhaps less surprising than the converse.

Example 6.6 Consider the chain complexes of sequences given by extending the following diagrams by zeros

$$\begin{array}{ccc} \mathbb{R} & \xrightarrow{\text{id}} & \mathbb{R} \\ {\scriptstyle \text{id}}\downarrow & & \downarrow \\ \mathbb{R} & \longrightarrow & 0 \end{array} \qquad \begin{array}{ccc} \mathbb{R} & \longrightarrow & 0 \\ \downarrow & & \downarrow \\ 0 & \longrightarrow & 0 \end{array}$$

These two diagrams are not isomorphic in **Kom**(**Seq**). After translating the left diagram up by one index, the row-wise homology of both diagrams is the same, it is simply the diagram at right above, so they are quasi-isomorphic.

Even if homology sequences for two chain complexes of sequences are isomorphic, there may not be a quasi-isomorphism that induces this isomorphism.

Example 6.7 Consider the two sequences

$$\begin{array}{ccc} 0 & \longrightarrow & \mathbb{R} \\ \downarrow & & \downarrow \\ \mathbb{R} & \longrightarrow & \mathbb{R} \\ \downarrow & & \downarrow \\ \mathbb{R} & \longrightarrow & 0 \end{array} \qquad \begin{array}{ccc} 0 & \longrightarrow & \mathbb{R} \\ \downarrow & & \downarrow \\ 0 & \longrightarrow & 0 \\ \downarrow & & \downarrow \\ \mathbb{R} & \longrightarrow & 0 \end{array}$$

The sequence on the right is the row-wise homology of the one on the left, so the two sequences have isomorphic homology.

On the other hand, there are no nontrivial chain maps between these two chain complexes in either direction. A chain map from the left to the right forces the component maps on the bottom to be trivial, while a chain map from the right to the left forces the component maps on the top to be trivial.

Exercise 6.12 Construct another example of a pair of chain complex of sequences that are quasi-isomorphic but not isomorphic.

6.3 Projective and Injective Sequences

When is it possible to extend a sequence map to a "larger" sequence in the domain? This kind of extension is not always possible, so to arrive at a sufficiently general extension property, this section establishes the preliminary tools to make this happen. Ultimately, we will obtain two general results in the subsequent sections (Propositions 6.10 and 6.16) for sequences and chain complexes. This section shows that there are two kinds of sequences, called *projective* and *injective* sequences, for which extensions are guaranteed. These sequences generalize a straightforward property of vector spaces, which is captured by Lemma 6.2 below.

Lemma 6.2 *Let V, W, X be vector spaces, $f : V \to W$ be a linear map, and $g : X \to W$ be a surjective linear map. Then there is another linear map $h : V \to X$ so that the following diagram commutes*

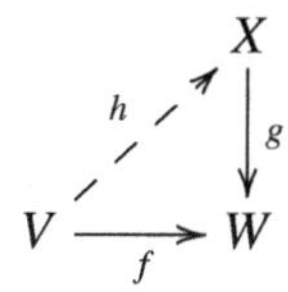

Proof Select a basis $\{v_1, \ldots, v_n\}$ for V. Since g is surjective, $g^{-1}(f(v_i))$ is nonempty for every i. Define $h(v_i)$ to be any element of $g^{-1}(f(v_i))$ and extend this definition by linearity. □

Remark 6.3 The proof of Lemma 6.2 relies on the finiteness of the dimension of V. Relaxing this hypothesis can be subtle and may involve the axiom of choice.

Example 6.8 It is relatively straightforward to find the h map in Lemma 6.2. For instance, consider the problem of finding h in the diagram

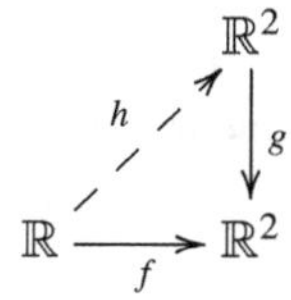

where

$$f = \begin{pmatrix} 0 \\ 2 \end{pmatrix}$$

and

$$g = \begin{pmatrix} 1 & 1 \\ 0 & 1 \end{pmatrix}.$$

The diagram effectively asks for a 2×1 matrix h, such that $f = gh$. This can be found by inspection to be

$$h = \begin{pmatrix} -2 \\ 2 \end{pmatrix}.$$

Exercise 6.13 Find the map h in Lemma 6.2, where

$$f = \begin{pmatrix} 1 \\ 0 \end{pmatrix}$$

and

$$g = \begin{pmatrix} 1 & 1 & 1 \\ -1 & 1 & 2 \end{pmatrix}.$$

Exercise 6.14 Find the map h in Lemma 6.2, where

$$f = \begin{pmatrix} 1 & 1 \\ 0 & 1 \end{pmatrix}$$

and

$$g = \begin{pmatrix} 1 & 1 & 1 \\ -1 & 1 & 2 \end{pmatrix}.$$

The dual property for vector spaces is also true, as is shown in the following exercise.

Exercise 6.15 Let V, W, X be vector spaces, $f : W \to V$ be a linear map, and $g : W \to X$ be an injective linear map. Show that there is another linear map $h : X \to V$ so that the following diagram commutes

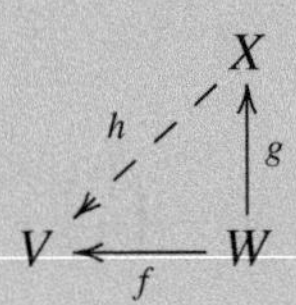

Lemma 6.2 for vector spaces can be generalized to sequences in a straightforward way.

Definition 6.5 A sequence $(P^\bullet, p^\bullet)$ is said to be *projective* if for every pair of sequence maps $m_1^\bullet : (A^\bullet, a^\bullet) \to (B^\bullet, b^\bullet)$ and $m_2^\bullet : (P^\bullet, p^\bullet) \to (B^\bullet, b^\bullet)$, where each component of $m_1^\bullet$ is surjective, there exists a sequence map $f^\bullet : (P^\bullet, p^\bullet) \to (A^\bullet, a^\bullet)$ such that $m_1^\bullet \circ f^\bullet = m_2^\bullet$. Diagrammatically, given the solid arrows below, the dashed arrow exists,

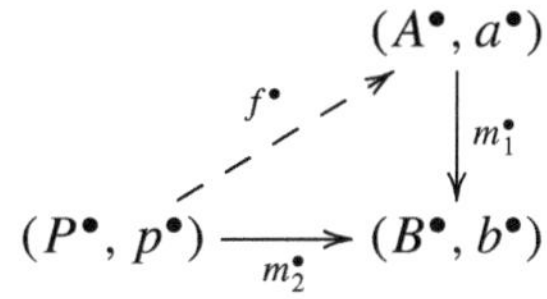

The reader is strongly cautioned that projectivity in **Seq** and **Kom** are quite distinct concepts, and that projective objects in these two categories are essentially unrelated.

Since there is only one sequence map from the zero sequence to any given sequence, the following is trivial.

Corollary 6.1 *The zero sequence is projective.*

Somewhat less trivially, certain kinds of infinite barcodes are projective. To characterize these barcodes, we begin with a few lemmas.

Lemma 6.3 *The sequence*

$$\cdots \to 0 \to \mathbb{R} \to \mathbb{R} \to \cdots$$

in which every arrow is of maximal rank is projective.

Extending the bar notation from Section 2.2 of Chap. 2, we will use the notation $\mathbb{R}^{(s,\infty)}$ to represent this kind of sequence, where s is the index of the first nonzero term.

Proof This requires a chase through the following diagram

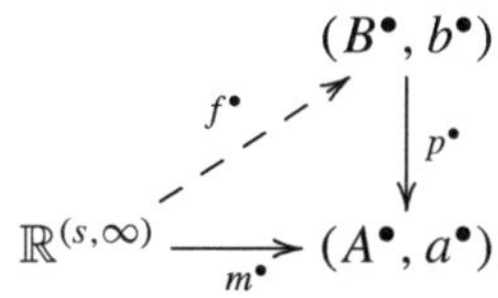

in which the components of the sequence map $p^\bullet$ are surjective linear maps. The goal is to define a sequence map $f^\bullet : \mathbb{R}^{(s,\infty)} \to (B^\bullet, b^\bullet)$ that makes the diagram commute.

Focusing on the diagram near index s, we have

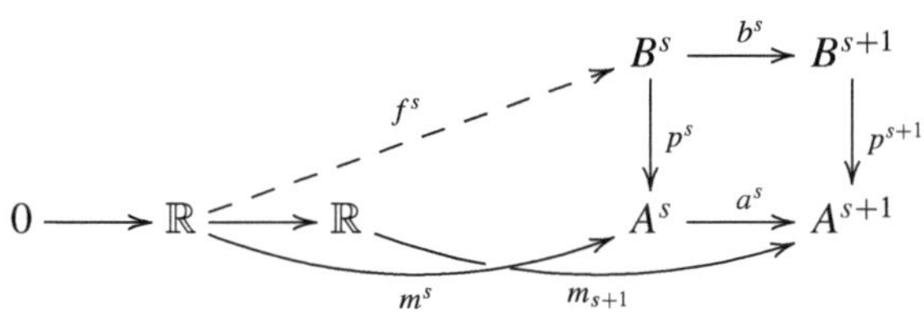

Define a map $f^s : \mathbb{R} \to B^s$ (dashed arrow above) by choosing $f^s(x)$ to be an element of $(p^s)^{-1}(m^s(x))$ for one nonzero $x \in \mathbb{R}$. That this set is nonempty follows by the surjectivity of the $p^\bullet$ component maps. The definition of f^s is completed by extending to the rest of $\mathbb{R}$ by linearity.

By commutativity of the diagram, the definition of f^{s+1} results in a definition

$$f^{s+1} := b^s \circ f^s .$$

This extends to define all the components f^k with $k > s$, that is, to the right of f^s.

To the left of f^s, we may simply take $f^k = 0$ for $k < s$. □

Exercise 6.16 Consider the diagram below, recognizing that Lemma 6.3 establishes that the bottom row is projective,

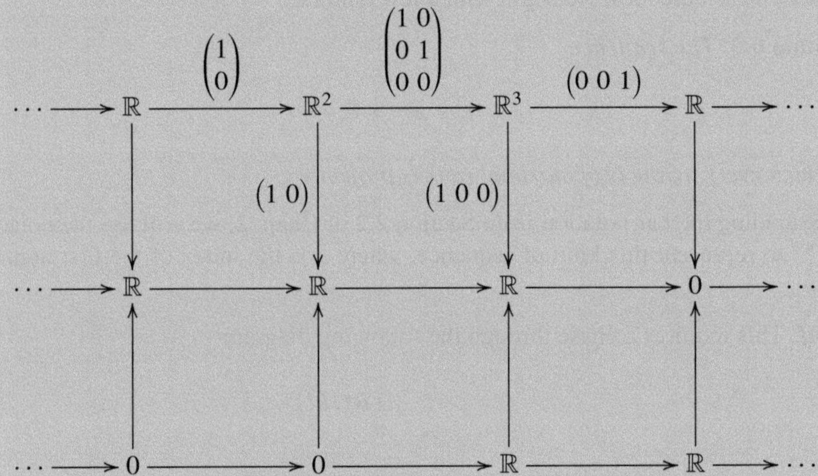

Construct a sequence map from the bottom row to the top row so that the resulting diagram commutes.

Exercise 6.17 Consider the diagram below, recognizing that Lemma 6.3 establishes that the bottom row is projective,

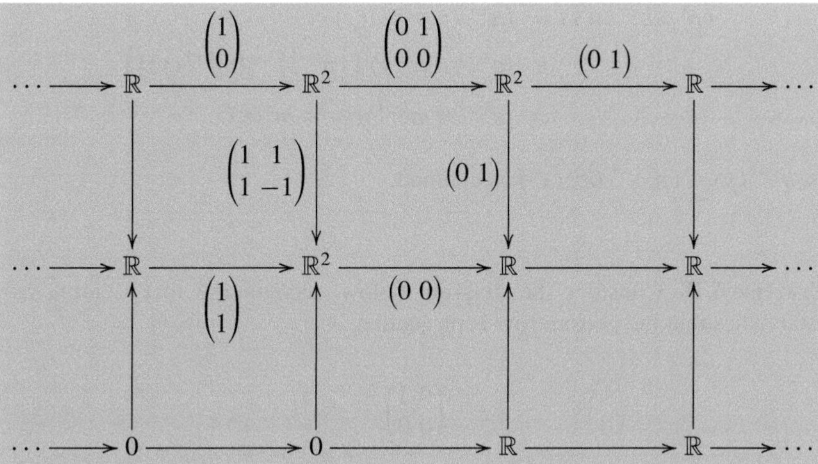

Construct a sequence map from the bottom row to the top row so that the resulting diagram commutes.

Lemma 6.4 *The sequence*

$$\cdots \to \mathbb{R} \to \mathbb{R} \to \cdots$$

in which every arrow is the identity map is projective.

We will use the notation $\mathbb{R}^{(-\infty,\infty)}$ to represent this kind of sequence.

Proof We begin by reusing the construction in the proof of Lemma 6.3 to partially construct the sequence map components f^k for $k \geq s$ starting at any index s we like.

We then need to establish that $f^{s-1}(x)$ can be defined consistently with this, where x is the same value as was used in the construction of f^s. The rest of the maps to the left of this can then be constructed inductively.

The appropriate diagram thus far is

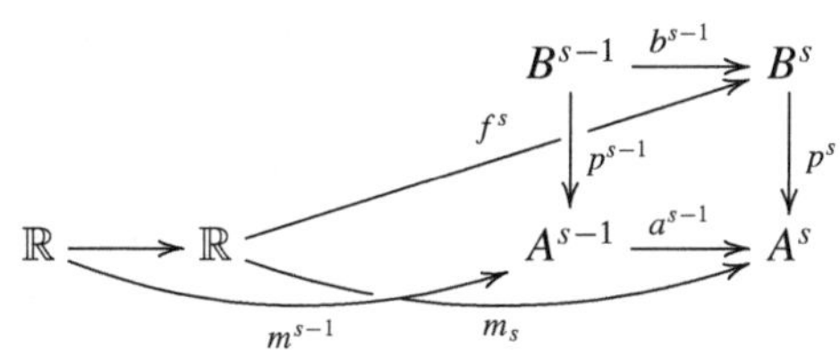

It suffices to show that the intersection of the preimages $\left(b^{s-1}\right)^{-1}(f^s(x))$ and $\left(p^{s-1}\right)^{-1}\left(m^{s-1}(x)\right)$ is nonempty. Suppose that $y \in \left(p^{s-1}\right)^{-1}\left(m^{s-1}(x)\right)$. We will show that $b^{s-1}(y) \in (p^s)^{-1}(m^s(x))$. To that end, compute

$$\begin{aligned}(p^s \circ b^{s-1})(y) &= (a^{s-1} \circ p^{s-1})(y) \\ &\in (a^{s-1} \circ p^{s-1}) \left(\left(p^{s-1}\right)^{-1} \left(m^{s-1}(x)\right) \right) \\ &= (a^{s-1} \circ m^{s-1})(x) = m^s(x).\end{aligned}$$

Thus, $b^{s-1}(y) \in (p^s)^{-1}(m^s(x))$ as claimed. □

Exercise 6.18 Consider the diagram below, recognizing that Lemma 6.4 establishes that the bottom row is projective,

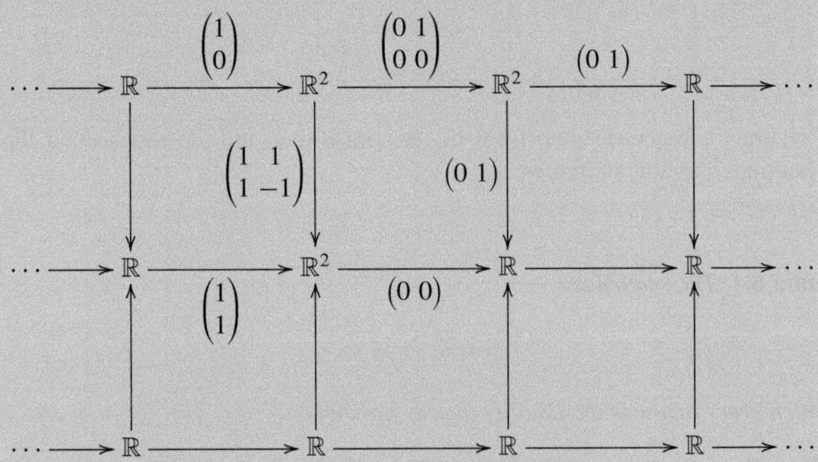

Construct a sequence map from the bottom row to the top row so that the resulting diagram commutes. Hint: infer the maps on the unlabeled arrows.

Example 6.9 The sequence $\cdots 0 \to \mathbb{R} \to 0 \to 0 \cdots$ is not projective in **Seq**. To see this consider the following commutative diagram

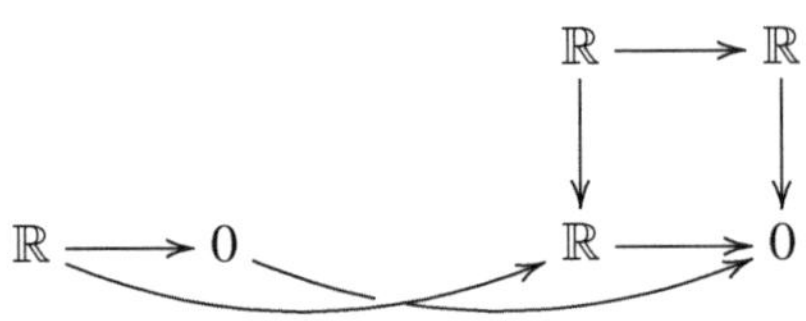

in which every arrow represents either an identity map $\mathbb{R} \to \mathbb{R}$ or a zero map. There is no way to insert two additional maps

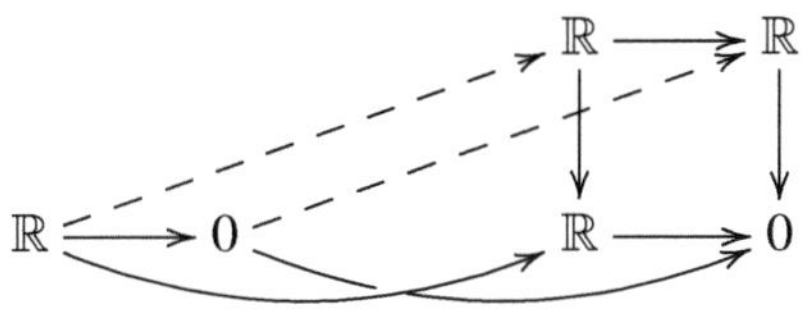

in a way that results in the diagram being commutative.

Exercise 6.19 Show that exact sequences are not projective in **Seq**. (Hint: use a sequence with a bar of length 3.)

Definition 6.6 A sequence $(E^\bullet, e^\bullet)$ is said to be *injective* if for every pair of sequence maps $m_1^\bullet : (B^\bullet, b^\bullet) \to (A^\bullet, a^\bullet)$ and $m_2^\bullet : (B^\bullet, b^\bullet) \to (E^\bullet, e^\bullet)$, where each component of $m_1^\bullet$ is injective, there exists a sequence map $s^\bullet : (A^\bullet, a^\bullet) \to (E^\bullet, e^\bullet)$ such that $s \circ m_1 = m_2$. Diagrammatically, given the solid arrows below, the dashed arrow exists,

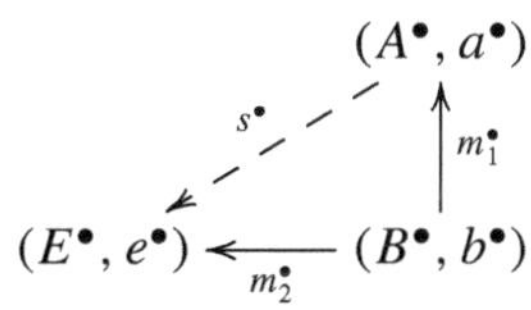

Lemma 6.5 *The sequence*

$$\cdots \to \mathbb{R} \to \mathbb{R} \to 0 \to \cdots$$

in which every arrow is of maximal rank is injective. We will use the notation $\mathbb{R}^{(-\infty,e)}$ *to represent this kind of sequence, where e is the index of the last nonzero term.*

This is the dual result to Lemma 6.3.

Proof This is a classic diagram chase. Since m_1^k is injective for each k, this means that each $x \in A^k$ either has zero or one preimage through m_1^k. We can therefore define

$$s^k(x) = \begin{cases} m_2^k\left((m_1^k)^{-1}(x)\right) & \text{if } x \in \text{image } m_1^k \\ 0 & \text{otherwise.} \end{cases}$$

It remains to show that this definition results in a commutative diagram. The final nonzero term is captured by the following diagram.

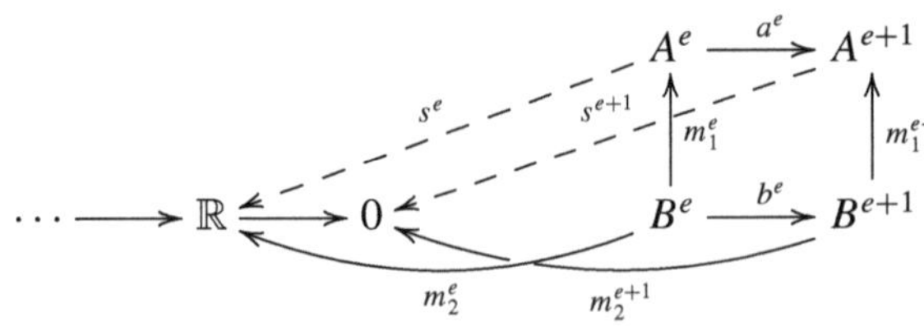

From the diagram, it is immediate that m_2^{e+1} is the zero map. The formula for s^{e+1} ensures that it is the zero map as well. Therefore, regardless of a^e, that portion of the diagram will commute.

For the lesser indices, if $s^k(x) \neq 0$ then $(m_1^k)^{-1}(x)$ is nonempty and nonzero. Moreover, this preimage consists of exactly one point, say y. The definition of s^k ensures that $m_2^k(y) \neq 0$. The hypothesis that $m_2^\bullet$ is a sequence map means that

$$m_2^k = m_2^{k+1} \circ b^k,$$

which implies that $b^k(y) \neq 0$ as well. Injectivity applied to m_1^{k+1} implies that $m_1^{k+1}(b^k(y)) \neq 0$. We must have that

$$0 \neq m_1^{k+1}(b^k(y)) = a^k(m_1^k(y)) = a^k(x).$$

which again by the definition of s^{k+1} ensures that $s^{k+1}(a^k(x)) = s^k(x)$. □

Exercise 6.20 Prove the analog of Lemma 6.4, that the infinite bar sequence is injective.

Lemma 6.6 *The direct sum of two projective (or injective) sequences is projective (or injective).*

Proof Suppose that $(P^\bullet, p^\bullet)$ and $(Q^\bullet, q^\bullet)$ are projective sequences and that $m_1^\bullet : (A^\bullet, a^\bullet) \to (B^\bullet, b^\bullet)$ is a sequence map with surjective components. The direct sum of the two projective sequences is a sequence $(P^\bullet \oplus Q^\bullet, p^\bullet \oplus q^\bullet)$, in which the direct sums are applied term-by-term. Let $m_2^\bullet : (P^\bullet \oplus Q^\bullet, p^\bullet \oplus q^\bullet) \to (B^\bullet, b^\bullet)$ be an arbitrary sequence map.

The sequence map $m_2^\bullet$ can be realized as the direct sum of two sequence maps: $m_{2,P}^\bullet : (P^\bullet, p^\bullet) \to (B^\bullet, b^\bullet)$ and $m_{2,Q}^\bullet : (Q^\bullet, q^\bullet) \to (B^\bullet, b^\bullet)$, since the elements of the two factors of the direct sum are linearly independent by construction.

Projectiveness of $(P^\bullet, p^\bullet)$ and $(Q^\bullet, q^\bullet)$ means that there are sequence maps $r_P^\bullet : (P^\bullet, p^\bullet) \to (A^\bullet, a^\bullet)$ and $r_Q^\bullet : (Q^\bullet, q^\bullet) \to (A^\bullet, a^\bullet)$ such that $m_1^\bullet \circ r_P^\bullet = m_{2,P}^\bullet$ and $m_1^\bullet \circ r_Q^\bullet = m_{2,Q}^\bullet$. It then follows that

$$\begin{aligned} m_2^\bullet &= m_{2,P}^\bullet \oplus m_{2,Q}^\bullet \\ &= (m_1^\bullet \circ r_P^\bullet) \oplus (m_1^\bullet \circ r_P^\bullet) \\ &= m_1^\bullet \oplus (r_P^\bullet \oplus r_Q^\bullet). \end{aligned}$$

The corresponding statement for injectiveness follows by duality. □

Proposition 6.8 *A sequence* $(P^\bullet, p^\bullet)$ *is projective in* **Seq** *if and only if its barcode decomposition is of the form*

$$(P^\bullet, p^\bullet) = \left(\bigoplus_{j=0}^{m} \mathbb{R}^{(s_j,\infty)} \right) \oplus \left(\bigoplus_{k=0}^{n} \mathbb{R}^{(-\infty,\infty)} \right).$$

Likewise, a sequence $(E^\bullet, e^\bullet)$ *is injective in* **Seq** *if and only if its barcode decomposition is of the form*

$$(E^\bullet, e^\bullet) = \left(\bigoplus_{j=0}^{m} \mathbb{R}^{(-\infty,s_j)} \right) \oplus \left(\bigoplus_{k=0}^{n} \mathbb{R}^{(-\infty,\infty)} \right).$$

Proof Let us handle the projective case directly. The injective case follows by duality.

For a projective sequence, we need only recall the Theorem 2.1 in Chap. 2 handle the left term in the direct sum decomposition. The reader is cautioned that Theorem 2.1 in Chap. 2 requires that the sequence be bounded from below. That presents no difficulties here since the only way a projective sequence might not be bounded from below is if it has direct summands of the form $\mathbb{R}^{(-\infty,\infty)}$. This explains the right term, which is projective according to Lemma 6.4. □

Example 6.10 The sequence

$$\left(\mathbb{R}^{(-\infty,\infty)}\right)^3 \oplus \mathbb{R}^{(1,\infty)},$$

which has diagram

$$\cdots \to \mathbb{R}^3 \to \mathbb{R}^3 \to \mathbb{R}^4 \to \mathbb{R}^4 \to \cdots$$

is projective. By the dual argument,

$$\cdots \to \mathbb{R}^4 \to \mathbb{R}^4 \to \mathbb{R}^3 \to \mathbb{R}^3 \to \cdots$$

is injective.

Exercise 6.21 If V is an n-dimensional vector space and W is an m-dimensional vector space, the space of all linear maps from V to W is a mn-dimensional vector space, whose elements are $m \times n$ matrices. Recalling Definition 1.12 in Chap.1 , write $\mathrm{Hom}_{\mathbf{Vec}}(V, W)$ for this space. Suppose that

$$0 \to A \to B \to C \to 0$$

is a short exact sequence of vector spaces. If X is projective, show that

$$0 \to \mathrm{Hom}_{\mathbf{Vec}}(X, A) \to \mathrm{Hom}_{\mathbf{Vec}}(X, B) \to \mathrm{Hom}_{\mathbf{Vec}}(X, C) \to 0$$

is also a short exact sequence.

With this preliminary result in hand, show that the same statement holds if A, B, C, and X are sequences.

6.4 Exact Resolutions

Exact sequences of vector spaces are useful tools for encoding constraints on the linear maps involved. Exact sequences of sequences are useful in a similar way. Because of the extension properties for projective and injective sequences, building an exact sequence from projective or injective sequences confers extension properties upon the entire exact sequence.

The first thing that needs to be established is that one can obtain such an exact sequence at all. This is guaranteed by the next result.

Proposition 6.9 *Suppose that*

$$A^\bullet = \bigoplus_k \mathbb{R}^{(s_k, e_k)}$$

is a sequence. Define

$$P^\bullet = \bigoplus_k \mathbb{R}^{(s_k, \infty)}$$

and

$$E^\bullet = \bigoplus_k \mathbb{R}^{(-\infty, e_k)}.$$

Then the natural projections (respectively, injections) induce exact sequences of sequences

$$P^\bullet \to A^\bullet \to 0 \text{ and } 0 \to A^\bullet \to E^\bullet.$$

Proof Observe that the diagram below is commutative, since every arrow is of maximal rank,

$$\begin{array}{ccccccccccccc}
\cdots & \longrightarrow & 0 & \longrightarrow & \mathbb{R} & \longrightarrow & \cdots & \longrightarrow & \mathbb{R} & \longrightarrow & \mathbb{R} & \longrightarrow & \cdots \\
 & & \downarrow & & \downarrow & & & & \downarrow & & \downarrow & & \\
\cdots & \longrightarrow & 0 & \longrightarrow & \mathbb{R} & \longrightarrow & \cdots & \longrightarrow & \mathbb{R} & \longrightarrow & 0 & \longrightarrow & \cdots
\end{array}$$

This can be expressed as an exact sequence of sequences $\mathbb{R}^{(s,\infty)} \to \mathbb{R}^{(s,e)} \to 0$ because each component map is surjective. To obtain the claimed statement, apply Lemma 6.6.

The corresponding statement about injectivity follows *mutatis mutandis.* □

Lemma 6.7 *For the sequences defined in Proposition 6.9, the term-by-term quotient $P^\bullet/A^\bullet$ is projective.*

Proof Observe that

$$\begin{aligned}
P^\bullet/A^\bullet &= \left(\bigoplus_k \mathbb{R}^{(s_k,\infty)}\right) / \left(\bigoplus_k \mathbb{R}^{(s_k,e_k)}\right) \\
&= \bigoplus_k \left(\mathbb{R}^{(s_k,\infty)}/\mathbb{R}^{(s_k,e_k)}\right) \\
&= \bigoplus_k \mathbb{R}^{(e_k,\infty)},
\end{aligned}$$

which is projective by Proposition 6.8. □

Exercise 6.22 Directly prove the analog of Lemma 6.7 for injective sequences. That is, show that if $E^\bullet$ is defined as in Proposition 6.9, then $E^\bullet/A^\bullet$ is injective by referring back to Definition 6.6.

Exact sequences of sequences that begin or end with a given sequence—a *resolution* of that given sequence—can help understand the given sequence.

Definition 6.7 A *projective resolution* of a sequence $A^\bullet$ is an object of **Kom**(**Seq**) given by the diagram

$$\cdots \longrightarrow P^{\bullet,}{}_1 \longrightarrow P^{\bullet,}{}_0 \longrightarrow A^\bullet \longrightarrow 0$$

in which the rows are exact, and each $P^{\bullet,}{}_k$ is a projective sequence. As a convention, we will define $P^{\bullet,}{}_k$ to be the zero sequence for $k < 0$.

Usually, we will say "$P^{\bullet,}{}_\diamond$ is a projective resolution of $A^\bullet$." This emphasizes that $P^{\bullet,}{}_\diamond$ (an object of $\mathbf{Kom}_\diamond(\mathbf{Seq}^\bullet)$) describes $A^\bullet$ (an object of $\mathbf{Seq}^\bullet$).

Dually, an *injective resolution* of $A^\bullet$ is an object of **Kom**(**Seq**) given by the diagram

$$0 \longrightarrow A^\bullet \longrightarrow E^{\bullet,0} \longrightarrow E^{\bullet,1} \longrightarrow \cdots$$

in which the rows are exact, and each $E^{\bullet,k}$ is an injective sequence. As a convention, we will define $E^{\bullet,k} := 0$ for $k < 0$.

We will usually express this by saying "$E^{\bullet,\diamond}$ is an injective resolution of $A^\bullet$."

Projective and injective resolutions are not unique and can be of different lengths.

Definition 6.8 The *projective dimension of* $A^\bullet$ is the largest integer n such that for all projective resolutions $P^{\bullet}{}_{;\diamond}$ of $A^\bullet$, $P^{\bullet}{}_{;k}$ is nontrivial for $0 \le k \le n$.
Similarly, the *injective dimension of* $A^\bullet$ is the largest integer n such that for all injective resolutions $E^{\bullet,k}$ of $A^\bullet$ $E^{\bullet,k}$ is nontrivial for $0 \le k \le n$.

Lemma 6.8 *A sequence has projective dimension zero if and only if it is a projective sequence.*

Proof Here is a projective resolution for a projective sequence $P^\bullet$,

$$0 \to P^\bullet \to P^\bullet \to 0,$$

which cannot be any shorter without violating exactness of the rows. This resolution has length 1, so the projective dimension of $P^\bullet$ is 0.

Conversely, an exact sequence

$$0 \to P^\bullet \to A^\bullet \to 0,$$

establishes an isomorphism between $P^\bullet$ and $A^\bullet$. If $P^\bullet$ is projective then $A^\bullet$ is also projective. □

Exercise 6.23 Show that a sequence has injective dimension zero if and only if it is an injective sequence.

Fortunately, although by no means obviously, all projective resolutions of a sequence are quasi-isomorphic to each other in **Kom**(**Seq**), as will be shown in Proposition 6.12. (The injective resolutions are all quasi-isomorphic to each other as well, though an injective and projective resolution will generally not be quasi-isomorphic to each other.) More is true, in that all projective resolutions of a sequence are homotopy equivalent to each other, as is established in Proposition 6.19. Identifying homotopy equivalent projective resolutions of a sequence is natural in the *homotopy category*, to be explored in Sect. 6.6. Identifying the quasi-isomorphism classes of **Kom**(**Seq**) results in another new category, the *derived category* D(**Seq**). The definition of morphisms of the derived category is somewhat subtle and is reminiscent of barycentric subdivision discussed in Sect. 5.4 of Chap. 5. We will explore the derived category in detail in Sect. 6.7.

Example 6.11 The sequence $A^\bullet$ consisting of a single bar of length 1, namely

$$\cdots \to 0 \to \mathbb{R} \to 0 \to \cdots$$

has multiple projective resolutions. One such resolution is given by the grid diagram,

$$\begin{array}{ccccccccccc}
 & & \vdots & & \vdots & & \vdots & & \vdots & & \vdots \\
 & & \downarrow & & \downarrow & & \downarrow & & \downarrow & & \downarrow \\
\cdots & \longrightarrow & 0 & \longrightarrow & 0 & \longrightarrow & 0 & \longrightarrow & A^{-1} = 0 & \longrightarrow & 0 \\
 & & \downarrow & & \downarrow & & \downarrow & & \downarrow & & \downarrow \\
\cdots & \longrightarrow & 0 & \longrightarrow & 0 & \longrightarrow & \mathbb{R} & \longrightarrow & A^{0} = \mathbb{R} & \longrightarrow & 0 \\
 & & \downarrow & & \downarrow & & \downarrow & & \downarrow & & \downarrow \\
\cdots & \longrightarrow & 0 & \longrightarrow & \mathbb{R} & \longrightarrow & \mathbb{R} & \longrightarrow & A^{1} = 0 & \longrightarrow & 0 \\
 & & \downarrow & & \downarrow & & \downarrow & & \downarrow & & \downarrow \\
\cdots & \longrightarrow & 0 & \longrightarrow & \mathbb{R} & \longrightarrow & \mathbb{R} & \longrightarrow & A^{2} = 0 & \longrightarrow & 0 \\
 & & \downarrow & & \downarrow & & \downarrow & & \downarrow & & \downarrow \\
 & & \vdots & & \vdots & & \vdots & & \vdots & & \vdots
\end{array}$$

where the original sequence is written as the second-to-rightmost column.

Another projective resolution is given by a different grid diagram,

$$\begin{array}{ccccccccccc}
 & & \vdots & & \vdots & & \vdots & & \vdots & & \vdots \\
 & & \downarrow & & \downarrow & & \downarrow & & \downarrow & & \downarrow \\
\cdots & \longrightarrow & 0 & \longrightarrow & 0 & \longrightarrow & 0 & \longrightarrow & A^{-1} = 0 & \longrightarrow & 0 \\
 & & \downarrow & & \downarrow & & \downarrow & & \downarrow & & \downarrow \\
\cdots & \longrightarrow & 0 & \longrightarrow & \mathbb{R} & \longrightarrow & \mathbb{R}^2 & \longrightarrow & A^{0} = \mathbb{R} & \longrightarrow & 0 \\
 & & \downarrow & & \downarrow & & \downarrow & & \downarrow & & \downarrow \\
\cdots & \longrightarrow & 0 & \longrightarrow & \mathbb{R}^2 & \longrightarrow & \mathbb{R}^2 & \longrightarrow & A^{1} = 0 & \longrightarrow & 0 \\
 & & \downarrow & & \downarrow & & \downarrow & & \downarrow & & \downarrow \\
\cdots & \longrightarrow & 0 & \longrightarrow & \mathbb{R}^2 & \longrightarrow & \mathbb{R}^2 & \longrightarrow & A^{2} = 0 & \longrightarrow & 0 \\
 & & \downarrow & & \downarrow & & \downarrow & & \downarrow & & \downarrow \\
 & & \vdots & & \vdots & & \vdots & & \vdots & & \vdots
\end{array}$$

Lemma 6.9 *Suppose that $f^\bullet : \mathbb{R}^{(a,\infty)} \to P^\bullet$ is a sequence map to a projective sequence, where $a \geq -\infty$. Using Proposition 6.8, suppose that the barcode decomposition of $P^\bullet$ is*

$$P^\bullet = \left(\bigoplus_{j=0}^{m} \mathbb{R}^{(s_j,\infty)} \right) \oplus \left(\bigoplus_{k=0}^{n} \mathbb{R}^{(-\infty,\infty)} \right).$$

The image of $f^\bullet$ is contained entirely within one of the terms in the decomposition above.

It is obvious that the image of $f^\bullet$ is contained within $P^\bullet$. The lemma establishes something stronger, namely that the image of $f^\bullet$ is contained in either $\mathbb{R}^{(s_j,\infty)}$ for some s_j, or it is contained in $\mathbb{R}^{(-\infty,\infty)}$.

Proof Suppose that $f^\ell \neq 0$ for some $\ell \geq a$. This means that there is some term $\mathbb{R}^{(s_j,\infty)}$ in the sum for $P^\bullet$ for which f^k has a nonzero image in the index ℓ term. (Without loss of generality, we can allow $s_j = -\infty$.) To establish the lemma, we need to show that the entire image of $f^\bullet$ is contained within $\mathbb{R}^{(s_j,\infty)}$.

Commutativity ensures that if $f^m = 0$ for some $m > \ell$, then $f^n = 0$ for all $n \geq m$. Conversely, let us assume that $f^{\ell+1} \neq 0$. This can be used to start an induction from index ℓ to the smallest index m where $f^m = 0$. Because the domain sequence $\mathbb{R}^{(a,\infty)}$ has maximum rank maps, and the bar $\mathbb{R}^{(s_j,\infty)}$ in the codomain also has maximum rank maps, then commutativity of $f^\bullet$ ensures that the image of $f^{\ell+1}$ is also in $\mathbb{R}^{(s_j,\infty)}$. □

Lemma 6.10 *Suppose that $m^\bullet : \mathbb{R}^{(a,\infty)} \to \mathbb{R}^{(b,c)}$ is a surjective sequence map. Then $a = b \leq c$.*

Proof Evidently, $b \leq c$. Commutativity ensures that $b \leq a$. To see this, suppose that $b > a$. This would imply a diagram of full rank linear maps of the form

$$\begin{array}{ccc} \mathbb{R} & \longrightarrow & 0 \\ \downarrow & & \downarrow \\ \mathbb{R} & \longrightarrow & \mathbb{R} \end{array}$$

which cannot commute. Finally, surjectivity requires that $b \geq a$, hence $a = b$. □

The above two results mean that projective resolutions have a recursive nesting structure characterized by iterative quotients.

Lemma 6.11 *Suppose that*

$$\cdots \longrightarrow P^{\bullet}_{1} \xrightarrow{p^{\bullet}_{1}} P^{\bullet}_{0} \xrightarrow{p^{\bullet}_{0}} A^\bullet \longrightarrow 0$$

is a projective resolution of an arbitrary sequence $A^\bullet$.

The induced map $(p^{\bullet,}{}_1)_* : P^{\bullet,}{}_1 \to P^{\bullet,}{}_0/A^\bullet$ is surjective, so that

$$\cdots \longrightarrow P^{\bullet,}{}_2 \xrightarrow{p^{\bullet,}{}_2} P^{\bullet,}{}_1 \xrightarrow{(p^{\bullet,}{}_1)_*} P^{\bullet,}{}_0/A^\bullet \to 0$$

is a projective resolution as well.

Moreover, $P^{\bullet,}{}_0/A^\bullet$ is a projective sequence.

Proof Lemmas 6.9 and 6.10 imply that $P^{\bullet,}{}_0/A^\bullet$ is projective, because the quotient operation can only "trim" the starts of the bars in $P^{\bullet,}{}_0$.

Now let us establish surjectivity of the induced map $P^{\bullet,}{}_1 \to P^{\bullet,}{}_0/A^\bullet$. If any term of $P^{k,}{}_0/A^k$ is nontrivial, it implies this term is not present in A^k. Therefore, the presence of each such term corresponds to a bar of the diagram

$$\cdots \to P^{k,}{}_1 \to P^{k,}{}_0 \to A^k \to 0$$

that ends at $P^{k,}{}_0$. Exactness implies that this same bar must start at $P^{k,}{}_1$. □

Exercise 6.24 Suppose that

$$0 \longrightarrow P^\bullet \xrightarrow{f^\bullet} Q^\bullet \xrightarrow{g^\bullet} R^\bullet \longrightarrow 0$$

is an exact sequence of projective sequences. Show that $Q^\bullet \cong P^\bullet \oplus R^\bullet$, which is to say that the exact sequence *splits*. We will discuss split and non-split exact sequences in a little more detail, later in Sect. 6.5.

If one considers a projective resolution of a projective sequence, one obtains a stronger special case of Lemma 6.11.

Proposition 6.10 *Suppose that $m^\bullet : A^\bullet \to B^\bullet$ is a sequence map and that*

$$P^{\bullet,}{}_\bullet \to A^\bullet \to 0,$$

$$Q^{\bullet,}{}_\bullet \to B^\bullet \to 0$$

are projective resolutions. There is a **Kom(Seq)** *morphism that extends $m^\bullet$ given by the diagram*

$$\begin{array}{ccccccc}
\cdots \longrightarrow & P^{\bullet,}{}_1 & \longrightarrow & P^{\bullet,}{}_0 & \longrightarrow & A^\bullet & \longrightarrow 0 \\
 & \downarrow f^{\bullet,}{}_1 & & \downarrow f^{\bullet,}{}_0 & & \downarrow m^\bullet & \\
\cdots \longrightarrow & Q^{\bullet,}{}_1 & \longrightarrow & Q^{\bullet,}{}_0 & \longrightarrow & B^\bullet & \longrightarrow 0
\end{array}$$

The reader is cautioned that the **Kom(Seq)** morphism extending $m^\bullet$ is by no means unique! Therefore this extension is not functorial. We *cannot* claim that this **Kom(Seq)** morphism is induced by $m^\bullet$. This defect will be remedied with the definition of the *homotopy category* in Sect. 6.6.

Proof We proceed by induction, though the base case and induction case are almost identical. For the base case, we merely need to recognize that the dashed sequence map $m^{\bullet,}_{0}$ in diagram below exists because $Q^{\bullet,}_{0} \to B^\bullet$ has surjective components (via exactness) and $P^{\bullet,}_{0}$ is projective,

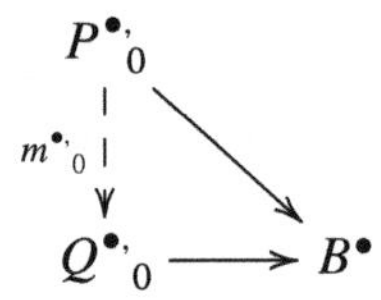

Lemma 6.11 applied to both resolutions implies that

$$\cdots \to P^{\bullet,}_{2} \to P^{\bullet,}_{1} \to P^{\bullet,}_{0}/A^\bullet \to 0$$

and

$$\cdots \to Q^{\bullet,}_{2} \to Q^{\bullet,}_{1} \to Q^{\bullet,}_{0}/B^\bullet \to 0$$

are both projective resolutions.

Moreover, the map $m^{\bullet,}_{0} : P^{\bullet,}_{0} \to Q^{\bullet,}_{0}$ descends to the quotients, inducing a map

$$(m^{\bullet,}_{0})_* : P^{\bullet,}_{0}/A^\bullet \to Q^{\bullet,}_{0}/B^\bullet.$$

Hence, the induction step is simply a repetition of the above construction (yielding $m^{\bullet,}_{k} : P^{\bullet,}_{k} \to Q^{\bullet,}_{k}$) with the modified resolutions and the sequence map $(m^{\bullet,}_{k-1})_*$. Specifically, we define

$$\begin{cases} Q'^{\bullet,}_{0} := Q^{\bullet,}_{0}/B^\bullet, \\ Q'^{\bullet,}_{k} := Q^{\bullet,}_{k}/Q'^{\bullet}_{k-1}. \end{cases}$$

The induction hypothesis implies that

$$\cdots \to Q^{\bullet,}_{k+1} \to Q^{\bullet,}_{k} \to Q'^{\bullet,}_{k-1} \to 0$$

is exact. Therefore, we construct the sequence map $m^{\bullet,}_{1}$

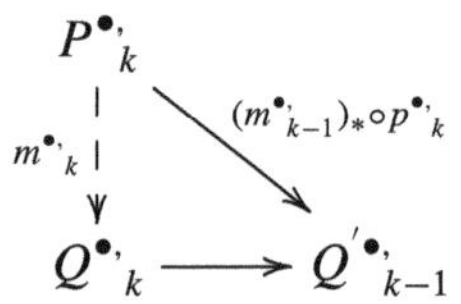

using the projectivity of $P^{\bullet,}{}_k$. □

The non-uniqueness of the extension from a sequence morphism to projective resolutions is quite severe. It even occurs for the simplest of cases!

Example 6.12 Consider the sequence morphism $m^\bullet : \mathbb{R}^{(1,1)} \to \mathbb{R}^{(1,1)}$ given by the identity sequence. If we call the domain $A^\bullet$ and the codomain $B^\bullet$, consistent with the notation of Proposition 6.10, then one possible pair of projective resolutions is given by the diagram below,

$$
\begin{array}{ccccccccc}
P^{\bullet,}{}_1 & & \cdots \longrightarrow & 0 & \longrightarrow & \mathbb{R} & \xrightarrow{\begin{pmatrix}1\\0\end{pmatrix}} & \mathbb{R}^2 & \longrightarrow \cdots \\
\downarrow & & & \downarrow & & \downarrow{\scriptstyle \begin{pmatrix}1\\0\end{pmatrix}} & & \downarrow{\scriptstyle \begin{pmatrix}1&0\\0&1\end{pmatrix}} & \\
P^{\bullet,}{}_0 & & \cdots \longrightarrow & 0 & \longrightarrow & \mathbb{R}^2 & \xrightarrow{\begin{pmatrix}1&0\\0&1\end{pmatrix}} & \mathbb{R}^2 & \longrightarrow \cdots \\
\downarrow & & & \downarrow & & \downarrow{\scriptstyle (1\ 0)} & & \downarrow & \\
A^\bullet & & \cdots \longrightarrow & 0 & \longrightarrow & \mathbb{R} & \longrightarrow & 0 & \longrightarrow \cdots \\
{\scriptstyle m^\bullet}\downarrow & & & \downarrow & & \downarrow{\scriptstyle \mathrm{id}} & & \downarrow & \\
B^\bullet & & \cdots \longrightarrow & 0 & \longrightarrow & \mathbb{R} & \longrightarrow & 0 & \longrightarrow \cdots \\
\uparrow & & & \uparrow & & \uparrow{\scriptstyle (1\ 0)} & & \uparrow & \\
Q^{\bullet,}{}_0 & & \cdots \longrightarrow & 0 & \longrightarrow & \mathbb{R}^2 & \xrightarrow{\begin{pmatrix}1&0\\0&1\end{pmatrix}} & \mathbb{R}^2 & \longrightarrow \cdots \\
\uparrow & & & \uparrow & & \uparrow{\scriptstyle \begin{pmatrix}1\\0\end{pmatrix}} & & \uparrow{\scriptstyle \begin{pmatrix}1&0\\0&1\end{pmatrix}} & \\
Q^{\bullet,}{}_1 & & \cdots \longrightarrow & 0 & \longrightarrow & \mathbb{R} & \xrightarrow{\begin{pmatrix}1\\0\end{pmatrix}} & \mathbb{R}^2 & \longrightarrow \cdots
\end{array}
$$

Proposition 6.10 asserts that there is a sequence morphism $f^{\bullet,}{}_0 : P^{\bullet,}{}_0 \to Q^{\bullet,}{}_0$ that makes the resulting diagram commute. Indeed, there are many such morphisms! In index 1 alone (the middle column), the map $f^{1,}{}_0$ must make the following diagram commute

$$
\begin{array}{ccc}
P^{1,}{}_{0} & \longrightarrow & A^{1} \\
{\scriptstyle f^{1,}{}_{0}}\big\downarrow & & \big\downarrow{\scriptstyle m^{1}} \\
Q^{1,}{}_{0} & \longrightarrow & B^{1}
\end{array}
\qquad
\begin{array}{ccc}
\mathbb{R}^2 & \xrightarrow{(1\ 0)} & \mathbb{R} \\
{\scriptstyle f^{1,}{}_{0}}\big\downarrow & & \big\downarrow{\scriptstyle \mathrm{id}} \\
\mathbb{R}^2 & \xrightarrow[(1\ 0)]{} & \mathbb{R}
\end{array}
$$

It is quickly seen that for any $x, y \in \mathbb{R}$, the definition

$$f^{1,}{}_{0} := \begin{pmatrix} 1 & 0 \\ x & y \end{pmatrix}$$

will work.

Exercise 6.25 Prove an analog of Proposition 6.10 for injective resolutions.

Proposition 6.11 *Projective resolutions exist for every sequence in* **Seq**.

Proof Consider an arbitrary sequence $A^\bullet$. Proposition 6.9 defines a projective sequence $P^\bullet$ such that

$$P^\bullet \xrightarrow{p^\bullet} A^\bullet \longrightarrow 0$$

is exact. Additionally, Lemma 6.7 ensures that quotients of projective sequences are projective. Hence $P^\bullet/A^\bullet$ is projective.

Consider the diagram formed from the terms of the sequences we have defined thus far,

$$0 \longrightarrow P^\ell/A^\ell \longrightarrow P^\ell \xrightarrow{p^\ell} A^\ell \longrightarrow 0,$$

where p^ℓ is the natural projection. This kind of sequence is exact for vector spaces (Proposition 2.8 in Chap. 2). Therefore, the sequence is a projective resolution of $A^\bullet$. □

Exercise 6.26 Prove that injective resolutions exist for every sequence in **Seq**.

Exercise 6.27 Use the procedure in the proof of Proposition 6.11 to construct a projective resolution for a sequence consisting of a single length 3 bar.

Remark 6.4 Proposition 6.11 and its analog for injective resolutions is usually called "Hilbert's syzygy theorem" (Eisenbud 1995). It generalizes neatly to sequences of sequences ... of sequences, namely arbitrary finite dimensional grids of vector spaces. If the grid in question is of dimension n, then the projective dimension of the original sequence is at most n.

Since sequences are 1-dimensional grids, this means that the projective dimension is at most $n = 1$. This explains why the construction in the proof of Proposition 6.11 has two terms to the left of A^{ℓ} (for dimensions 0 and 1).

Projective and injective resolutions extend homology to arbitrary sequences, not just chain complexes. Because projective and injective resolutions are exact sequences in **Kom(Seq)**, some care is needed to avoid obtaining trivial homology. The trick is to leave off the original sequence once a resolution of it has been constructed.

Specifically, consider a projective resolution of a sequence $A^{\bullet}$,

$$P^{\bullet,}{}_{\bullet} \to A^{\bullet} \to 0.$$

If we replace the $A^{\bullet}$ with the zero sequence, we have a diagram of the form

$$\cdots \longrightarrow P^{\bullet,}{}_{1} \longrightarrow P^{\bullet,}{}_{0} \longrightarrow 0,$$

which is no longer exact. However, this diagram is still an object of **Kom(Seq)**, and we can compute row-wise homology.

Example 6.13 Recall the two projective resolutions of the sequence $A^{\bullet} = \mathbb{R}^{(0,0)}$ from Example 6.11. We can write the first one as

$$\cdots \to 0 \to \mathbb{R}^{(1,\infty)} \to \mathbb{R}^{(0,\infty)} \to A^{\bullet} = \mathbb{R}^{(0,0)} \to 0$$

If we remove the length 1 bar sequence at the rightmost end above from consideration, the grid diagram resulting from the first resolution is

$$\begin{array}{ccccccccc}
 & & \vdots & & \vdots & & \vdots & & \vdots \\
 & & \downarrow & & \downarrow & & \downarrow & & \downarrow \\
\cdots & \longrightarrow & 0 & \longrightarrow & 0 & \longrightarrow & 0 & \longrightarrow & 0 \\
 & & \downarrow & & \downarrow & & \downarrow & & \downarrow \\
\cdots & \longrightarrow & 0 & \longrightarrow & 0 & \longrightarrow & \mathbb{R} & \longrightarrow & 0 \\
 & & \downarrow & & \downarrow & & \downarrow & & \downarrow \\
\cdots & \longrightarrow & 0 & \longrightarrow & \mathbb{R} & \longrightarrow & \mathbb{R} & \longrightarrow & 0 \\
 & & \downarrow & & \downarrow & & \downarrow & & \downarrow \\
\cdots & \longrightarrow & 0 & \longrightarrow & \mathbb{R} & \longrightarrow & \mathbb{R} & \longrightarrow & 0 \\
 & & \downarrow & & \downarrow & & \downarrow & & \downarrow \\
 & & \vdots & & \vdots & & \vdots & & \vdots
\end{array}$$

Applying homology row-wise yields a new grid diagram

$$\begin{array}{ccccccccc}
 & & \vdots & & \vdots & & \vdots & & \vdots \\
 & & \downarrow & & \downarrow & & \downarrow & & \downarrow \\
\cdots & \longrightarrow & 0 & \longrightarrow & 0 & \longrightarrow & 0 & \longrightarrow & 0 \\
 & & \downarrow & & \downarrow & & \downarrow & & \downarrow \\
\cdots & \longrightarrow & 0 & \longrightarrow & 0 & \longrightarrow & \mathbb{R} & \longrightarrow & 0 \\
 & & \downarrow & & \downarrow & & \downarrow & & \downarrow \\
\cdots & \longrightarrow & 0 & \longrightarrow & 0 & \longrightarrow & 0 & \longrightarrow & 0 \\
 & & \downarrow & & \downarrow & & \downarrow & & \downarrow \\
\cdots & \longrightarrow & 0 & \longrightarrow & 0 & \longrightarrow & 0 & \longrightarrow & 0 \\
 & & \downarrow & & \downarrow & & \downarrow & & \downarrow \\
 & & \vdots & & \vdots & & \vdots & & \vdots
\end{array}$$

which can be briefly written

$$\cdots \to 0 \to 0 \to \mathbb{R}^{(0,0)} \to 0.$$

The other projective resolution from Example 6.11, namely

$$\cdots \to 0 \to \mathbb{R}^{(1,\infty)} \oplus \mathbb{R}^{(0,\infty)} \to \mathbb{R}^{(0,\infty)} \oplus \mathbb{R}^{(0,\infty)} \to A^\bullet = \mathbb{R}^{(0,0)} \to 0$$

results in the same row-wise homology, as is easily verified from its grid diagram.

Proposition 6.12 *All projective resolutions of an arbitrary sequence $A^\bullet$ have isomorphic row-wise homology sequences. Because the columns of the row-wise homology sequences are all trivial except for the last column, the homology sequence is precisely $A^\bullet$.*

Proof For concreteness, suppose that the projective resolution is

$$\cdots \to P^{\bullet,}{}_{2} \to P^{\bullet,}{}_{1} \to P^{\bullet,}{}_{0} \to A^\bullet \to 0.$$

The row-wise homology sequence is the homology of the sequence of a single row without $A^\bullet$, namely for row k,

$$\cdots \to P^{k,}{}_{2} \to P^{k,}{}_{1} \to P^{k,}{}_{0} \to 0.$$

Because this sequence was extracted from a resolution, the sequence is exact at all indices except 0. This means that all bars have length 2, except any bars that start at index 0. Hence, the only place where nontrivial homology appears is at index 0. On the other hand, because the original projective resolution was exact when $A^\bullet$ was included, this means that any bars starting at index 0 correspond to nontrivial terms in $A^\bullet$. Hence the last column of row-wise homology has the same terms as $A^\bullet$.

It remains to establish that the bars of the last column of row-wise homology are the bars of $A^\bullet$, namely that the maps are the same. If the last column has a nontrivial bar in its row-wise homology, we will have a diagram like the following in index 0 in row k,

$$\cdots \longrightarrow P^{\bullet,}{}_{1} \longrightarrow P^{\bullet,}{}_{0} \longrightarrow A^\bullet$$

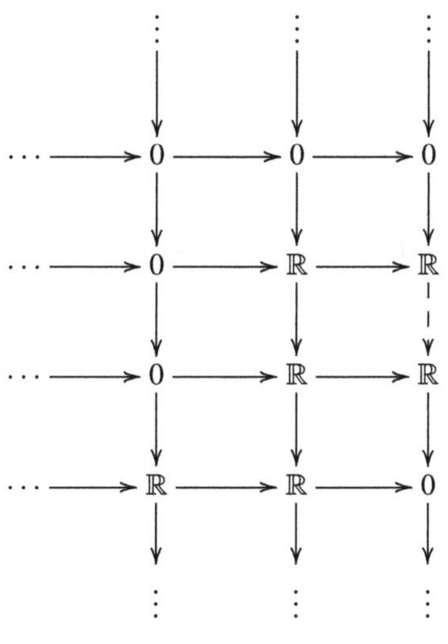

where the solid arrows are of full rank. Evidently, commutativity means that the dashed arrow must also be of full rank. Longer bars can be handled *mutatis mutandis*. □

Example 6.14 Consider the sequence

$$0 \longrightarrow \mathbb{R} \xrightarrow{\begin{pmatrix}1\\0\\0\end{pmatrix}} \mathbb{R}^3 \xrightarrow{\begin{pmatrix}1&0&0\\0&0&1\end{pmatrix}} \mathbb{R}^2 \xrightarrow{(0\ 1)} \mathbb{R} \longrightarrow 0$$

We can construct many different projective resolutions of this sequence, for instance

$$\begin{array}{ccccccccc}
\cdots & \longrightarrow & 0 & \longrightarrow & 0 & \longrightarrow & 0 & \longrightarrow & 0 \\
 & & \downarrow & & \downarrow & & \downarrow & & \downarrow \\
\cdots & \longrightarrow & 0 & \longrightarrow & 0 & \longrightarrow & \mathbb{R} & \longrightarrow & \mathbb{R} \\
 & & \downarrow & & \downarrow & & \downarrow & & \downarrow{\scriptstyle \begin{pmatrix}1\\0\\0\end{pmatrix}} \\
\cdots & \longrightarrow & 0 & \longrightarrow & 0 & \longrightarrow & \mathbb{R}^3 & \longrightarrow & \mathbb{R}^3 \\
 & & \downarrow & & \downarrow & & \downarrow & & \downarrow{\scriptstyle \begin{pmatrix}1&0&0\\0&0&1\end{pmatrix}} \\
\cdots & \longrightarrow & 0 & \longrightarrow & \mathbb{R} & \longrightarrow & \mathbb{R}^3 & \longrightarrow & \mathbb{R}^2 \\
 & & \downarrow & & \downarrow & & \downarrow & & \downarrow{\scriptstyle (0\ 1)} \\
\cdots & \longrightarrow & \mathbb{R} & \longrightarrow & \mathbb{R}^2 & \longrightarrow & \mathbb{R}^3 & \longrightarrow & \mathbb{R} \\
 & & \downarrow & & \downarrow & & \downarrow & & \downarrow \\
\cdots & \longrightarrow & \mathbb{R} & \longrightarrow & \mathbb{R}^3 & \longrightarrow & \mathbb{R}^3 & \longrightarrow & 0
\end{array}$$

If we set the rightmost column to the zero sequence, we obtain the row-wise chain complex given by the diagram

$$\begin{array}{ccccccccc}
\cdots & \longrightarrow & 0 & \longrightarrow & 0 & \longrightarrow & 0 & \longrightarrow & 0 \\
 & & \downarrow & & \downarrow & & \downarrow & & \downarrow \\
\cdots & \longrightarrow & 0 & \longrightarrow & 0 & \longrightarrow & \mathbb{R} & \longrightarrow & 0 \\
 & & \downarrow & & \downarrow & & \downarrow & & \downarrow \\
\cdots & \longrightarrow & 0 & \longrightarrow & 0 & \longrightarrow & \mathbb{R}^3 & \longrightarrow & 0 \\
 & & \downarrow & & \downarrow & & \downarrow & & \downarrow \\
\cdots & \longrightarrow & 0 & \longrightarrow & \mathbb{R} & \longrightarrow & \mathbb{R}^3 & \longrightarrow & 0 \\
 & & \downarrow & & \downarrow & & \downarrow & & \downarrow \\
\cdots & \longrightarrow & \mathbb{R} & \longrightarrow & \mathbb{R}^2 & \longrightarrow & \mathbb{R}^3 & \longrightarrow & 0 \\
 & & \downarrow & & \downarrow & & \downarrow & & \downarrow \\
\cdots & \longrightarrow & \mathbb{R} & \longrightarrow & \mathbb{R}^3 & \longrightarrow & \mathbb{R}^3 & \longrightarrow & 0
\end{array}$$

This clearly only has row-wise homology in the rightmost column. Taking quotients recovers the original sequence, which is to say that the homology of this projective resolution is exactly the sequence we started with, in accordance with Proposition 6.12.

Exercise 6.28 Construct a projective resolution of the following sequence, and then compute its homology,

$$0 \longrightarrow \mathbb{R} \xrightarrow{(0)} \mathbb{R} \xrightarrow{(0)} \mathbb{R} \longrightarrow 0$$

Exercise 6.29 Construct a projective resolution of the following sequence, and then compute its homology,

$$0 \to \mathbb{R}^3 \xrightarrow{\begin{pmatrix} -1 & 1 & 1 \\ -10 & -2 & 0 \\ -31 & -11 & -3 \end{pmatrix}} \mathbb{R}^3 \xrightarrow{\begin{pmatrix} 0 & 3 & -3 \\ -9 & 6 & 0 \end{pmatrix}} \mathbb{R}^2 \xrightarrow{(6\ 6)} \mathbb{R} \to 0$$

Exercise 6.30 Construct a projective resolution of the following sequence, and then compute its homology,

$$0 \longrightarrow \mathbb{R} \xrightarrow{\begin{pmatrix}-1\\-2\end{pmatrix}} \mathbb{R}^2 \xrightarrow{\begin{pmatrix}1 & 4\\0 & 3\\-6 & -3\end{pmatrix}} \mathbb{R}^3 \longrightarrow 0$$

Example 6.15 Consider the same sequence as in Example 6.14, namely

$$0 \longrightarrow \mathbb{R} \xrightarrow{\begin{pmatrix}1\\0\\0\end{pmatrix}} \mathbb{R}^3 \xrightarrow{\begin{pmatrix}1&0&0\\0&0&1\end{pmatrix}} \mathbb{R}^2 \xrightarrow{\begin{pmatrix}0&1\end{pmatrix}} \mathbb{R} \longrightarrow 0$$

While in Example 6.14, we constructed a projective resolution, we can also construct an injective resolution, namely

$$\begin{array}{ccccccccc}
0 & \longrightarrow & \mathbb{R}^3 & \longrightarrow & \mathbb{R}^3 & \longrightarrow & \mathbb{R}^3 & \longrightarrow & \cdots \\
\downarrow & & \downarrow & & \downarrow & & \downarrow & & \\
\mathbb{R} & \longrightarrow & \mathbb{R}^3 & \longrightarrow & \mathbb{R}^2 & \longrightarrow & 0 & \longrightarrow & \cdots \\
\begin{pmatrix}1\\0\\0\end{pmatrix}\downarrow & & \downarrow & & \downarrow & & \downarrow & & \\
\mathbb{R}^3 & \longrightarrow & \mathbb{R}^3 & \longrightarrow & 0 & \longrightarrow & 0 & \longrightarrow & \cdots \\
\begin{pmatrix}1&0&0\\0&0&1\end{pmatrix}\downarrow & & \downarrow & & \downarrow & & \downarrow & & \\
\mathbb{R}^2 & \longrightarrow & \mathbb{R}^2 & \longrightarrow & 0 & \longrightarrow & 0 & \longrightarrow & \cdots \\
\begin{pmatrix}0&1\end{pmatrix}\downarrow & & \downarrow & & \downarrow & & \downarrow & & \\
\mathbb{R} & \longrightarrow & \mathbb{R} & \longrightarrow & 0 & \longrightarrow & 0 & \longrightarrow & \cdots \\
\downarrow & & \downarrow & & \downarrow & & \downarrow & & \\
0 & \longrightarrow & 0 & \longrightarrow & 0 & \longrightarrow & 0 & \longrightarrow & \cdots
\end{array}$$

In exactly the same way as in Example 6.14, if we set the leftmost column to be the zero sequence, then row-wise homology of the resulting chain complex is exactly the sequence we started with.

Exercise 6.31 Show that all injective resolutions of an arbitrary sequence $A^\bullet$ have isomorphic row-wise homology sequences.

Exercise 6.32 Prove "Schanuel's theorem," which states that if

$$0 \to K \to P \to M \to 0$$

and

$$0 \to K' \to P' \to M \to 0$$

are short exact sequences of objects in **Seq**, and both P and P' are projective, then

$$K \oplus P' \cong K' \oplus P.$$

6.5 Double Complexes

Now having explored **Kom**(**Seq**), one can reasonably ask how this relates to the subcategory **Kom**(**Kom**).

Definition 6.9 A *double complex* is an object of **Kom**(**Kom**), or in other words, a chain complexes of chain complexes.

The categories **Kom**(**Seq**) and **Kom**(**Kom**) are quite different in terms of injectives and projectives. Because the structure of morphisms of **Kom** is substantially more constrained than that of **Seq**, it is not the case that projective (or injective) sequences in **Seq** remain projective (or injective) in **Kom**. Moreover, although projective and injective sequences are distinct concepts in **Seq**, these concepts coincide in **Kom**.

Exercise 6.33 Choose any index a. Why is $\mathbb{R}^{(a,\infty)}$ not projective in **Kom**, even though it is projective in **Seq**?

Proposition 6.13 *The following are equivalent in* **Kom**:

1. *A chain complex is an exact sequence,*
2. *A chain complex is projective, and*
3. *A chain complex is injective.*

Proof As a preliminary step, surjectivity for chain maps implies that the chain map decomposition can contain only bar chain maps of Type 4, 5, 7, 8, and 9. Because this eliminates bar chain maps of Type 10, we need only consider two consecutive terms at a time.

Let us assume that we have an exact sequence. We must show that it is projective. Schematically, we can assume that the vertical arrows in the commutative diagram below are surjective,

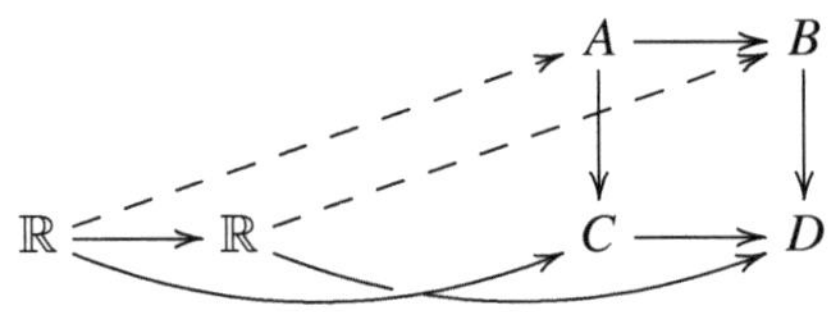

Surjectivity implies that the right side of the diagram is a bar chain map of Type 4, 5, 7, 8, or 9. In contrast, the curved horizontal arrows fill out a bar chain map that need not be surjective. We need to establish that the dashed arrows can be filled out regardless of what these two bar chain maps happen to be. Notice that since the domain on the left is a bar of length 2, this means that the horizontal bar chain map must be of Type 7, 8, 9, or 10. At this point, one needs to check all possible cases for vertical and horizontal bar chain maps. Some of these cases cannot occur under the constraint that the vertical chain map be both surjective and a *bar* chain map. Table 6.2 shows the bar chain map Type of the bar chain map that completes the above diagram (dashed arrows) for each possible combination.

Conversely, a non-exact sequence has at least one length 1 bar. In this case, Example 6.9 shows that the presence of a length 1 bar precludes the sequence being projective.

To establish the statement that injective chain complexes must also be exact, we must dualize everything in the diagram. That is, we reverse all the arrows. We can re-run the same calculation as before, by simply following the translation that

- Type 4 becomes Type 2 (transforms a surjection into an injection),
- Type 5 is unchanged,
- Type 7 becomes Type 3 (transforms a surjection into an injection),
- Type 8 becomes Type 6 (transforms a surjection into an injection),
- Type 9 is unchanged, and
- Type 10 is unchanged. □

Table 6.2 is a bit subtle, but the following two examples—which are paired—should help explain what is happening.

Table 6.2 Bar chain map types for the projective extension in Proposition 6.13. Impossible scenarios are shown as blank entries

Vertical map	Horizontal map			
	Type 7	Type 8	Type 9	Type 10
Type 4	Zero map			
Type 5		Type 8		
Type 7	Zero map			
Type 8		Type 9		
Type 9			Type 9	Type 10

Example 6.16 Consider the situation where the vertical map mentioned in Table 6.2 is a Type 5, and the horizontal map is a Type 8. This corresponds to a diagram of the form

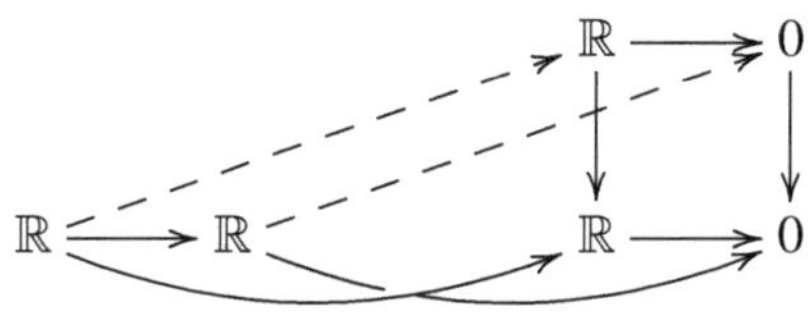

It is clear that the only possibility for the extension (the dashed arrows) is a Type 8.

Example 6.17 Consider the case in which the vertical map in Table 6.2 is a Type 8. If the horizontal map is also Type 8, the diagram takes the form

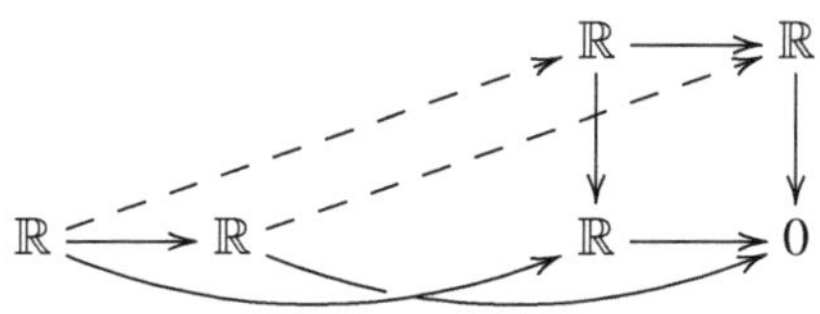

Notice that the only possibility for the extension (the dashed arrows) is a Type 9.

Exercise 6.34 Show that an exact sequence is neither projective nor injective in **Seq**. (Hint: rework Example 6.9.)

Proposition 6.13 notwithstanding, we will usually distinguish between projective and injective chain complexes both because they differ in **Seq**, and because projective and injective resolutions of chain complexes are rather different. This difference is also visible if we apply a bound to the indices in all chain complexes under consideration.

Proposition 6.14 *A chain complex in* $\mathbf{Kom}^+$ *is injective if and only if its only nontrivial homology is in index* 0.

Notice in particular the difference between this statement and Proposition 6.13.

Proof Proposition 6.13 handles the case where the homology is completely trivial. Without loss of generality we merely need to show that $\mathbb{R}^{(0,0)}$ is injective in $\mathbf{Kom}^+$.

According to Definition 6.6, we need to show that for arbitrary chain complexes $(A^\bullet, a^\bullet)$ and $(B^\bullet, b^\bullet)$ in $\mathbf{Kom}^+$, if $m_1 : (B^\bullet, b^\bullet) \to (A^\bullet, a^\bullet)$ is a chain map with injective components and $m_2 : (B^\bullet, b^\bullet) \to \mathbb{R}^{(0,0)}$ is completely arbitrary, then there is a chain map $s : (A^\bullet, a^\bullet) \to \mathbb{R}^{(0,0)}$ making the diagram below commute,

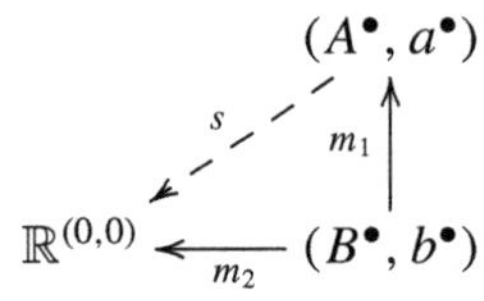

Considering only the situation at index 0, the diagram is of the form

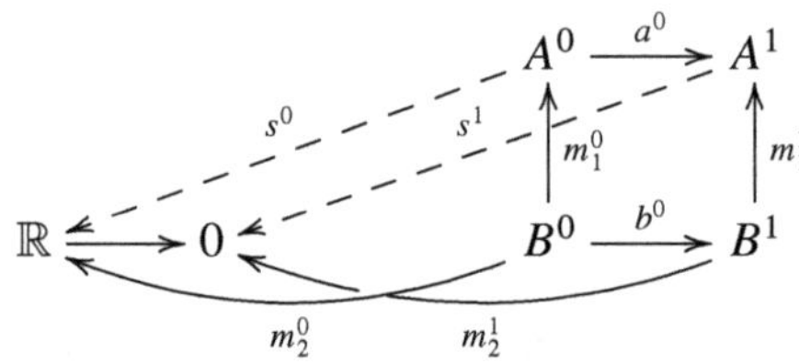

Following the same line of reasoning as in the proof of Lemma 6.5, we define

$$s^0(x) = \begin{cases} m_2^0\left((m_1^0)^{-1}(x)\right) & \text{if } x \in \text{image}\, m_1^0 \\ 0 & \text{otherwise.} \end{cases}$$

This is well defined because m_1 is injective. Moreover, this ensures that the diagram commutes in index 0. In the remaining indices, $s^k = 0$ for $k > 0$ is clearly the only (and correct) definition that ensures commutativity. □

The bar length restriction imposed on chain complexes ensures that there are many projective and injective sequences. This makes proving an analog for Proposition 6.9 for **Kom** straightforward.

Proposition 6.15 *If $A_\bullet$ is an arbitrary chain complex, then there exist exact sequences $P_\bullet$ and $E_\bullet$ such that*

$$P_\bullet \to A_\bullet \to 0$$

and

$$0 \to A_\bullet \to E_\bullet$$

are exact sequences.

Proof The barcode decomposition of $A_\bullet$ consists of bars of length 1 and 2 since it is a chain complex. To construct the required exact sequence $P_\bullet$, we merely need a chain map $P_\bullet \to A_\bullet$ that has surjective component maps. The easiest way to arrange this is to construct $P_\bullet$ and the chain map simultaneously. A suitable chain map can be constructed as the direct sum of bar chain maps. Specifically, we can use a bar chain map of Type 8 for each length 1 bar of $A_\bullet$ and a bar chain map of Type 9 for each length 2 bar of $A_\bullet$.

The same idea works to construct $E_\bullet$. We merely need to use a bar chain map of Type 6 for each length 1 bar of $A_\bullet$ and a bar chain map of Type 9 for each length 2 bar of $A_\bullet$. □

Repeated application of Proposition 6.15 is all that is necessary to construct projective and injective resolutions.

Example 6.18 Consider the chain complex $A_\bullet$ with a single bar at index 2, in other words $\mathbb{R}^{(2,2)}$, given by the diagram

$$\cdots \to 0 \to \mathbb{R} \to 0 \to 0.$$

Using Proposition 6.15 repeatedly, we can construct a projective resolution of $A_\bullet$, namely an exact sequence

$$\cdots \to P_{3,\bullet} \to P_{2,\bullet} \to P_{1,\bullet} \to P_{0,\bullet} \to A_\bullet \to 0.$$

Recognizing that projective sequences must themselves be exact and writing the sequences as columns, here is one such resolution (compare with Example 6.11 in the previous section)

$$\cdots \longrightarrow P_{3,\bullet} \longrightarrow P_{2,\bullet} \longrightarrow P_{1,\bullet} \longrightarrow P_{0,\bullet} \longrightarrow A_{\bullet}$$

$$\begin{array}{ccccccccccc}
 & & \vdots & & \vdots & & \vdots & & \vdots & & \vdots \\
 & & \downarrow & & \downarrow & & \downarrow & & \downarrow & & \downarrow \\
\cdots & \longrightarrow & 0 & \longrightarrow & 0 & \longrightarrow & 0 & \longrightarrow & 0 & \longrightarrow & 0 = A_3 \\
 & & \downarrow & & \downarrow & & \downarrow & & \downarrow & & \downarrow \\
\cdots & \longrightarrow & 0 & \longrightarrow & 0 & \longrightarrow & 0 & \longrightarrow & \mathbb{R} & \longrightarrow & \mathbb{R} = A_2 \\
 & & \downarrow & & \downarrow & & \downarrow & & \downarrow & & \downarrow \\
\cdots & \longrightarrow & 0 & \longrightarrow & 0 & \longrightarrow & \mathbb{R} & \longrightarrow & \mathbb{R} & \longrightarrow & 0 = A_1 \\
 & & \downarrow & & \downarrow & & \downarrow & & \downarrow & & \downarrow \\
\cdots & \longrightarrow & 0 & \longrightarrow & \mathbb{R} & \longrightarrow & \mathbb{R} & \longrightarrow & 0 & \longrightarrow & 0 = A_0 \\
 & & \downarrow & & \downarrow & & \downarrow & & \downarrow & & \\
 & & \vdots & & \vdots & & \vdots & & \vdots & &
\end{array}$$

Going the other direction with injective sequences, here is an injective resolution of $A_{\bullet}$,

$$A_{\bullet} \longrightarrow E_{0,\bullet} \longrightarrow E_{1,\bullet} \longrightarrow \cdots$$

$$\begin{array}{ccccccc}
\vdots & & \vdots & & \vdots & & \\
\downarrow & & \downarrow & & \downarrow & & \\
A_3 = 0 & \longrightarrow & \mathbb{R} & \longrightarrow & \mathbb{R} & \longrightarrow & \cdots \\
\downarrow & & \downarrow & & \downarrow & & \\
A_2 = \mathbb{R} & \longrightarrow & \mathbb{R} & \longrightarrow & 0 & \longrightarrow & \cdots \\
\downarrow & & \downarrow & & \downarrow & & \\
A_1 = 0 & \longrightarrow & 0 & \longrightarrow & 0 & \longrightarrow & \cdots \\
\downarrow & & \downarrow & & \downarrow & & \\
A_0 = 0 & \longrightarrow & 0 & \longrightarrow & 0 & \longrightarrow & \cdots
\end{array}$$

Example 6.18 is merely an expression of the general situation.

Corollary 6.2 *Every chain complex of vector spaces has at least one projective (and at least one injective) resolution.*

There is also a variant of Proposition 6.10 for projective resolutions of chain complexes as well. The proof works a little differently because the structure of projectives in **Kom** is different from that for projectives in **Seq**.

Proposition 6.16 *Suppose that $m^\bullet : A^\bullet \to B^\bullet$ is a chain map in* **Kom**, *and that*

$$P^{\bullet,}{}_{\bullet} \to A^\bullet \to 0$$

and

$$Q^{\bullet,}{}_{\bullet} \to B^\bullet \to 0$$

are projective resolutions. There is a **Kom(Kom)** *morphism that extends $m^\bullet$ given by the diagram*

$$\begin{array}{ccccccccc} \cdots & \longrightarrow & P^{\bullet,}{}_1 & \longrightarrow & P^{\bullet,}{}_0 & \longrightarrow & A^\bullet & \longrightarrow & 0 \\ & & \big\downarrow f^{\bullet,}{}_1 & & \big\downarrow f^{\bullet,}{}_0 & & \big\downarrow m^\bullet & & \\ \cdots & \longrightarrow & Q^{\bullet,}{}_1 & \longrightarrow & Q^{\bullet,}{}_0 & \longrightarrow & B^\bullet & \longrightarrow & 0 \end{array}$$

Proof We would like to mimic the construction used in the proof of Proposition 6.10, but the details of taking the quotients are different because $Q^{\bullet,}{}_k$ is not a subspace of $Q^{\bullet,}{}_{k+1}$. We still take quotients recursively according to the following recipe,

$$\begin{cases} Q'^{\bullet,}{}_0 := Q^{\bullet,}{}_0 / B^\bullet \\ Q'^{\bullet,}{}_k := Q^{\bullet,}{}_k / Q'^{\bullet,}{}_{k-1}. \end{cases}$$

However, for this proof, we need to show that this construction is well defined, namely that each $Q'^{\ell,}{}_{k-1}$ is a subspace of $Q^{\ell,}{}_k$. The motivation for this construction is that the induced maps $Q^{\bullet,}{}_{k+1} \to Q'^{\bullet,}{}_k$ are surjective on each term, and this will enable us to construct the desired **Kom(Kom)** morphism using projectivity of the $P^{\bullet,}{}_k$ sequences.

Towards the end of showing well definedness of the $Q'^{\bullet,}{}_k$, we assume, as an inductive hypothesis, that the following sequence is exact,

$$\cdots \longrightarrow Q^{\bullet,}{}_{k+1} \xrightarrow{q^{\bullet,}{}_{k+1}} Q^{\bullet,}{}_k \xrightarrow{q'^{\bullet,}{}_k} Q'^{\bullet,}{}_{k-1} \longrightarrow 0$$

Exactness implies that $q'^{\bullet,}{}_k$ is surjective, and so this chain map can only contain bar chain maps of Types 4, 5, 7, 8, and 9. Given that observation, $Q'^{\bullet,}{}_k := Q^{\bullet,}{}_k / Q'^{\bullet,}{}_{k-1}$ can only contain bars of length 1. These length 1 bars only arise from bar chain maps of Types 4 and 8 in $q'^{\bullet,}{}_k$. Exactness implies that the horizontal bars in the grid diagram are all of length 2. Therefore, the length 1 bars in $Q'^{\bullet,}{}_k$ must be at the right ends of these length 2 horizontal bars, since they correspond to bar chain maps of Types 4 and 8 in $q'^{\bullet,}_{k}$. Thus, we must conclude that

$$\cdots \longrightarrow Q^{\bullet,}_{k+1} \xrightarrow{q'^{\bullet,}_{k+1}} Q'^{\bullet,}_{k} \longrightarrow 0$$

is also exact, where $q'^{\bullet,}_{k+1}$ is the map induced by $q^{\bullet,}_{k+1}$ on the quotient.

Given the construction of the $Q'^{\bullet,}_{k}$ sequences and the surjective chain maps $q'^{\bullet,}_{k+1}$, we now may apply projectivity to obtain the dashed chain maps below

$$\begin{array}{ccc} P^{\bullet,}_{k+1} & & \\ {\scriptstyle f^{\bullet,}_{k+1}}\,\downarrow & \searrow^{\pi^{\bullet,}_{k}\circ f^{\bullet,}_{k}\circ p^{\bullet,}_{k+1}} & \\ Q^{\bullet,}_{k+1} & \longrightarrow & Q'^{\bullet,}_{k} \end{array}$$

where $\pi^{\bullet,}_{k} : Q^{\bullet,}_{k} \to Q'^{\bullet,}_{k}$ is the canonical projection. □

Example 6.19 Let us extend the following chain map into projective resolutions,

$$\begin{array}{ccccccccc} 0 & \longrightarrow & \mathbb{R} & \xrightarrow{(0)} & \mathbb{R} & \xrightarrow{(1)} & \mathbb{R} & \longrightarrow & 0 \\ & & \downarrow & & \downarrow{\scriptstyle (1)} & & \downarrow{\scriptstyle (1)} & & \\ 0 & \longrightarrow & 0 & \longrightarrow & \mathbb{R} & \xrightarrow[(1)]{} & \mathbb{R} & \longrightarrow & 0 \end{array}$$

Here is a projective resolution for the domain chain complex (the top row in the above is the rightmost column below),

$$\begin{array}{ccccccc} 0 & \longrightarrow & 0 & \longrightarrow & 0 & \longrightarrow & 0 \\ \downarrow & & \downarrow & & \downarrow & & \downarrow \\ 0 & \longrightarrow & 0 & \longrightarrow & \mathbb{R} & \longrightarrow & \mathbb{R} \\ \downarrow & & \downarrow & & \downarrow{\scriptstyle \binom{1}{0}} & & \downarrow{\scriptstyle (0)} \\ 0 & \longrightarrow & \mathbb{R} & \xrightarrow{\binom{1}{0}} & \mathbb{R}^2 & \xrightarrow[(0\,1)]{} & \mathbb{R} \\ \downarrow & & \downarrow & & \downarrow{\scriptstyle (0\,1)} & & \downarrow{\scriptstyle (1)} \\ \mathbb{R} & \longrightarrow & \mathbb{R} & \xrightarrow{(0)} & \mathbb{R} & \xrightarrow[(1)]{} & \mathbb{R} \\ \downarrow & & \downarrow & & \downarrow & & \downarrow \\ \mathbb{R} & \longrightarrow & 0 & \longrightarrow & 0 & \longrightarrow & 0 \end{array}$$

and here is a projective resolution for the codomain chain complex (the rightmost column is the bottom row in the original chain map),

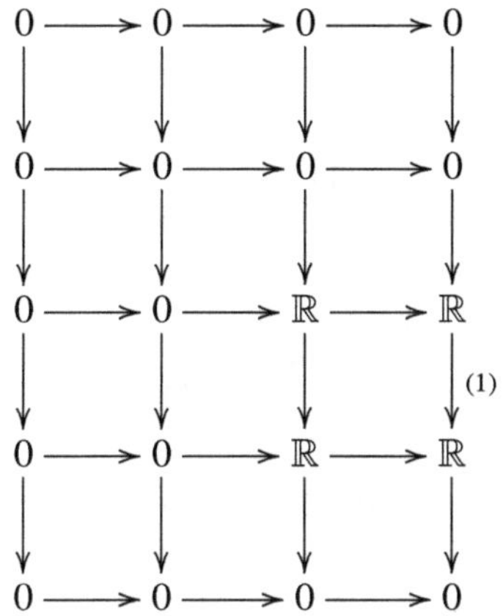

Notice that these are both not unique; many other projective resolutions exist for each chain complex. We will consider a different resolution for the codomain in Example 6.20.

The original chain map takes the rightmost column of the domain's resolution to the rightmost column of the codomain's resolution. The nice thing about lifting the original chain map into the resolutions is that we can simply proceed term-by-term within each column. To that end, we next work on the second column from the right, starting at the top. Since both of the terms in question are zero spaces, the lifted component map is a zero map. There is similarly no option for the second map. For the third component map, we need to make the diagram below commute by defining the dashed arrow

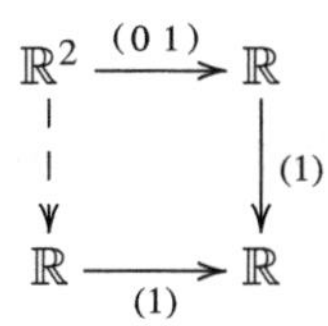

There is only one possibility! Finally, the last component map only has one possibility as well.

Given what we have defined thus far for the lifts of the previous maps, there is only one possibility for the entire chain map, namely the one below:

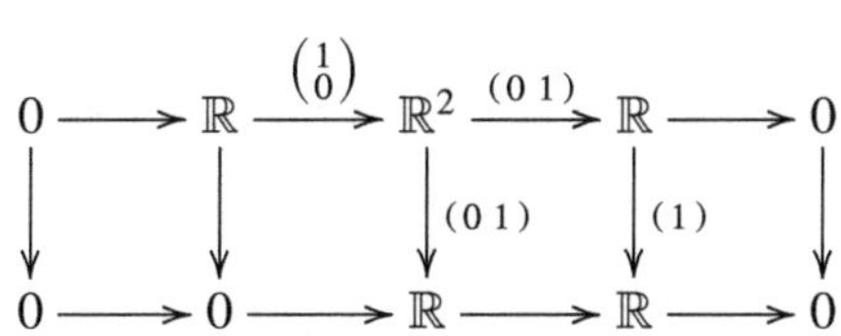

All of the other lifted maps must be zero maps because the remaining codomain chain complexes are all trivial.

Exercise 6.35 Extend the chain map

$$
\begin{array}{ccccccccc}
0 & \longrightarrow & \mathbb{R} & \xrightarrow{(1)} & \mathbb{R} & \longrightarrow & 0 & \longrightarrow & 0 \\
 & & \downarrow & & \downarrow{\scriptstyle (0)} & & \downarrow & & \\
0 & \longrightarrow & 0 & \longrightarrow & \mathbb{R} & \xrightarrow[(1)]{} & \mathbb{R} & \longrightarrow & 0
\end{array}
$$

to a chain map between projective resolutions.

Exercise 6.36 Extend the chain map

$$
\begin{array}{ccccccccc}
0 & \longrightarrow & \mathbb{R} & \xrightarrow{(1)} & \mathbb{R} & \longrightarrow & 0 & \longrightarrow & 0 \\
 & & \downarrow{\scriptstyle (1)} & & \downarrow & & \downarrow & & \\
0 & \longrightarrow & \mathbb{R} & \longrightarrow & 0 & \longrightarrow & \mathbb{R} & \longrightarrow & 0
\end{array}
$$

to a chain map between projective resolutions.

Exercise 6.37 Prove the dual statement to Proposition 6.16 for injective resolutions. Hint: see Exercise 6.25.

Two different resolutions of the same chain complex may not be isomorphic in **Kom**(**Kom**). We will delay discussion of isomorphisms between resolutions until we recall chain homotopies in Sect. 6.6.

Example 6.20 Consider the chain map given in Example 6.19,

$$
\begin{array}{ccccccccc}
0 & \longrightarrow & \mathbb{R} & \xrightarrow{(0)} & \mathbb{R} & \xrightarrow{(1)} & \mathbb{R} & \longrightarrow & 0 \\
 & & \downarrow & & \downarrow{\scriptstyle (1)} & & \downarrow{\scriptstyle (1)} & & \\
0 & \longrightarrow & 0 & \longrightarrow & \mathbb{R} & \xrightarrow[(1)]{} & \mathbb{R} & \longrightarrow & 0
\end{array}
$$

Let us consider a projective resolution of the codomain chain complex that is different from the one we used in Example 6.19, namely

$$
\begin{array}{ccccccc}
0 & \longrightarrow & 0 & \longrightarrow & 0 & \longrightarrow & 0 \\
\downarrow & & \downarrow & & \downarrow & & \downarrow \\
0 & \longrightarrow & 0 & \longrightarrow & 0 & \longrightarrow & 0 \\
\downarrow & & \downarrow & & \downarrow & & \downarrow \\
0 & \longrightarrow & \mathbb{R} & \xrightarrow{\binom{1}{0}} & \mathbb{R}^2 & \xrightarrow{(0\ 1)} & \mathbb{R} \\
\downarrow & & \downarrow & & \downarrow & & \downarrow{\scriptstyle (1)} \\
0 & \longrightarrow & \mathbb{R} & \xrightarrow{\binom{1}{0}} & \mathbb{R}^2 & \xrightarrow{(0\ 1)} & \mathbb{R} \\
\downarrow & & \downarrow & & \downarrow & & \downarrow \\
0 & \longrightarrow & 0 & \longrightarrow & 0 & \longrightarrow & 0
\end{array}
$$

The chain map still induces chain maps with this new resolution, though they are clearly different from the ones obtained in Example 6.19. It is helpful to draw the entire diagram in the three rightmost columns and the third and fourth rows, namely

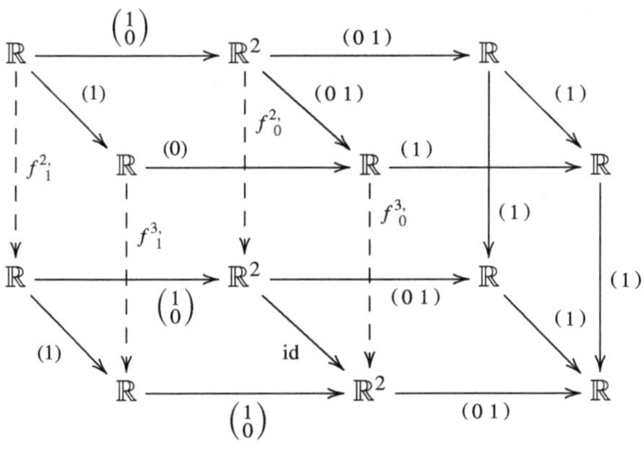

From this, the nontrivial component maps can be read off in the following order

$$
f^{3,}{}_{0} = \begin{pmatrix} 0 \\ 1 \end{pmatrix},
$$

$$
f^{2,}{}_{0} = \begin{pmatrix} 0 & 0 \\ 0 & 1 \end{pmatrix},
$$

$$
f^{3,}{}_{1} = (0),
$$

and

$$f^{2,}{}_{1} = (0)\,.$$

For a double complex, it is possible to compute homology along the rows *and* the columns in a double complex. These homologies will be different in general. Moreover, computing homology along the columns does not disrupt the chain complex structure in the rows and *vice versa*, as the next lemma states.

Lemma 6.12 *Suppose that* $(A^{\bullet,\bullet}, a^{\bullet,\bullet}, b^{\bullet,\bullet})$ *is a double complex. Both homology along the columns* $(H^{\square}(A^{\diamond,\bullet}, b^{\diamond,\bullet}), a_*^{\diamond,\square}, 0)$*, and homology along the rows* $(H^{\square}(A^{\bullet,\diamond}, a^{\bullet,\diamond}), 0, b_*^{\square,\diamond})$ *yield double complexes.*

Proof If taking homology along the rows of a double complex yields another double complex, then the same holds for taking homology along the columns as well. Without loss of generality, we need only establish the statement for row-wise homology.

Each row and each column of the grid diagram for a double complex is itself a chain complex. This means that the barcode decomposition of each row and each column contains length 1 and length 2 bars only. In each row, the homology sequence will consist of only the length 1 bars, removing the length 2 bars via quotienting. The rows therefore remain chain complexes.

The column-wise maps in the row-wise homology sequence will become the induced maps on homology, which is to say that they are the maps descended to the quotients. Since this merely reduces the dimension of the domains and codomains of each column-wise map, any bars in the columns can only become shorter after taking homology. Hence, the column-wise barcode decompositions in the grid diagram after row-wise homology has been taken still contain bars of length 1 or 2. Therefore, the columns in the grid diagram after row-wise homology are still chain complexes. □

The lemma establishes that for a double complex, it is possible to compute homology sequentially: first rows and then columns, or *vice versa*. Perhaps surprisingly, the order in which this is done matters. That is, taking the homology along the rows and then the columns is not the same as taking the homology along the columns and then the rows.

Example 6.21 Consider a grid diagram for a double complex made from a Type 6 bar chain map. That is,

$$\begin{array}{ccc} 0 & \longrightarrow & \mathbb{R} \\ \downarrow & & \downarrow \\ \mathbb{R} & \longrightarrow & \mathbb{R} \end{array}$$

If we take homology along the columns first, we obtain the grid diagram

$$\begin{array}{ccc} 0 & \longrightarrow & 0 \\ \downarrow & & \downarrow \\ \mathbb{R} & \longrightarrow & 0 \end{array}$$

Taking homology along the rows next does not change the diagram.

On the other hand, if we take homology of a Type 6 bar chain map along the rows first, we obtain a different grid diagram

$$
\begin{array}{ccc}
0 & \longrightarrow & \mathbb{R} \\
\downarrow & & \downarrow \\
0 & \longrightarrow & 0
\end{array}
$$

which is also unchanged by then taking homology along the columns.

Exercise 6.38 For which bar chain map Types does the order matter for taking homology along rows and columns?

Double complexes can be used to relate two chain complexes $(C^\bullet, c^\bullet)$ and $(D^\bullet, d^\bullet)$. If the two chain complexes are bounded from below, so that $C^k = 0$ and $D^k = 0$ for $k < 0$, then one can consider a grid diagram extending both of them

$$
\begin{array}{ccccccc}
0 & \longrightarrow & C^0 & \xrightarrow{c^0} & C^1 & \xrightarrow{c^1} & \cdots \\
\downarrow & & \downarrow & & \downarrow & & \\
D^0 & \longrightarrow & A^{1,1} & \longrightarrow & A^{1,2} & \longrightarrow & \cdots \\
\downarrow{\scriptstyle d^0} & & \downarrow & & \downarrow & & \\
D^1 & \longrightarrow & A^{2,1} & \longrightarrow & A^{2,2} & \longrightarrow & \cdots \\
\downarrow{\scriptstyle d^1} & & \downarrow & & \downarrow & & \\
\vdots & & \vdots & & \vdots & &
\end{array}
$$

With careful choice of $A^{\bullet,\bullet}$, a correspondence between the two chain complexes $(C^\bullet, c^\bullet)$ and $(D^\bullet, d^\bullet)$ can be established. For instance, if the diagram has exact rows and columns (excepting the first of each), then the two chain complexes are quasi-isomorphic.

Theorem 6.1 (the Čech double complex theorem) *Suppose that* $(C^\bullet, c^\bullet)$ *and* $(D^\bullet, d^\bullet)$ *are chain complexes.*

If $(A^{\bullet,\bullet}, a^{\bullet,\bullet}, b^{\bullet,\bullet})$ *is a double complex that satisfies*

1. $(A^{0,k}, a^{0,k}) = (C^{k-1}, c^{k-1})$ *for all* k,
2. $(A^{j,0}, b^{j,0}) = (D^{j-1}, d^{j-1})$ *for all* j,
3. $(A^{\bullet,\bullet}, a^{\bullet,\bullet}, b^{\bullet,\bullet})$ *has exact rows for indices* 1 *and higher,*
4. $(A^{\bullet,\bullet}, a^{\bullet,\bullet}, b^{\bullet,\bullet})$ *has exact columns for indices* 1 *and higher, and*

5. $A^{j,k} = 0$ *if* $j < 0$ *or* $k < 0$,

then $H^\ell(C^\bullet, c^\bullet) \cong H^{\ell+1}(A^{0,\bullet}, a^{0,\bullet}) \cong H^{\ell+1}(A^{\bullet,0}, b^{\bullet,0}) \cong H^\ell(D^\bullet, d^\bullet)$*forall*$\ell \geq 0$.

The proof uses the construction in Example 6.18 to "connect" length 1 bars in the first row to length 1 bars in the first column by a "staircase" path between them.

Proof Suppose that there is a length 1 bar in C^ℓ for some ℓ. Exactness of column $\ell + 1$ in the grid diagram implies that this bar has a nontrivial image in $A^{1,\ell+1}$. Since it is a length 1 bar in C^ℓ, this means that its image in $C^{\ell+1}$ is zero. Commutativity implies that it cannot have a nontrivial image in $A^{1,\ell+2}$. However, exactness of the rows of the grid diagram implies that there is a nontrivial preimage of this bar in $A^{1,\ell}$. This argument is summarized by a diagram chase on

$$\begin{array}{ccccc}
0 & \longrightarrow & \mathbb{R} \subseteq C^\ell & \longrightarrow & 0 \\
\downarrow & & \downarrow & & \\
\mathbb{R} \subseteq A^{1,\ell} & \longrightarrow & \mathbb{R} \subseteq A^{1,\ell+1} & \longrightarrow & 0
\end{array}$$

At this point, we can repeat the argument starting at $A^{1,\ell}$ with one small change. Although the image of the original bar was zero in $C^{\ell+1}$, it is not zero in $A^{1,\ell+1}$. However, exactness of column $\ell + 1$ ensures that the image in $A^{2,\ell+1}$ must be zero. Thus the diagram extends to

$$\begin{array}{ccccccc}
 & & 0 & \longrightarrow & \mathbb{R} \subseteq C^\ell & \longrightarrow & 0 \\
 & & \downarrow & & \downarrow & & \\
 & & \mathbb{R} \subseteq A^{1,\ell} & \longrightarrow & \mathbb{R} \subseteq A^{1,\ell+1} & \longrightarrow & 0 \\
 & & \downarrow & & \downarrow & & \\
\mathbb{R} \subseteq A^{2,\ell-1} & \longrightarrow & \mathbb{R} \subseteq A^{2,\ell} & \longrightarrow & 0 & &
\end{array}$$

This can be iterated until we arrive at column 0, which occurs at D^ℓ. The result is a length 1 bar in exactly index D^ℓ.

Hence, we have defined a correspondence that takes every length 1 bar of $C^\bullet$ at index ℓ to a length 1 bar of $D^\bullet$ at the same index ℓ. The construction is symmetric under transposition of the grid diagram. We also have a correspondence taking every length 1 bar of $D^\bullet$ to a length 1 bar of $C^\bullet$. Because the indices are preserved, this ensures that these two correspondences are inverses of each other. We have a bijection between the length 1 bars of $C^\bullet$ and $D^\bullet$. These correspondences may not be chain maps, so $C^\bullet$ and $D^\bullet$.

There was no freedom in the above construction. We must conclude that any other double complex $(A^{\bullet,\bullet}, a^{\bullet,\bullet}, b^{\bullet,\bullet})$ satisfying the hypotheses will include the structure above. While there may be length 2 bars present in various places in the diagram, they do not impact homology. □

Exercise 6.39 Using Theorem 6.1, show that the simplicial chain complexes for an abstract simplicial complex X and its barycentric subdivision BX are quasi-isomorphic. Hint: the simplicial chain complexes for X and for BX should be $C^\bullet$ and $D^\bullet$, respectively. Your task is to construct a double complex so that the rows and columns are exact, as is required by Theorem 6.1. (Extra hint: If you get stuck, consult the appendix of Hubbard 2006, where a similar construction is used.)

We end this section with an application to *split exact sequences* of sequences. Recall that we say that a short exact sequence

$$0 \longrightarrow A \longrightarrow B \longrightarrow C \longrightarrow 0$$

splits if

$$B \cong A \oplus C.$$

Split exact sequences were a mere footnote in the case of vector spaces (Sect. 2.4 of Chap. 2) but now are more interesting.

Proposition 6.17 *If $P^\bullet$ is a projective sequence in either* **Kom** *or* **Seq** *then every exact sequence of the form*

$$0 \longrightarrow A^\bullet \longrightarrow B^\bullet \longrightarrow P^\bullet \longrightarrow 0$$

splits. Similarly, if $E^\bullet$ is an injective sequence, then every exact sequence of the form

$$0 \longrightarrow E^\bullet \longrightarrow B^\bullet \longrightarrow C^\bullet \longrightarrow 0$$

splits.

A special case of this for **Seq** is considered in Exercise 6.24 in Sect. 6.4. To prove these statements in general, we do not require any specifics about **Kom** or **Seq** beyond projectivity or injectivity.

Proof We can handle both categories at once, so by "map" we mean "sequence map" or "chain map" as appropriate.

Suppose that we have an exact sequence

$$0 \longrightarrow A^\bullet \longrightarrow B^\bullet \xrightarrow{m^\bullet} P^\bullet \longrightarrow 0$$

in which $P^\bullet$ is projective. We can augment the diagram with an identity map for $P^\bullet$, which induces the dashed map $f^\bullet$ via projectivity

$$\begin{array}{ccccccccc} & & & & & & P^\bullet & & \\ & & & & & \swarrow^{f^\bullet} & \downarrow \mathrm{id}^\bullet & & \\ 0 & \longrightarrow & A^\bullet & \longrightarrow & B^\bullet & \xrightarrow[m^\bullet]{} & P^\bullet & \longrightarrow & 0 \end{array}$$

such that the diagram commutes.

Notice that commutativity means that

$$m^\bullet \circ f^\bullet = \mathrm{id}^\bullet$$

which is to say that $f^\bullet$ is injective.

Because the map $A^\bullet \to B^\bullet$ is injective as well, its image (of $A^\bullet$) in $B^\bullet$ is in the kernel of $m^\bullet$. Hence the images of $A^\bullet$ and $P^\bullet$ within $B^\bullet$ must only intersect along the zero sequence. Thus $A^\bullet \oplus P^\bullet$ is a subsequence of $B^\bullet$.

On the other hand, exactness in each term ensures that we have $A^k \oplus P^k \cong B^k$ for each k, which completes the argument.

The statement for injectivity follows by dualization, *mutatis mutandis*. □

Proposition 6.17 generalizes slightly to projective resolutions of projective sequences (and to injective resolutions of injective sequences).

Corollary 6.3 *If $P^\bullet$ is a projective sequence in either* **Kom** *or* **Seq** *then an exact sequence of the form*

$$\cdots \longrightarrow A^{\bullet,}{}_1 \longrightarrow A^{\bullet,}{}_0 \longrightarrow P^\bullet \longrightarrow 0$$

implies that $P^\bullet$ is a subsequence of $A^{\bullet,}{}_0$. Namely, there exists a sequence $Q^\bullet$ such that $A^{\bullet,}{}_0 \cong P^\bullet \oplus Q^\bullet$. Similarly, if $E^\bullet$ is an injective sequence, then every exact sequence of the form

$$0 \longrightarrow E^\bullet \longrightarrow B^{\bullet,}{}_0 \longrightarrow B^{\bullet,}{}_1 \longrightarrow \cdots$$

implies that $E^\bullet$ is a subsequence of $B^{\bullet,}{}_0$.

It is tempting to think that the resolution shown in Example 6.18 contradicts Proposition 6.17, at least in the case of **Kom**. However, pruning out three consecutive columns does not yield a short exact sequence of the kind demanded in the hypothesis of Proposition 6.17. The essential obstruction is the presence of a Type 10 bar chain map. Type 10 bar chain maps play a key role in creating exact sequences that do not split (compare with Example 3.8 in Chap. 3).

Proposition 6.18 *A short exact sequence of chain complexes splits if and only if its grid diagram contains no Type* 10 *bar chain maps.*

Proof Notice that the only way a Type 10 bar chain map can appear in a grid diagram with exact rows is

$$\begin{array}{ccccc} 0 & \longrightarrow & \mathbb{R} & \longrightarrow & \mathbb{R} \\ \downarrow & & \downarrow & & \downarrow \\ \mathbb{R} & \longrightarrow & \mathbb{R} & \longrightarrow & 0 \end{array}$$

Rotating the diagram would violate exactness. Similarly, the only allowable bar chain maps in an exact sequence of chain complexes are Types 3, 7, 9, 10, since each row needs to have bars of length 0 or 2. Of these, in a short exact sequence, Type 10 guarantees that a sequence does not split (Example 3.8 in Chap. 3), but the others easily split. □

6.6 The Homotopy Category

Chain maps in **Kom**(**Seq**) or **Kom**(**Kom**) induced by a sequence map (Proposition 6.10) or a chain map (Proposition 6.16) are not unique on their own. Chain homotopies, defined in Definition 3.4 in Chap. 3, provide an equivalence between chain maps that removes this non-uniqueness. One can construct a new category called the *homotopy category* in which the morphisms are chain homotopy classes. In the homotopy category, projective and injective resolutions are unique up to homotopy equivalence.

Additionally in Sect. 3.4 of Chap. 3, we noted that for chain complexes of vector spaces, homotopy equivalence and quasi-isomorphism is the same. This does not hold for **Kom**(**Seq**) or **Kom**(**Kom**) as Example 6.22 below shows. Using quasi-isomorphism instead of homotopy equivalence yields the *derived category*, to be discussed in Sect. 6.7.

Example 6.22 Homotopy equivalence and quasi-isomorphism are different for chain complexes of sequences. Consider a Type 10 bar chain map extended to make each column into a separate sequence,

$$
\begin{array}{ccccc}
\vdots & & \vdots & & \vdots \\
\downarrow & & \downarrow & & \downarrow \\
0 & \longrightarrow & 0 & \longrightarrow & 0 \\
\downarrow & & \downarrow & & \downarrow \\
0 & \longrightarrow & \mathbb{R} & \longrightarrow & \mathbb{R} \\
\downarrow & & \downarrow & & \downarrow \\
\mathbb{R} & \longrightarrow & \mathbb{R} & \longrightarrow & 0 \\
\downarrow & & \downarrow & & \downarrow \\
0 & \longrightarrow & 0 & \longrightarrow & 0 \\
\downarrow & & \downarrow & & \downarrow \\
\vdots & & \vdots & & \vdots
\end{array}
$$

This can be thought of as a short exact sequence of sequences

$$0 \to S_2 \to S_1 \to S_0 \to 0,$$

where we have suppressed the superscript (row) index for clarity. This short exact sequence of sequences has trivial (row-wise) homology. Therefore, the zero chain map (of sequences!) induces an isomorphism on row-wise homology, which means that $S_\bullet$ is quasi-isomorphic to the zero chain complex.

On the other hand, $S_\bullet$ is not homotopy equivalent to the zero chain complex. The reason is a bit subtle, so let us prove the inequivalence by contradiction. A homotopy equivalence would require there to be two chain maps $f_\bullet : S_\bullet \to 0_\bullet$ and $g_\bullet : 0_\bullet \to S_\bullet$, such that $g_\bullet \circ f_\bullet \simeq \mathrm{id}_\bullet$. Evidently both $f_\bullet$ and $g_\bullet$ are zero chain maps. What must really be shown is that $\mathrm{id}_\bullet : S_\bullet \to S_\bullet$ is nullhomotopic. This is where the contradiction lies. We would need to construct the homotopy $h_\bullet$ shown as the dashed arrows in the grid diagram below

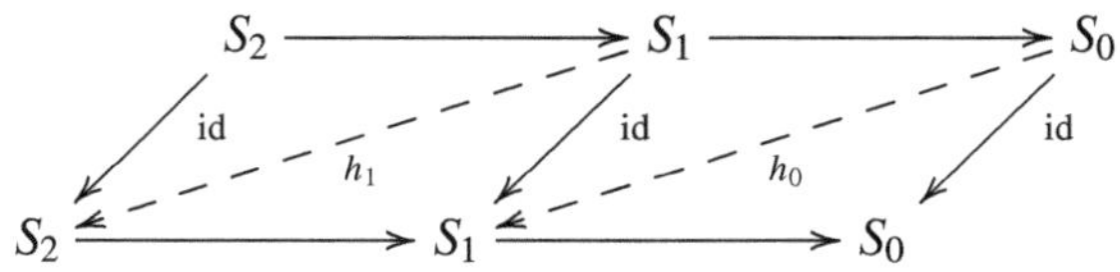

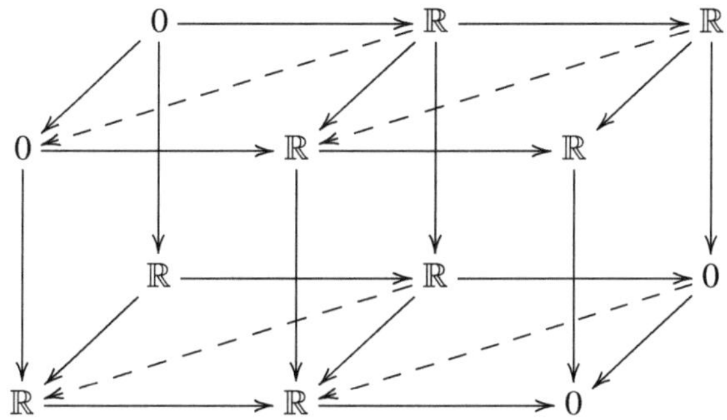

where each solid map is of full rank. Notice that while the homotopy $h_\bullet$ is not a chain map (a morphism of **Kom**(**Seq**)), each component h_k of the homotopy must be a sequence map (a morphism of **Seq**). What this means is that in the grid diagram above, the two vertical squares that involve the dashed arrows must both commute. This is impossible! Therefore, the identity chain map is not nullhomotopic. We conclude that the chain complex $S_\bullet$ is not homotopy equivalent to the zero chain complex.

Suppose that **C** is one of the following categories: **Vec**, **Seq**, or **Kom**. Then **Kom**(**C**) is a well defined category of chain complexes, where we observe that **Kom** = **Kom**(**Vec**) has occupied our attention throughout most of this book.

The proliferation of indices can become cumbersome, especially because **Vec** needs no indices but the other two categories do. Additionally, the indices on objects of **Seq** and **Kom** become ambiguous when we consider **Kom**(**Seq**) or **Kom**(**Kom**). In these last three sections of this book, we will suppress the indices when referring to an object in **C**. That is, we write A for an object of **C** even if A is itself a sequence (an object of **Seq**) or a chain complex (an object of **Kom**). Similarly, we will leave off indices for **C** morphisms as well. We emphasize the chain complex index for objects of **Kom**(**C**) by using only one index, writing such an object as $A_\bullet$ given our (slight) preference for projective resolutions. In the cases where this causes ambiguity in calculations, we will reintroduce the relevant indices.

Theorem 6.2 (morphisms extend into sequence maps between resolutions) *Suppose that* **C** *is one of the following categories:* **Vec**, **Seq**, *or* **Kom**. *Suppose that A and B are both objects in* **C**. *As Propositions 6.10 and 6.16 show, any* **C**-*morphism $m : A \to B$ extends to a* **Kom**(**C**)-*morphism $m_\bullet$ between any two projective resolutions $P_\bullet \to A \to 0$ and $Q_\bullet \to B \to 0$. All of these extensions are chain homotopic to each other.*

See also Gelfand and Manin (2003, III.1.3), which shows that the same statement is true at greater generality.

Proof Suppose that there are two chain map extensions $m_\bullet$ and $w_\bullet$ of m. We need to construct a chain homotopy between them. Recall that this means that there are **C** morphisms h_k, which may not be the components of a **Kom**(**C**) morphism, such that in the (non-commutative) diagram

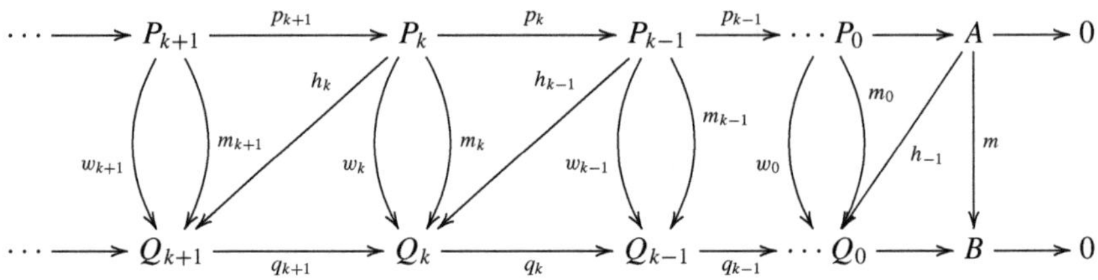

we have that

$$q_{k+1} \circ h_k + h_{k-1} \circ p_k = m_k - w_k. \tag{6.1}$$

We will construct the h_k inductively. To start an induction, notice that since both $m_\bullet$ and $w_\bullet$ extend m, it is the case that

$$m_{-1} = w_{-1} = m.$$

(We do not need to handle indices less than -1 in Eq. (6.1), because all of the **C** objects in question—which would appear off the right side of the diagram above—are zero.) We can simply define h_k to be the zero morphism for $k \leq -1$, because then Eq. (6.1) is trivially satisfied.

Now let us attend to the induction step. Recall that the arguments from the proofs of both Propositions 6.10 and 6.16 used iterative quotients of the form

$$\begin{cases} Q'_0 := Q_0/B \\ Q'_k := Q_k/Q'_{k-1}. \end{cases}$$

The proofs of Propositions 6.10 and 6.16 established that each of the Q'_k is well defined, and that

$$\cdots \longrightarrow Q_{k+2} \xrightarrow{q_{k+2}} Q_{k+1} \xrightarrow{q'_{k+1}} Q'_k \longrightarrow 0 \tag{6.2}$$

is a projective resolution. While the q_{k+1} map might not be surjective, the induced map q'_{k+1} is surjective. It will be useful in calculations below to define $\pi_k : Q_k \to Q'_k$ to be the canonical projection.

Let $(f_*)_k : P_k \to Q'_k$ be the map induced by

$$f_k = m_k - w_k - h_{k-1} \circ p_k.$$

With this in place, we can construct h_k using the projectivity of P_k, such that

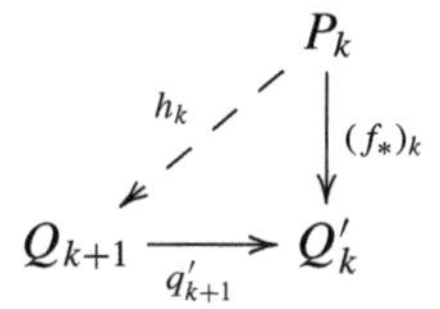

A slight subtlety remains. Commutativity of the above diagram states that

$$q'_{k+1} \circ h_k = (f_*)_k, \tag{6.3}$$

while the chain homotopy condition requires instead that

$$q_{k+1} \circ h_k = f_k.$$

In short, we need to show that the dashed triangle in the diagram below commutes

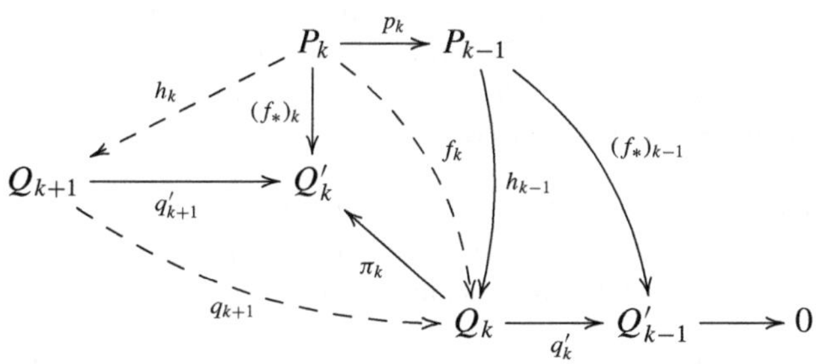

Exactness of Eq. (6.2), a projective resolution, implies that the image of q_{k+1} does not intersect Q'_{k-1} in Q_k. Hence the image of $q_{k+1} \circ h_k$ cannot intersect Q'_{k-1} either. Equation (6.3) is equivalent to

$$\begin{aligned} \pi_k \circ f_k &= \pi_k \circ q_{k+1} \circ h_k \\ 0 &= \pi_k \circ (q_{k+1} \circ h_k - f_k) \end{aligned}$$

Notice that the kernel of π_k is precisely Q'_{k-1}. If the image of $(q_{k+1} \circ h_k - f_k)$ intersects Q'_{k-1}, this is entirely because the image of f_k does.

The induction hypothesis implies that

$$q_k \circ h_{k-1} = f_{k-1},$$

and by descending to the quotients

$$\pi_{k-1} \circ q_k \circ h_{k-1} = \pi_{k-1} \circ f_{k-1},$$
$$q'_k \circ h_{k-1} = (f_*)_{k-1}.$$

This and the fact that $m_\bullet$ and $w_\bullet$ are both chain maps, allows us to compute,

$$\begin{aligned}
q'_k \circ f_k &= q'_k \circ (m_k - w_k - h_{k-1} \circ p_k) \\
&= q'_k \circ (m_k - w_k) - q'_k \circ h_{k-1} \circ p_k \\
&= q'_k \circ (m_k - w_k) - (f_*)_{k-1} \circ p_k \\
&= \pi_{k-1} \circ (m_{k-1} - w_{k-1}) \circ p_k - \pi_{k-1} \circ f_{k-1} \circ p_k \\
&= \pi_{k-1} \circ (m_{k-1} - w_{k-1} - f_{k-1}) \circ p_k \\
&= \pi_{k-1} \circ (h_{k-2} \circ p_{k-1}) \circ p_k \\
&= \pi_{k-1} \circ h_{k-2} \circ (p_{k-1} \circ p_k) \\
&= 0.
\end{aligned}$$

This calculation ensures that the image of f_k cannot intersect Q'_{k-1}. Therefore, we conclude that our construction of h_k satisfies the chain homotopy condition. □

Propositions 3.7 and 3.8 in Chap. 3 establish that chain homotopies are an equivalence relation between chain maps, and that homotopy equivalence for chain complexes is an equivalence relation. The proofs given in Chap. 3 for these two propositions did not rely upon any particulars of chain complexes of vector spaces, and are both valid for **Kom**(**Seq**) and **Kom**(**Kom**).

Definition 6.10 Suppose that **C** is one of the following categories: **Vec**, **Seq**, or **Kom**. The *homotopy category* K(**C**) has chain complexes in **C** as its objects, and has homotopy classes of chain maps as its morphisms.

The homotopy category of vector spaces K(**Vec**) is closely related to **Kom** = **Kom**(**Vec**), with the sole distinction that we get one morphism in K(**Vec**) for each homotopy class of morphisms in **Kom**(**Vec**). This has the effect of promoting each homotopy equivalence between chain complexes in **Kom**(**Vec**) into an isomorphism in K(**Vec**).

Since Theorem 3.4 in Chap. 3 establishes that quasi-isomorphism and homotopy equivalence in **Kom**(**Vec**) are the same, this means that K(**Vec**) is the most natural place to study homology for chain complexes of vector spaces, since its isomorphism classes of objects and morphisms are completely characterized by homology. The reader is cautioned that quasi-isomorphic objects of **Kom**(**Seq**) or **Kom**(**Kom**) are not homotopy equivalent in general, since one need only consider Example 6.22. Because of this, row-wise homology in these categories does not exactly characterize the objects of K(**Seq**) and K(**Kom**).

Exercise 6.40 Show that the identity **Kom**(**Kom**) morphism for the grid diagram below is not null homotopic.

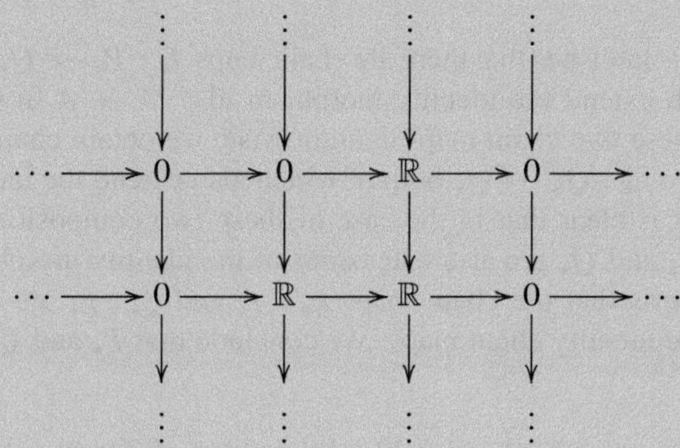

Remark 6.5 It would seem to be useful to try to define a category in which objects are chain complexes up to homotopy equivalence. In such a category, it would seem natural that morphisms should be homotopy classes of chain maps. Unfortunately, composition of morphisms in such a category is not well defined. Suppose that $m_\bullet : A_\bullet \to B_\bullet$ and $w_\bullet : B'_\bullet \to C_\bullet$ are two chain maps. If $B_\bullet$ and $B'_\bullet$ are homotopy equivalent chain complexes, we ought to be able to compose these two chain maps as morphisms in our category, since the objects corresponding to $B_\bullet$ and $B'_\bullet$ are identical. However, if $B_\bullet$ and $B'_\bullet$ are not identical as chain complexes, then the composition $c_\bullet := w_\bullet \circ m_\bullet$, which should be constructed by composing the component maps

$$c_k := w_k \circ m_k,$$

is not well defined.

Recall that projective resolutions give a way to ascribe homological information to categories—most notably **Seq**—that lack it. If this is to succeed, we need to make sure that the projective resolutions of a given object are really a feature of the original object, not some additional information that has been introduced by the non-uniqueness of resolutions. Fortunately, in $K(\mathbf{C})$, we do not need to worry.

Proposition 6.19 *Suppose that* **C** *is one of the following categories:* **Vec**, **Seq**, *or* **Kom**. *All projective resolutions of a given object in* **C** *are homotopy equivalent chain complexes. In other words, all projective resolutions of a given object in* **C** *are isomorphic in* $K(\mathbf{C})$.

Proof Suppose that A is an object in **C** with two projective resolutions

$$\cdots P_1 \to P_0 \to A,$$

and

$$\cdots Q_1 \to Q_0 \to A.$$

Proposition 6.10 establishes that there are chain maps $f_\bullet : P_\bullet \to Q_\bullet$ and $g_\bullet : Q_\bullet \to P_\bullet$ both of which extend the identity morphism $\mathrm{id}_A : A \to A$ in **C**. Observe that if we compose these two chain maps column-wise, we obtain chain maps $g_\bullet \circ f_\bullet : P_\bullet \to P_\bullet$ and $f_\bullet \circ g_\bullet : Q_\bullet \to Q_\bullet$ both of which also extend the identity morphism id_A. Moreover, it is clear that in the case of these two compositions, the identity chain maps on $P_\bullet$ and $Q_\bullet$ are also extensions of the identity morphism id_A. Hence Theorem 6.2 asserts that the chain maps $g_\bullet \circ f_\bullet$ and $f_\bullet \circ g_\bullet$ are both homotopic to their respective identity chain maps. We conclude that $P_\bullet$ and $Q_\bullet$ are homotopy equivalent. □

Recall that Examples 6.19 and 6.20 exhibited two different extensions to **Kom** (**Kom**) morphisms from the same chain map. Not only are the component maps different, the resolutions are different, which means that Theorem 6.2 does not apply directly. Instead, we must first apply Proposition 6.19 to the codomain resolutions.

Example 6.23 Consider the two resolutions of the codomain complex in Examples 6.19 and 6.20. We will show that they are homotopy equivalent directly by constructing a homotopy equivalence: two morphisms in **Kom**(**Kom**) whose compositions are homotopic to their respective identity morphisms.

Let us start with the codomain complex in Example 6.19. The two morphisms are shown as the dashed vertical lines below, in which the horizontal layers are the two resolutions. Note that this particular diagram starts and ends at the same resolution.

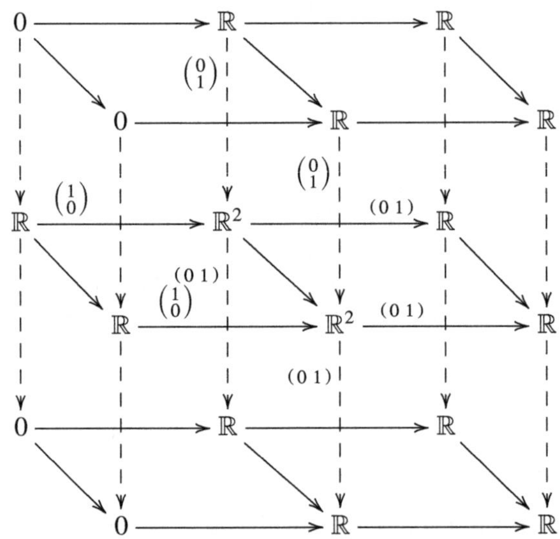

In this diagram, vertical composition yields an identity from the top layer to the bottom. This establishes the first homotopy equation for demonstrating homotopy equivalence.

In the other diagram, shown below,

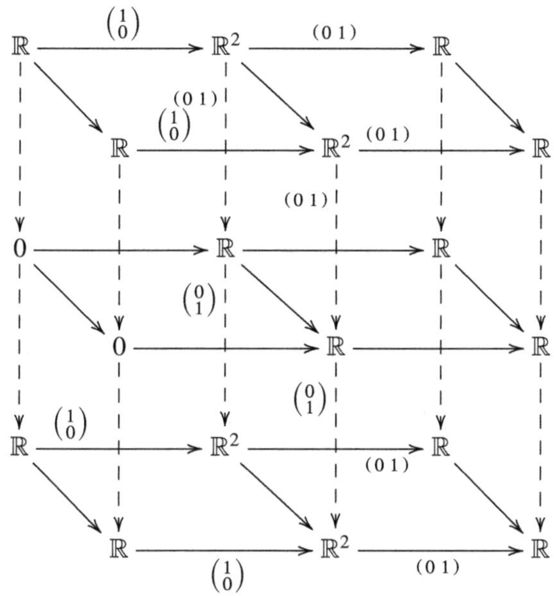

vertical composition does not yield an identity from the top layer to the bottom, so more work is required.

Composing the two morphisms yields the following smaller diagram,

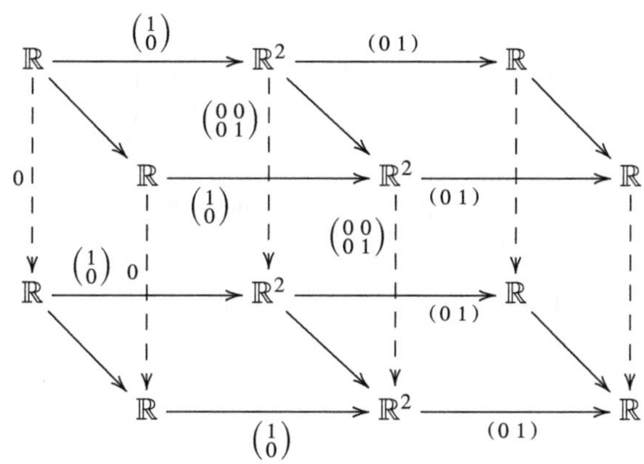

which in order to show that this is homotopic to the identity, we take the difference between the diagram above and the same diagram with identities for the vertical maps. This yields the following diagram,

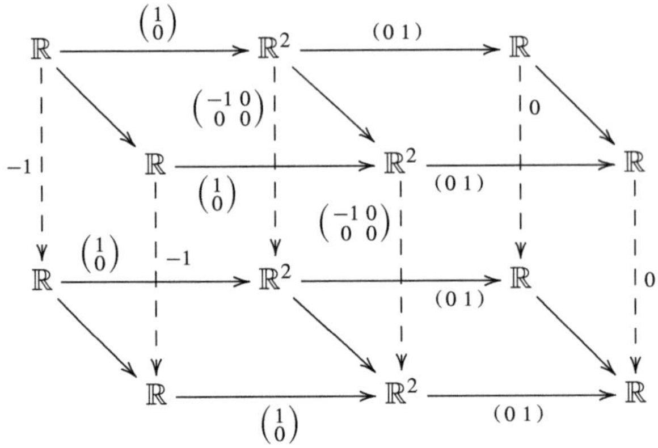

Notice that both the front and back faces of the above diagram are the same. If we can establish that the chain map expressed on the front face is null homotopic, then we are done. The front face is given by the diagram

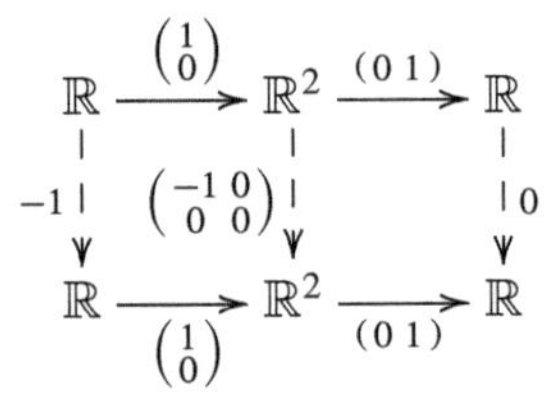

This is a chain map that can be decomposed as follows,

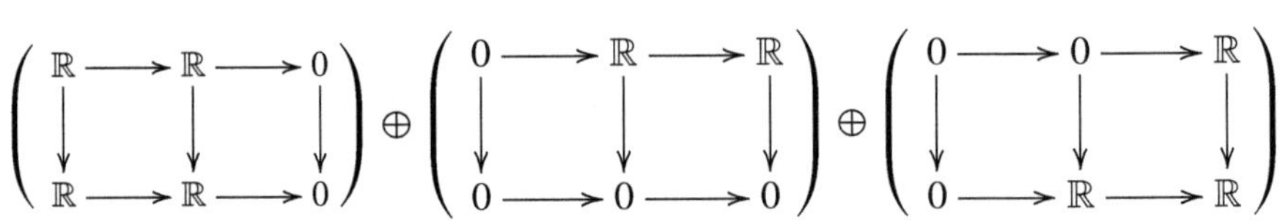

Observe that since there are no Type 5 bar chain maps present, Lemma 3.2 in Chap. 3 establishes that all three of these are null homotopic.

Now that we have a homotopy equivalence on the codomains, we may now apply Theorem 6.2 to the induced maps.

Example 6.24 The induced morphisms constructed in Examples 6.19 and 6.20 are chain homotopic by Theorem 6.2.

Let us show this directly. Consider the two induced **Kom**(**Kom**) morphisms constructed in Examples 6.19 and 6.20. From Example 6.19, the third and fourth rows are the most salient, and are summarized by the diagram below, where the induced morphism is shown as vertical dashed arrows,

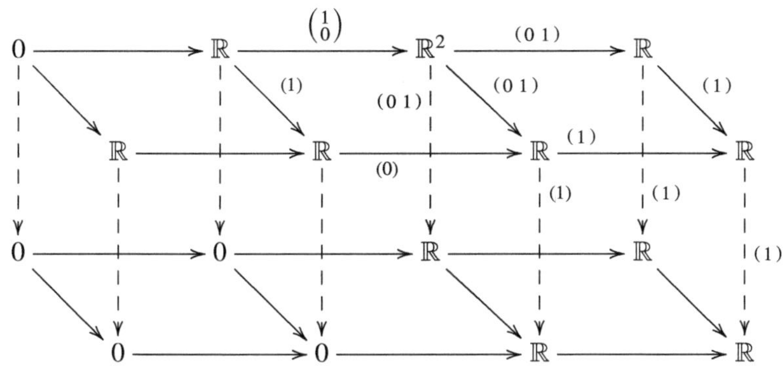

while from Example 6.20, we have

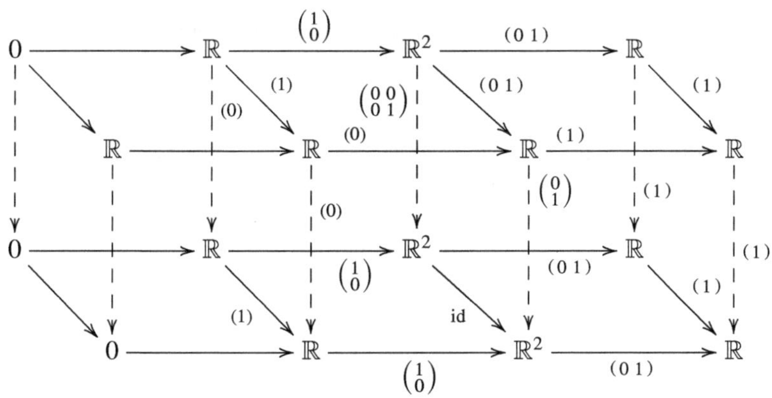

According to Example 6.23, the two codomain complexes (the bottom faces above) are homotopy equivalent. By composing with one of the morphisms that effect this homotopy equivalence, we can modify the second diagram so that it has the same codomain as the first diagram,

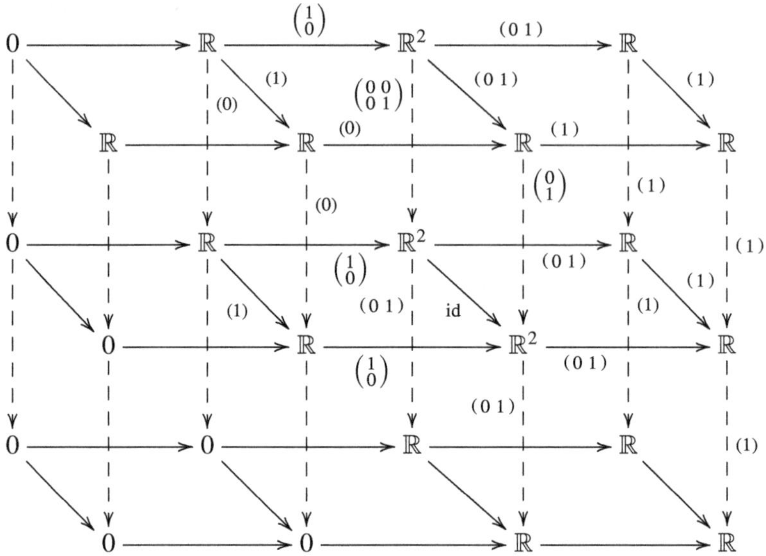

Composing the vertical maps yields exactly the same morphism as the one induced in Example 6.19 (the first diagram shown in the present example).

Exercise 6.41 Continuing Exercise 6.35, construct two different extensions of the chain map below to a chain map between projective resolutions

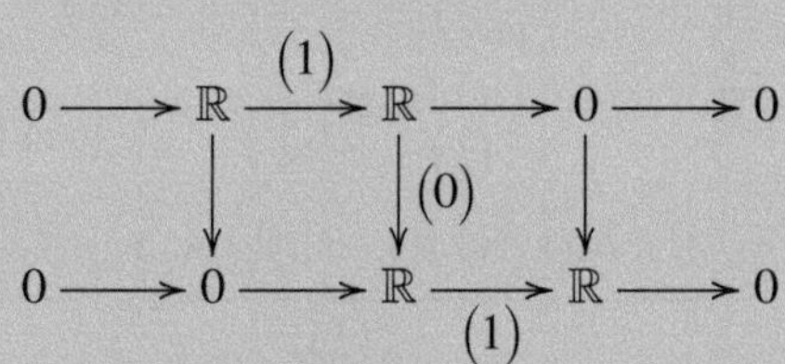

Then construct the relevant homotopy between them.

Exercise 6.42 Redo Example 6.24 with injective resolutions. That is,

1. Construct an extension of the original chain map in Example 6.19 into a chain map between injective resolutions,
2. Construct a second such extension,
3. Show that the domain and/or codomain resolutions are homotopy equivalent, and finally
4. Show that the two chain maps between resolutions are homotopic.

Proposition 6.20 *Suppose that* **C** *is one of the following categories:* **Vec**, **Seq**, *or* **Kom**. *There are two well defined canonical functors* $P : \mathbf{C} \to K(\mathbf{C})$ *and* $I : \mathbf{C} \to K(\mathbf{C})$ *that transform a sequence into its projective and injective resolutions, respectively.*

Proof We treat the case of projective resolutions, the functor P, leaving the injective resolutions as an exercise.

We have already discussed that there are many potential projective resolutions of a given sequence. well definedness of P means that all of these resolutions are homotopy equivalent chain complexes, since then they all correspond to the isomorphic objects in $K(\mathbf{C})$. This is precisely what Proposition 6.19 states.

The case of the morphisms is already handled by Proposition 6.10, a morphism $m : A \to B$ in **C** extends to a chain map $m_\bullet : A_\bullet \to B_\bullet$ on any two projective resolutions $A_\bullet$ and $B_\bullet$ of A and B, respectively. While there are many possible extensions, Theorem 6.2 establishes that they are all homotopic, and hence correspond to the same morphism in $K(\mathbf{C})$. Thus we can define $P(m)$ to be the homotopy class of $m_\bullet$. □

Exercise 6.43 Show that Proposition 6.20 holds for vector spaces by explicitly constructing the two functors $P, I : \mathbf{Vec} \to K(\mathbf{Vec})$.

6.7 The Derived Category

Homology characterizes chain complexes of vectors spaces up to chain homotopy equivalence. For chain complexes of vector spaces, which we already discussed in Chap. 3, this is the end of the story. But for chain complexes of sequences or of chain complexes, there is a bit more to say. In this section, we establish that there is a category, called the *derived category*, for which quasi-isomorphisms characterize all isomorphisms.

We would like to say that in the derived category, "objects are chain complexes" and "morphisms are chain maps up to homology." Unfortunately, this notion of morphism is not well defined. The correct notion comes from the roof construction (Definition 5.17 in Chap. 5), suitably adjusted to handle chain complexes instead of simplicial complexes.

To start this discussion, recall that chain homotopies are more restrictive than quasi-isomorphisms. As in Sect. 6.6, let **C** be any one of the following categories: **Vec**, **Seq**, or **Kom**.

Proposition 6.21 *If two chain complexes in* **C** *are homotopy equivalent, then they are also quasi-isomorphic (as objects of* $\mathbf{Kom}(\mathbf{C})$*).*

Proof The earlier proof of Proposition 3.8 in Chap. 3 that establishes homotopy equivalence implies quasi-isomorphism still holds at this new level of generality. It does not rely upon any particulars of **Kom**(**Vec**) and specifically does rely upon reasoning about bar chain maps. □

Suppose $C_\bullet$, $D_\bullet$, $E_\bullet$ are objects of **Kom**(**C**), which is to say that they are chain complexes of objects in **C**.

Definition 6.11 A *roof* consists of a pair of **Kom**(**C**) morphisms (chain maps)

$$C_\bullet \xleftarrow[\simeq]{f_\bullet} D_\bullet \xrightarrow{g_\bullet} E_\bullet,$$

where $f_\bullet$ is a quasi-isomorphism.

We sometimes write a roof suggestively as

$$\begin{array}{ccccc} & & D_\bullet & & \\ & \overset{f_\bullet}{\swarrow} \simeq & & \overset{g_\bullet}{\searrow} & \\ C_\bullet & & & & E_\bullet \end{array}$$

where unless otherwise stated, the morphism on the left is assumed to be a quasi-isomorphism, even if we do not mark it with $\simeq$. This convention turns out to be helpful when thinking about how rooves can be composed. The composition rule is reminiscent of Definition 5.18 in Chap. 5. Before we can make the corresponding definition here, we need to assemble some algebraic machinery.

Lemma 6.13 *Suppose that U, V, and W are three objects in* **C**, *and that $f : U \to W$ and $g : V \to W$ are two* **C** *morphisms. We can define*

$$Z := \{(u, v) \in U \oplus V : f(u) = g(v)\}$$

so that the following diagram commutes

$$\begin{array}{ccc} Z & \xrightarrow{\mathrm{pr}_1} & U \\ {\scriptstyle \mathrm{pr}_2}\downarrow & & \downarrow{\scriptstyle f} \\ V & \xrightarrow[g]{} & W \end{array}$$

If f is an isomorphism, then pr_2 *is also an isomorphism.*

Moreover, if X is any object in **C** *for which the diagram*

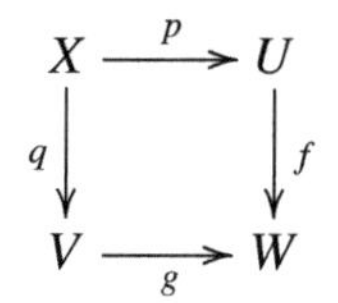

commutes, then there is a unique morphism $h : X \to Z$ *such that the larger diagram below commutes*

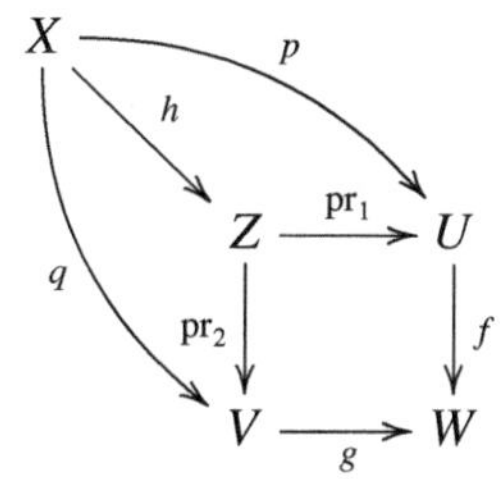

One often says that Z *is the* pullback *of* f *and* g*.*

Proof We first establish the result for $\mathbf{C} = \mathbf{Vec}$, and then extend term-by-term. That is, let us start by assuming that U, V, and W are vector spaces. We will show that pr_2 is an injective linear map. Suppose that $z_1 = (u_1, v)$ and $z_2 = (u_2, v)$ are elements of Z. By construction, z_1 and z_2 have the same image through pr_2. Therefore,

$$
\begin{aligned}
(g \circ \mathrm{pr}_2)(z_1) &= (g \circ \mathrm{pr}_2)(z_2) \\
(f \circ \mathrm{pr}_1)(z_1) &= (f \circ \mathrm{pr}_1)(z_2) \\
f(u_1) &= f(u_2),
\end{aligned}
$$

but since f is an isomorphism, $u_1 = u_2$. Hence pr_2 is injective.

Now to show that pr_2 is a surjective linear map, suppose that $v \in V$. Since f is an isomorphism,

$$u := f^{-1}(g(v))$$

is an element of U. On the other hand, this means that

$$f(u) = f\left(f^{-1}(g(v))\right) = g(v),$$

so $z := (u, v)$ is an element of Z, and moreover, $\mathrm{pr}_2(z) = v$.

Now consider the X posited in the statement of the lemma. We need to construct a $h : X \to Z$. We must have that for each $x \in X$,

$$p(x) = \mathrm{pr}_1(h(x))$$

and

$$q(x) = \mathrm{pr}_2(h(x)).$$

Because Z was constructed as a subset of $U \oplus V$, there is only one option, which is that

$$h(x) = (p(x), q(x)).$$

For the generalization to sequences and chain complexes, we argue term-by-term. For instance, if $(U_\bullet, u_\bullet)$, $(V_\bullet, v_\bullet)$, $(W_\bullet, w_\bullet)$ are sequences, we apply the above argument to each U_k, V_k, and W_k to obtain a space Z_k. To this end, the maps $z_k : Z_k \to Z_{k-1}$ are defined as the restriction of $(u_k \oplus v_k) : (U_k \oplus V_k) \to (U_{k-1} \oplus V_{k-1})$ to Z_k. We need to verify that this is well defined, which is to say that the image of z_k defined in this way is within Z_{k-1}. Suppose that $z := (a, b)$ is an element of Z_k. That means that $f_k(a) = g_k(b)$. On the other hand, because $f_\bullet$ and $g_\bullet$ are **C** morphisms, we have that

$$\begin{aligned} f_k(a) &= g_k(b) \\ (w_k \circ f_k)(a) &= (w_k \circ g_k)(b) \\ (f_{k-1} \circ u_k)(a) &= (g_{k-1} \circ v_k)(b) \\ (f_{k-1} \circ \mathrm{pr}_1 \circ z_k)(z) &= (g_{k-1} \circ \mathrm{pr}_2 \circ z_k)(z), \end{aligned}$$

which completes the argument. □

It is easy to see that the well definedness argument at the end of the proof of Lemma 6.13 applies not only to sequences and chain complexes, but to grid diagrams. This is a bit tedious to check.

Corollary 6.4 *The statement of Lemma 6.13 applies to* **Kom(C)** *as well. Briefly, one can always construct pullbacks in* **Kom(C)**.

Example 6.25 One should not be too blithe about the uniqueness of pullbacks, since they are only well defined up to isomorphism. For instance, consider the diagram

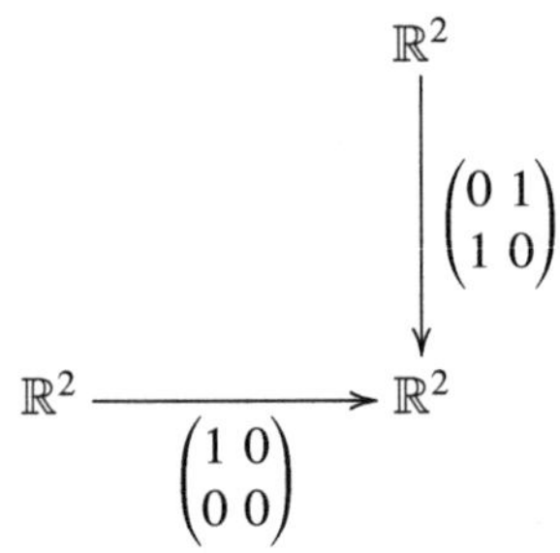

The construction in Lemma 6.13 yields the space

$$\begin{aligned} Z &:= \left\{ \left(\begin{pmatrix} u_1 \\ u_2 \end{pmatrix}, \begin{pmatrix} v_1 \\ v_2 \end{pmatrix} \right) \in \mathbb{R}^2 \oplus \mathbb{R}^2 : \begin{pmatrix} 0 & 1 \\ 1 & 0 \end{pmatrix} \begin{pmatrix} u_1 \\ u_2 \end{pmatrix} = \begin{pmatrix} 1 & 0 \\ 0 & 0 \end{pmatrix} \begin{pmatrix} v_1 \\ v_2 \end{pmatrix} \right\} \\ &= \left\{ \left(\begin{pmatrix} u_1 \\ u_2 \end{pmatrix}, \begin{pmatrix} v_1 \\ v_2 \end{pmatrix} \right) \in \mathbb{R}^2 \oplus \mathbb{R}^2 : \begin{pmatrix} u_2 \\ u_1 \end{pmatrix} = \begin{pmatrix} v_1 \\ 0 \end{pmatrix} \right\} \\ &= \left\{ \left(\begin{pmatrix} 0 \\ v_1 \end{pmatrix}, \begin{pmatrix} v_1 \\ v_2 \end{pmatrix} \right) \in \mathbb{R}^2 \oplus \mathbb{R}^2 \right\} \\ &\cong \mathbb{R}^2, \end{aligned}$$

which means that we have the diagram

$$\begin{array}{ccc} \mathbb{R}^2 & \xrightarrow{\begin{pmatrix} 0 & 0 \\ 0 & 1 \end{pmatrix}} & \mathbb{R}^2 \\ {\scriptstyle \begin{pmatrix} 0 & 1 \\ 1 & 0 \end{pmatrix}} \Big\downarrow & & \Big\downarrow {\scriptstyle \begin{pmatrix} 0 & 1 \\ 1 & 0 \end{pmatrix}} \\ \mathbb{R}^2 & \xrightarrow[\begin{pmatrix} 1 & 0 \\ 0 & 0 \end{pmatrix}]{} & \mathbb{R}^2 \end{array}$$

On the other hand, we could have also written the diagram as

$$\begin{array}{ccc} \mathbb{R}^2 & \xrightarrow{\begin{pmatrix} 0 & 0 \\ 0 & \mathbf{2} \end{pmatrix}} & \mathbb{R}^2 \\ {\scriptstyle \begin{pmatrix} 0 & \mathbf{2} \\ \mathbf{2} & 0 \end{pmatrix}} \Big\downarrow & & \Big\downarrow {\scriptstyle \begin{pmatrix} 0 & 1 \\ 1 & 0 \end{pmatrix}} \\ \mathbb{R}^2 & \xrightarrow[\begin{pmatrix} 1 & 0 \\ 0 & 0 \end{pmatrix}]{} & \mathbb{R}^2 \end{array}$$

which satisfies exactly the same properties and so is *also* a pullback.

The statement of Corollary 6.4 is too strong to be of use when working with rooves. This is because rooves contain *quasi-isomorphisms*, which are not necessarily isomorphisms. Fortunately, an appropriately weaker statement still holds.

Proposition 6.22 *Suppose that U, V, and W are three objects in* $\mathbf{Kom}(\mathbf{C})$*, and that $f : U \to W$ and $g : V \to W$ are two* $\mathbf{Kom}(\mathbf{C})$ *morphisms. We can define*

$$Z := \{(u, v) \in U \oplus V : f(u) = g(v)\}$$

so that the following diagram commutes

$$\begin{array}{ccc} Z & \xrightarrow{\mathrm{pr}_1} & U \\ {\scriptstyle \mathrm{pr}_2}\downarrow & & \downarrow{\scriptstyle f} \\ V & \xrightarrow[g]{} & W \end{array}$$

If f is a quasi-isomorphism, then pr_2 *is also a quasi-isomorphism.*

Proof Given that the above commutative diagram is in **Kom**(**C**), this means that there is an induced diagram

$$\begin{array}{ccc} H_k(Z) & \xrightarrow{(\mathrm{pr}_1)_*} & H_k(U) \\ {\scriptstyle (\mathrm{pr}_2)_*}\downarrow & & \downarrow{\scriptstyle f_*} \\ H_k(V) & \xrightarrow[g_*]{} & H_k(W) \end{array}$$

where we are assuming that the induced map f_* is an isomorphism. If we assume for the moment that $H_k(Z)$ is of the form stipulated in Lemma 6.13, then it follows immediately that $(pr_2)_*$ is an isomorphism.

Now we establish that $H_k(Z)$ is of the correct form. By hypothesis, we have already constructed

$$Z_k := \{(u, v) \in U_k \oplus V_k : f_k(u) = g_k(v)\},$$

and z_k by the restriction of $(u_k \oplus v_k)$ to Z_k. Therefore,

$$\begin{aligned} \ker z_k &= \ker(u_k \oplus v_k) \cap Z_k \\ &= \{(u, v) \in \ker u_k \oplus \ker v_k : f_k(u) = g_k(v)\}. \end{aligned}$$

On the other hand,

$$\begin{aligned} \text{image}\, z_k &= (u_{k+1}(U_{k+1}) \oplus v_{k+1}(V_{k+1})) \\ &= \{(u_{k+1}(u), v_{k+1}(v)) \in U_k \oplus V_k : \text{ for } u \in U_{k+1} \text{ and } v \in V_{k+1}\}. \end{aligned}$$

In short, this means that

$$\{([u], [v]) \in H_k(U) \oplus H_k(V) : f_*([u]) = g_*([v])\}$$

is a subspace of $H_k(Z)$.

On the other hand, if we have an element $[z]$ of $H_k(Z)$, this necessarily is in the same coset as an element z of $\ker z_k$. By definition of z_k, this must also be in the kernel of $(u_k \oplus v_k)$, and therefore has the form $([u], [v])$ in $H_k(U) \oplus H_k(V)$.

Notice, that in virtue of being an element of $\ker z_k$, this means that z is an element of Z_k. This means that $f_k(\mathrm{pr}_1(z)) = g_k(\mathrm{pr}_2(z))$. Therefore

$$\begin{aligned} f_*([u]) &= \left(f_* \circ (\mathrm{pr}_1)_*\right)([(z)]) \\ &= f_*([\mathrm{pr}_1(z)]) \\ &= f_k(\mathrm{pr}_1(z)) \\ &= g_k(\mathrm{pr}_2(z)) \\ &= g_*([\mathrm{pr}_2(z)]) \\ &= \left(g_* \circ (\mathrm{pr}_2)_*\right)([(z)]) = g_*([v]). \end{aligned}$$

This completes the proof. □

Now we are in a position to define roof composition. Suppose that

$$A_1 \xleftarrow{f_1} A_2 \xrightarrow{g_1} A_3,$$

and

$$A_3 \xleftarrow{f_2} A_4 \xrightarrow{g_2} A_5$$

are two rooves, in which f_1 and f_2 are quasi-isomorphisms. Proposition 6.22 ensures that there is a unique object B and associated morphisms such that

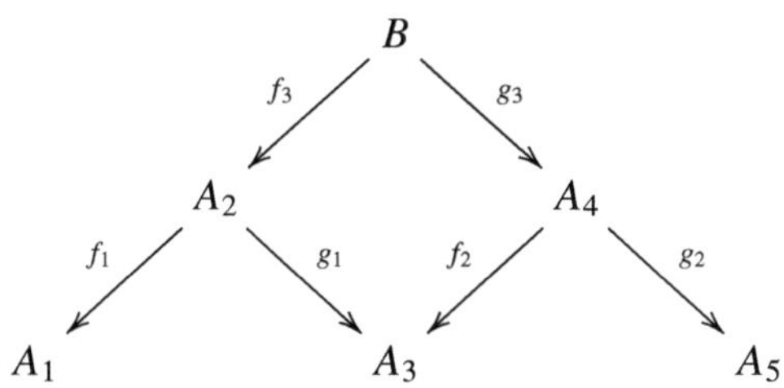

commutes, and f_3 is a quasi-isomorphism.

Definition 6.12 The *composition* of the two rooves above is the roof

$$A_1 \xleftarrow{f_1 \circ f_3} B \xrightarrow{g_2 \circ g_3} A_5.$$

Exercise 6.44 Construct the composition of the following two rooves in **Vec**,

$$\mathbb{R}^2 \xleftarrow{\begin{pmatrix}1 & -1\\ 1 & 1\end{pmatrix}} \mathbb{R}^2 \xrightarrow{\begin{pmatrix}1 & 2\\ 3 & 4\\ 5 & 6\end{pmatrix}} \mathbb{R}^3$$

and

$$\mathbb{R}^3 \xleftarrow{\begin{pmatrix}1 & 0 & 0\\ 1 & 1 & 0\\ 1 & 1 & 1\end{pmatrix}} \mathbb{R}^3 \xrightarrow{\begin{pmatrix}6 & 5 & 4\\ 3 & 2 & 1\end{pmatrix}} \mathbb{R}^2$$

Then verify that the resulting diagram is still a roof, which is to say that the left map is still an isomorphism.

Exercise 6.45 Construct the composition of the following two rooves in **Kom**,

$$\begin{array}{ccccccccc}
0 & \longrightarrow & \mathbb{R} & \xrightarrow{\begin{pmatrix}0\\0\end{pmatrix}} & \mathbb{R}^2 & \xrightarrow{\begin{pmatrix}-6 & -2\\ 16 & 6\end{pmatrix}} & \mathbb{R}^2 & \longrightarrow & 0\\
 & & \uparrow{\scriptstyle(-3/4)} & & \uparrow{\scriptstyle\begin{pmatrix}-1/4\\1\end{pmatrix}} & & \uparrow{\scriptstyle\begin{pmatrix}1/2\\-2\end{pmatrix}} & & \\
0 & \longrightarrow & \mathbb{R} & \xrightarrow{(0)} & \mathbb{R} & \xrightarrow{(-1)} & \mathbb{R} & \longrightarrow & 0\\
 & & \downarrow & & \downarrow{\scriptstyle(0)} & & \downarrow{\scriptstyle(0)} & & \\
0 & \longrightarrow & 0 & \longrightarrow & \mathbb{R} & \xrightarrow[(3)]{} & \mathbb{R} & \longrightarrow & 0
\end{array}$$

and

$$\begin{array}{ccccccccc}
0 & \longrightarrow & 0 & \longrightarrow & \mathbb{R} & \xrightarrow{(3)} & \mathbb{R} & \longrightarrow & 0\\
 & & \uparrow & & \uparrow{\scriptstyle(0)} & & \uparrow & & \\
0 & \longrightarrow & \mathbb{R} & \xrightarrow{(-2)} & \mathbb{R} & \longrightarrow & 0 & \longrightarrow & 0\\
 & & \downarrow{\scriptstyle(2)} & & \downarrow & & \downarrow & & \\
0 & \longrightarrow & \mathbb{R} & \longrightarrow & 0 & \longrightarrow & \mathbb{R} & \longrightarrow & 0
\end{array}$$

For rooves to form the morphisms in a category, roof composition has to satisfy the usual axioms. Identity rooves clearly exist, since the identity morphism in **C** extends to an identity morphism in **Kom**(**C**), and that is a quasi-isomorphism.

Unfortunately, the roof composition we have defined is not generally associative, even in **Vec**. The problem is that associativity requires *exact equality*, but roof composition is only uniquely defined up to isomorphism, as was shown in Example 6.25.

To remedy the lack of associativity for roof composition, recall that in Sect. 5.4 of Chap. 5, we treated all subdivisions of a simplicial complex as equivalent. The same idea works for rooves and even uses a similar diagram. Roof composition under this equivalence is associative, as will be shown in Lemma 6.16 after a lengthy and technical preparation.

Definition 6.13 Two rooves

$$X \xleftarrow{f} P \xrightarrow{g} Y \qquad \text{and} \qquad X \xleftarrow{f'} Q \xrightarrow{g'} Y$$

are *roof homotopic* if there is another roof

$$P \xleftarrow{p} Z \xrightarrow{q} Q$$

such that

$$f \circ p \simeq f' \circ q$$

and

$$g \circ p \simeq g' \circ q.$$

(Since a roof requires either p or q to be a quasi-isomorphism, the above two equations imply that both p and q must be quasi-isomorphisms.) We will write $(f, g) \simeq (f', g')$ in this case, and will also write $(f, g) : X \rightsquigarrow Y$ to represent a roof homotopy equivalence class of rooves.

Notice that the roof homotopy equivalence condition ensures that the diagram below commutes in $K(\mathbf{C})$,

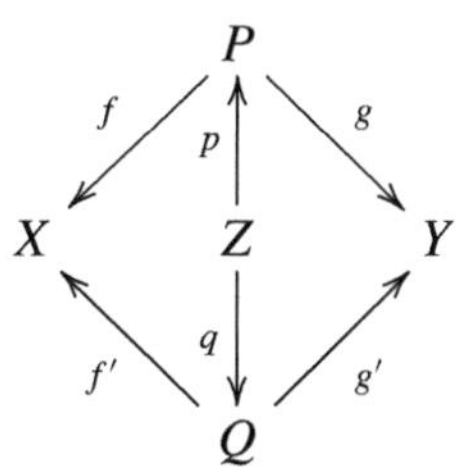

even though it may not commute in **Kom**(**C**). The two pairs of maps from $Z \to X$ and $Z \to Y$ are only required to be homotopic instead of being equal.

It might appear that Proposition 6.22 makes roof homotopies trivial, since we can always construct Z as a pullback. Fortunately, this is not a problem. Such a pullback is not necessarily a roof homotopy. Given two rooves

$$X \xleftarrow{f} P \xrightarrow{g} Y \qquad \text{and} \qquad X \xleftarrow{f'} Q \xrightarrow{g'} Y,$$

Proposition 6.22 allows us to construct *two* pullback diagrams. The first is

$$\begin{array}{ccc} Z & \xrightarrow{p} & P \\ {\scriptstyle q}\downarrow & & \downarrow{\scriptstyle f} \\ Q & \xrightarrow[f']{} & X \end{array}$$

in which p and q are quasi-isomorphisms, but they do not necessarily yield

$$g \circ p \simeq g' \circ q.$$

The second pullback is

$$\begin{array}{ccc} Z' & \xrightarrow{p'} & P \\ {\scriptstyle q'}\downarrow & & \downarrow{\scriptstyle g} \\ Q & \xrightarrow[g']{} & Y \end{array}$$

though p' and q' may not be quasi-isomorphisms, and it might not be the case that

$$f \circ p \simeq f' \circ q.$$

Moreover, the example below ensures that roof homotopy is a nontrivial condition.

Example 6.26 Consider the following two rooves in **Kom**(**Vec**), where we only show the nontrivial terms,

$$\begin{array}{ccccc} \vdots & & \vdots & & \vdots \\ \downarrow & & \downarrow & & \downarrow \\ \mathbb{R} & \xleftarrow{(1)} & \mathbb{R} & \xrightarrow{(1)} & \mathbb{R} \\ \downarrow & & \downarrow & & \downarrow \\ \vdots & & \vdots & & \vdots \end{array} \qquad \text{and} \qquad \begin{array}{ccccc} \vdots & & \vdots & & \vdots \\ \downarrow & & \downarrow & & \downarrow \\ \mathbb{R} & \xleftarrow{(1)} & \mathbb{R} & \xrightarrow{(0)} & \mathbb{R} \\ \downarrow & & \downarrow & & \downarrow \\ \vdots & & \vdots & & \vdots \end{array}$$

A roof homotopy would require a commutative diagram of vector spaces of the form (sliced along the nontrivial row in the two diagrams above),

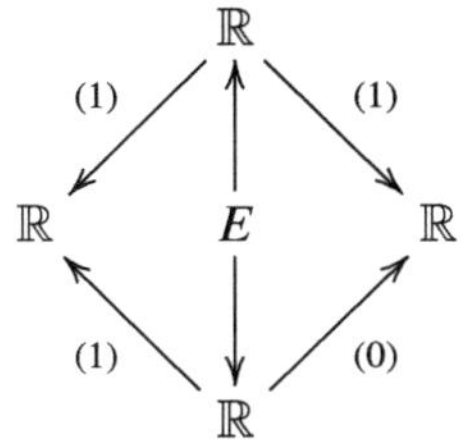

but the only way that this diagram will commute is if E is trivial or the vertical maps are not isomorphisms. Therefore, the vertical maps cannot be the components of a roof homotopy (nor even a roof). Such a roof homotopy cannot exist.

Lemma 6.14 *Roof homotopy is an equivalence relation between rooves.*

Proof Reflexivity Follows from using the identity roof,

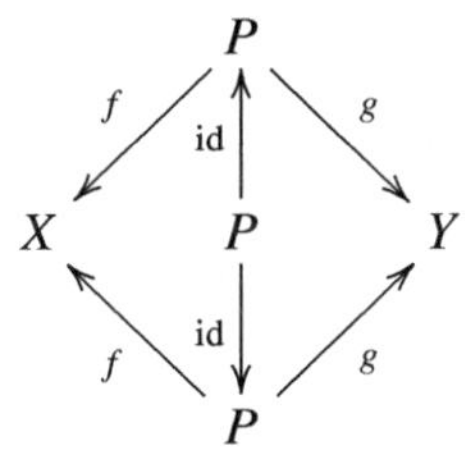

Symmetry Is immediate from the definition.
Transitivity Consider the diagram of two roof homotopies involving three rooves,

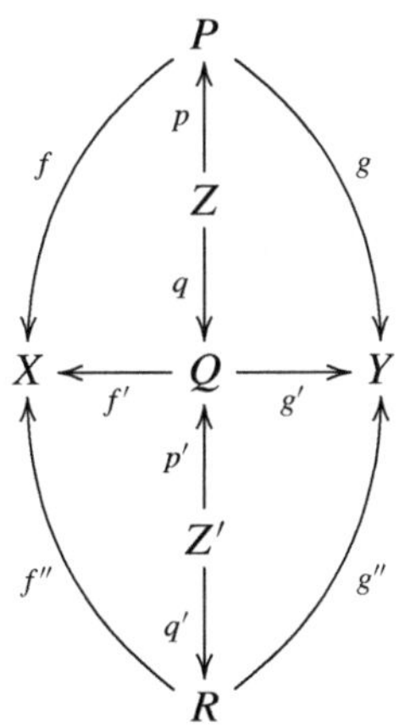

Noting that the center column is a pair of rooves, their composition is the roof homotopy from the top roof to the bottom. □

Lemma 6.15 *In* **Kom**(**C**), *two rooves*

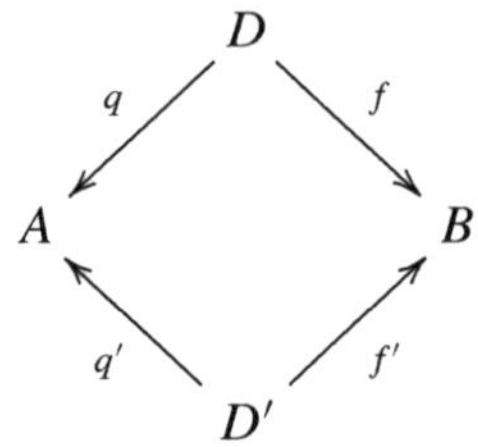

are roof homotopic if and only if they induce the same map on homology.

Because of Example 6.7 in Sect. 6.2, two chain complexes having isomorphic homology does not mean that they are quasi-isomorphic. There are rooves that induce the same maps on homology without there being a possibility of a roof homotopy, in virtue of the fact that their domains are not quasi-isomorphic. It is therefore essential that the domains and codomains of the two rooves be identical.

Proof First, suppose that there is a roof homotopy

$$D \xleftarrow{Q} E \xrightarrow{Q'} D'.$$

Because roof homotopies make the combined diagram below commute up to homotopy,

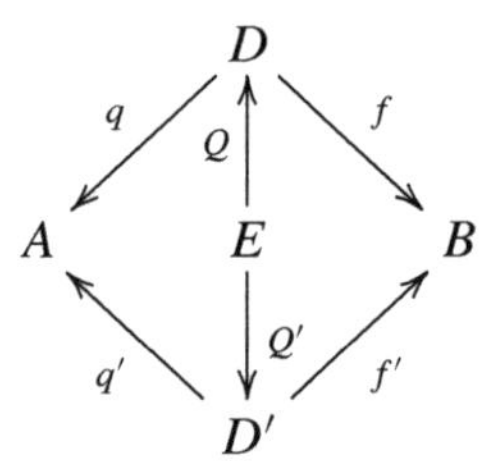

and because at least one of Q or Q' must be a quasi-isomorphism by assumption, it is immediate that the other of Q' and Q must also be a quasi-isomorphism.

Passing to homology, the following commutative diagram is induced,

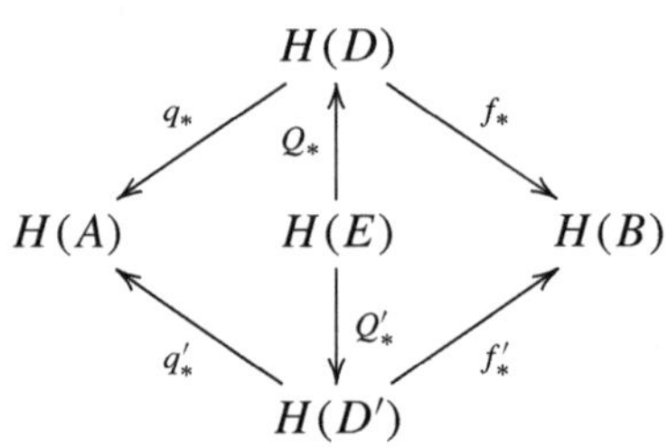

where q_*, q'_*, Q_*, and Q'_* are all isomorphisms. This means that we have the following

$$\begin{aligned} f_* \circ (q_*)^{-1} &= \left(f'_* \circ Q'_* \circ (Q_*)^{-1}\right) \circ (q_*)^{-1} \\ &= f'_* \circ \left(Q'_* \circ (Q_*)^{-1} \circ (q_*)^{-1}\right) \\ &= f'_* \circ (q'_*)^{-1}, \end{aligned}$$

which is simply the statement that the maps induced on homology by both rooves are equal.

Conversely, suppose that the two rooves induce the same map on homology. We must construct a roof homotopy from this situation, by constructing the object E and the morphisms Q and Q' so that the diagram

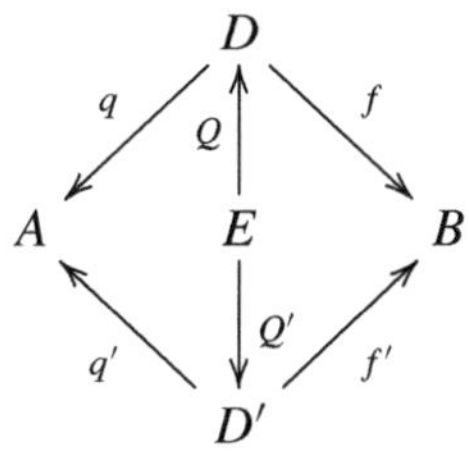

commutes up to homotopy. The reader is cautioned that although we can make such a construction in general, it is not necessarily the case that Q and Q' will be quasi-isomorphisms as is required for a roof homotopy. In our case, this fact will follow from the hypothesis that the original two rooves induce the same map on homology.

First, let us construct E. Since our diagrams are in $\mathbf{Kom(C)}$, the objects are chain complexes. Therefore, we construct E term-by-term. For instance at index k, let

$$E_k := \{(x, y) \in D_k \oplus D'_k : q_k(x) = q'_k(y) \text{ and } f_k(x) = f'_k(y)\}.$$

This ensures that the diagram above commutes.[1]

Passing to homology, notice that the defining equations for E_k remain the same, namely that

$$\begin{aligned} H(E)_k = \{([x], [y]) \in H(D)_k \oplus H(D')_k : (q_*)_k([x]) = (q'_*)_k([y]) \\ \text{and } (f_*)_k([x]) = (f'_*)_k([y])\}, \end{aligned}$$

where $[x]$ is the homology class represented by $x \in D$. This means that we have the following commutative diagram (rearranged from the earlier ones slightly),

[1] *Actually* commutes; not just up to homotopy!

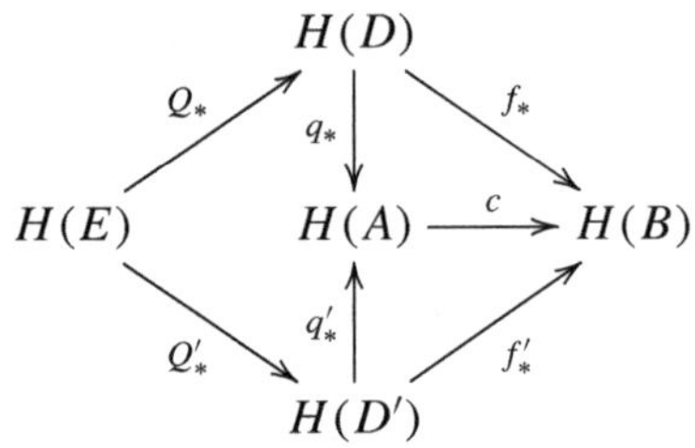

where c is the common map induced by both rooves. As a result, we have that

$$f_* = c \circ q_* \text{ and } f'_* = c \circ q'_*.$$

If $(x, y) \in D_k \oplus D'_k$ are such that

$$q_k(x) = q'_k(y),$$

it follows that

$$(q_*)_k([x]) = (q'_*)_k([y]).$$

Using this equation and the two for f_* and f'_*, we have that

$$\begin{aligned}(f_*)_k([x]) &= (c \circ q_*)_k([x]) \\ &= c_k((q_*)_k([x])) \\ &= c_k((q'_*)_k([y])) \\ &= (c \circ q'_*)_k([y]) = (f'_*)_k([y]).\end{aligned}$$

We conclude that the second condition for elements of $H(E)$ is unnecessary. We merely need to state

$$H(E)_k = \{([x], [y]) \in H(D)_k \oplus H(D')_k : (q_*)_k([x]) = (q'_*)_k([y])\}.$$

Since q and q' are quasi-isomorphisms, this means that q_* and q'_* are isomorphisms. Therefore, $H(E) \cong H(D) \cong H(D')$, which means that the construction of E constructs a roof homotopy. □

Given the fact that Lemma 6.15 is a complete characterization of roof homotopies in terms of induced maps on homology, it is instructive to see what happens when the induced maps differ.

Example 6.27 In Example 6.26, the two rooves presented do not induce the same maps on homology, and so Lemma 6.15 provides a justification for why there is no roof homotopy.

We are finally in a position to prove that roof composition is associative under roof homotopy equivalence!

Lemma 6.16 *Roof composition satisfies the usual morphism axioms. That is, an identity roof exists for each chain complex $C^\bullet$ in* $\mathbf{Kom}(\mathbf{C})$*, namely* $(\mathrm{id}_{C^\bullet}, \mathrm{id}_{C^\bullet})$*, and roof composition is associative.*

Proof Suppose that

$$A_1 \xleftarrow{q_1} A_2 \xrightarrow{f_1} A_3,$$

$$A_3 \xleftarrow{q_2} A_4 \xrightarrow{f_2} A_5,$$

and

$$A_5 \xleftarrow{q_3} A_6 \xrightarrow{f_3} A_7$$

are three rooves.

It suffices to show that there exists a roof

$$B_{\ell 2} \xleftarrow{t} C \xrightarrow{u} B_{r2},$$

where $B_{\ell 2}$ and B_{r2} arise from the two possible ways of parenthesizing the composition of the three rooves in the (not completely commutative) diagram below,

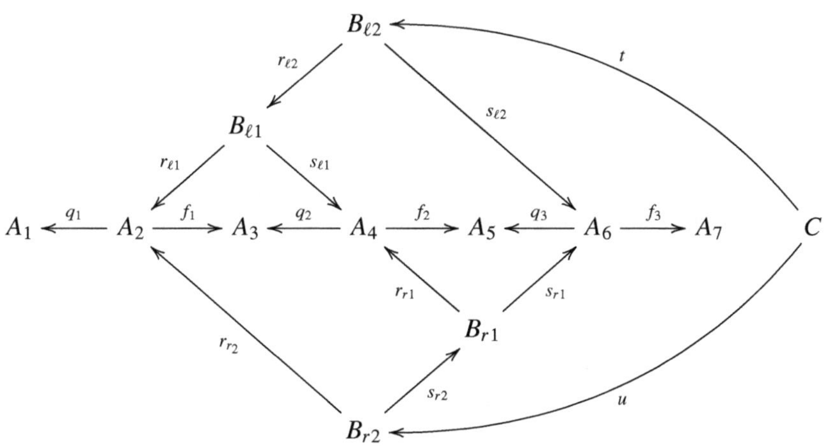

(6.4)

such that

$$r_{\ell 1} \circ r_{\ell 2} \circ t \simeq r_{r2} \circ u$$

and

$$s_{\ell 2} \circ t \simeq s_{r1} \circ s_{r2} \circ u.$$

We can obtain nearly what we want by employing Proposition 6.22 to construct the object C_{26} in the diagram below

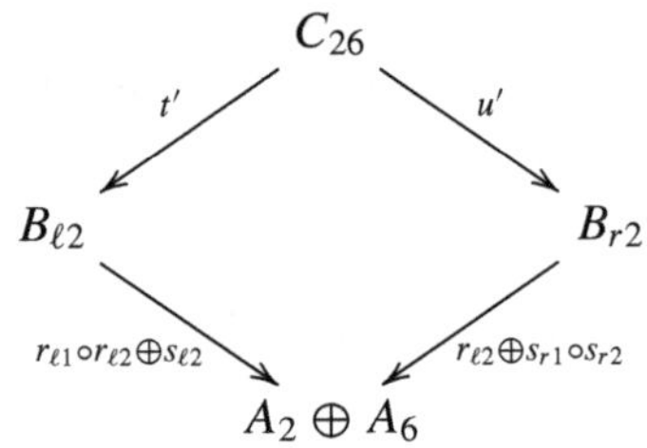

because then both homotopy equations are satisfied by exact equality. What remains to be shown is that we have actually obtained a roof by this construction, namely that either t' or u' are quasi-isomorphisms.

As a result of this construction, the diagram below commutes

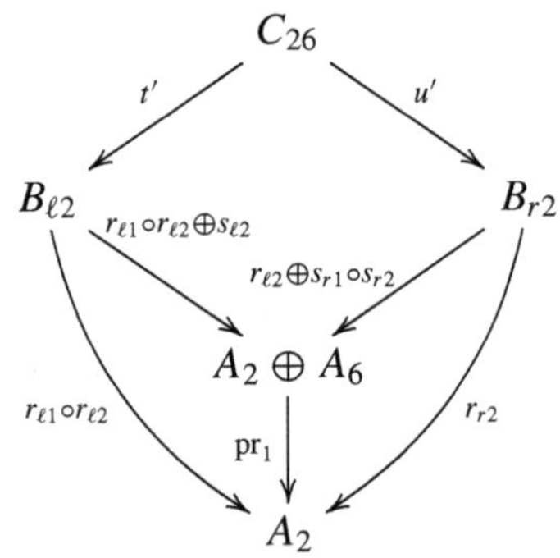

In the diagram, both of the bottom morphisms, $r_{\ell 1} \circ r_{\ell 2}$ and r_{r_2} are quasi-isomorphisms. Thus, if C_{26} were to have the form specified in Proposition 6.22, then we could conclude that t' and u' are also quasi-isomorphisms.

On the other hand, suppose that C_2 is defined as in Proposition 6.22. Then the following diagram commutes,

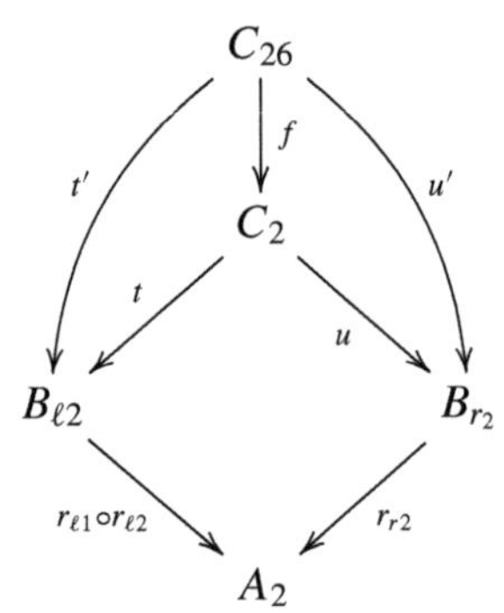

and moreover t and u are quasi-isomorphisms. Notice that the vertical morphism f must exist by Corollary 6.4 (see the last part of the statement of Lemma 6.13).

Referring back to the first (large) diagram in Eq. (6.4) on page 453, we can compute,

$$
\begin{aligned}
A_2: &\quad r_{\ell 1} \circ r_{\ell 2} \circ t = r_{r2} \circ u \\
A_3: &\quad f_1 \circ r_{\ell 1} \circ r_{\ell 2} \circ t = f_1 \circ r_{r2} \circ u \\
A_3: &\quad q_2 \circ s_{\ell 1} \circ r_{\ell 2} \circ t = q_2 \circ r_{r1} \circ s_{r2} \circ u \\
A_4: &\quad s_{\ell 1} \circ r_{\ell 2} \circ t = r_{r1} \circ s_{r2} \circ u + k_1 \\
A_5: &\quad f_2 \circ s_{\ell 1} \circ r_{\ell 2} \circ t = f_2 \circ r_{r1} \circ s_{r2} \circ u + f_2 \circ k_1 \\
A_5: &\quad q_3 \circ s_{\ell 2} \circ t = q_3 \circ s_{r1} \circ s_{r2} \circ u + f_2 \circ k_1 \\
A_6: &\quad s_{\ell 2} \circ t = s_{r1} \circ s_{r2} \circ u + k_2
\end{aligned}
$$

in which k_1 and k_2 are morphisms that induce zero maps on homology. Accordingly, this calculation means that while the diagram below does not commute,

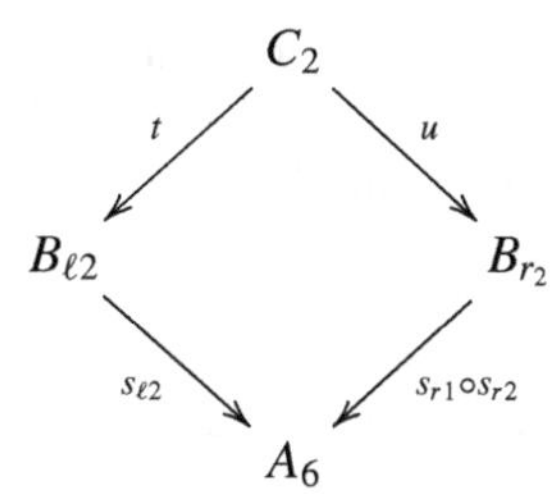

its induced diagram on homology does commute.

Thus, the maps induced on homology by the following diagram commute

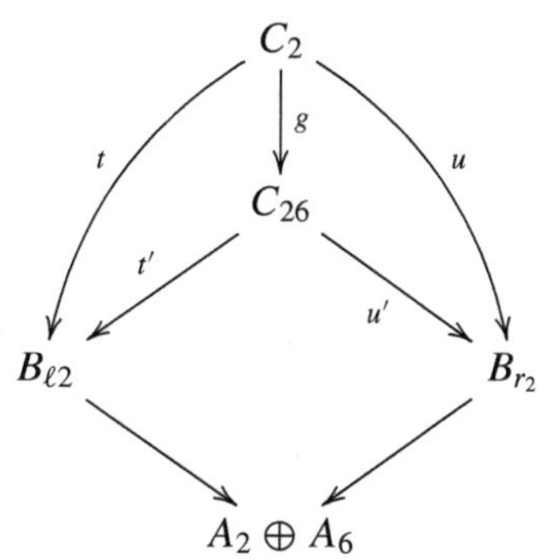

Since the vertical map g is unique, it induces an inverse to the map induced by f on homology. As a result, both t' and u' are compositions of quasi-isomorphisms. They are also quasi-isomorphisms. □

Lemma 6.17 *Roof composition preserves roof homotopy classes. If* $(f_1, g_1) \simeq (f_2, g_2)$, $(h_1, k_1) \simeq (h_2, k_2)$, *and* $(f_1, g_1) \circ (h_1, k_1)$ *is well defined, then*

1. $(f_2, g_2) \circ (h_2, k_2)$ *is well defined, and*

2. $((f_1, g_1) \circ (h_1, k_1)) \simeq ((f_2, g_2) \circ (h_2, k_2))$.

Proof Because $(f_1, g_1) \circ (h_1, k_1)$ is well defined, and we have the two roof homotopies in the hypothesis, this means that we have the following commutative diagram

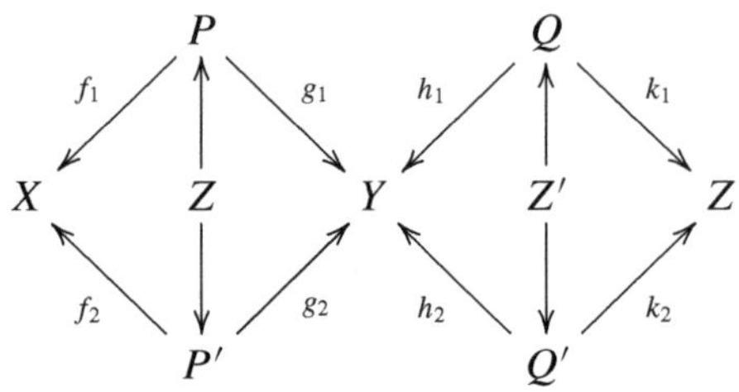

This establishes that $(f_2, g_2) \circ (h_2, k_2)$ is well defined, because the bottom two rooves must share the object Y.

Since the two pairs of rooves are roof homotopic, according to Lemma 6.15, they induce the same maps $H(X) \to H(Y)$ and $H(Y) \to H(Z)$ on homology. This implies that the composition of the corresponding induced maps is well defined and unambiguously equal to the map $H(X) \to H(Z)$ induced by the roof composition. □

Given Lemmas 6.16 and 6.17, we can finally write the definition of the derived category.

Definition 6.14 The *derived category* $D(\mathbf{C})$ has chain complexes of objects in $\mathbf{C}$ for its objects and roof homotopy classes as its morphisms. The *bounded derived categories* $D^+(\mathbf{C})$ and $D^-(\mathbf{C})$ are the full subcategories obtained by restricting to chain complexes with trivial terms to the left and right (respectively) of index 0.

Theorem 6.3 (quasi-isomorphism is equivalent to derived category isomorphism) *Two objects are isomorphic in the derived category $D(\mathbf{C})$ if and only if they are quasi-isomorphic in* $\mathbf{Kom}(\mathbf{C})$.

Proof If two chain complexes A and B (objects in $\mathbf{Kom}(\mathbf{C})$) are quasi-isomorphic, then we have rooves

$$A \xleftarrow{\mathrm{id}_A} A \xrightarrow{f} B$$

and

$$B \xleftarrow{f} A \xrightarrow{\mathrm{id}_A} A$$

where f induces an isomorphism on homology. As a result of Lemma 6.17, the two compositions of these two rooves are roof homotopy equivalent to the identity rooves on A and B, so they are isomorphic in the derived category.

Conversely, if two chain complexes are isomorphic in the derived category, then there are rooves

$$A \xleftarrow{g} C \xrightarrow{f} B$$

and

$$B \xleftarrow{h} D \xrightarrow{k} A$$

that compose to rooves that are roof homotopy equivalent to identity rooves on A and B. By Lemma 6.17, the chain maps f, g, h, and k must be quasi-isomorphisms. □

Theorem 6.3 finally completes Table 6.1 shown at the beginning of this chapter. Isomorphism classes of objects and morphisms become progressively larger as we move from $\mathbf{Kom}(\mathbf{C})$ to $K(\mathbf{C})$ to $D(\mathbf{C})$. The derived category $D(\mathbf{C})$ is the place where one treats chain complexes of objects in $\mathbf{C}$ up to homology, both because isomorphism classes of objects of $D(\mathbf{C})$ are precisely quasi-isomorphism classes, and because morphisms are identified up to induced maps on homology.

Most of this book has therefore been describing properties of the derived category $D(\mathbf{Vec})$. The objects are chain complexes of vector spaces, and the morphisms are equivalence classes of chain maps, identified by the usual induced maps on homology.

Exercise 6.46 Prove that the derived category $D(\mathbf{Vec})$ is equivalent to the subcategory of $\mathbf{Kom}(\mathbf{Vec})$ consisting of chain complexes whose barcodes only have length 1 bars. (Hint: if two chain complexes of vector spaces have the same homology, does it follow that they are quasi-isomorphic? Recall Sect. 3.4 in Chap. 3.)

The derived category $D(\mathbf{Seq})$ is something that is a bit more subtle due to the richer structure within each column of the grid diagrams. Objects in $D(\mathbf{Seq})$ are the same as in $\mathbf{Kom}(\mathbf{Seq})$, namely grid diagrams in which bar length in rows is 2 or less. Isomorphism classes are induced by grid morphisms that induce isomorphisms on row-wise homology.

Example 6.28 Consider the $\mathbf{Kom}(\mathbf{Seq})$ objects defined by the following two grid diagrams,

$$\begin{array}{ccccc} 0 & \rightarrow & 0 & \rightarrow & 0 \\ \downarrow & & \downarrow & & \downarrow \\ 0 & \rightarrow & \mathbb{R} & \rightarrow & 0 \\ \downarrow & & \downarrow & & \downarrow \\ \mathbb{R} & \rightarrow & \mathbb{R} & \rightarrow & 0 \end{array} \qquad \begin{array}{ccccc} 0 & \rightarrow & \mathbb{R} & \rightarrow & \mathbb{R} \\ \downarrow & & \downarrow & & \downarrow \\ 0 & \rightarrow & \mathbb{R} & \rightarrow & 0 \\ \downarrow & & \downarrow & & \downarrow \\ 0 & \rightarrow & 0 & \rightarrow & 0 \end{array}$$

Notice that these two objects have the same row-wise homology. Moreover, there is a morphism from the object on the left to the object on the right given by the maximal rank component maps. This morphism has exactly one nontrivial component map, namely the one in the middle row and middle column. Since this morphism induces an isomorphism on row-wise homology, by Theorem 6.3 both objects are isomorphic in $D(\mathbf{Seq})$.

It is not the case that row-wise homology always induces quasi-isomorphisms in $D(\mathbf{Seq})$, though.

Example 6.29 Consider the objects in $\mathbf{Kom}(\mathbf{Seq})$ defined by the grid diagrams,

$$
\begin{array}{ccccc}
0 & \longrightarrow & \mathbb{R} & \longrightarrow & \mathbb{R} \\
\downarrow & & \downarrow & & \downarrow \\
0 & \longrightarrow & \mathbb{R} & \longrightarrow & 0 \\
\downarrow & & \downarrow & & \downarrow \\
\mathbb{R} & \longrightarrow & \mathbb{R} & \longrightarrow & 0
\end{array}
\qquad\qquad
\begin{array}{ccccc}
0 & \longrightarrow & 0 & \longrightarrow & 0 \\
\downarrow & & \downarrow & & \downarrow \\
0 & \longrightarrow & \mathbb{R} & \longrightarrow & 0 \\
\downarrow & & \downarrow & & \downarrow \\
0 & \longrightarrow & 0 & \longrightarrow & 0
\end{array}
$$

These grid diagrams have isomorphic row-wise homology, and moreover it is the same row-wise homology as the grid diagrams in Example 6.28. However, there are only trivial $\mathbf{Kom}(\mathbf{Seq})$ morphisms from one grid diagram to the other, so they cannot be quasi-isomorphic. Thus in the derived category $D(\mathbf{Seq})$, these two grid diagrams are not isomorphic.

Exercise 6.47 Which of the remaining pairs of objects in Examples 6.28 and 6.29 are isomorphic in $D(\mathbf{Seq})$?

Exercise 6.48 Describe the derived category $D(\mathbf{Kom})$. (Hint: what data are needed to characterize the induced maps on homology? If two double complexes have the same row-wise homology, does it follow that they are quasi-isomorphic?)

The derived category is a way to ascribe homology to objects in $\mathbf{C}$, even if they do not intrinsically have homology. We can use resolutions (encountered in Sect. 6.4) to transform an object of $\mathbf{C}$ into one of $D(\mathbf{C})$. According to Proposition 6.12 all the projective (and by duality, injective) resolutions of a given sequence will be isomorphic.

Proposition 6.23 *There are two functors* $P : \mathbf{C} \to D^-(\mathbf{C})$ *and* $I : \mathbf{C} \to D^+(\mathbf{C})$ *that translate each object of* $\mathbf{C}$ *into its projective and injective resolutions (respectively). Explicitly, if* A *is an object of* $\mathbf{C}$*, it has a projective resolution*

$$\cdots P_2 \to P_1 \to P_0 \to A \to 0$$

and an injective resolution

$$0 \to A \to E^0 \to E^1 \to E^2 \to \cdots,$$

both of which are unique up to isomorphism in $D(\mathbf{C})$. *We define* $P(A)$ *to be the chain complex* $P_\bullet$ *and* $I(A)$ *to be the chain complex* $E^\bullet$, *where we note that A is not present in either of these chain complexes. These are only defined up to isomorphism in their respective derived categories, namely quasi-isomorphism in* $\mathbf{Kom}(\mathbf{C})$.

Proof The identity functor on $\mathbf{Kom}(\mathbf{C})$ also induces a covariant functor $\mathbf{Kom}(\mathbf{C}) \to K(\mathbf{C})$. This means that projective and injective resolutions are best thought of as compositions of functors

$$\mathbf{C} \to \mathbf{Kom}(\mathbf{C}) \to K(\mathbf{C}) \to D(\mathbf{C}),$$

since resolutions are well defined up to quasi-isomorphism. □

Exercise 6.49 Consider the sequence

$$\mathbb{R}^{(1,3)} \oplus \mathbb{R}^{(2,2)}.$$

Apply P and I to this sequence—constructing two resolutions—and then compute row-wise homology of the resulting objects in $D(\mathbf{C})$. (Hint: you should find that there is only nontrivial homology in index 0.)

Exercise 6.50 Prove or give a counterexample: if A and B are two sequences in **Seq** for which the row-wise homology of $P(A)$ is isomorphic to the row-wise homology $P(B)$ in index 0, then $A \cong B$.

6.8 Derived Functors

We end with a taste of how the enriched algebraic structure of the derived category can be useful. If we consider an arbitrary functor $F : \mathbf{C} \to \mathbf{D}$ (where both $\mathbf{C}$ and $\mathbf{D}$ are one of **Vec**, **Seq**, or **Kom**), we can apply F termwise to a chain complex whose terms are objects in $\mathbf{C}$. The result is certainly a sequence in $\mathbf{D}$, but might no longer be a chain complex. If we can guarantee that F yields chain complexes when applied termwise, then we can obtain homological invariants for F.

For a suitable functor F, the process involves three steps:

1. Transform objects in $\mathbf{C}$ into their resolutions,
2. Apply the functor F termwise, and then
3. Examine the result within the context of the derived category $D(\mathbf{C})$.

Since the derived category is characterized by homology, this process can be seen to ascribe homological invariants to F by "passing through" the derived category. Naturally enough, the resulting homological invariants are called *derived functors* of F.

As noted above, we need to restrict our attention to those functors that preserve chain complexes. For simplicity, we will additionally require that the functors transform exact sequences into chain complexes. The interested reader can find many advanced books, for instance (Gelfand and Manin 2003; Weibel 1994), that remove these restrictions.

Definition 6.15 Suppose that $\mathbf{C}$ and $\mathbf{D}$ are one of **Vec**, **Seq**, or **Kom**, with no requirement that they be the same.
A covariant functor $F : \mathbf{C} \to \mathbf{D}$ is *left exact* if it transforms every short exact sequence in $\mathbf{C}$

$$0 \to A \to B \to C \to 0,$$

into an exact sequence

$$0 \to F(A) \to F(B) \to F(C).$$

Notice that the right trivial object has been removed, indicating that the rightmost morphism may not be surjective (or may not have surjective components). If instead, F induces an exact sequence

$$F(A) \to F(B) \to F(C) \to 0,$$

then F is said to be *right exact*.

Dually, if $G : \mathbf{C} \to \mathbf{D}$ is a contravariant functor, it is *left exact* if

$$0 \to G(C) \to G(B) \to G(A)$$

is exact. G is said to be *right exact* if

$$G(C) \to G(B) \to G(A) \to 0$$

is an exact sequence.

Example 6.30 The dualization functor T : **Vec** $\to$ **Vec** is a contravariant functor. According to Lemma 3.5 in Sect. 3.5 in Chap. 3, if

$$0 \to A \to B \to C \to 0,$$

is a short exact sequence, then

$$0 \leftarrow A^T \leftarrow B^T \leftarrow C^T \leftarrow 0$$

is also a short exact sequence. Therefore, T : **Vec** $\to$ **Vec** is both a left exact functor and a right exact functor.

Exercise 6.51 Recall Proposition 3.9 in Sect. 3.5 in Chap. 3, which establishes that the dualization functor T : **Kom** $\to$ **Kom** is a contravariant functor. Is it left exact? Right exact?

Example 6.31 Consider the zeroth homology functor H^0 : **Kom**$^+$ $\to$ **Vec** that takes a chain complex $(C^\bullet, c^\bullet)$ (with $C^k = 0$ for all $k < 0$) to the vector space $\ker c^0$. As has been established earlier in this book (Proposition 3.2 in Chap. 3), this is a covariant functor. It is also left exact. This fact follows from the Snake Lemma (Theorem 3.2 of Chap. 3). Specifically, if

$$0 \to A^\bullet \to B^\bullet \to C^\bullet \to 0$$

is an exact sequence of chain complexes in **Kom**$^+$, then it induces an exact sequence of vector spaces,

$$0 \to H^0(A^\bullet) \to H^0(B^\bullet) \to H^0(C^\bullet) \to H^1(A^\bullet) \to \cdots$$

In particular, notice that left exactness follows by stopping the sequence above after $H^0(C^\bullet)$.

Seeing as this is the final section of the book, we will only consider covariant functors. The process of passing through the derived category is well defined for left (and right) exact functors, because left (and right) exact functors transform exact sequences into chain complexes.

Proposition 6.24 *If F : **C** $\to$ **D** is a covariant left or right exact exact functor and*

$$\cdots \longrightarrow A^0 \xrightarrow{a^0} A^1 \longrightarrow \cdots$$

is an exact sequence in **C**, *then*

$$\cdots \longrightarrow F(A^0) \xrightarrow{F(a^0)} F(A^1) \longrightarrow \cdots$$

is a chain complex in **C**.

Proof Because the long exact sequence in the hypothesis is exact at each index, we can focus our attention on a single index k to form a short exact sequence

$$0 \longrightarrow \ker a^k \xrightarrow{i^k} A^k \xrightarrow{a^k} \operatorname{image} a^k \longrightarrow 0,$$

where i^k is an inclusion map. But since exactness holds at all indices, we also have that

$$\operatorname{image} a^{k-1} = \ker a^k$$

for each k. Therefore, the short exact sequence above can also be written as

$$0 \longrightarrow \operatorname{image} a^{k-1} \xrightarrow{i^k} A^k \xrightarrow{a^k} \operatorname{image} a^k \longrightarrow 0.$$

We can assemble each of these into a larger commutative diagram,

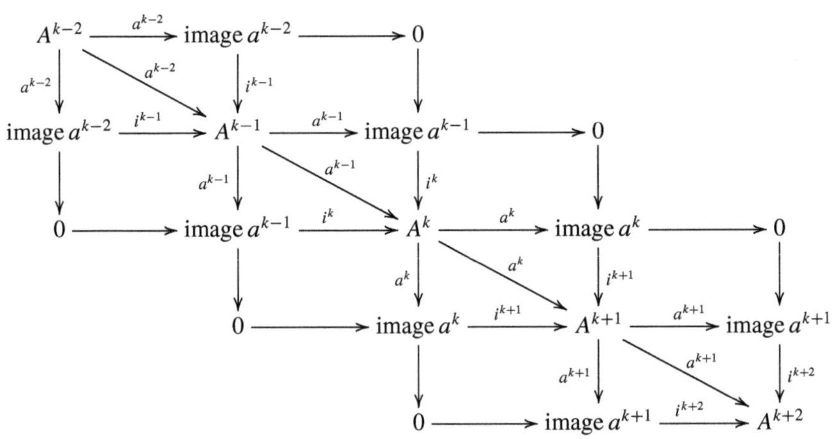

in which every row and every column is a short exact sequence. Notice that the original long exact sequence runs along the diagonal.

Since the covariant functor F is either left exact or right exact, each row and column in the diagram below (which is induced by F) is also exact,

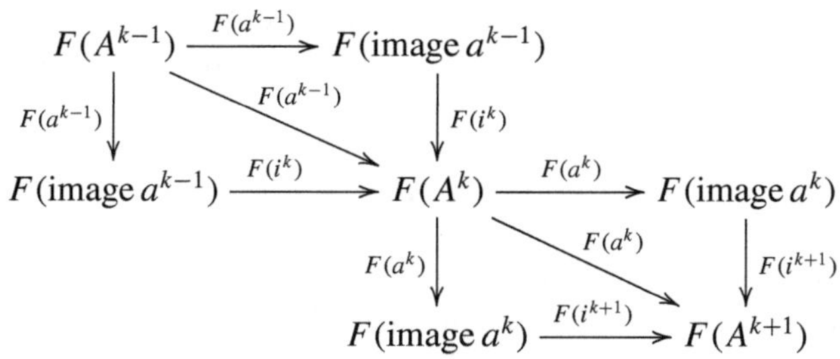

The exactness of the middle row implies that ker $F(a^k) =$ image $F(i^k)$, so we have that

$$\text{image } F(a^{k-1}) = \text{image } \big(F(i^k) \circ F(a^{k-1})\big) \subseteq \text{image } F(i^k) = \ker F(a^k).$$

$\square$

Recall the P and I functors that transform objects into projective and injective resolutions, respectively, shown to be well defined up to homotopy in Proposition 6.20 in Sect. 6.6.

Definition 6.16 Suppose $\mathbf{C}$ and $\mathbf{D}$ are each one of the following categories: **Vec**, **Seq**, or **Kom**, and that $F : \mathbf{C} \to \mathbf{D}$ is a covariant right exact functor. The *left*[2] *derived functors* of F are covariant functors $L_k F : \mathbf{C} \to \mathbf{D}$ defined by

$$L_k F(X) := H_k\left(F\left((P(X))_\bullet\right)\right),$$

where we have abused notation by regarding F as also being a functor $\mathbf{Kom(C)} \to \mathbf{Kom(D)}$, by applying F to each term in a chain complex.

Dually, the *right derived functors* of a left exact covariant functor F are the covariant functors $R^k F : \mathbf{C} \to \mathbf{D}$ defined by

$$R^k F(X) := H^k\left(F\left((I(X))^\bullet\right)\right).$$

Let us unpack the definition of the derived functors. Start with a covariant right exact functor F and an object X from one of our three categories $\mathbf{C}$. We form a projective resolution of X

$$\cdots \to P_2 \to P_1 \to P_0 \to X \to 0.$$

This is an exact sequence in $K(\mathbf{C})$. Since F is right exact, applying F termwise yields

$$\cdots \to F(P_2) \to F(P_1) \to F(P_0) \to F(X) \to 0$$

[2] Watch out! **Right** exact functors have **left** derived functors and *vice versa*!.

which is a chain complex in **Kom**(**D**) by Proposition 6.24. The right derived functors of F compute the row-wise homology in each index of this chain complex with the last term removed, namely

$$\cdots \to F(P_2) \to F(P_1) \to F(P_0) \to 0.$$

Each column of row-wise homology for the above chain complex is an object in **D**. Since projective resolutions are unique up to chain homotopy (Proposition 6.19), using a different projective resolution of X will not alter the homology of the above chain complexes. As a result, the right derived functors are well defined. They do not depend on the projective resolution chosen for X. Dually, one can obtain left derived functors by the same procedure.

Proposition 6.25 *Suppose* **C** *and* **D** *are each one of the following categories:* **Vec**, **Seq**, *or* **Kom**, *and that* $F : \mathbf{C} \to \mathbf{D}$ *is a covariant right exact functor. Then for every object* X *of* **C**,

$$(L_0 F)(X) \cong F(X).$$

Dually, if $G : \mathbf{C} \to \mathbf{D}$ *is a covariant left exact functor, then*

$$\left(R^0 G\right)(X) \cong G(X).$$

Notice that no guarantees are provided about the left derived functors of a left exact functor, nor about the right derived functors of a right exact functor. This is reflected in the mechanics of the proof.

Proof Consider any projective resolution of X in **C**

$$\cdots \longrightarrow P_2 \xrightarrow{p_2} P_1 \xrightarrow{p_1} P_0 \xrightarrow{p_0} X \longrightarrow 0.$$

Since F is right exact, it takes the short exact sequence

$$0 \longrightarrow \ker p_0 \longrightarrow P_0 \xrightarrow{p_0} X \longrightarrow 0,$$

where the first map is the inclusion, to the exact sequence

$$F(\ker p_0) \to F(P_0) \to F(X) \to 0,$$

which means that the morphism $F(P_0) \to F(X)$ is row-wise surjective.

The value of the derived functor $(L_0 F)(X)$ is the row-wise homology of

$$\cdots \to F(P_2) \to F(P_1) \to F(P_0) \to 0$$

in index $k = 0$, which consists of length 1 bars in the grid diagram along the right edge. Each of these length 1 bars correspond to length 2 bars in

$$\cdots \to F(P_2) \to F(P_1) \to F(P_0) \to F(X) \to 0$$

in virtue of the surjectivity of $F(P_0) \to F(X)$.

Duality establishes the statement about $(R^0 G)(X)$, *mutatis mutandis.* □

Example 6.32 Let us explore the right derived functors of the zeroth homology functor $H^0 : \mathbf{Kom}^+ \to \mathbf{Vec}$, given by

$$H^0\left((C^\bullet, c^\bullet)\right) := \ker c^0,$$

which was shown to be left exact in Example 6.31.

Consider $(C^\bullet, c^\bullet)$ given by $\mathbb{R}^{(\ell,\ell)}$. Using Proposition 6.14, one realization of $I(\mathbb{R}^{(\ell,\ell)})$ is the grid diagram

$$\begin{array}{ccccccccccccc}
C^\bullet & \longrightarrow & I^{0,\bullet} & \longrightarrow & I^{1,\bullet} & \longrightarrow & \cdots & \longrightarrow & I^{\ell-1,\bullet} & \longrightarrow & I^{\ell,\bullet} & \longrightarrow & \cdots \\
\\
C^0 = 0 & \longrightarrow & 0 & \longrightarrow & 0 & \longrightarrow & \cdots & \longrightarrow & \mathbb{R} & \longrightarrow & \mathbb{R} & \longrightarrow & \cdots \\
\downarrow & & \downarrow & & \downarrow & & & & \downarrow & & \downarrow & & \\
C^1 = 0 & \longrightarrow & 0 & \longrightarrow & 0 & \longrightarrow & \cdots & \longrightarrow & \mathbb{R} & \longrightarrow & 0 & \longrightarrow & \cdots \\
\downarrow & & \downarrow & & \downarrow & & & & \downarrow & & \downarrow & & \\
\vdots & & \vdots & & \vdots & & & & \vdots & & \vdots & & \\
\downarrow & & \downarrow & & \downarrow & & & & \downarrow & & \downarrow & & \\
C^{\ell-1} = 0 & \longrightarrow & \mathbb{R} & \longrightarrow & \mathbb{R} & \longrightarrow & \cdots & \longrightarrow & 0 & \longrightarrow & 0 & \longrightarrow & \cdots \\
\downarrow & & \downarrow & & \downarrow & & & & \downarrow & & \downarrow & & \\
C^{\ell} = \mathbb{R} & \longrightarrow & \mathbb{R} & \longrightarrow & 0 & \longrightarrow & \cdots & \longrightarrow & 0 & \longrightarrow & 0 & \longrightarrow & \cdots \\
\downarrow & & \downarrow & & \downarrow & & & & \downarrow & & \downarrow & & \\
C^{\ell+1} = 0 & \longrightarrow & 0 & \longrightarrow & 0 & \longrightarrow & \cdots & \longrightarrow & 0 & \longrightarrow & 0 & \longrightarrow & \cdots \\
\downarrow & & \downarrow & & \downarrow & & & & \downarrow & & \downarrow & & \\
\vdots & & \vdots & & \vdots & & & & \vdots & & \vdots & &
\end{array}$$

The H^0 functor applies column-wise, yielding the sequence

$$\text{(omitted)} \longrightarrow 0 \longrightarrow 0 \longrightarrow \cdots \longrightarrow 0 \longrightarrow \mathbb{R} \longrightarrow \cdots$$

in which the only nonzero term is in index ℓ, which means that the only nontrivial row-wise homology is in index $k = \ell$. Hence

$$R^k(H^0(\mathbb{R}^{(\ell,\ell)})) = H^k(H^0((I(\mathbb{R}^{(\ell,\ell)}))^\bullet))$$
$$= H^k(\mathbb{R}^{(\ell,\ell)}) \cong \begin{cases} \mathbb{R} & \text{if } k = \ell \\ 0 & \text{otherwise.} \end{cases}$$

Following this line of calculation, if $(C^\bullet, c^\bullet)$ is an arbitrary chain complex in $\mathbf{Kom}^+$, then

$$R^k(H^0((C^\bullet, c^\bullet))) \cong H^k(C^\bullet, c^\bullet).$$

Notice that this is consistent with Proposition 6.25.

In the above example, the right derived functor we computed depends only on the homology of the chain complex. The following theorem should not be surprising.

Theorem 6.4 (derived functors are independent of resolution) *The left (respectively, right) derived functors are independent of the projective (respectively, injective) resolution, up to isomorphism.*

Proof The way we have defined the left derived functors of $F : \mathbf{C} \to \mathbf{D}$, they involve the homotopy category, being defined by the composition of several functors

$$\mathbf{C} \xrightarrow{P} K(\mathbf{C}) \xrightarrow{F} K(\mathbf{D}) \xrightarrow{H_k} \mathbf{D}.$$

As such, Proposition 6.19 implies that a different resolution will be homotopy equivalent.

On the other hand, Proposition 3.8 in Chap. 3 establishes that homotopy equivalences imply quasi-isomorphisms. Thus the last step (of taking homology) means that the composition

$$\mathbf{C} \xrightarrow{P} D(\mathbf{C}) \xrightarrow{F} D(\mathbf{D}) \xrightarrow{H_k} \mathbf{D}, \tag{6.5}$$

is an isomorphic definition for the left derived functors. Therefore, a different projective resolution will not change the isomorphism classes of the left derived functors of F.

The argument for the right derived functors is identical, *mutatis mutandis*. □

As is apparent from Equation (6.5) in the proof of Theorem 6.4, taking row-wise homology of the projective (or injective) resolution of an object in $\mathbf{C}$ corresponds to the left (or right) derived functor of the identity functor $\mathrm{id}_\mathbf{C}$. This provides an alternative interpretation of Proposition 6.12.

Proposition 6.26 *If A is an object of $\mathbf{C}$ with projective dimension n, then $L_k \,\mathrm{id}_\mathbf{C}(A) = 0$ for all $k > n$.*

Proof Since the projective dimension of A is n, this means that at least one projective resolutions of A has length $n+1$ (including A),

$$\cdots \to 0 \to P_n \to \cdots \to P_0 \to A \to 0.$$

Since the derived functors are defined by omitting A from its projective resolution,

$$\cdots \to 0 \to P_n \to \cdots \to P_0 \to 0,$$

this particular chain complex has trivial row-wise homology for $k > n$. Theorem 6.4 establishes that this result holds for all projective resolutions of A, which is to say that $L_k \,\mathrm{id}_{\mathbf{C}}(A) = H_k(P_\bullet) = 0$ for $k > n$. □

Because of Remark 6.4, the projective dimension of objects in **Kom** or **Seq** is either 0 (the sequence is projective) or 1 (the sequence is not projective). Proposition 6.26 only provides a bound on $L_k \,\mathrm{id}_{\mathbf{C}}(A)$ based upon projective dimension. Nevertheless, bounding the projective dimension directly via Proposition 6.26 is illustrative.

Example 6.33 If we start with a projective sequence, for instance $\mathbb{R}^{(0,\infty)}$, Proposition 6.26 asserts that $L_k \,\mathrm{id}_{\mathbf{Seq}}(\mathbb{R}^{(0,\infty)}) = 0$ for all $k > 0$. To demonstrate this directly, start with a projective resolution of $\mathbb{R}^{(0,\infty)}$, for instance

$$\begin{array}{ccccccccc}
 & & \vdots & & \vdots & & \vdots & & \vdots \\
 & & \downarrow & & \downarrow & & \downarrow & & \downarrow \\
\cdots & \to & 0 & \to & 0 & \to & 0 & \to & 0 \\
 & & \downarrow & & \downarrow & & \downarrow & & \downarrow \\
\cdots & \to & 0 & \to & \mathbb{R} & \to & \mathbb{R} & \to & 0 \\
 & & \downarrow & & \downarrow & & \downarrow & & \downarrow \\
\cdots & \to & 0 & \to & \mathbb{R} & \to & \mathbb{R} & \to & 0 \\
 & & \downarrow & & \downarrow & & \downarrow & & \downarrow \\
 & & \vdots & & \vdots & & \vdots & & \vdots
\end{array}$$

remove the original sequence, yielding

$$\begin{array}{ccccccc}
 & & \vdots & & \vdots & & \vdots \\
 & & \downarrow & & \downarrow & & \downarrow \\
\cdots & \longrightarrow & 0 & \longrightarrow & 0 & \longrightarrow & 0 \\
 & & \downarrow & & \downarrow & & \downarrow \\
\cdots & \longrightarrow & 0 & \longrightarrow & \mathbb{R} & \longrightarrow & 0 \\
 & & \downarrow & & \downarrow & & \downarrow \\
\cdots & \longrightarrow & 0 & \longrightarrow & \mathbb{R} & \longrightarrow & 0 \\
 & & \downarrow & & \downarrow & & \downarrow \\
 & & \vdots & & \vdots & & \vdots
\end{array}$$

and then compute row-wise homology. This is trivial everywhere except in index 0, which is consistent with the projective dimension being 0.

Example 6.34 Consider a projective resolution of $\mathbb{R}^{(0,0)}$, for instance

$$\begin{array}{ccccccccccc}
 & & \vdots & & \vdots & & \vdots & & \vdots & & \vdots \\
 & & \downarrow & & \downarrow & & \downarrow & & \downarrow & & \downarrow \\
\cdots & \longrightarrow & 0 & \longrightarrow & 0 & \longrightarrow & \mathbb{R} & \longrightarrow & \mathbb{R} & \longrightarrow & 0 \\
 & & \downarrow & & \downarrow & & \downarrow & & \downarrow & & \downarrow \\
\cdots & \longrightarrow & 0 & \longrightarrow & \mathbb{R} & \longrightarrow & \mathbb{R} & \longrightarrow & 0 & \longrightarrow & 0 \\
 & & \downarrow & & \downarrow & & \downarrow & & \downarrow & & \downarrow \\
\cdots & \longrightarrow & 0 & \longrightarrow & \mathbb{R} & \longrightarrow & \mathbb{R} & \longrightarrow & 0 & \longrightarrow & 0 \\
 & & \downarrow & & \downarrow & & \downarrow & & \downarrow & & \downarrow \\
 & & \vdots & & \vdots & & \vdots & & \vdots & & \vdots
\end{array}$$

and remove the original sequence, yielding

$$
\begin{array}{ccccccccc}
 & & \vdots & & \vdots & & \vdots & & \vdots \\
 & & \downarrow & & \downarrow & & \downarrow & & \downarrow \\
\cdots & \longrightarrow & 0 & \longrightarrow & 0 & \longrightarrow & \mathbb{R} & \longrightarrow & 0 \\
 & & \downarrow & & \downarrow & & \downarrow & & \downarrow \\
\cdots & \longrightarrow & 0 & \longrightarrow & \mathbb{R} & \longrightarrow & \mathbb{R} & \longrightarrow & 0 \\
 & & \downarrow & & \downarrow & & \downarrow & & \downarrow \\
\cdots & \longrightarrow & 0 & \longrightarrow & \mathbb{R} & \longrightarrow & \mathbb{R} & \longrightarrow & 0 \\
 & & \downarrow & & \downarrow & & \downarrow & & \downarrow \\
 & & \vdots & & \vdots & & \vdots & & \vdots
\end{array}
$$

There are nontrivial terms in index 0 and 1 (columns), because $\mathbb{R}^{(0,0)}$ has projective dimension 1; it is not itself projective in **Seq**. However, row-wise homology is nontrivial only in index 0. This is consistent with Proposition 6.26, which only claims that the row-wise homology in indices greater than 1 will be trivial.

Exercise 6.52 Bound the projective dimension using Proposition 6.26 for two chain complexes in **Kom**: $\mathbb{R}^{(0,0)}$ and $\mathbb{R}^{(0,1)}$. Does your answer change if you instead consider **Kom**$^+$? **Kom**$^-$?

Proposition 6.27 *If F is both a left exact functor and a right exact functor, then $L_k F = 0$ for all $k > 0$.*

Proof Because F is both left exact and right exact, it turns every projective resolution into an exact sequence, not just a chain complex. The homology of an exact sequence is trivial. □

In short, derived functors measure how far a left or right exact functor deviates from exactness in much the same way that homology measures how much a chain complex deviates from exactness. Continuing this thought, we end this section with a discussion about the derived functor $\mathrm{Ext}^k_{\mathbf{C}}$, which measures how far a given object deviates from being projective or injective.

Suppose that U and V are vector spaces. Recall (all the way back in Definition 1.12 in Chap. 1) that $\mathrm{Hom}_{\mathbf{Vec}}(U, V)$ is the class of all linear maps $U \to V$. Moreover, $\mathrm{Hom}_{\mathbf{Vec}}(U, V)$ is itself a vector space. For all the categories $\mathbf{C}$ we will consider here, $\mathrm{Hom}_{\mathbf{C}}(U, V)$ can also be considered to be a vector space. $\mathrm{Ext}^k_{\mathbf{C}}$ is constructed as a right derived functor of $\mathrm{Hom}_{\mathbf{C}}(U, V)$. It detects projective and injective objects by examining the class of morphisms with a given domain or codomain.

Example 6.35 If $U = \mathbb{R}^m$ and $V = \mathbb{R}^n$, then $\mathrm{Hom}_{\mathbf{Vec}}(U, V)$ is the vector space of $n \times m$ matrices, namely

$$\mathrm{Hom}_{\mathbf{Vec}}(U, V) = \mathrm{Hom}_{\mathbf{Vec}}(\mathbb{R}^m, \mathbb{R}^n) \cong \mathbb{R}^{n\times m} \cong \mathbb{R}^{nm}.$$

Leaving the first argument of $\mathrm{Hom}_{\mathbf{C}}(U, V)$ fixed while allowing the second to vary, we can treat $\mathrm{Hom}_{\mathbf{C}}(U, V)$ as defining a function from the set of objects in **C** to **Vec**,

$$(\mathrm{Hom}_{\mathbf{C}}(U, -))\,(V) := \mathrm{Hom}_{\mathbf{C}}(U, V),$$

which acts on morphisms by post-composition. Explicitly, if $f : V \to W$ is a morphism, this induces a morphism $\mathrm{Hom}_{\mathbf{Vec}}(U, V) \to \mathrm{Hom}_{\mathbf{Vec}}(U, W)$ defined by taking each $m : U \to V$ to $f \circ m$. This results in a functor $\mathrm{Hom}_{\mathbf{C}}(U, -) : \mathbf{C} \to \mathbf{Vec}$.

Example 6.36 Suppose that $U = V = W = \mathbb{R}^2$, and $f : V \to W$ is the linear map given by

$$f = \begin{pmatrix} 1 & 1 \\ 0 & 1 \end{pmatrix}.$$

Then since both

$$\mathrm{Hom}_{\mathbf{Vec}}(U, V) = \mathrm{Hom}_{\mathbf{Vec}}(\mathbb{R}^2, \mathbb{R}^2) \cong \mathbb{R}^{2\times 2}$$

and

$$\mathrm{Hom}_{\mathbf{Vec}}(U, W) = \mathrm{Hom}_{\mathbf{Vec}}(\mathbb{R}^2, \mathbb{R}^2) \cong \mathbb{R}^{2\times 2}$$

are the set of 2×2 matrices, the linear map

$$(\mathrm{Hom}_{\mathbf{Vec}}(U, -))\,(f) : \mathrm{Hom}_{\mathbf{Vec}}(U, V) \to \mathrm{Hom}_{\mathbf{Vec}}(U, W)$$

induced by f is given by

$$\begin{aligned} ((\mathrm{Hom}_{\mathbf{Vec}}\,(U, -))\,(f)) \begin{pmatrix} a_{11} & a_{12} \\ a_{21} & a_{22} \end{pmatrix} &= f \begin{pmatrix} a_{11} & a_{12} \\ a_{21} & a_{22} \end{pmatrix} \\ &= \begin{pmatrix} 1 & 1 \\ 0 & 1 \end{pmatrix} \begin{pmatrix} a_{11} & a_{12} \\ a_{21} & a_{22} \end{pmatrix} \\ &= \begin{pmatrix} a_{11} + a_{21} & a_{12} + a_{22} \\ a_{12} & a_{22} \end{pmatrix}, \end{aligned}$$

which can also be expressed by the 4×4 matrix equation

$$((\mathrm{Hom}_{\mathbf{Vec}}\,(U, -))\,(f)) \begin{pmatrix} a_{11} \\ a_{12} \\ a_{21} \\ a_{22} \end{pmatrix} = \begin{pmatrix} 1 & 0 & 1 & 0 \\ 0 & 1 & 0 & 1 \\ 0 & 1 & 0 & 0 \\ 0 & 0 & 0 & 1 \end{pmatrix} \begin{pmatrix} a_{11} \\ a_{12} \\ a_{21} \\ a_{22} \end{pmatrix},$$

which itself means that we can make the interpretation that

$$(\mathrm{Hom}_{\mathbf{Vec}}(U, -))(f) = \begin{pmatrix} 1 & 0 & 1 & 0 \\ 0 & 1 & 0 & 1 \\ 0 & 1 & 0 & 0 \\ 0 & 0 & 0 & 1 \end{pmatrix}.$$

Exercise 6.53 Compute the matrix representation of $\left(\mathrm{Hom}_{\mathbf{Vec}}\left(\mathbb{R}^2, -\right)\right)(f)$ when

$$f = \begin{pmatrix} f_{11} & f_{12} \\ f_{21} & f_{22} \end{pmatrix}.$$

Exercise 6.54 Show that $\mathrm{Hom}_{\mathbf{Vec}}(U, -)$ is a covariant functor that is both left and right exact.

Exercise 6.55 We can define a contravariant Hom functor $\mathrm{Hom}_{\mathbf{Vec}}(-, W) : \mathbf{Vec} \to \mathbf{Vec}$ by dualizing the construction of $\mathrm{Hom}_{\mathbf{Vec}}(U, -)$. Make the appropriate definition, and then show that $\mathrm{Hom}_{\mathbf{Vec}}(-, W)$ is a contravariant functor that is both left and right exact.

As a consequence of Proposition 6.27, the left derived functors of $\mathrm{Hom}_{\mathbf{Vec}}(U, -)$ are trivial except for the one for index 0. This is not the case for our other two categories, **Kom** and **Seq**. The fact that $\mathrm{Hom}_{\mathbf{Vec}}(U, -)$ is both left and right exact is special to **Vec**.

Exercise 6.56 Show that if **C** is either **Seq** and **Kom**, then $\mathrm{Hom}_{\mathbf{C}}(U, -)$ is still a covariant functor, but it is only left exact.

We can compute the vector space $\mathrm{Hom}_{\mathbf{Kom}}(U, V)$ using the Escolar-Hiraoka decomposition, as the next example shows.

Example 6.37 Let $(A^\bullet, a^\bullet)$ be the chain complex

$$\cdots \to 0 \to \mathbb{R} \to 0 \to 0 \to \cdots$$

and let $(B^\bullet, b^\bullet)$ be the chain complex

$$\cdots \to 0 \to \mathbb{R} \to \mathbb{R} \to 0 \to \cdots$$

Assume that the linear maps above are all of full rank. $\mathrm{Hom}_{\mathbf{Kom}}((A^\bullet, a^\bullet), (B^\bullet, b^\bullet))$ is the class of all chain maps $m^\bullet : (A^\bullet, a^\bullet) \to (B^\bullet, b^\bullet)$, namely every diagram of the form

$$\begin{array}{ccccccc} \cdots \longrightarrow & \mathbb{R} & \longrightarrow & 0 & \longrightarrow & 0 & \longrightarrow \cdots \\ & \downarrow{\scriptstyle m^0} & & \downarrow{\scriptstyle m^1} & & \downarrow{\scriptstyle m^2} & \\ \cdots \longrightarrow & \mathbb{R} & \longrightarrow & \mathbb{R} & \longrightarrow & 0 & \longrightarrow \cdots \end{array}$$

In this particular case $m^k = 0$ for $k > 0$, so $\mathrm{Hom}_{\mathbf{Kom}}((A^\bullet, a^\bullet), (B^\bullet, b^\bullet))$ is therefore parameterized by the single real number m^0. However, commutativity forces $m^0 = 0$, so $\mathrm{Hom}_{\mathbf{Kom}}((A^\bullet, a^\bullet), (B^\bullet, b^\bullet))$ must be trivial.

Conversely, $\mathrm{Hom}_{\mathbf{Kom}}((B^\bullet, b^\bullet), (A^\bullet, a^\bullet))$ is the class of all chain maps of the form

$$\begin{array}{ccccccc} \cdots \longrightarrow & \mathbb{R} & \longrightarrow & \mathbb{R} & \longrightarrow & 0 & \longrightarrow \cdots \\ & \downarrow{\scriptstyle m^0} & & \downarrow{\scriptstyle m^1} & & \downarrow{\scriptstyle m^2} & \\ \cdots \longrightarrow & \mathbb{R} & \longrightarrow & 0 & \longrightarrow & 0 & \longrightarrow \cdots \end{array}$$

where now any choice of m^0 will result in a commutative diagram, so

$$\mathrm{Hom}_{\mathbf{Kom}}((B^\bullet, b^\bullet), (A^\bullet, a^\bullet)) \cong \mathbb{R}.$$

The Escolar-Hiraoka decomposition for chain maps means that the structure of $\mathrm{Hom}_{\mathbf{Kom}}(U, V)$ can be deduced from the barcode decompositions of U and V. Each of the possible options for bar chain map decompositions of elements of $\mathrm{Hom}_{\mathbf{Kom}}(U, V)$ acts as a basis element. Even though the indices also matter, so $\mathrm{Hom}_{\mathbf{Kom}}(U, V)$ actually has the structure of a sequence, we will not use this information here.

Example 6.38 Let $(A^\bullet, a^\bullet)$ be the chain complex

$$\cdots \to 0 \to \mathbb{R} \to \mathbb{R} \to 0 \to \cdots$$

and let $(B^\bullet, b^\bullet)$ be the chain complex

$$\cdots \to 0 \to \mathbb{R} \to \mathbb{R} \to 0 \to \cdots$$

Assume that the linear maps above are all of full rank. $\mathrm{Hom}_{\mathbf{Kom}}((A^\bullet, a^\bullet), (B^\bullet, b^\bullet))$ is the class of all chain maps $m^\bullet : (A^\bullet, a^\bullet) \to (B^\bullet, b^\bullet)$, namely every diagram of the form

$$\begin{array}{ccccccccc}
\cdots & \longrightarrow & \mathbb{R} & \longrightarrow & \mathbb{R} & \longrightarrow & 0 & \longrightarrow & \cdots \\
 & & \big\downarrow{\scriptstyle m^0} & & \big\downarrow{\scriptstyle m^1} & & \big\downarrow & & \\
\cdots & \longrightarrow & \mathbb{R} & \longrightarrow & \mathbb{R} & \longrightarrow & 0 & \longrightarrow & \cdots
\end{array}$$

Since we assumed that the maps in the chain complexes (rows) are all of full rank, there are two possibilities for the chain map decompositions:

- Type 3 and Type 7 ($m^0 = m^1 = 0$), or
- Type 9 (both m^0 and m^1 are nonzero, but only one is free to be chosen).

Therefore, $\mathrm{Hom}_{\mathbf{Kom}}\left((A^\bullet, a^\bullet), (B^\bullet, b^\bullet)\right)$ is isomorphic to a 1-dimensional vector space.

Exercise 6.57 Let $(A^\bullet, a^\bullet)$ be the chain complex

$$\cdots \to 0 \to \mathbb{R}^2 \to \mathbb{R} \to 0 \to \cdots$$

and let $(B^\bullet, b^\bullet)$ be the chain complex

$$\cdots \to 0 \to \mathbb{R} \to \mathbb{R} \to 0 \to \cdots$$

Assume that the linear maps above are all of full rank, and characterize

$$\mathrm{Hom}_{\mathbf{Kom}}\left((A^\bullet, a^\bullet), (B^\bullet, b^\bullet)\right)$$

as a vector space. (Challenge: give $\mathrm{Hom}_{\mathbf{Kom}}\left((A^\bullet, a^\bullet), (B^\bullet, b^\bullet)\right)$ the structure of a sequence.)

Exercise 6.58 Let $(A^\bullet, a^\bullet)$ be the sequence

$$\cdots \to 0 \to \mathbb{R} \to \mathbb{R} \to \mathbb{R} \to 0 \to \cdots$$

and let $(B^\bullet, b^\bullet)$ be the sequence

$$\cdots \to 0 \to \mathbb{R} \to \mathbb{R} \to 0 \to 0 \to \cdots$$

Assume that the linear maps above are all of full rank, and characterize

$$\mathrm{Hom}_{\mathbf{Seq}}\left((A^\bullet, a^\bullet), (B^\bullet, b^\bullet)\right).$$

(Hint: although you cannot use the Escolar-Hiraoka decomposition, $\text{Hom}_{\mathbf{Seq}}(U, V)$ is still finitely generated in this particular exercise.)

Definition 6.17 The right derived functors of $\text{Hom}_{\mathbf{C}}(U, -)$ are called $\text{Ext}^k_{\mathbf{C}}$, and are defined as

$$\text{Ext}^k_{\mathbf{C}}(U, V) := \left(R^k(\text{Hom}_{\mathbf{C}}(U, -))\right)(V).$$

Example 6.39 Let $U := \mathbb{R}^3$ and $V := \mathbb{R}^2$. Let us verify that $\text{Ext}^k_{\mathbf{Vec}}(U, V) = 0$ by direct computation. By definition,

$$\begin{aligned}\text{Ext}^k_{\mathbf{Vec}}(U, V) &= \left(R^k(\text{Hom}_{\mathbf{Vec}}(U, -))\right)(V) \\ &= H^k\left((\text{Hom}_{\mathbf{Vec}}(U, -))((I(V))^\bullet)\right) \\ &= H^k\left((\text{Hom}_{\mathbf{Vec}}(U, I(V)^\bullet))\right) \\ &= H^k\left(\text{Hom}_{\mathbf{Vec}}(\mathbb{R}^3, I(\mathbb{R}^2)^\bullet)\right).\end{aligned}$$

We need an injective resolution $I(\mathbb{R}^2)^\bullet$ of $\mathbb{R}^2$, for instance

$$\mathbb{R}^2 \to \mathbb{R}^2 \to 0 \to \cdots.$$

Notice that $\text{Hom}(\mathbb{R}^3, -)$ is the set of all matrices that have 3 columns, and that $\text{Hom}(\mathbb{R}^3, \mathbb{R}^2) \cong \mathbb{R}^6$ is the vector space of all 2×3 matrices. Moreover,

$$\text{Hom}_{\mathbf{Vec}}(\mathbb{R}^3, I(\mathbb{R}^2)^\bullet) = \mathbb{R}^6 \to \mathbb{R}^6 \to 0 \to \cdots$$

is exact, because the first map is induced by the identity map. Therefore, the homology is trivial.

In Examples 6.40–6.44 below, we will use three chain complexes U, V, and W. The chain complex U is defined by

$$\cdots \to 0 \to \mathbb{R} \to 0 \to 0 \to \cdots$$

where index 0 is the only nontrivial term, the chain complex V is defined by

$$\cdots \to 0 \to 0 \to \mathbb{R} \to 0 \to \cdots,$$

where index 1 is the only nontrivial term, and finally, the chain complex W is defined by

$$\cdots \to 0 \to \mathbb{R} \to \mathbb{R} \to 0 \to \cdots,$$

which has nontrivial terms in both index 0 and index 1. We will use the following injective resolutions (in **Kom**), for U:

$$0 \longrightarrow U \longrightarrow I(U)^0 \longrightarrow I(U)^1 \longrightarrow \cdots$$

$$\begin{array}{ccccccccc}
\vdots & & \vdots & & \vdots & & \vdots & & \\
\downarrow & & \downarrow & & \downarrow & & \downarrow & & \\
0 & \longrightarrow & 0 & \longrightarrow & \mathbb{R} & \longrightarrow & \mathbb{R} & \longrightarrow & \cdots \\
\downarrow & & \downarrow & & \downarrow & & \downarrow & & \\
0 & \longrightarrow & \mathbb{R} & \longrightarrow & \mathbb{R} & \longrightarrow & 0 & \longrightarrow & \cdots \\
\downarrow & & \downarrow & & \downarrow & & \downarrow & & \\
0 & \longrightarrow & 0 & \longrightarrow & 0 & \longrightarrow & 0 & \longrightarrow & \cdots \\
\downarrow & & \downarrow & & \downarrow & & \downarrow & & \\
0 & \longrightarrow & 0 & \longrightarrow & 0 & \longrightarrow & 0 & \longrightarrow & \cdots \\
\downarrow & & \downarrow & & \downarrow & & \downarrow & & \\
\vdots & & \vdots & & \vdots & & \vdots & &
\end{array}$$

for V:

$$0 \longrightarrow V \longrightarrow I(V)^0 \longrightarrow I(V)^1 \longrightarrow \cdots$$

$$\begin{array}{ccccccccc}
\vdots & & \vdots & & \vdots & & \vdots & & \\
\downarrow & & \downarrow & & \downarrow & & \downarrow & & \\
0 & \longrightarrow & 0 & \longrightarrow & 0 & \longrightarrow & \mathbb{R} & \longrightarrow & \cdots \\
\downarrow & & \downarrow & & \downarrow & & \downarrow & & \\
0 & \longrightarrow & 0 & \longrightarrow & \mathbb{R} & \longrightarrow & \mathbb{R} & \longrightarrow & \cdots \\
\downarrow & & \downarrow & & \downarrow & & \downarrow & & \\
0 & \longrightarrow & \mathbb{R} & \longrightarrow & \mathbb{R} & \longrightarrow & 0 & \longrightarrow & \cdots \\
\downarrow & & \downarrow & & \downarrow & & \downarrow & & \\
0 & \longrightarrow & 0 & \longrightarrow & 0 & \longrightarrow & 0 & \longrightarrow & \cdots \\
\downarrow & & \downarrow & & \downarrow & & \downarrow & & \\
\vdots & & \vdots & & \vdots & & \vdots & &
\end{array}$$

and for W:

$$0 \longrightarrow W \longrightarrow I(W)^0 \longrightarrow I(W)^1 \longrightarrow \cdots$$

$$\begin{array}{ccccccccc}
\vdots & & \vdots & & \vdots & & \vdots & & \\
\downarrow & & \downarrow & & \downarrow & & \downarrow & & \\
0 & \longrightarrow & 0 & \longrightarrow & 0 & \longrightarrow & 0 & \longrightarrow & \cdots \\
\downarrow & & \downarrow & & \downarrow & & \downarrow & & \\
0 & \longrightarrow & \mathbb{R} & \longrightarrow & \mathbb{R} & \longrightarrow & 0 & \longrightarrow & \cdots \\
\downarrow & & \downarrow & & \downarrow & & \downarrow & & \\
0 & \longrightarrow & \mathbb{R} & \longrightarrow & \mathbb{R} & \longrightarrow & 0 & \longrightarrow & \cdots \\
\downarrow & & \downarrow & & \downarrow & & \downarrow & & \\
0 & \longrightarrow & 0 & \longrightarrow & 0 & \longrightarrow & 0 & \longrightarrow & \cdots \\
\downarrow & & \downarrow & & \downarrow & & \downarrow & & \\
\vdots & & \vdots & & \vdots & & \vdots & &
\end{array}$$

Example 6.40 Let us compute

$$\mathrm{Ext}^k_{\mathbf{Kom}}(U, V) = H^k(\mathrm{Hom}_{\mathbf{Kom}}(U, I(V)^\bullet)).$$

That is, we compute $\mathrm{Hom}_{\mathbf{Kom}}(U, I(V)^\ell)$ for each ℓ, which uses the diagram

$$\mathrm{Hom}_{\mathbf{Kom}}(U, I(V)^0) \longrightarrow \mathrm{Hom}_{\mathbf{Kom}}(U, I(V)^1) \longrightarrow \cdots$$

Each object in the Hom sets fits into a larger diagram of the form,

$$\begin{array}{ccccccc}
U & & & U & & & \cdots \\
& \searrow^{m^0} & & & \searrow^{m^1} & & \\
& & I(V)^0 & \longrightarrow & & I(V)^1 & \longrightarrow \cdots
\end{array}$$

which expands into the following diagram based upon the injective resolutions,

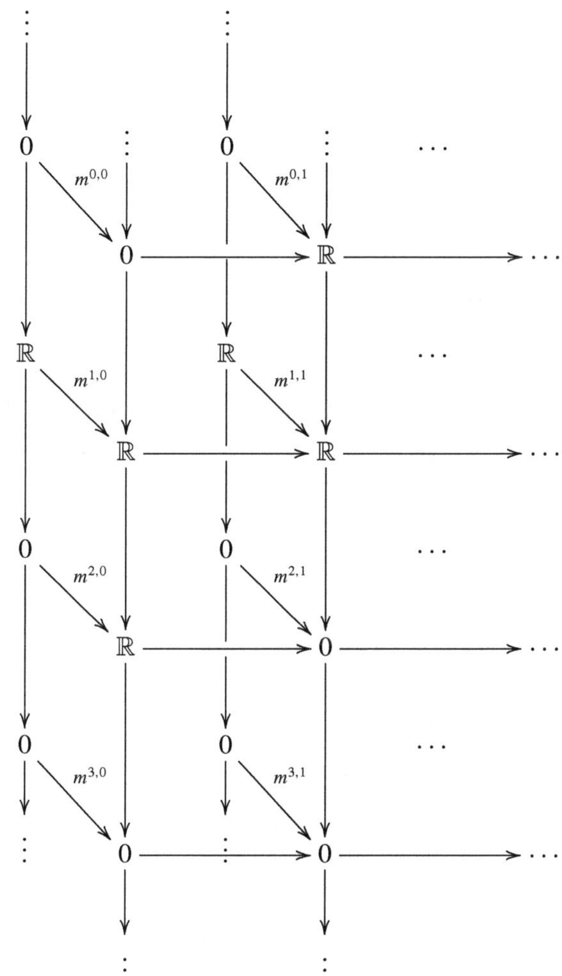

Commutativity requires that $m^{\bullet,0} = 0$, and $m^{k,1} = 0$ if $k \neq 1$. Therefore, the induced maps on the Hom sets together yield the chain complex

$$0 \longrightarrow \mathbb{R} \longrightarrow 0 \longrightarrow \cdots$$

so

$$\mathrm{Ext}^k_{\mathbf{Kom}}(U, V) = \begin{cases} \mathbb{R} & \text{if } k = 1 \\ 0 & \text{otherwise.} \end{cases}$$

Example 6.41 Using the same procedure as in Example 6.40, we can compute $\mathrm{Ext}^k_{\mathbf{Kom}}(V, U)$. Along the way, we use the chain complex

$$\mathrm{Hom}_{\mathbf{Kom}}(V, I(U)^0) \longrightarrow \mathrm{Hom}_{\mathbf{Kom}}(V, I(U)^1) \longrightarrow \mathrm{Hom}_{\mathbf{Kom}}(V, I(U)^2) \longrightarrow \cdots$$

and compute

$$\mathrm{Hom}_{\mathbf{Kom}}(V, I(U)^0) = 0,$$
$$\mathrm{Hom}_{\mathbf{Kom}}(V, I(U)^1) = 0,$$
$$\mathrm{Hom}_{\mathbf{Kom}}(V, I(U)^2) = 0, \ldots$$

so

$$\mathrm{Ext}^k_{\mathbf{Kom}}(V, U) = H^k(\mathrm{Hom}_{\mathbf{Kom}}(V, I(U)^\bullet)) = 0.$$

Example 6.42 To compute $\mathrm{Ext}^k_{\mathbf{Kom}}(W, U)$, we use the chain complex

$$\mathrm{Hom}_{\mathbf{Kom}}(W, I(U)^0) \longrightarrow \mathrm{Hom}_{\mathbf{Kom}}(W, I(U)^1) \longrightarrow \mathrm{Hom}_{\mathbf{Kom}}(W, I(U)^2) \longrightarrow \cdots$$

and compute

$$\mathrm{Hom}_{\mathbf{Kom}}(W, I(U)^0) \cong \mathbb{R},$$
$$\mathrm{Hom}_{\mathbf{Kom}}(W, I(U)^1) \cong \mathbb{R},$$
$$\mathrm{Hom}_{\mathbf{Kom}}(W, I(U)^2) = 0, \ldots$$

so that the chain complex is exact, so

$$\mathrm{Ext}^k_{\mathbf{Kom}}(W, U) = H^k(\mathrm{Hom}_{\mathbf{Kom}}(W, I(U)^\bullet)) = 0.$$

Example 6.43 To compute $\mathrm{Ext}^k_{\mathbf{Kom}}(U, W)$, we use the chain complex

$$\mathrm{Hom}_{\mathbf{Kom}}(U, I(W)^0) \longrightarrow \mathrm{Hom}_{\mathbf{Kom}}(U, I(W)^1) \longrightarrow \mathrm{Hom}_{\mathbf{Kom}}(U, I(W)^2) \longrightarrow \cdots$$

and compute

$$\mathrm{Hom}_{\mathbf{Kom}}(U, I(W)^0) = 0,$$
$$\mathrm{Hom}_{\mathbf{Kom}}(U, I(W)^1) = 0,$$
$$\mathrm{Hom}_{\mathbf{Kom}}(U, I(W)^2) = 0, \ldots$$

so that

$$\mathrm{Ext}^k_{\mathbf{Kom}}(U, W) = H^k(\mathrm{Hom}_{\mathbf{Kom}}(U, I(W)^\bullet)) = 0.$$

Example 6.44 Finally, to compute $\mathrm{Ext}^k_{\mathbf{Kom}}(V, W)$, we use the chain complex

$$\mathrm{Hom}_{\mathbf{Kom}}(V, I(W)^0) \longrightarrow \mathrm{Hom}_{\mathbf{Kom}}(V, I(W)^1) \longrightarrow \mathrm{Hom}_{\mathbf{Kom}}(V, I(W)^2) \longrightarrow \cdots$$

and compute

$$\mathrm{Hom}_{\mathbf{Kom}}(V, I(W)^0) = \mathbb{R},$$
$$\mathrm{Hom}_{\mathbf{Kom}}(V, I(W)^1) = \mathbb{R},$$
$$\mathrm{Hom}_{\mathbf{Kom}}(V, I(W)^2) = 0, \ldots$$

so the chain complex is exact, so

$$\mathrm{Ext}^k_{\mathbf{Kom}}(V, W) = H^k(\mathrm{Hom}_{\mathbf{Kom}}(V, I(W)^\bullet)) = 0.$$

Exercise 6.59 Consider the chain complex U

$$\cdots 0 \to \mathbb{R} \to 0 \to 0 \to \cdots$$

where index 0 is the only nontrivial term. Compute $\mathrm{Ext}^k_{\mathbf{Kom}}(U, U)$.

Exercise 6.60 Given chain complexes U

$$\cdots \to 0 \to \mathbb{R}^2 \to \mathbb{R} \to 0 \to \cdots$$

and V

$$\cdots \to 0 \to \mathbb{R} \to 0 \to 0 \to \cdots$$

in **Kom**, compute $\mathrm{Ext}^k_{\mathbf{Kom}}(U, V)$. Assume all maps shown above are of full rank.

Exercise 6.61 Given sequences U

$$\cdots \to 0 \to \mathbb{R}^2 \to \mathbb{R} \to \mathbb{R} \to 0 \to \cdots$$

and V

$$\cdots \to 0 \to \mathbb{R} \to \mathbb{R} \to 0 \to 0 \to \cdots$$

in **Seq**, compute $\mathrm{Ext}^k_{\mathbf{Seq}}(U, V)$. Assume all maps shown above are of full rank.

Exercise 6.62 Show if either U is projective or V is injective, then $\mathrm{Ext}^k_{\mathbf{C}}$ $(U, V) = 0$ for all $k > 0$.

Examples 6.42–6.44 are special cases of the following two exercises.

Exercise 6.63 Show that if for all V it follows that $\mathrm{Ext}^k_{\mathbf{C}}(U, V) = 0$ for all $k > 0$, then U is projective.

Exercise 6.64 Show that if for all U it follows that $\mathrm{Ext}^k_{\mathbf{C}}(U, V) = 0$ for all $k > 0$, then V is injective.

6.9 Further Reading

For purposes of concreteness, we did not consider the homotopy (or derived) category of categories other than **Vec**, **Seq**, and **Kom**. *Abelian categories* are those for which the homotopy category makes sense. They are of considerable interest in advanced texts on homological algebra (Atiyah and Macdonald 2018; Bruns and Herzog 1998; Hilton and Stammbach 1997; Eisenbud 1995; Weibel 1994). These books are now approachable to the reader who has completed the present text. The reader might also consider (Schapira 2001), whose exposition begins roughly where this book ends. Another algebraic class of invariants for grid diagrams are *spectral sequences* (Chow 2006).

Grid diagrams of vector spaces naturally arise when one starts with a grid of simplicial complexes, and then applies the simplicial chain complex functor to each. In the literature, this is usually called *multi-parameter persistence* (Botnan et al. 2024; Lesnick and Wright 2022). Multi-parameter persistence is stable under perturbations of these simplicial chain complexes (Botnan et al. 2024). There are algorithms for efficiently computing these grid diagrams and invariants of them (Bauer et al. 2023; Fugacci et al. 2023). The algebra of multi-parameter persistence modules is quite intricate. In addition to the somewhat global approach taken in this book, one can also consider the local structure of grid diagrams, and in this way arrive at the notion of a *sheaf*. The Ext functor described in this book then acquires rich local structure (Berkouk and Ginot 2022; Bubenik and Milićević 2021).

References

Atiyah, M.F., Macdonald, I.G.: Introduction to Commutative Algebra. CRC Press (2018)
Bauer, U., Lenzen, F., Lesnick, M.: Efficient two-parameter persistence computation via cohomology. In: Chambers, E.W., Gudmundsson, J., (eds.) 39th International Symposium on Computational Geometry (SoCG 2023), Leibniz International Proceedings in Informatics (LIPIcs), vol. 258, pp. 15:1–15:17. Schloss Dagstuhl – Leibniz-Zentrum für Informatik, Dagstuhl, Ger-

many (2023). https://doi.org/10.4230/LIPIcs.SoCG.2023.15, https://drops.dagstuhl.de/entities/document/10.4230/LIPIcs.SoCG.2023.15

Bauer, U., Scoccola, L.: Multi-parameter persistence modules are generically indecomposable. Int. Math. Res. Notices **2025**(5), rnaf034 (2025)

Berkouk, N., Ginot, G.: A derived isometry theorem for sheaves. Adv. Math. **394**, 108033 (2022). https://doi.org/10.1016/j.aim.2021.108033, https://www.sciencedirect.com/science/article/pii/S0001870821004722

Botnan, M.B., Oppermann, S., Oudot, S., Scoccola, L.: On the bottleneck stability of rank decompositions of multi-parameter persistence modules. Adv. Math. **451**, 109780 (2024). https://doi.org/10.1016/j.aim.2024.109780, https://www.sciencedirect.com/science/article/pii/S0001870824002950

Botnan, M.B., Oppermann, S., Oudot, S.: Signed barcodes for multi-parameter persistence via rank decompositions and rank-exact resolutions. Found. Comput. Math. (2024). https://doi.org/10.1007/s10208-024-09672-9

Bruns, W., Herzog, H.J.: Cohen-Macaulay Rings, vol. 39. Cambridge University Press (1998)

Bubenik, P., Milićević, N.: Homological algebra for persistence modules. Found. Comput. Math. **21**, 1233–1278 (2021). https://doi.org/10.1007/s10208-020-09482-9

Chow TY (2006) You could have invented spectral sequences. Notices AMS 53(1):15–19

Crawley-Boevey W (2015) Decomposition of pointwise finite-dimensional persistence modules. J. Algebra Appl. 14(05):1550066

Eisenbud, D.: Commutative algebra with a view toward algebraic geometry. In: Graduate Texts in Mathematics. Springer, New York (1995)

Fugacci, U., Kerber, M., Rolle, A.: Compression for 2-parameter persistent homology. Comput. Geometry **109**, 101940 (2023)

Gelfand, S.I., Manin, Y.I.: Methods of Homological Algebra. Springer (2003)

Hilton, P., Stammbach, U.: A course in homological algebra. In: Graduate Texts in Mathematics. Springer, New York (1997). https://doi.org/10.1007/978-1-4419-8566-8

Hubbard, J.H.: Teichmüller Theory, vol. 1. Matrix Editions (2006)

Lesnick M, Wright M (2022) Computing minimal presentations and bigraded betti numbers of 2-parameter persistent homology. SIAM J. Appl. Algebra Geometry 6(2):267–298. https://doi.org/10.1137/20M1388425

Oudot S, Scoccola L (2024) On the stability of multigraded Betti numbers and Hilbert functions. SIAM J. Appl. Algebra Geometry 8(1):54–88

Schapira, P.: Categories and Homological Algebra. Société mathétique de France (2001)

Simson, D.: Linear Representations of Partially Ordered Sets and Vector Space Categories, vol. 4. CRC Press (1993)

Weibel, C.A.: An Introduction to Homological Algebra. Cambridge Studies in Advanced Mathematics. Cambridge University Press (1994). https://doi.org/10.1017/CBO9781139644136

Appendix
Linear Algebra Basics

Homological algebra is built on linear algebra, the study of vector spaces. This appendix is a reference to remind the reader of the main ideas of linear algebra. It also establishes the notational conventions used throughout the book. No proofs are included in this appendix, since the material is properly prerequisite knowledge for our study of homological algebra.

Nearly all of the vector spaces described in this book are variants of the vector space $\mathbb{R}^n$, which—as a set—is the collection of ordered n-tuples $(x_1, \cdots, x_n)$ of real numbers. We call these ordered n-tuples *vectors*. Addition for real numbers gives rise to an addition operation for vectors, by adding the numbers in corresponding slots,

$$(x_1, \cdots, x_n) + (y_1, \cdots, y_n) := (x_1 + y_1, \cdots, x_n + y_n).$$

Commutativity of addition of the real numbers leads to commutativity of vector addition in $\mathbb{R}^n$. Vector addition has an additive identity, which is just the n-tuple of all zeros. Vectors can also be *scaled*, by multiplying a real number (called a *scalar* in this case) to each slot,

$$a \cdot (x_1, \cdots, x_n) := (ax_1, \cdots, ax_n).$$

Abstracting these operations to an arbitrary set instead of $\mathbb{R}^n$ leads to a general definition for a real vector space.

Definition A.1 A *vector space* consists of a set V with two operations: *vector addition* $+ : V \times V \to V$ and *scalar multiplication* $\cdot : \mathbb{R} \times V \to V$ that satisfies the following axioms:

Associativity of addition $u + (v + w) = (u + v) + w$ for all $u, v, w \in V$,
Existence of additive identity There is an element $0 \in V$ such that $u + 0 = 0 + u = u$ for all $u \in V$,
Existence of additive inverses For each $v \in V$, there is a $(-v) \in V$ such that $v + (-v) = 0$,

M. Robinson, *Computational Homological Algebra*, Mathematical Engineering,
https://doi.org/10.1007/978-3-032-08634-1

Commutativity of addition $v + w = w + v$ for all $v, w \in V$,
Multiplicative identity For all $v \in V$, $v = 1 \cdot v$,
Compatibility $a \cdot (b \cdot v) = (ab) \cdot v$ for all $a, b \in \mathbb{R}$ and $v \in V$,
Distributivity I $a \cdot (v + w) = (a \cdot v) + (a \cdot w)$ for all $a \in \mathbb{R}$ and $v, w \in V$, and
Distributivity II $(a + b) \cdot v = (a \cdot v) + (b \cdot v)$, for all $a, b \in \mathbb{R}$ and $v \in V$.

Whenever it is unambiguous, we will usually write av instead of $a \cdot v$ for scalar multiplication. All elements of V that are not the additive identity 0 are called *nonzero*.

Exercise A.1 Prove that all of the axioms of a vector space are satisfied by the operations on $\mathbb{R}^n$ given above.

Definition A.2 A *linear combination* of a subset S of a vector space V is a sum of the form

$$\sum_{x \in S} a_x x$$

where the a_x are in $\mathbb{R}$, only finitely many of which are nonzero. We call the situation where $a_x = 0$ for all $x \in S$ a *trivial linear combination*.

Definition A.3 A *linear map* (or *linear transformation*) is a function $\ell : V \to W$ from a vector space V to another W that preserves linear combinations. Specifically, for all $a, b \in \mathbb{R}$ and all $v, w \in V$,

$$\ell(av + bw) = a\ell(v) + b\ell(w).$$

Definition A.4 The *identity function* $\mathrm{id}_V : V \to V$ is given by

$$id_V(x) := x,$$

and is a linear map by the definition of a vector space.

Proposition A.1 *The composition of two linear maps is also a linear map.*

Definition A.5 The set of all linear combinations of a subset S is called its *span* and is denoted span$S$.

If we have a vector v that is in the span of $S \subseteq V$, then this means that there is a way to write

$$v = \sum_{s \in S} a_s s.$$

we call the set $\{a_s\}$ the *coefficients* of v in S. The reader is cautioned that coefficients are generally not unique!

Exercise A.2 Consider the subset of $\mathbb{R}^2$ given by

$$S := \left\{ \begin{pmatrix} 1 \\ -1 \end{pmatrix}, \begin{pmatrix} 1 \\ 1 \end{pmatrix}, \begin{pmatrix} 2 \\ 1 \end{pmatrix} \right\}.$$

First, show that $v := \begin{pmatrix} 4 \\ 0 \end{pmatrix}$ is in the span of S by finding a set of coefficients for v in S. Then, find a different set of coefficients for v in S.

This problem of non-uniqueness of coefficients can be resolved if we only consider a more restricted class of subsets, those that are *linearly independent*.

Definition A.6 A subset S of a vector space V is called *linearly independent* if every nontrivial linear combination of elements of S is nonzero.

Lemma A.1 *If $v \in V$ is a vector in the span of a linearly independent set S in a vector space V, then there is exactly one set of coefficients for v in S.*

Lemma A.2 *Suppose that I and S are both finite subsets of a vector space V with the same cardinality. If I is linearly independent and S spans V, then I also spans V.*

Definition A.7 A *basis* of a vector space V is a linearly independent set whose span is V.

Proposition A.2 *Every vector space V has at least one basis.*

Proposition A.3 *Every linearly independent set is a subset of some basis.*

The proof of this proposition is a construction that extends an arbitrary linearly independent set into a basis. With a little work, it can be rendered into an algorithm that is useful in computations.

Proposition A.4 *All bases of a vector space V have the same cardinality.*

Although true in general, we only need the case for finite cardinality in this book.

Definition A.8 We call the cardinality of a basis its *dimension*, and refer to the *dimension of a vector space* as the cardinality of any of its bases.

If we have bases for V and W, we can write a linear map as an array of numbers. Suppose that $\{v_1, \ldots, v_n\}$ is a basis for V and $\{w_1, \ldots, w_m\}$ is a basis for W. Then we can write each $\ell(v_i)$ in the basis for W

$$\ell(v_i) = \sum_{j=1}^{m} (l_{j,i}) w_j,$$

where the set $L = (l_{j,i})$ is called the *matrix representation* or *matrix* for ℓ. For an arbitrary vector $x \in V$ written in the V basis

$$x = \sum_{i=1}^{n} (x_i) v_i,$$

it is easy to compute $\ell(x)$ using its matrix

$$\begin{aligned}
\ell(x) &= L\left(\sum_{i=1}^{n} (x_i) v_i\right) \\
&= \sum_{i=1}^{n} (x_i) L(v_i) \\
&= \sum_{i=1}^{n} (x_i) \sum_{j=1}^{m} (l_{j,i}) w_j \\
&= \sum_{j=1}^{m} \left(\sum_{i=1}^{n} (l_{j,i})(x_i)\right) w_j.
\end{aligned}$$

This is easy to remember if we write the $(l_{j,i})$ in an $m \times n$ rectangular array

$$\begin{pmatrix} l_{1,1} & l_{1,2} & \cdots & l_{1,n} \\ l_{2,1} & l_{2,2} & \cdots & l_{2,n} \\ \vdots & \vdots & & \vdots \\ l_{m,1} & l_{m,2} & \cdots & l_{m,n} \end{pmatrix},$$

and the coefficients for x in a $n \times 1$ column

$$\begin{pmatrix} x_1 \\ x_2 \\ \vdots \\ x_n \end{pmatrix},$$

because the formula $(l_{j,i})(x_i)$ has us multiply the ith column of the matrix with ith row, obtaining a new array

$$\begin{pmatrix} l_{1,1}x_1 & l_{1,2}x_2 & \cdots & l_{1,n}x_n \\ l_{2,1}x_1 & l_{2,2}x_2 & \cdots & l_{2,n}x_n \\ \vdots & \vdots & & \vdots \\ l_{m,1}x_1 & l_{m,2}x_2 & \cdots & l_{m,n}x_n \end{pmatrix},$$

which we then sum along rows to obtain the coefficients for $\ell(x)$ in the W basis.

Example A.1 Matrices whose nonzero entries only appear along the *diagonal*—where the row and column indices are the same—play a special role. These matrices are called *diagonal matrices*. For instance, the following diagonal matrix when multiplied by a typical 4-dimensional vector

$$\begin{pmatrix} 1 & 0 & 0 & 0 \\ 0 & 2 & 0 & 0 \\ 0 & 0 & 3 & 0 \\ 0 & 0 & 0 & 4 \end{pmatrix} \begin{pmatrix} v_1 \\ v_2 \\ v_3 \\ v_4 \end{pmatrix} = \begin{pmatrix} v_1 \\ 2v_2 \\ 3v_3 \\ 4v_4 \end{pmatrix}$$

has the behavior of independently scaling each of the components. Thus, the diagonal matrix with ones along the diagonal

$$\begin{pmatrix} 1 & 0 & \cdots & 0 \\ 0 & 1 & \cdots & 0 \\ \vdots & \vdots & \ddots & \vdots \\ 0 & 0 & \cdots & 1 \end{pmatrix}$$

is the matrix for the identity map $\mathrm{id}_{\mathbb{R}^n}$, and is called the *identity matrix*.

Proposition A.1 showed that the composition $(b \circ a) : V \to X$ of two linear maps $a : V \to W, b : W \to X$ is itself a linear map. When the linear maps are written with respect to bases, the matrix representation has a convenient form. Suppose that V has a basis $\{e_1, \ldots, e_n\}$, W has a basis $\{f_1, \ldots, f_m\}$, and X has a basis $\{g_1, \ldots, g_p\}$. When written with respect to these bases, a has a matrix representation of size $m \times n$

$$a := \begin{pmatrix} a_{1,1} & a_{1,2} & \cdots & a_{1,n} \\ a_{2,1} & a_{2,2} & \cdots & a_{2,n} \\ \vdots & \vdots & & \vdots \\ a_{m,1} & a_{m,2} & \cdots & a_{m,n} \end{pmatrix}$$

and b has a matrix representation of size $p \times m$

$$b := \begin{pmatrix} b_{1,1} & b_{1,2} & \cdots & b_{1,m} \\ b_{2,1} & b_{2,2} & \cdots & b_{2,m} \\ \vdots & \vdots & & \vdots \\ b_{p,1} & b_{p,2} & \cdots & b_{p,m} \end{pmatrix}.$$

Exercise A.3 Show that $(b \circ a)$ has a matrix representation of size $p \times n$ given by the usual *matrix multiplication*

$$(b \circ a) = \begin{pmatrix} b_{1,1} & b_{1,2} & \cdots & b_{1,m} \\ b_{2,1} & b_{2,2} & \cdots & b_{2,m} \\ \vdots & \vdots & & \vdots \\ b_{p,1} & b_{p,2} & \cdots & b_{p,m} \end{pmatrix} \begin{pmatrix} a_{1,1} & a_{1,2} & \cdots & a_{1,n} \\ a_{2,1} & a_{2,2} & \cdots & a_{2,n} \\ \vdots & \vdots & & \vdots \\ a_{m,1} & a_{m,2} & \cdots & a_{m,n} \end{pmatrix}$$
$$= \begin{pmatrix} \sum_{j=1}^{m} b_{1,j}a_{j,1} & \sum_{j=1}^{m} b_{1,j}a_{j,2} & \cdots & \sum_{j=1}^{m} b_{1,j}a_{j,n} \\ \sum_{j=1}^{m} b_{2,j}a_{j,1} & \sum_{j=1}^{m} b_{2,j}a_{j,2} & \cdots & \sum_{j=1}^{m} b_{2,j}a_{j,n} \\ \vdots & \vdots & & \vdots \\ \sum_{j=1}^{m} b_{p,j}a_{j,1} & \sum_{j=1}^{m} b_{p,j}a_{j,2} & \cdots & \sum_{j=1}^{m} b_{p,j}a_{j,n} \end{pmatrix},$$

or more briefly that

$$(b \circ a)_{i,k} = \sum_{j=1}^{n} b_{i,j}a_{j,k}.$$

In order for this composition to be well defined, it is necessary that the dimensions of the matrices for a and b be compatible: the number of rows of a must be the number of columns of b. But beware! Unlike multiplication in $\mathbb{R}$, matrix multiplication is not commutative, even if the dimensions are compatible.

Recall that *bijective* functions are both *surjective* (onto) and *injective* (one-to-one).

Proposition A.5 *If $\ell : V \to W$ is a bijective linear map, then there exists another linear map $\ell^{-1} : W \to V$ with $\ell \circ \ell^{-1} = \mathrm{id}_W$ and $\ell^{-1} \circ \ell = \mathrm{id}_V$.*

Definition A.9 A bijective linear map $\ell : V \to W$ is called an *isomorphism* between V and W. The map ℓ^{-1} guaranteed in Proposition A.5 is called the *inverse* of ℓ, and ℓ is said to be *invertible*.

Since every basis $\{v_1, \ldots, v_n\}$ of a vector space V is a linearly independent set, if we form the matrix

$$a := \begin{pmatrix} \vdots & \vdots & & \vdots \\ v_1 & v_2 & \cdots & v_n \\ \vdots & \vdots & & \vdots \end{pmatrix}$$

this is both an isomorphism $a : V \to V$ and a way to re-express a given vector in basis

$$\begin{pmatrix} 1 \\ 0 \\ \vdots \\ 0 \end{pmatrix}, \begin{pmatrix} 0 \\ 1 \\ \vdots \\ 0 \end{pmatrix}, \ldots, \begin{pmatrix} 0 \\ 0 \\ \vdots \\ 1 \end{pmatrix},$$

in terms of the $\{v_1, \ldots, v_n\}$ basis. Conversely, every isomorphism $V \to V$ can be interpreted as a basis change for V.

Consider a linear map $\ell : V \to W$, for which we have two bases for V (related by the isomorphism $a : V \to V$) and two bases for W (related by the isomorphism $b : W \to W$). Given this, if we have a matrix representation for ℓ in the original bases for V and W, then in the new bases, it has matrix representation

$$b^{-1} \circ \ell \circ a.$$

This can also be expressed by the commutative diagram

$$\begin{array}{ccc} V & \xrightarrow{\;b^{-1}\circ\ell\circ a\;} & W \\ {\scriptstyle a}\downarrow & & \downarrow{\scriptstyle b} \\ V & \xrightarrow[\;\ell\;]{} & W \end{array}$$

Definition A.10 A subset W of a vector space V is called a *subspace* whenever every linear combination of elements in W is itself an element of W.

A subspace of a vector space is a vector space in its own right.

Proposition A.6 *The span of a subset S of a vector space V is a subspace $W =$ span S of V. If S is linearly independent, then S is a basis for W.*

Example A.2 Suppose that V is a 3-dimensional vector space with basis

$$v_1 := \begin{pmatrix} 1 \\ 1 \\ 0 \\ 1 \end{pmatrix}, \quad v_2 := \begin{pmatrix} 2 \\ -1 \\ 0 \\ 0 \end{pmatrix}, \quad v_3 := \begin{pmatrix} 0 \\ 0 \\ 1 \\ 0 \end{pmatrix}$$

and that W is a 2-dimensional vector space with basis

$$w_1 := \begin{pmatrix} 0 \\ -3 \\ 0 \\ -2 \end{pmatrix}, \quad w_2 := \begin{pmatrix} 3 \\ 0 \\ 0 \\ 1 \end{pmatrix}.$$

Notice that because

$$\begin{pmatrix} 1 & 2 & 0 \\ 1 & -1 & 0 \\ 0 & 0 & 1 \\ 1 & 0 & 0 \end{pmatrix} \begin{pmatrix} -2 & 1 \\ 1 & 1 \\ 0 & 0 \end{pmatrix} = \begin{pmatrix} 0 & 3 \\ -3 & 0 \\ 0 & 0 \\ -2 & 1 \end{pmatrix},$$

this means that each basis element of W can be expressed as a linear combination of basis elements of V. Thus, W is a subspace of V.

Index

M. Robinson, *Computational Homological Algebra*, Mathematical Engineering,
https://doi.org/10.1007/978-3-032-08634-1

MIX
Papier aus verantwortungsvollen Quellen
Paper from responsible sources
FSC® C105338

If you have any concerns about our products,
you can contact us on
ProductSafety@springernature.com

In case Publisher is established outside the EU,
the EU authorized representative is:
Springer Nature Customer Service Center GmbH
Europaplatz 3, 69115 Heidelberg, Germany

Printed by Libri Plureos GmbH
in Hamburg, Germany